ESO ASTROPHYSICS SYMPOSIA
European Southern Observatory

Series Editor: Bruno Leibundgut

Springer-Verlag
Berlin Heidelberg GmbH

Physics and Astronomy ONLINE LIBRARY

http://www.springer.de/phys/

ESO ASTROPHYSICS SYMPOSIA
European Southern Observatory

Series Editor: Bruno Leibundgut

G. Meylan (Ed.), **QSO Absorption Lines**
Proceedings, 1994. XXIII, 471 pages. 1995.

D. Minniti, H.-W. Rix (Eds.),
Spiral Galaxies in the Near-IR
Proceedings, 1995. X, 350 pages. 1996.

H. U. Käufl, R. Siebenmorgen (Eds.),
The Role of Dust in the Formation of Stars
Proceedings, 1995. XXII, 461 pages. 1996.

P. A. Shaver (Ed.),
Science with Large Millimetre Arrays
Proceedings, 1995. XVII, 408 pages. 1996.

J. Bergeron (Ed.),
The Early Universe with the VLT
Proceedings, 1996. XXII, 438 pages. 1997.

F. Paresce (Ed.),
Science with the VLT Interferometer
Proceedings, 1996. XXII, 406 pages. 1997.

D. L. Clements, I. Pérez-Fournon (Eds.),
Quasar Hosts
Proceedings, 1996. XVII, 336 pages. 1997.

L. N. da Costa, A. Renzini (Eds.),
Galaxy Scaling Relations: Origins, Evolution and Applications
Proceedings, 1996. XX, 404 pages. 1997.

L. Kaper, A. W. Fullerton (Eds.),
Cyclical Variability in Stellar Winds
Proceedings, 1997. XXII, 415 pages. 1998.

R. Morganti, W. J. Couch (Eds.),
Looking Deep in the Southern Sky
Proceedings, 1997. XXIII, 336 pages. 1999.

J. R. Walsh, M. R. Rosa (Eds.), **Chemical Evolution from Zero to High Redshift**
Proceedings, 1998. XVIII, 312 pages. 1999.

J. Bergeron, A. Renzini (Eds.), **From Extrasolar Planets to Cosmology: The VLT Opening Symposium**
Proceedings, 1999. XXVIII, 575 pages. 2000.

A. Weiss, T. G. Abel, V. Hill (Eds.),
The First Stars
Proceedings, 1999. XIII, 355 pages. 2000.

A. Fitzsimmons, D. Jewitt, R. M. West (Eds.),
Minor Bodies in the Outer Solar System
Proceedings, 1998. XV, 192 pages. 2000.

L. Kaper, E. P. J. van den Heuvel, P. A. Woudt (Eds.), **Black Holes in Binaries and Galactic Nuclei: Diagnostics, Demography and Formation**
Proceedings, 1999. XXIII, 378 pages. 2001.

G. Setti, J.-P. Swings (Eds.), **Quasars, AGNs and Related Research Across 2000**
Proceedings, 2000. XVII, 220 pages. 2001.

A. J. Banday, S. Zaroubi, M. Bartelmann (Eds.),
Mining the Sky
Proceedings, 2000. XV, 705 pages. 2001.

E. Costa, F. Frontera, J. Hjorth (Eds.),
Gamma-Ray Bursts in the Afterglow Era
Proceedings, 2000. XIX, 459 pages. 2001.

S. Cristiani, A. Renzini, R. E. Williams (Eds.),
Deep Fields
Proceedings, 2000. XXVI, 379 pages. 2001.

J. F. Alves, M. J. McCaughrean (Eds.), **The Origins of Stars and Planets: The VLT View**
Proceedings, 2001. XXVII, 515 pages. 2002.

Series homepage – http://www.springer.de/phys/books/eso/

João F. Alves Mark J. McCaughrean (Eds.)

The Origins of Stars and Planets: The VLT View

Proceedings of the ESO Workshop
Held in Garching, Germany, 24–27 April 2001

Springer

Volume Editors

João F. Alves
European Southern Observatory
Karl-Schwarzschild-Strasse 2
85748 Garching, Germany

Mark J. McCaughrean
Astrophysikalisches Institut Potsdam
An der Sternwarte 16
14482 Potsdam, Germany

Series Editor

Bruno Leibundgut
European Southern Observatory
Karl-Schwarzschild-Strasse 2
85748 Garching, Germany

Library of Congress Cataloging-in-Publication Data

The origins of stars and planets : the VLT view : proceedings of the ESO workshop held in Garching, Germany, 24-27 April 2001 / Joao F. Alves, Mark J. McCaughrean (eds.).
p. cm. -- (ESO astrophysics symposia)
Includes bibliographical references.

ISBN 978-3-662-39255-3 ISBN 978-3-662-40277-1 (eBook)
DOI 10.1007/978-3-662-40277-1

1. Stars--Formation--Congresses. 2. Planets--Origin--Congresses. 3. Very Large Telescope--Congresses. I. Alves, Joao F., 1968- II. McCaughrean, Mark J., 1961- III. Series.

QB806 .O77 2002
523.8'8--dc21

2002066928

ISBN 978-3-662-39255-3

Additional material to this book can be downloaded from http://extra.springer.com

http://www.springer.de

Softcover reprint of the hardcover 1st edition 2002

Typesetting: Camera-ready by the authors/editors
Cover design: Erich Kirchner, Heidelberg

Printed on acid-free paper SPIN: 10856518 55/3141/du - 5 4 3 2 1 0

Preface

Understanding how stars and planets form in the cold interiors of molecular clouds presents one of the most formidable challenges of modern astrophysics. Many important observational clues concerning this fundamental process have been amassed only during the last two decades and almost all associated with major technological developments. The recent advent of ESO's Very Large Telescope (VLT) promises to be a major step in our ability to investigate stellar nurseries and infant stars, and so the time seemed right to bring the star and planet formation community together and sketch the scientific breakthroughs we can hope to obtain in the next decade or two.

The response from the community was overwhelming, and even before the second announcement we had filled all 80 seats of the ESO auditorium. As a result, the conference was held in the Hörsaal of the Max Planck Institute for Extraterrestrial Physics in Garching from April 24th to 27th 2001, and was attended by almost 200 scientists who were native speakers of more than 20 languages.

Best of all, we had a magnificent and lively participation, with superb talks and poster contributions, impromptu debates, and "name that object" polls. Part of what happened during the conference can be found in this book, and we hope that it attests well to quite how exciting star formation research is and will be in the VLT era. We thank all participants for making the "The Origins of Stars and Planets" meeting such a memorable event.

This conference would not have been possible without the outstanding organisational skills of Christina Stoffer who inherited the wheel of an overcrowded boat and smoothly steered it to its destination. The proceedings book was carefully composed by the mathematical eye of Pamela Bristow who never avoided extra hard work to follow a good suggestion, usually her own. The organisers are grateful to Reinhard Genzel for facilitating the move to the brand new auditorium of the MPE, and of course to the European Southern Observatory, which fully sponsored the conference.

Garching, Potsdam
September 2001

João Alves
Mark McCaughrean

The Origins of Stars and Planets
The VLT View

ESO Workshop, Garching 24 - 27 April 2001

SOC : J.Alves (co-chair), S.Beckwith, F.Bertoldi, F.Boulanger, C.Clarke, C.Cesarsky, E. vanDishoeck, F.Malbet, M.Mayor, R.Genzel, T.Henning, C.Lada, M.McCaughrean (co-chair), A.Moorwood, A.Natta, F.Paresce, B.Reipurth, M.Romaniello, E.Tolstoy

LOC : J.Alves (chair), M.Romaniello, C.Stoffer ▯ For more information - http://www.eso.org/starplanet2001

Contact Address: European Southern Observatory Karl-Schwarzschild-Str. 2, D-85748 Garching bei München
Telephone: (+49 89) 320-060 Fax (+49 89) 3200-6480 Email: starplan@eso.org

Contents

Posters available on the CD-ROM

VLT/ISAAC Study of Four Massive Star Forming Regions
D. Apai, B. Stecklum, Th. Henning

Search for Pre-Main Sequence Stars in Open Clusters: NGC 6871
Z. Balog, S. Kenyon

Detailed Study of the Very Young Accreting Protostar IRAM 04191: Infall, Rotation and Outflow
A. Belloche, P. André, D. Despois, A. Bacmann, F. Motte

Submm Study of the Protostellar Envelopes Around Very Young Stellar Objects in L1634
M.T. Beltran, R. Estalella, P.T.P. Ho, N. Calvet, G. Anglada

The Stellar Population of Ultra-Compact H II Regions: The Case of IRAS 10049-5657
A. Bik, L. Kaper, M.M. Hanson, F. Comerón

High-Resolution Spectra Toward Massive Protostars: Probing the Hot Core
A.M.S. Boonman, J.H. Lacy, N.J. II Evans, M.J. Richter, E.F. van Dishoeck, R. Stark, F.F.S. van der Tak

High-Mass Protostars
J. Brand, R. Cesaroni, F. Palla, Q. Zhang, M.A. Kramer, T.R. Hunter, S. Molinari, T.K. Shridharan

Status of the Hobby-Eberly 9.2 Meter Telescope at the McDonald Observatory
M.L. Broyles

Star Formation and High Velocity Outflows in OMC1 seen at High Spatial Resolution
Y. Clenet, E. Le Coarer, L. Vannier, J.L. Lemaire, D. Field, G. Pineau des Forêts, D. Rouan

List of Participants

Adams, Fred
Univ. of Michigan, Physics Dept.
fca@umich.edu

Ageorges, Nancy
ESO-Chile
nageorge@eso.org

Agostinho, Rui Jorge
Astronomy Observatory of Lisbon
ruiag@oal.ul.pt

Alves, João
ESO, Garching
jalves@eso.org

Andersen, Morten
Astrophysikalisches Institut Potsdam
mortena@aip.de

André, Philippe
CEA – Saclay, Service d'Astrophysique
pandre@cea.fr

Apai, Daniel
Astrophysikalisches Institut &
Univ.-Sternwarte, Jena
apai@astro.uni-jena.de

Bacciotti, Francesca
Osservatorio Astrofisico di Arcetri
fran@arcetri.astro.it

Balog, Zoltan
Harvard Smithsonian Center
for Astrophysics
zbalog@cfa.harvard.edu

Baraffe, Isabelle
CRAL – ENS, Lyon
ibaraffe@ens-lyon.fr

Barrado y Navascués, David
Universidad Autonoma de Madrid
barrado@pollux.ft.uam.es

Bate, Matthew
University of Exeter
mbate@ast.cam.ac.uk

Beaulieu, Jean-Philippe
Institut d'Astrophysique de Paris
beaulieu@iap.fr

Belloche, Arnaud
CEA Saclay, Service d'Astrophysique
belloche@discovery.saclay.cea.fr

Beltran, Maite
Harvard Smithsonian Center
for Astrophysics
mbeltran@cfa.harvard.edu

Bertoldi, Frank
MPI für Radioastronomie, Bonn
bertoldi@mpifr-bonn.mpg.de

Bik, Arjan
University of Amsterdam,
Astronomical Institute
bik@astro.uva.nl

Blitz, Leo
University of California, Berkeley
blitz@gmc.berkeley.edu

Bockelee-Morvan, Dominique
Observatoire de Paris, Meudon
dominique.bockelee@obspm.fr

Boden, Andy
Caltech, ISCI/IPAC, Pasadena
bode@ipac.caltech.edu

Bontemps, Sylvain
Observatoire de Bordeaux
bontemps@observ.u-bordeaux.fr

Boonman, Annemieke
Leiden University Observatory
boonman@strw.leidenuniv.nl

Bourke, Tyler
Harvard Smithsonian Center
for Astrophysics
tbourke@cfa.harvard.edu

Bouvier, Jérôme
Lab. Astrophysique de Grenoble
jbouvier@obs.ujf-grenoble.fr

Bouwman, Jeroen
University of Amsterdam,
Astronomical Institute
jeroenb@astro.uva.nl

Brand, Jan
Istituto di Radioastronomia, Bologna
brand@ira.bo.cnr.it

Brandner, Wolfgang
ESO, Garching
brandner@eso.org

Broyles, Michael
Univ. of Texas, Dept. of Astronomy
mbroyles@astro.as.utexas.edu

Burkert, Andreas
MPI für Astronomie, Heidelberg
burkert@mpia-hd.mpg.de

Cesarsky, Catherine
ESO, Garching
ccesarsk@eso.org

Chini, Rolf
Ruhr-Universität Bochum,
Astronomisches Institut
chini@astro.ruhr-uni-bochum.de

Clarke, Cathie
University of Cambridge,
Institute of Astronomy
cclarke@ast.cam.ac.uk

Cox, Pierre
Institut d'Astrophysique Spatiale,
Université de Paris XI
Pierre.Cox@ias.u-psud.fr

de Koter, Alex
University of Amsterdam,
Astronomical Institute
dekoter@astro.uva.nl

De Marchi, Guido
STScI, Baltimore
demarchi@stsci.edu

Delgado, Eduardo
University of Cambridge,
Institute of Astronomy
edelgado@ast.cam.ac.uk

Dominik, Carsten
University of Amsterdam,
Astronomical Institute
dominik@astro.uva.nl

Donatowicz, Jadzia
University of Vienna,
Institute for Astronomy
JDonatowicz@solar.Stanford.EDU

Dullemond, Cornelis
MPI für Astrophysik, Garching
dullemon@mpa-garching.mpg.de

Edgar, Richard
University of Cambridge,
Institute of Astronomy
rge21@ast.cam.ac.uk

Eiroa, Carlos
Universidad Autonoma de Madrid
carlos@xiada.ft.uam.es

Eislöffel, Jochen
Thüringer Landessternwarte,
Tautenburg
eisloeffel@tls-tautenburg.de

Els, Sebastian
Institut für Theoretische Astrophysik,
Heidelberg
sels@ita.uni-heidelberg.de

Field, David
University of Aarhus,
Inst. of Physics and Astronomy
dfield@ifa.au.dk

Fujiyoshi, Takuya
Cavendish Laboratory
tak@mrao.cam.ac.uk

Galli, Daniele
Osservatorio Astrofisico di Arcetri
galli@arcetri.astro.it

Garay, Guido
Universidad de Chile
guido@das.uchile.cl

Garcia, Paulo
Centre for Astrophysics,
University of Porto
pgarcia@astro.up.pt

Giannini, Teresa
Osservatorio Astronomico di Roma,
Monteporzio
teresa@coma.mporzio.astro.it

Gilmozzi, Roberto
ESO, Paranal
rgilmozz@eso.org

Gittins, David
University of Cambridge,
Institute of Astronomy
dgittins@ast.cam.ac.uk

Gonçalves, José
Astronomy Observatory of Lisbon
jgoncalv@oal.ul.pt

Greene, Tom
NASA Ames Research Center
tgreene@arc.nasa.gov

Guenther, Eike
Thüringer Landessternwarte,
Tautenburg
guenther@tls-tautenburg.de

Habart, Emilie
Institut d'Astrophysique Spatiale,
Université de Paris XI
emilie.habart@ias.u-psud.fr

Haisch Jr., Karl
NASA Ames Research Center
haisch@astro.ufl.edu

Harvey, Paul
University of Texas at Austin
pmh@astro.as.utexas.edu

Harvey, Daniel
Harvard Smithsonian Center
for Astrophysics
dharvey@cfa.harvard.edu

Hatzes, Artie
Thüringer Landessternwarte,
Tautenburg
artie@tls-tautenburg.de

Heitsch, Fabian
MPI für Astronomie, Heidelberg
heitsch@mpia-hd.mpg.de

Henning, Thomas
AIU Jena
henning@astro.uni-jena.de

Hogerheijde, Michiel
University of California, Berkeley
michiel@astro.berkeley.edu

Huélamo, Nuria
MPI für Extraterrestrische Physik, Garching
huelamo@mpe.mpg.de

Ibrahimov, Mansur
Ulugh Beg Astronomical Observatory, Tashkent
mansur@astrin.uzsci.net

Jayawardhana, Ray
University of California, Berkeley
rayjay@astro.berkeley.edu

Jewitt, David
University of Hawaii, Institute for Astronomy
jewitt@ifa.hawaii.edu

Joergens, Viki
MPI für Extraterrestrische Physik, Garching
viki@mpe.mpg.de

Jørgensen, Jes
Leiden University Observatory
joergens@strw.leidenuniv.nl

Kaper, Lex
University of Amsterdam
lexk@eso.org, lexk@astro.uva.nl

Kendall, Tim
Astronomical Observatory, Lisbon
miguelm@oal.ul.pt

Khanzadyan, Tigran
Armagh Observatory, Northern Ireland
tig@star.arm.ac.uk

Klessen, Ralf
University of California, Lick Observatory
ralf@ucolick.org

König, Brigitte
MPI für Extraterrestrische Physik, Garching
bkoenig@mpe.mpg.de

Kramer, Carsten
Universität zu Köln, KOSMA
kramer@ph1.uni-koeln.de

Krivova, Natalia
Max-Planck-Institut für Aeronomie, Katlenburg-Lindau
natalie@linmpi.mpg.de

Lachaume, Régis
Lab. Astrophysique de Grenoble
lachaume@laog.obs.ujf-grenoble.fr

Lada, Charles
Harvard Smithsonian Center for Astrophysics
clada@cfa.harvard.edu

Lagage, Pierre-Olivier
CEA Saclay, DAPNIA/SAp
lagage@cea.fr

Lamzin, Sergei
Sternberg Astronomical Institute, Moscow
lamzin@sai.msu.ru

Lang, Bernd
MPIA Heidelberg
lang@mpia.de

Launhardt, Ralf
Caltech, Pasadena
rl@astro.caltech.edu

Lawson, Warrick
Australian Defence Force Academy
UNSW, Canberra
wal@ph.adfa.edu.au

Lefloch, Bertrand
Lab. Astrophysique de Grenoble
Bertrand.Lefloch
@obs.ujf-grenoble.fr

Lemaire, Jean-Louis
Observatoire de Paris, Dépt. DAMAp,
Meudon
lemaire@obspm.fr

Li, Zhi-Yun
University of Virginia, Charlottesville
zl4h@virginia.edu

Linz, Hendrik
Thüringer Landessternwarte,
Tautenburg
linz@tls-tautenburg.de

Lodato, Giuseppe
Scuola Normale Superiore, Pisa
lodato@cibs.sns.it

Lodieu, Nicolas
Astrophysikalisches Institut Potsdam
nlodieu@aip.de

Lombardi, Marco
Universität Bonn, Astronomie
lombardi@astro.uni-bonn.de

Looney, Leslie
MPI für Extraterrestrische Physik,
Garching
lwl@mpe.mpg.de

Lopez, Rosario
Universidad de Barcelona
rosario@am.ub.es

Malbet, Fabien
Observatoire de Grenoble,
Laboratoire d'Astrophysique
Fabien.Malbet
@obs.ujf-grenoble.fr

Marcaide, Jon
Universidad de Valencia
J.M.Marcaide@uv.es

Mayor, Michel
Observatoire de Genève
Michel.Mayor@obs.unige.ch

McCaughrean, Mark
Astrophysikalisches Institut Potsdam
mjm@aip.de

Meyer, Michael R.
The University of Arizona,
Steward Observatory
mmeyer@as.arizona.edu

Miao, Jingqi
University of Kent
j.miao@ukc.ac.uk

Monin, Jean-Louis
Observatoire de Grenoble,
Laboratoire d'Astrophysique
Jean-Louis.Monin
@obs.ujf-grenoble.fr

Montmerle, Thierry
CEA Saclay, DAPNIA/SAP
montmerle@cea.fr

Mookerjea, Bhaswati
Universität zu Köln, KOSMA
bhaswati@ph1.uni-koeln.de

Moorwood, Alan
ESO, Garching
amoor@eso.org

Mora, Alcione
Universidad Autonoma de Madrid
alcione.mora@uam.es

Moraux, Estelle
Lab. Astrophysique de Grenoble
moraux@laog.obs.ujf-grenoble.fr

Moreira, Miguel
Astronomical Observatory, Lisbon
miguelm@oal.ul.pt

Muecke, Anita
Université de Montréal
muecke@astro.umontreal.ca

Mundt, Reinhard
MPI für Astronomie, Heidelberg
mundt@mpia.de

Natta, Antonella
Osservatorio Astrofisico di Arcetri
natta@arcetri.astro.it

Neuhäuser, Ralph
MPI für Extraterrestrische Physik,
Garching
rne@mpe.mpg.de

Nielbock, Markus
Ruhr-Universität Bochum
nielbock
@astro.ruhr-uni-bochum.de

Nürnberger, Dieter
IRAM
nurnberg@astro.uni-wuerzburg.de

Ossenkopf, Volker
Universität zu Köln, KOSMA
ossk@ph1.uni-koeln.de

Palla, Francesco
Osservatorio Astrofisico di Arcetri
palla@arcetri.astro.it

Paresce, Francesco
ESO, Garching
fparesce@eso.org

Pascucci, Ilaria
Astrophysikalisches Institut, Jena
ilaria@astro.uni-jena.de

Pavlovski, Georgi
Armagh Observatory,
Northern Ireland
gbp@star.arm.ac.uk

Petr-Gotzens, Monika
MPI für Radioastronomie, Bonn
mpetr@mpifr-bonn.mpg.de

Pontoppidan, Klaus
Leiden University Observatory
pontoppi@strw.leidenuniv.nl

Preibisch, Thomas
MPI für Radioastronomie, Bonn
preib@mpifr-bonn.mpg.de

Prusti, Timo
ESTEC, Astrophysics Division
tprusti@astro.estec.esa.nl

Qualtrough, Catherine
Cavendish Astrophysics Group,
Cambridge
cjq20@mrao.cam.ac.uk

Ray, Tom
Dublin Institute for Advanced Studies
tr@cp.dias.ie

Reipurth, Bo
University of Colorado, CASA
reipurth@casa.colorado.edu

Rengel Lamus, Miriam
Thüringer Landessternwarte,
Tautenburg
mrengel@tls-tautenburg.de

Richer, John
Cavendish Laboratory
jsr@mrao.cam.ac.uk

Richichi, Andrea
ESO, Garching
arichich@eso.org

Robberto, Massimo
STScI
robberto@stsci.edu

Roche, Patrick
University of Oxford
pfr@astro.ox.ac.uk

Romaniello, Martino
ESO, Garching
mromanie@eso.org

Rosen, Alex
Armagh Observatory,
Northern Ireland
rar@star.arm.ac.uk

Rubio, Monica
Universidad de Chile, Santiago
mrubio@das.uchile.cl

Santos, Carlos
Astronomical Observatory, Lisbon
csantos@oal.ul.pt

Santos, Nuno
Observatoire de Genève
Nuno.Santos@obs.unige.ch

Sargent, Anneila
Caltech
afs@astro.caltech.edu

Schneider, Nicola
Observatoire de Bordeaux
schneider@observ.u-bordeaux.fr

Schulz, Rita
ESA Space Science Department
Rita.Schulz@esa.int

Siebenmorgen, Ralf
ESO, Garching
rsiebenm@eso.org

Smith, Michael
Armagh Observatory,
Northern Ireland
mds@star.arm.ac.uk

Stamatellos, Dimitris
Cardiff University
Dimitrios.Stamatellos
@astro.cf.ac.uk

Stecklum, Bringfried
Thüringer Landessternwarte,
Tautenburg
stecklum@tls-tautenburg.de

Sterzik, Michael
ESO, Santiago
msterzik@eso.org

Stolte, Andrea
MPI für Astronomie, Heidelberg
stolte@mpia-hd.mpg.de

Tachihara, Kengo
MPI für Extraterrestrische Physik,
Garching
tatihara@mpe.mpg.de

Testi, Leonardo
Osservatorio Astrofisico di Arcetri
lt@arcetri.astro.it

Thi, Wing-Fai
Leiden University Observatory
thi@strw.leidenuniv.nl

Tieftrunk, Achim
Universität zu Köln, KOSMA
tieftrunk@ph1.uni-koeln.de

Torrelles, José-Maria
Inst. de Ciencias del Espacio (CSIC), Barcelona
torrelles@ieec.fcr.es

Trimble, Virginia
University of California, Irvine
vtrimble@uci.edu

van Boekel, Roy
ESO, Garching
rvboekel@eso.org

van der Tak, Floris
MPI für Radioastronomie, Bonn
vdtak@mpifr-bonn.mpg.de

van Dishoeck, Ewine
Leiden Observatory
ewine@strw.LeidenUniv.nl

Vavrek, Roland
Konkoly Observatory / OCA
vavrek@konkoly.hu

Vitali, Fabrizio
Osservatorio Astronomico di Roma, Monteporzio
vitali@coma.mporzio.astro.it

Walmsley, Malcolm
Osservatorio Astrofisico di Arcetri
walmsley@arcetri.astro.it

Waters, Rens
University of Amsterdam, Astronomical Institute
rensw@astro.uva.nl

Weigelt, Gerd
MPI für Radioastronomie, Bonn
weigelt@mpifr-bonn.mpg.de

White, Glenn
University of Kent
g.j.white@ukc.ac.uk

Whitworth, Anthony
Cardiff University
Anthony.Whitworth@astro.cf.ac.uk

Wiedemann, Günter
University Observatory, Munich
gwiedema@eso.org

Wilner, David
Harvard Smithsonian Center for Astrophysics
dwilner@cfa.harvard.edu

Woitas, Jens
Thüringer Landessternwarte, Tautenburg
woitas@tls-tautenburg.de

Wolf, Sebastian
Thüringer Landessternwarte, Tautenburg
wolf@tls-tautenburg.de

Wright, Christopher
Australian Defence Force Academy, UNSW, Canberra
wright@ph.adfa.edu.au

Wuchterl, Günther
MPI für Extraterrestrische Physik, Garching
wuchterl@mpe.mpg.de

Yun, João
Astronomical Observatory, Lisbon
yun@oal.ul.pt

Zavagno, Annie
Institut d'Astrophysique de Marseille
zavagno@observatoire.cnrs-mrs.fr

Ziegler, Mareike
MPI für Astronomie, Heidelberg
ziegler@mpia-hd.mpg.de

Zinnecker, Hans
Astrophysikalisches Institut Potsdam
hzinnecker@aip.de

Catherine Cesarsky, Anneila Sargent, and Antonella Natta.
A 1977 connection.

As Eyes See Young Stars Assemble: Star and Planet Formation in the VLT Era

Mark J. McCaughrean

Astrophysikalisches Institut Potsdam,
An der Sternwarte 16, 11482 Potsdam, Germany

Abstract. The river of time leading to this meeting on star and planet formation with the VLT can apparently be traced back to somewhere in the vicinity of Lake Geneva almost 25 years ago, in 1977. In this introductory talk, I attempt to navigate back upstream and locate the source of the river and in doing so, discover a rather surprising link to musical happenings elsewhere on the lake at the same time. The reader is free to decide whether they find the connection more than mere coincidence, let alone interesting.

1 Introduction and Thanks

We are here this week in Garching to celebrate the European Southern Observatory Very Large Telescope in Chile, and to discuss its impact on the field of star and planet formation (Figure 1). Right at the outset, before things get too irredeemably silly, I'd like to pay a serious tribute to the many, many people who have made the VLT a physical reality and such an immediate success across a broad range of modern astronomical problems. They are far too numerous to name individually, but include ESO staff and contractors over the years who actually designed and built the VLT and operate it today, the many outside astronomers who were involved in detailed planning and building of the VLT and its instrumentation, and the broad community of European and Chilean astronomers whose long-term support was fundamental in promoting this vision of the world's largest telescope for optical and infrared astronomy. In particular, as a frequent user of the VLT myself, I'd like to thank the astronomers who sit on Paranal and help ensure that high quality scientific data are taken at the sharp end, be it in service or visitor mode. Hopefully the data shown in my talk and others at this meeting will be ample testament to their work. This talk itself is all downhill from here ...

2 Inspiration

While I was preparing for this introductory talk, João Alves kept sending emails reminding me that he wanted me to inspire the participants, to get them all fired up about the meeting. A sort of P. T. Barnum start to things. Just in case I wasn't getting the message, he also sent me dictionary definitions of the word 'inspiration', as follows:

Inspiration \In'spi*ra*tion\, n. [F. inspiration, L. inspiratio. See Inspire.]

The act or power of exercising an elevating or stimulating influence upon the intellect or emotions; the result of such influence which quickens or stimulates; as the inspiration of occasion, of art, etc.

A supernatural divine influence on the prophets, apostles, or sacred writers, by which they were qualified to communicate moral or religious truth with authority; a supernatural influence which qualifies men to receive and communicate divine truth.

While I suspect many of us fancy (secretly or otherwise) that we possess the latter talents of *ex cathedra* persuasion, João probably had more the former definition in mind. In any case, as I pondered the task ahead of me, I wondered where my own inspiration would come from, which thread could be used to draw together the many aspects of star and planet formation with the VLT in a way that would stimulate both the intellect and the emotions. At some point, it became blindingly obvious: Switzerland 1977.

3 1977 – The Swiss Connection

3.1 How Big Was Your Hair?

Ok, perhaps it's not *that* obvious. Let's start with 1977, in the middle of the 'joke decade' as Martin Amis (2001) calls it, wondering quite how anything intelligent can have happened when we were all patently stupid enough to wear flared trousers. I should immediately set the scene here personally as well, with a photo from that era (Figure 2), in which the reader is left to identify me. Clue: I had more hair then.

Perhaps the defining event of 1977 for many people of my generation and nerdish disposition, was the release of the first *Star Wars* film (Figure 3), George Lucas' bazillion dollar romp across another galaxy and the laws of physics. I'm tempted to ask who can forget the panning screech of a TIE fighter, the asthmatic wheeze and CNN intonations of Darth Vader/James Earl Jones, and the curious headphone hairdos of Princess Leia Organa, but to my continued amazement, I *still* meet astronomers who have never seen any of these films.

3.2 Music Makes the World Go Around

In the same year, The (Burger) King, Elvis Presley (Figure 4), died in Memphis. A testament to his long-lived iconic status is that he lives on, both literally in the National Enquirer and metaphorically in the guise of 50,000 professional Elvis impersonators (according to an article in The Economist, August 16 1997, as cited by Christian & Kinney 1999). When you consider that this is roughly five times the number of active professional astronomers in the world, it certainly begs the question "And you think *they* have a strange job?". Of course, Elvis also lives on through his recordings, including his 1970s Las Vegas rendition of Frank Sinatra's classic, *My Way*. It's of passing amusement to listen to this dirge-like version alongside the altogether more spirited, snarling punk take by Sid Vicious in *The Great Rock and Roll Swindle*: 1977 was also the year the Sex Pistols expectorated onto the musical scene with their Silver Jubilee paean to Elizabeth Windsor, *God Save The Queen* (Figure 5), adding a another nail to Elvis' coffin.

3.3 To Boldly Go

Moving to something a little more directly relevant, albeit briefly, 1977 saw the first flights of the Space Shuttle test vehicle, OV-101, better known as the trekkily-named Enterprise, off the back of NASA's modified Boeing 747 carrier aircraft (Figure 6). It took another thirteen years (unlucky for some, perhaps optical test technicians in particular) before the Hubble Space Telescope was launched on a full-up shuttle, Discovery. The Enterprise never made it into space, of course, and has long since been in various forms of storage, latterly at the end of a runway at Washington Dulles International Airport, awaiting the completion of the briskly-named Steven F. Udvar-Hazy Center of the Smithsonian National Air and Space Museum. It would be nice to think that NASA will close that circle and bring back the HST at the end of its mission, and install it in place of the full-scale model that presently resides in the main museum building in Washington. However, its current long-term fate is uncertain, and less palatable options also including boosting it to high orbit from which it would take half a millennium to decay, or dropping it in the Pacific (Leckrone, personal communication).

3.4 Papers You Should Have Read

Which sort of brings us to astronomy, and in 1977 in particular, two seminal papers in star formation which still serve to illustrate the breadth of the problem. Frank Shu's paper (Figure 7; Shu 1977) examined the small scale, how a single isolated molecular core would evolve via inside-out collapse to form a protostar, while Bruce Elmegreen and Charlie Lada's paper (Figure 8; Elmegreen & Lada 1977) looked at the other end, how the formation of one generation of stars in a giant molecular cloud may sweep up and compress more gas, triggering another generation in its turn. Capturing, integrating, and understanding the whole star formation process, over at least these nine orders of magnitude in scale (from 1 kiloparsec to 1 solar radius), remains a real challenge today, both observationally and theoretically. Things get worse by yet another fifteen orders of magnitude when we try to fold in the evolution of small particles in circumstellar disks to form planets. Speaking of which, 1977 was also the year in which Michel Mayor co-authored his first paper (Baranne, Mayor, & Poncet 1977) on measuring radial velocities with an echelle spectrograph. This instrument was to become CORAVEL, the direct predecessor of ELODIE, which made its landmark achievement almost twenty years later with the first confirmed detection of an extrasolar planet (Mayor & Queloz 1995).

Fig. 1. The ESO VLT UT1 Antu

Fig. 2. Part of the Royal Latin School, Buckingham, school photograph, 1976

Fig. 3. The favourite film of cosmologists and Matthew Bate

Fig. 4. The King

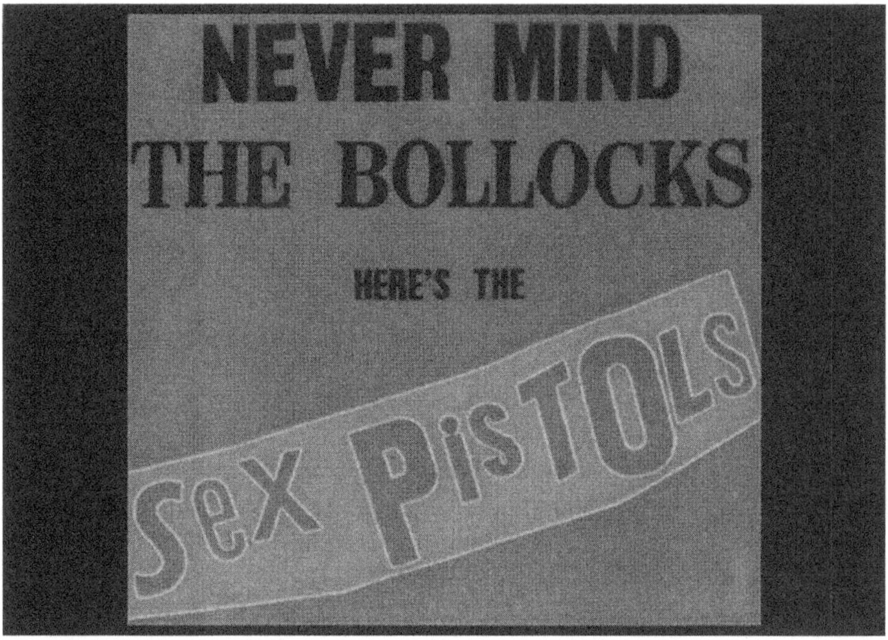

Fig. 5. Johnny Rotten and friends' debut album

Fig. 6. The Shuttle test vehicle Enterprise

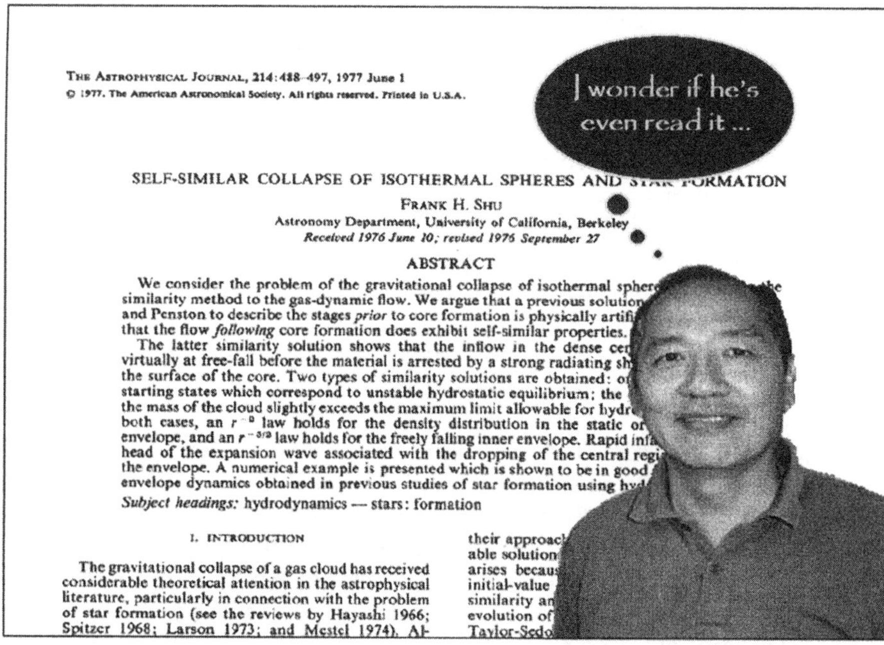

Fig. 7. Frank Shu and his inside-out protostellar collapse

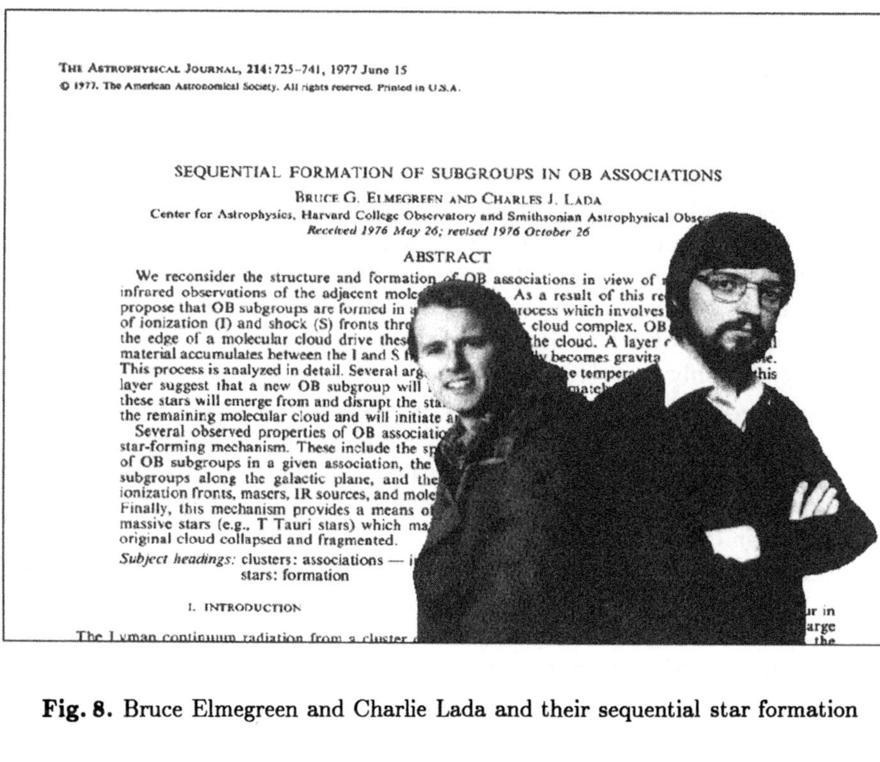

Fig. 8. Bruce Elmegreen and Charlie Lada and their sequential star formation

3.5 Through a Looking Glass

Of course, Michel Mayor has long been associated with the Observatoire de Genève, which conveniently (if long-windedly) brings us to Switzerland in 1977. At that time, Geneva (if you'll pardon the switch to the Anglicised spelling) was the location for two important conferences with direct relevance to ours today. Admittedly, the first was actually held in 1976, but its proceedings are dated 1977, so I'll take a little artistic license here for the sake of continuity. This meeting, IAU Symposium 75, was entitled simply *Star Formation* (de Jong & Maeder 1977) and could take this rather presumptuous approach since it was, in fact, more or less the first conference on the subject (Figure 9, left). It was attended by a goodly number of our present participants, whose names I shall withhold for obvious reasons: at the other extreme, some of the students in our audience were still in nappies at the time.

The direct line from that meeting to ours today is clear, as it is from the second, held in Geneva in 1977, and titled *Optical Telescopes of the Future* (Pacini, Richter, & Wilson 1978) (Figure 9, right). It was at this meeting that Lo Woltjer introduced a concept that is remarkably similar to the VLT we are using today, namely an array of four 8-metre diameter telescopes (Woltjer 1978; Woltjer, personal communication). At the same meeting, Jerry Nelson first outlined his concept of a segmented 10-metre diameter mirror telescope that would later become the Keck(s) (Nelson 1978). If nothing else, this underlines that long-term, multi-decade tenacity is required to bring big astronomy projects to fruition, well outlasting any fashions in trouser width and hair length.

3.6 Death and Taxes

The star formation meeting was held at the University of Geneva, in conjunction with the IAU General Assembly in Geneva that year, while the telescopes meeting was held at ESO, before its move to the bucolic Kuhfelder of Garching. The presence of important international organisations and meetings in neutral Switzerland is hardly unusual, but another important consideration for many individuals and organisations is the relatively low maximum taxation rate for high earners in Switzerland (e.g., a single person earning $250,000 per annum presently sees only ~ 40% disappear in federal, cantonal, and municipal taxes; `www.taxation.ch`). This long-standing Swiss generosity, combined with the fact that people resident there for less than half a year pay no taxes, and the exorbitant UK *Supertax* rate of 98% in 1977 (on earnings over £20,000 per annum, ~ £80,000 or $120,000 today; `www.accountingweb.co.uk`), leads to the next connection in our story.

Namely this. If you're old enough to own one, take a flip through your LP collection from the 1970s, and notice how many albums were recorded in Switzerland by very wealthy and thus tax-exiled Brits. Indeed, many were recorded at the Mountain Studios in Montreux, at the other end of Lake Geneva from the eponymous city (Figure 10). Among others, David Bowie (who was, of course, a *Space Oddity*), Led Zeppelin, Deep Purple, the Rolling Stones, and Queen all

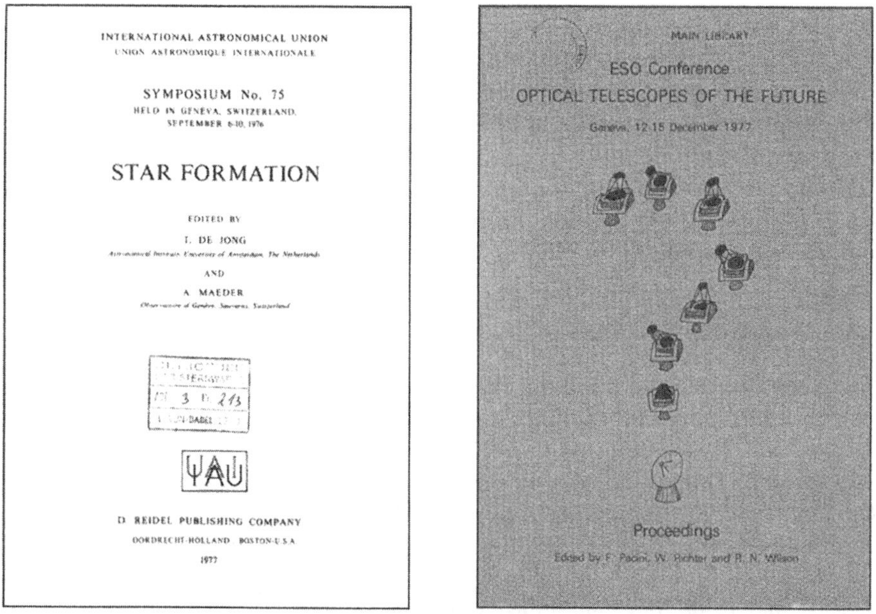

Fig. 9. Two seminal conference proceedings

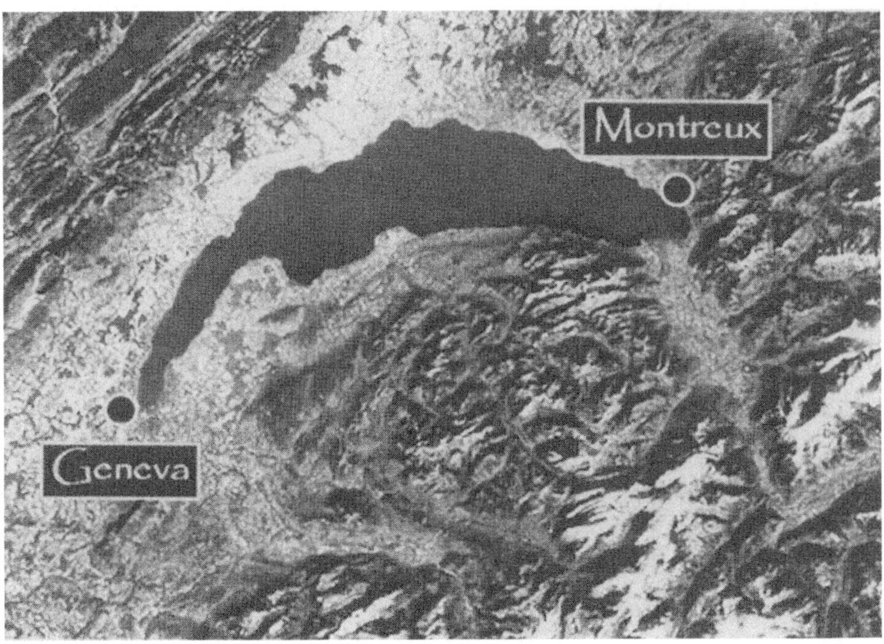

Fig. 10. Lake Geneva in Switzerland

recorded there in the 1970s. In fact, Queen liked the studio so much, they bought it in 1978 and there's even a statue of Freddie Mercury in Montreux as a result. This is an important epicycle in our trail of celestial conjunctions, since Brian May worked for four years on a PhD in infrared astronomy at Imperial College in London, before deciding to get rich and famous as Queen's lead guitarist. Although he never finished his thesis, he did publish at least two papers relevant to our field before quitting, on the motions of zodiacal dust particles in our remnant circumstellar disk (Hicks, May, & Reay 1972, 1974).

3.7 Wondrous Stories

This diversion aside, of more direct interest to us here is the English progressive rock group Yes and their album *Going For The One*, recorded at the Mountain Studios in 1976 and released in 1977 (Figure 11). At the time, I was a 16 year old schoolboy (Figure 2) living in rural Buckinghamshire, listening to *Going For The One* every day for a whole summer of lunchbreaks from working on a farm. On the album, my favourite song was a fifteen minute symphony called *Awaken*, and only years later did I realise quite how much this song predestined my career, in particular these four verses near the song's apex:

Master of images
Songs cast a light on you
Hark thru dark ties
That tunnel us out of sane existence
In challenge as direct
As eyes see young stars assemble

Master of light
All pure chance
As exists cross divided
In all encircling mode
Oh closely guided plan
Awaken in our heart

Master of soul
Set to touch
All impenetrable youth
Ask away
That thought be contact
With all that's clear
Be honest with yourself
There's no doubt no doubt

Master of time
Setting sail
Over all our lands
And as we look
Forever closer
Shall we now bid
Farewell
Farewell ...

Err, what on earth is all *that* about, I hear you asking. Quite obvious really; it's about star and planet formation with the VLT, isn't it? No? Well, let me see if I can explain.

3.8 Master of Images

The opening line sets the scene: almost any modern observing project, whether imaging or spectroscopy, involves the use of a two-dimensional detector array. Thus, in order to make the most of your data, you had better be a *master of images*. Of course, image processing can be tedious, routine work, and at least in my institute, many a student and post-doc will be plugged into a CD-player listening to *songs*, as they work through their data.

I see the next phrase as a cautionary note, as it seems to refer to the infamous NICMOS images of TMR-1C, purported to be a protoplanet being ejected from a young binary system, down a filament or *tunnel* of gas (Figure 12; Terebey *et al.* 1998). The race to obtain the first direct images of an exoplanet is exciting and important, garnering great public interest, and I have no doubt it will be won within a decade, perhaps much sooner. Nevertheless, we should keep sceptical and self-critical hats firmly on our heads as we run, to avoid a damaging boy-who-cried-wolf effect in the media. After all, pure imaging can be misleading: spectroscopy has since shown TMR-1C to be a background star as many suspected (Terebey *et al.* 2000), and the same is true for another candidate source, TWA-7B (Neuhäuser *et al.* 2000). We also need to avoid conveniently moving the goalposts in terms of what we define as a planet, as in the case of free-floating, possibly sub-13 M_{Jup} sources in Orion (Zapatero-Osorio *et al.* 1998; Lucas & Roche 1999). Above all, let's stay *sane* as we chase the brass ring.

The verse ends on a much more positive note, reminding us what our whole business is about: the *direct challenge* we're posed, to try and understand which processes and physical principles are at work *as [our] eyes see young stars assemble* in the beautiful images returned by the VLT. In particular, this phrase seems to refer to young embedded clusters where young low- and high-mass stars congregate together. One of the nearest regions which such clusters exist is, of course, Orion, including the very appropriately numbered (and strangely otherwise nameless) NGC 1977 (Figure 13) and just to its south, the far-more-famous NGC 1976 or Orion Nebula (Figure 14).

3.9 Master of Light

The second verse starts with the obvious statement that astronomy is, by and large, all about *master*ing the analysis of *light*, since this is all we can collect from our targets, modulo a few cosmic rays, neutrinos, and as-yet ethereal gravitational waves that we might one day be able to utilise.

While the next two lines remain completely inscrutable, we then move straight to circumstellar disks (Figure 15), which we believe to surround all young stars in some *encircling mode*, and from which we believe planetary systems form. Disks are also implicated in the next phrase, since they are believed to help drive and collimate the *closely guided* jets and outflows, which are also thought to *awaken in [the] heart* of every young low-mass protostellar system (Figure 16).

3.10 Master of Soul

In the third verse, there is but one readily decipherable reference to star and planet formation, albeit a crucial one. After all, it is the *impenetrable youth* of the situation that often pushes us out to infrared wavelengths to reduce the effects of dust extinction. For example, the much improved transparency at infrared

Fig. 11. Recorded in 1976 at the Mountain Studios, Montreux

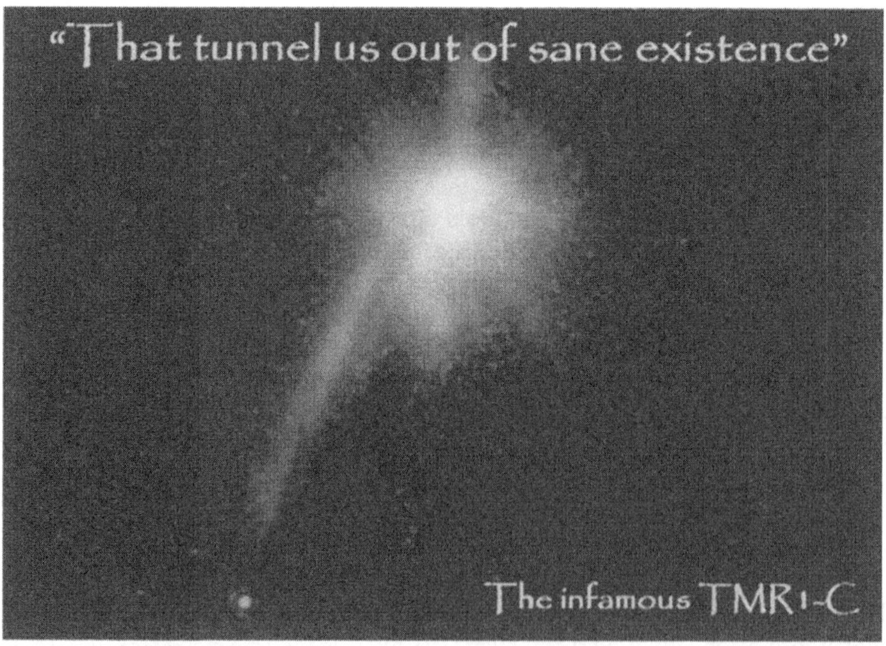

Fig. 12. NICMOS image of TMR-1C (Terebey *et al.* 1998)

Fig. 13. Optical photo of NGC 1977 (Robert Bickel)

Fig. 14. ISAAC image of NGC 1976, the Orion Nebula

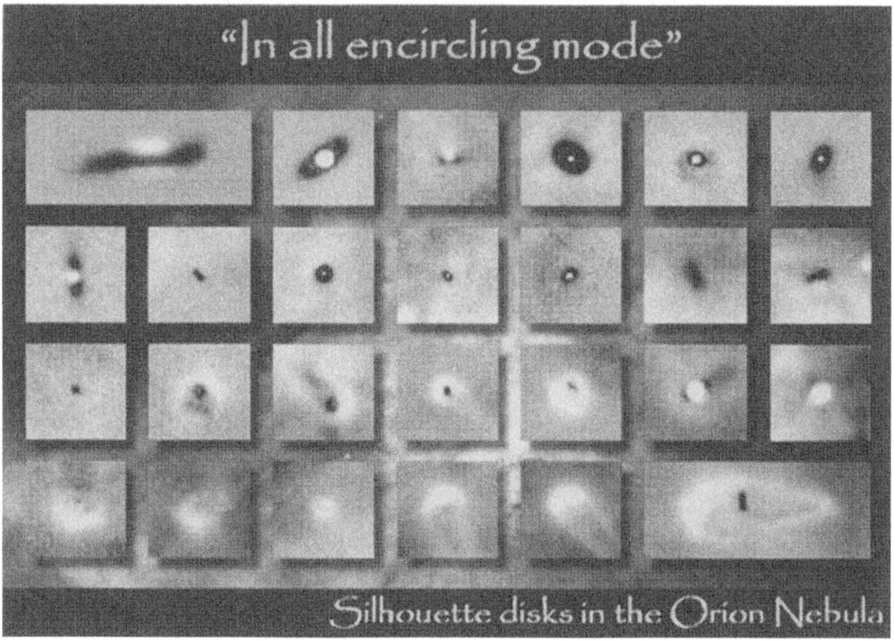

Fig. 15. Silhouette disks in the Orion Nebula

Fig. 16. ISAAC image of HH 212

Fig. 17. FORS/SOFI image of Barnard 68 (Alves *et al.* 2001)

Fig. 18. ISAAC/HST image covering the head of one elephant trunk in M 16

Fig. 19. Typical European spring weather a few days before the meeting

Fig. 20. The VLT Interferometer

wavelengths of a dense molecular cloud allows us to probe its structure and gas-to-dust ratio before it forms stars, by imaging background stars shining through it (Figure 17; Alves, Lada, & Lada 2001).

Similarly, infrared imaging can be used to penetrate the placental gas and dust in which recently born stars are found. Just one example of a region where this can be fruitful is M 16, the well-known Eagle Nebula, of which the HST's visible-wavelength images have become iconic (Hester *et al.* 1996). The famous 'pillars of creation' are covered in many small ($\sim$1 arcsec or 2000 AU), opaque EGGs (for evaporating gaseous globules), which Hester *et al.* conjectured might contain young sources, about to be revealed by the nearby NGC 6611 OB stars. This hypothesis has been tested directly via VLT infrared imaging (Figure 18), with the preliminary conclusion that in fact the EGGs may be empty, although for once counter to the song, quite some analysis will be needed before we can say *there's no doubt, no doubt.*

3.11 Master of Time

The final verse moves on from the actual mechanics of star and planet formation, to describe how it is we go about this business in a day-to-day sense. It starts by making reference to the all-important ESO Observing Programme Committee, who have control over the allocation of observing *time* on the VLT. Without their approval, you won't be *setting sail* (or taking a flight) to Chile to use the telescopes. The phrase *all our lands* then implies the implicit and subtly-administered policy that the observing time should be distributed amongst all the ESO member states in proportion to their financial contribution (Figure 19).

The next part of the verse is a look forward to adaptive optics and then the VLTI (Figure 20), which we fervently hope will soon enable us to *look forever closer*, at higher and higher resolution into the places where stars are being made, and to search for indirect (via astrometry) and direct (via imaging) evidence for the ongoing birth of planets. Finally, the last three lines are self-explanatory, and indicate that we are at last nearing the end of our journey.

4 Coda

Having completed this over-imaginative, Nostradamus-like divination of Yes' lyrics, it nevertheless may be worth asking ourselves whether there's more than pure coincidence at work here. After all, Montreux is less than 100 kilometres from Geneva, and perhaps one September afternoon in 1976, the band got bored and drove into the big city to seek some of that inspiration thing. It at least amuses me to think of Jon Anderson of Yes sitting in on the IAU Symposium, leaning against the back wall of the lecture theatre, taking a toke of his herbal tobacco, and grooving along as Charlie Lada explained his new theory of sequential star formation. Or maybe it was just a case of long-distance karmic resonance (or some such twaddle), with Jon just tapping into a hidden channel of new ideas as he took a look out the window, trying to write a punchline. After all, Rick

Wakeman, peripatetic Yes keyboard player, once said that Jon Anderson was the only person he knew who was trying to understand this universe, while living in another one.

In any case, that's perhaps an apposite quote on which to end. I very much hope that you will all enjoy and profit from the real scientific talks that follow in this conference on *The Origins of Stars and Planets—The VLT View.*

Acknowledgements The lyrics from *Awaken* by Jon Anderson of Yes are copyright Topographic Music, and I would like to thank the band for composing, playing, and recording this beautiful song. I would also like to thank *en-masse* the proprietors of the many websites I plundered in order to put together my presentation. Finally, I would like to thank the participants of the meeting for entertaining this somewhat unusual opening talk in its full multimedia version, and especially Leo Blitz, who was left with the thankless task of following it with a proper scientific presentation.

References

1. J. Alves, C. J. Lada, E. A. Lada: Nature **409**, 159 (2001)
2. M. Amis: *Experience* (Random House, London, 2001), p26
3. A. Baranne, M. Mayor, J.-L. Poncet: Comptes Rendus, Serie B **285**, (Academie des Sciences, Paris, 1977) p117
4. A. Baranne, M. Mayor, J.-L. Poncet: Vistas in Astronomy **23**, 279 (1979)
5. C. A. Christian, A. Kinney: *The Public Impact of Hubble Space Telescope*, STScI OPO monograph series **102** (STScI, Baltimore, 1999), p4
6. B. G. Elmegreen, C. J. Lada: Astrophys. J. **214**, 725 (1977)
7. J. J. Hester *et al.*: Astron. J. **111**, 2349 (1996)
8. T. R. Hicks, B. H. May, N. K. Reay: Nature **240**, 401 (1972)
9. T. R. Hicks, B. H. May, N. K. Reay: Mon. Not. R. astr. Soc. **166**, 439 (1977)
10. T. de Jong, A. Maeder, editors: *Star Formation, proceedings of the symposium, Université de Genève, Geneva, Switzerland, September 6–10, 1976*, IAU Symposium **75** (Reidel, Dordrecht, 1977)
11. P. W. Lucas, P. F. Roche: Mon. Not. R. astr. Soc. **314**, 526 (2000)
12. M. Mayor, D. Queloz: Nature **378**, 355 (1995)
13. R. Neuhäuser *et al.*: Astron. Astrophys. (Letts) **354**, L9 (2000)
14. J. Nelson: 'The proposed University of California 10 meter telescope'. In: *Optical Telescopes of the Future, Geneva, Switzerland, December 12–15, 1977*, ed. by F. Pacini, W. Richter, R. N. Wilson (ESO, Geneva, 1978), pp133–140
15. F. Pacini, W. Richter, R. N. Wilson, editors: *Optical Telescopes of the Future, Geneva, Switzerland, December 12–15, 1977* (ESO, Geneva, 1978)
16. F. H. Shu: Astrophys. J. **214**, 488 (1977)
17. S. Terebey, D. van Buren, D. L. Padgett, T. Hancock, M. Brundage: Astrophys. J. (Letts) **507**, L71 (1998)
18. S. Terebey, D. van Buren, K. Matthews, D. L. Padgett: Astron. J. **119**, 2341 (2000)
19. L. Woltjer: 'The case for large optical telescopes'. In: *Optical Telescopes of the Future, Geneva, Switzerland, December 12–15, 1977*, ed. by F. Pacini, W. Richter, R. N. Wilson (ESO, Geneva, 1978), pp5–12
20. M. R. Zapatero-Osorio *et al.*: Science **290**, 103 (2000)

Kueyen structure. View of mirror cell from behind

Clockwise from lower left: Marco Lombardi, Martino Romaniello, David Silva, Anthony Brown, and Francesca Primas

Mapping the Interstellar Dust with Near-Infrared Observations: An Optimized Multi-Band Technique

Marco Lombardi[1] and João Alves[2]

[1] Institut für Astrophysik und Extraterrestrische Forschung, Auf dem Hügel 71, D-53121 Bonn, Germany
[2] European Southern Observatory, Karl-Schwarzschild-Straße 2, D-85748 Garching bei München, Germany

Abstract. We generalize Lada et al. (1994) NICE technique to map dust column density through a molecular cloud to an optimized multi-band technique that can be applied to any multi-band survey of molecular clouds. We present a first application to a $\sim 625\ \mathrm{deg}^2$ subset of the Two Micron All Sky Survey (2MASS) data and show that when compared to NICE, the optimized NICER (i) achieves the same extinction peak values, (ii) improves the noise variance of the map by a factor of 2 and (iii) is able to reach $3\,\sigma$ dust extinction measurements as low as $A_V \simeq 0.5$ magnitudes, better than or equivalent to classical optical star count techniques and below the threshold column density for the formation of CO.

1 Introduction

Dark clouds remain one of the least understood objects in the Universe. Although they are ubiquitous in galaxies like the Milky Way, we still do not have a basic understanding of how they relate to the more diffuse interstellar medium (ISM), of how some of them form stars and planets, or how they vanish. Almost everything we know about molecular cloud structure has been derived from radio observations of tracers (e.g., CO, CS, NH_3) of the undetectable H_2 (Lada 1996; Myers 1999). However, the interpretation of results derived from radio observations is not always straightforward (e.g., Alves et al. 1999).

It has long been recognized that infrared color excess can be used to obtain a reliable estimate of the extinction through a molecular cloud (see, e.g., Bok 1977; Frerking et al. 1982; Jones et al. 1984). Such estimates are free from the complications that plague molecular-line or dust emission data and enable detailed maps of cloud density structure to be constructed. Lada et al. (1994) pioneered a technique, the Near-Infrared Color Excess (NICE), for measuring and mapping the distribution of dust through a molecular cloud using data obtained in large-scale, multi-wavelength, infrared imaging surveys. This method combines measurements of near-infrared color excess to directly measure extinctions and map the dust column density distribution through a cloud.

We generalize Lada's NICE technique to an optimized multi-band technique, the Near-Infrared Color Excess Revisited technique (NICER). Although inspired by the Two Micron All Sky Survey (2MASS) (Kleinmann et al. 1994), and by

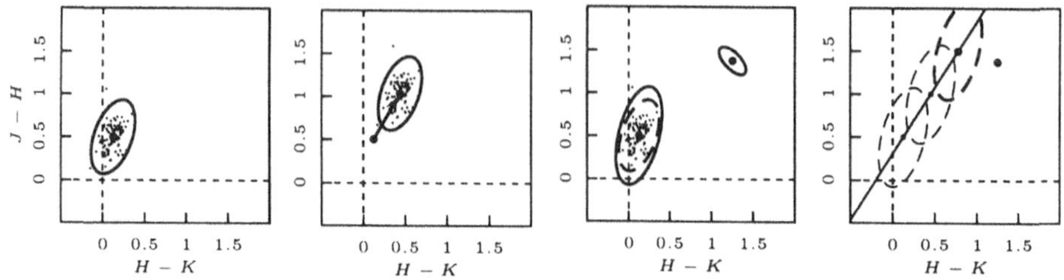

Fig. 1. From the left to the right: **(a)** JHK color-color of a hundred stars subject to no extinction (sketch); **(b)** The same stars as observed through a cloud with $A_V = 5$ magnitudes; **(c)** NICER obtains an estimate of the extinction A_V of each star using a maximum-likelihood technique, taking into account both the intrinsic color scatter (dashed ellipse) and photometric errors (small ellipse); **(d)** Method used to infer the column density of a single star (see text)

the possibility of all sky dust extinction mapping, this technique can be applied to any multi-band survey of molecular clouds. We present a first application to 2MASS data and show that NICER improves the noise variance of a map by a factor of 2 when compared to NICE, and is able to reach $3\,\sigma$ dust extinction measurements of $A_V \simeq 0.5$ magnitudes.

2 The Optimized Method

NICER has been designed by considering separately the two steps needed to make an extinction map:

- Local extinction estimate for each star;
- Spatial smoothing of individual stars estimates.

2.1 Local Absorption

The cloud column density A_V is normally measured by comparing the $H - K$ color for stars observed through the cloud with the same color for stars for which no reddening is expected:

$$\hat{A}_V = 15.87 \left[\frac{1}{N} \sum_{n=1}^{N} (H - K)_i^{\mathrm{obs}} - \langle (H - K)^{\mathrm{tr}} \rangle \right] . \tag{1}$$

Clearly, other possibilities are also viable. For instance, we could choose the $J - H$ color and use the expression

$$\hat{A}_V = 9.35 \left[\frac{1}{N} \sum_{n=1}^{N} (J - H)_i^{\mathrm{obs}} - \langle (J - H)^{\mathrm{tr}} \rangle \right] . \tag{2}$$

Should we use Eq. (2) or (3) to infer the cloud column density? In both cases, we expect some error on the estimate of A_V arising from three factors:

1. The scatter of the intrinsic colors: Normally, stars presents a scatter larger in $J - H$ than in $H - K$ (see Fig. 1a).
2. The photometric error on the individual bands: Typically errors on J are smaller than errors on K (except along line-of-sights through dense clouds, where extinction makes stars fainter in J).
3. The numerical coefficient used to convert color excess into A_V extinction: This coefficient is larger for Eq. (2) (15.87 vs. 9.35).

Hence, on the one hand we are encouraged to use Eq. (2) because of the smaller scatter in colors; on the other hand we want to avoid this estimator because of the larger numerical coefficient and larger photometric errors. Thus, this heuristic discussion does not lead to a definitive answer regarding a choice between the two estimators (2) and (3); a more quantitative approach is needed.

With three bands at our disposal we can make two independent colors, $c_1 = J - H$ and $c_2 = H - K$. For each c_i (with $i = 1, 2$), we can write the relationship between the observed color c_i^{obs} and the true one c_i^{tr} (i.e., the color that would be observed if no extinction were present) as

$$c_i^{\mathrm{obs}} = c_i^{\mathrm{tr}} + k_i A_V + \epsilon_i \ . \tag{3}$$

Here $k_i = E_i / A_V$ is the ratio between the color excess on the band i and the extinction on the V band; for the colors considered here we have $k_1 = 1/9.35$ and $k_2 = 1/15.87$. The extra term ϵ_i above represents the noise on the colors, i.e. the result of photometric error on the estimate of $J - H$ and $H - K$. Let us restrict the discussion to estimators of A_V which are linear on the observed colors c_i^{obs}, i.e. of the form

$$\hat{A}_V = a + b_1 c_1^{\mathrm{obs}} + b_2 c_2^{\mathrm{obs}} \ . \tag{4}$$

The coefficients a, b_1, and b_2 need to be determined by satisfying two conditions:

1. The estimator is *unbiased*, i.e. its expected value is the true extinction A_V.
2. The estimator has *minimum variance*.

These two conditions lead to a *unique* estimator, given by

$$\hat{A}_V = b_1 \left[c_1^{\mathrm{obs}} - \langle c_1^{\mathrm{tr}} \rangle \right] + b_2 \left[c_2^{\mathrm{obs}} - \langle c_2^{\mathrm{tr}} \rangle \right] \ , \tag{5}$$

with $(b_1, b_2) = \boldsymbol{b}$ given by (in matrix notation)

$$\boldsymbol{b} = \frac{C^{-1} \cdot \boldsymbol{k}}{\boldsymbol{k} \cdot C^{-1} \cdot \boldsymbol{k}} \ . \tag{6}$$

Here $C = \mathrm{Cov}(c^{\mathrm{tr}}) + \mathrm{Cov}(\epsilon)$ is a "total" covariance matrix, sum of the intrinsic color scatter $\mathrm{Cov}(c^{\mathrm{tr}})$ and of the individual photometric errors for the star considered $\mathrm{Cov}(\epsilon)$.

A nice feature of this technique is that it can be applied without significant modifications in the case where one of the bands is not observed. Suppose, for example, that the J band is not available. Then we can assume an arbitrary

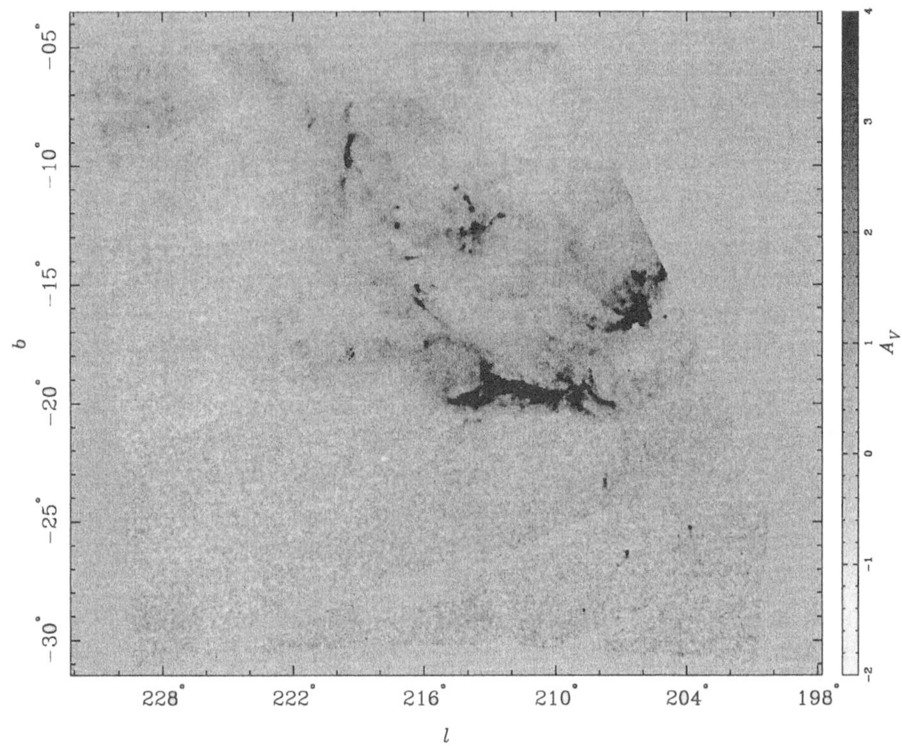

Fig. 2. The total region mapped using NICER. The cuts used in A_V emphasize the faint, diffuse halos around the main cores. The higher noise observed in the southern part is due to the smaller star density

value for this band and a set the corresponding error σ_J to an extremely large value. If we "blindly" apply Eqs. (6) and (5), we obtain a matrix C^{-1} with only a single non-vanishing element at the position $(2, 2)$. In other words, the use of a large error on J automatically suppresses the use of this band on the evaluation of $\hat{A}_V$. Interestingly, the same technique can be used if the missing band is H, which contributes to both colors $c_1 = J - H$ and $c_2 = H - K$: In this case the estimator will be composed only of the combination $c_1 + c_2 = J - K$, thus avoiding the use of the missing H band.

2.2 Spatial Smoothing

So far we have considered only a single star. In order to obtain a smooth extinction map with high signal-to-noise ratio, NICER applies a *spatial smoothing* to angularly close stars. The smoothing is performed using three different techniques, namely weighted mean, sigma-clipped mean, and weighted median.

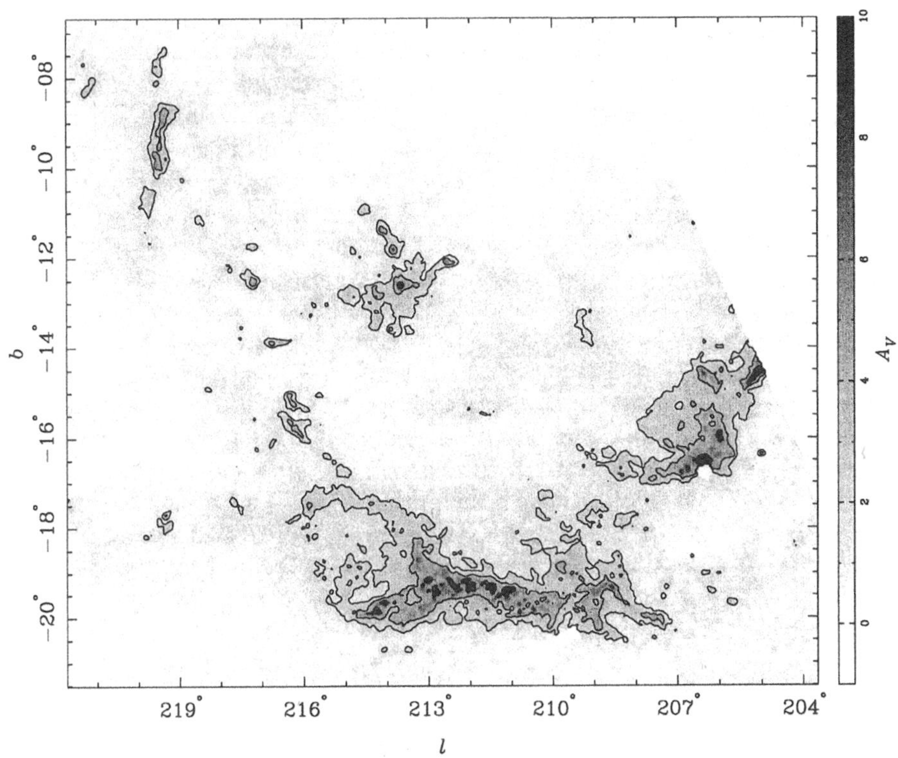

Fig. 3. Zoom of the central region of the NICER extinction map. Cuts in this map emphasize the faint, diffuse halos around the main cores. Contours are displayed at $A_V = \{2, 4, 8\}$ magnitudes; the maximum value obtained for A_V is 17.7 magnitudes

The choice of the smoothing technique is important for two main reasons: (i) The final map has a signal-to-noise ratio that strongly depends on the smoothing used; (ii) The smoothing is responsible for the selection of background stars. In particular, sigma-clipped mean and median are able to effectively remove foreground stars and are, in general, more robust than the simple mean.

3 Nicer at Work: The Orion Region from 2MASS Data

The Two Micron All Sky Survey (2MASS) offers a unique opportunity to test our algorithm. We have selected from the 2MASS archive a large area around the Orion and MonR2 nebulae (see also Carpenter 2000), with galactic coordinates

$$200^\circ < l < 230^\circ \,, \qquad -30^\circ < b < -5^\circ \,. \tag{7}$$

Given the area on the sky and the number of stars, the pipeline has suggested to perform the analysis using a grid of 820 × 700 points, with scale 2.4′ per pixel, and with Gaussian smoothing characterized by FWHM = 4.8′. We then have made a preliminary analysis using standard values for the average colors and color scatters, thus obtaining a first extinction map for the region. The map obtained is only used as a first check for the basic parameters chosen, to identify a control field, and to obtain the photometric properties of stars in our sample. In particular, inside the boundaries (7), we have found a region in the South-East of the map which does not show any significant extinction and which thus has been selected as control field (see Fig. 2). We then have used the color of stars in this control field to set the photometric parameters needed in our algorithm, namely the average colors of stars (i.e., $\langle J-H \rangle$ and $\langle H-K \rangle$) and the scatter in colors. Note that here only the 2MASS photometric data are used, and not the preliminary map (this map is used only to select the control field). As a result, an inaccurate choice for the initial constants of Eq. (1) to make the preliminary map has no influence on the statistical parameters obtained.

Figures 2 and 3 shows the final result obtained using a Gaussian smoothing characterized by FWHM $= 5'$ with σ-clipping (other smoothing schemes give similar results). Note that use of our optimized method, combined with the quality of 2MASS data, allows us to detect at an unprecedented level of detail the faint, extended halos on which the main cores are embedded. At the smoothing size used, we reach the excellent sensitivity $A_V = 0.13$ at $1\,\sigma$ level.

4 Conclusions

We have presented NICER, an optimized technique to produce highly accurate extinction maps from multi-band near-infrared photometric data. A first example of application of this new technique to 2MASS data has shown an improvement with respect to the standard NICE algorithm of a factor 2 on the noise variance. This way, we have been able to detect extended diffuse halos down to $A_V \simeq 0.5$ magnitudes.

References

1. J. Alves, C.J. Lada, E.A. Lada: ApJ 515, 265 (1999)
2. B.J. Bok: PASP 89, 597 (1977)
3. J.M. Carpenter: AJ 120, 3139 (2000)
4. M.A. Frerking, W.D. Langer, R.W. Wilson: ApJ 262, 590 (1982)
5. T.J. Jones, A.R. Hyland, J. Bailey: ApJ 282, 590 (1982)
6. S.G. Kleinmann, M.G. Lysaght et al.: Experimental Astronomy 3, 65 (1994)
7. C.J. Lada: IAU Symp. 170: CO: Twenty-Five Years of Millimeter-Wave Spectroscopy, Vol. 170, 387 (1996)
8. C.J. Lada, E.A. Lada et al.: ApJ 429, 694 (1994)
9. P.C. Myers in: NATO ASIC Proc. 540: The Origin of Stars and Planetary Systems, 67 (1999)

Centaurus A (detail)
(Kueyen/UT2 + FORS2)

Malcolm Walmsley and David Field

Physical Properties of Prestellar Cores

Malcolm Walmsley[1], Paola Caselli[1], Antonio Zucconi[2], and Daniele Galli[1]

[1] Osservatorio Astrofisico di Arcetri, Largo E. Fermi 5, I-50125 Firenze, Italy
[2] Dipartimento di Astronomia e Scienza dello Spazio, Università di Firenze, Largo E. Fermi 5, I-50125 Firenze, Italy

Abstract. We present a brief summary of current progress on research into the physical and chemical properties of pre–protostellar cores. In particular, we summarize recent investigations into the thermal structure and the degree of molecular depletion in the high density ($\sim 10^6$ cm^{-3}) nucleus of such cores.

1 Temperature and Density Structure

A brief study of models of protostellar evolution convinces one that in order to make advances in this subject, it is essential to understand the initial conditions. The initial density and temperature structure to a considerable extent defines the time dependence of the mass accretion rate and this in turn determines many of the observable properties of a young protostar. This is not the whole story of course and there are many *caveats* which one should apply. Still, it is clear that in most scenarios of star formation, the nascent star passes through something close to a hydrostatic (or magnetohydrostatic) equilibrium state just prior to the moment at which collapse ensues. Li & Shu (1996) introduced the concept of a pivotal state, with the scale-free magnetostatic density distribution approaching $\rho \propto r^{-2}$ for an isothermal equation of state, when the mass-to-flux ratio has a spatially constant value. More in general, gravitationally bound, starless, dense molecular cloud cores are usually referred to as "pre-protostellar cores" (e.g. Ward-Thompson et al. 1994) and are thought to evolve towards the pivotal state through ambipolar diffusion and/or the dissipation of turbulence (see Shu, Adams & Lizano 1987).

Magnetized self-gravitating equilibria tend to be somewhat flattened, unless high levels of toroidal fields are present. When the support against self-gravity is dominated by poloidal magnetic fields and/or rotation, the configuration becomes disk-like, but not necessarily axisymmetric (Galli et al. 2001). Also, if turbulent support is modeled as a polytropic or logatropic scalar pressure with a relatively soft equation of state (e.g. Lizano & Shu 1989, Galli et al. 1999), then all magnetostatic configurations with mass-to-flux ratio less than the critical value are highly flattened in the direction perpendicular to the magnetic field.

Pre–protostellar cores must clearly be cold, of stellar mass, and of column density considerably higher than their surroundings. Given that most molecular clouds have column densities corresponding to several magnitudes of visual extinction, this latter condition suggests that a pre–protostellar core will be an object with visual extinction A_V in the range 10-100 magnitudes and that this might be observable in either dust emission at long wavelengths or due to dust extinction in the infrared. Indeed as discussed by Alves elsewhere in this volume, sensitive near infrared photometry of giants to the rear of dust concentrations in molecular clouds has been a very successful way of determining the structure of condensations with visual extinctions as high as 50 magnitudes. Somewhat higher extinctions have been probed in the mid IR by both ISOCAM and the MSX experiment (e.g. Bacmann et al. 2000, Carey et al. 2000) but the background emission in this case is that of small grains heated by the UV radiation from nearby O-B stars. But probably most effort has gone into mapping the millimeter and sub–millimeter emission from pre–protostellar cores due to the availability of instruments such as SCUBA on the JCMT and MAMBO on the IRAM 30-m (e.g. Shirley et al. 2000, Ward–Thompson et al. 1999).

These latter observations are sensitive to the assumed dust temperature structure of the pre–stellar core. As a rule, isothermality has been assumed and indeed an isothermal fit gives reasonable agreement with the observed spectral energy distribution in several objects (e.g. André et al. 2000 for L1544). However, it seems improbable that real cores will be isothermal if only because their main heating source is the external interstellar radiation field. Indeed recently, Evans et al. 2001 and Zucconi et al. 2001 have considered the temperature structure which one would expect for plausible density distributions and conclude that one expects a dust temperature gradient such that the central temperature is about half the value in the external molecular cloud. While these results do not negate conclusions drawn on the basis of assumed isothermality, they do show that in certain cases the temperature gradient causes the dense center of the core to be invisible to sub-mm observers.

To illustrate the type of results one obtains from this sort of analysis, we show in Fig. 1 the temperature distribution and expected mm intensities one would expect for a Bonnor–Ebert sphere density distribution such as that which Alves et al. (2001) have used to fit their NIR data toward B68. We note here that the densities within about 4000 AU from the center of this globule should be sufficiently large that gas and dust temperatures are coupled (Goldsmith 2001). Thus inside 4000 AU, one expects the gas temperature to be close to that shown in Fig. 1 whereas at higher radii, the gas temperature will be independent of the dust temperature. It will depend on both the cosmic ray ionization rate and the degree of depletion of important coolants such as CO (see next section).

An interesting feature of Fig. 1 is the relatively flat behavior of the computed mm/sub-mm intensity as a function of radius relative to the visual extinction A_V (which is a measure of the column density). One sees that the temperature gradient causes the millimeter intensity to reflect the hot exterior of the globule rather than the cold nucleus and that therefore interpreting the millimeter data requires care. Another way of putting this is to say that the extinction (measured either in the mid or near IR) is in principle a more reliable way of deducing the density distribution than using mm/sub-mm emission.

From these studies, some solid evidence for "equilibrium structures" has emerged. In particular, we note the excellent fit obtained by Alves et al. (2001, see also Alves this volume) to a Bonnor–Ebert density distribution. In more general, evidence from both ISOCAM and sub-millimeter mapping (André et al. 2000, Bacmann et al. 2000) suggests a flattened density distribution in the inner core regions ($\rho \propto r^{-0.4}$ at radii r less than 4000 AU) with a steeper fall off (roughly $\rho \propto r^{-2}$) at higher radii. These distributions have been fit with some success to expectations based on models of quasistatic contraction of magnetically supported cores by Ciolek and Mouschovias (1994) and Ciolek and Basu (2000). Non–axisymmetric models have also had some success (see e.g. Basu 2000, Galli et al. 2001) and indeed many of the observed millimeter emission maps suggest departures from axial symmetry. All of this is qualitatively consistent with models of low mass star formation due to ambipolar diffusion across field lines (see e.g. Shu et al. 1997) and raises the question of the magnitude of the degree of ionization in the central regions of the core. It also raises the question of the kinematics of these prestellar cores which is briefly discussed in the next section.

2 Depletion and its Consequences

As mentioned above, the temperature distribution in the core nucleus depends on the gas cooling and this in turn depends on the gas phase abundances of

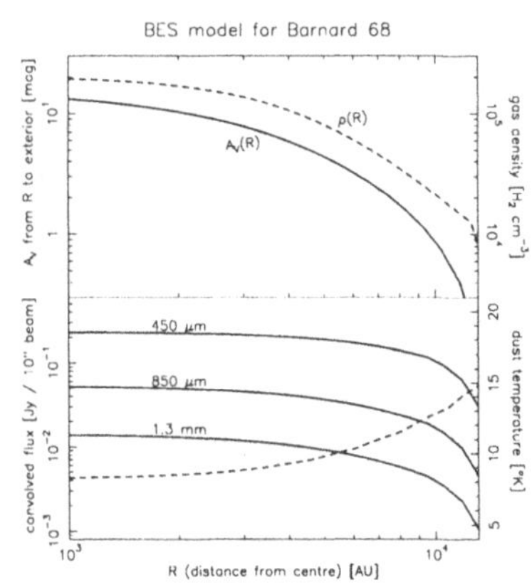

Fig. 1. Model for the dust temperature distribution in the core B68 based on the results of Zucconi et al. (2001). In the top panel, we show the density distribution and distribution of visual extinction (from the center of the core) for the model used by Alves et al. (2001) to fit their NIR extinction data. In the bottom panel, we show the corresponding dust temperature distribution (dashed) computed using the mean interstellar radiation field proposed by Black (1994) and optical constants from Ossenkopf and Henning (1994). We also show the intensities as a function of offset which one would expect on the basis of this model at some typical millimeter and sub-millimeter frequencies

important coolants such as CO. There is now abundant evidence that at low dust temperatures and high density (probably below 15 K and above 10^5 cm^{-3}), most CO becomes deposited as ice upon the surface of dust grains. The evidence for this is both the solid-CO ice absorption features seen in the near IR (see e.g. Langer et al. 2000) and the anti–correlation between CO isotopomer emission and dust emission seen towards many nearby molecular clouds (see Caselli et al. 1999, Kramer et al. 1999, Mitchell et al. 2001). These observations show that CO is quite commonly depleted by at least an order of magnitude and probably much more in many cores. As well as CO, many other species such as CS and HCO^+ have the characteristic that their column density appears to anti–correlate with that of the dust. In such cases, it seems likely that the observed molecular emission in the direction of the dust maxima comes from foreground or background layers at lower densities and higher temperatures than the dense depleted pre–stellar core.

If this was true for all species, the unfortunate consequence would be that we are denied the possibility to examine the kinematics of the dense gas. Fortunately, this appears not to be the case and a number of studies have shown that at least N_2H^+ and probably ammonia have their column density maxima close to the dust peaks (see Caselli and Walmsley 2001, Bergin et al. 2001). Bergin et al. for example find towards IC5146 that CO and CS are underabundant towards cores of visual extinctions greater than 10 magnitudes whereas N_2H^+ is relatively undepleted. Thus, the kinematics of the dense gas can be traced with N_2H^+ but not with CS. Why is there this dichotomy?

One plausible explanation which goes back to Bergin and Langer (1997) is that N_2H^+ is basically a tracer for molecular nitrogen which is known to be extremely volatile with a binding energy to plausible dust grain surfaces smaller than that of CO. A relatively small difference in the dissociation energy of N_2 from grain surfaces with respect to CO suffices to explain a large difference in the gas phase abundances of N_2 and CO. Indeed, it is quite plausible that there could be layers where N_2 is the most abundant gas phase species containing heavy elements (C,N,O, ...). This has important consequences for the gas cooling rate and one can have situations where the gas temperature rises considerably (Goldsmith 2001) if the density at which CO depletion takes place is less than the critical density at which collisions couple dust and gas temperatures. These two critical densities however are rather close to one another and their relative magnitudes depend on the composition of the icy grain surfaces.

Fig. 2 illustrates some of the properties we expect in such pre–protostellar cores. Parameters have been chosen to allow an approximate fit to observations of the nearby core L1498. The density distribution shown has been obtained by fitting to the observed millimeter intensity (ignoring departures from isothermality and spherical symmetry). The CO depletion has been chosen to match the observed intensity of $C^{17}O$(1-0) at the cloud center (see Caselli et al. 1999, 2001b for a discussion of this) and the N_2 abundance has been computed assuming a binding energy of N_2 to the grain surface of 800 K (as compared to 1200 K for CO). The result is a relatively constant N_2H^+ abundance as a function of radius and an electron abundance which varies slightly more rapidly with density than the $x(e)$ proportional to $n^{-0.5}$ dependence usually assumed in theoretical analyses.

More detailed modelling and observations have been carried out for L 1544 (Caselli et al. 2001a, 2001b) which differs from L 1498 in that the central column density (as measured by the dust emission) is a factor roughly 5 larger than for L1498. It also shows evidence on large scales for infall which has been studied in detail by Tafalla et al. (1998) and by Williams et al. (1999). A model interpreting this behavior in terms of a disk–like core contracting across magnetic field lines due to the effects of ambipolar diffusion has been put forward by Ciolek and Basu (2000). This manages to reproduce some features of the observed kinematics (see Caselli et al. 2001a) without however fitting completely the observed behavior. The ionization degree derived in the central regions of L1544 by Caselli et al. (2001b) is consistent moreover with an ambipolar diffusion timescale of the same order as the free–fall time. In terms of these models, L 1544 is very close to the pivotal state from which dynamical collapse will take place assuming that it has not already commenced. There is presently however no evidence for dynamical collapse in the sense of infall onto a point mass.

Future work in this field must contend with the fact that millimeter emission lines from the cold nuclei of dense cores at densities higher than a few times 10^5 cm^{-3} may be very difficult or impossible to detect. The reason for this is that both the decline of the temperature in the central density peak and the increase in depletion cause the emission from high density regions to be weak relative to that from the warmer less depleted surroundings. As a consequence, absorption measurements in the near and mid infrared may be the most successful approach to probing such regions. Of particular interest will be observations of the ice bands in the 2-5 micron region where one might reasonably expect to see a

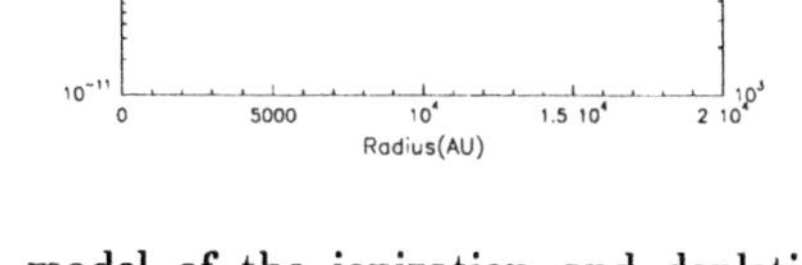

Fig. 2. Results from a model of the ionization and depletion in the nearby pre–protostellar core L1498. We show as a function of radius the density $n(H_2)$ and the fractional abundance of electrons $x(e)$, of N_2H^+ and of $C^{17}O$. The fall-off of the $C^{17}O$ abundance at small radii reflects the depletion of CO onto dust grains (in this model the $C^{17}O$ abundance is proportional to that of CO). The N_2H^+ abundance in contrast remains relatively constant with radius due partially to the fact that molecular nitrogen is less depleted than CO

greater proportion of "apolar" ices (ices constituted of species such as CO and N_2) than in the lines of sight probed to date.

References

1. J. Alves, C.J. Lada, E.A. Lada: Nature 409, 159 (2001)
2. P. André, D. Ward–Thompson, M. Barsony: p. 59 in *Protostars and Planets IV*, edited by V.Mannings, A.P. Boss, S.S.Russell, publ. U. of Arizona press (2000)
3. A. Bacmann, P.André, J.-L. Puget, A. Abergel, S. Bontemps, D. Ward–Thompson: A&A 361, 555 (2000)
4. S. Basu: ApJ, 540, L103 (2000)
5. E.A. Bergin, W.D. Langer: ApJ, 486, 316 (1997)
6. E.A. Bergin, D.R. Ciardi, C.J. Lada, J. Alves, E.A. Lada: ApJ (in press), (2001)
7. J.H. Black: in *The first symposium on the infrared Cirrus and diffuse interstellar clouds*, ASP Conference Series 58, edited by R.M. Cutri and W.B. Latter
8. S.J. Carey, P.A. Feldman, R.O. Redman, M.P. Egan, J.M. MacLeod, S.D. Price: ApJ, 543, L157 (2000)
9. P. Caselli, C.M. Walmsley, M. Tafalla, L. Dore, P.C. Myers: ApJ, 523, L165 (1999)
10. P. Caselli, C.M. Walmsley: in *From Darkness to Light*, edited by T.Montmerle, P. André, in press
11. P. Caselli, C.M. Walmsley, A. Zucconi, M. Tafalla, L. Dore, P.C. Myers: ApJ, in press
12. P. Caselli, C.M. Walmsley, A. Zucconi, M. Tafalla, L. Dore, P.C. Myers: ApJ (submitted) (2001b)
13. G. Ciolek, T. Mouschovias: ApJ, 425, 142 (1994)
14. G. Ciolek, S.Basu: ApJ, 529, 925 (2000)
15. N.J. Evans II, J.M.C. Rawlings, Y.L. Shirley, L.G. Mundy: ApJ, in press
16. D. Galli, S. Lizano, Z.-Y. Li, F.C. Adams, F.H. Shu: ApJ, 521, 630 (1999)
17. D. Galli, F.H. Shu, G. Laughlin, S. Lizano: ApJ, 551, 367 (2001)
18. P.F. Goldsmith: ApJ, in press
19. C. Kramer, J. Alves, C.J.Lada, E.A. Lada, A. Sievers, H. Ungerechts, C.M. Walmsley: A&A 342, 257 (1999)
20. W.D. Langer, E.F. van Dishoeck, E.A. Bergin, G.A. Blake, A.G.G.M. Tielens, T. Velusamy, D.C.B. Whittet: p. 29 in *Protostars and Planets IV*, edited by V. Mannings, A.P. Boss, S.S. Russell, publ. U. Arizona (2000)
21. Z.-Y. Li, F.H. Shu: ApJ, 472, 211 (1996)
22. S. Lizano, F.H. Shu: ApJ, 342, 834 (1989)
23. G.F. Mitchell, D. Johnstone, G. Moriarty-Schieven, M. Fich, N.F.H. Tothill: ApJ, in press
24. V. Ossenkopf, T. Henning: A&A 291,943 (1994)
25. Shirley Y.L., Evans N.J. III, Rawlings J.M.C., Gregersen E.M.: ApJS, 131, 249 (2000)
26. F.H. Shu, F.C. Adams, S. Lizano: Ann. Rev. Astron. Astrophys. 25,31 (1987)
27. M. Tafalla, D. Mardones, P.C. Myers, P. Caselli, R. Bachiller, P.J. Benson: ApJ, 504, 900 (1998)
28. Ward-Thompson, D., Scott, P.F., Hills, R.E. et al.: MNRAS, 268, 276 (1994)
29. D. Ward–Thompson, F. Motte, P. André: MNRAS, 268, 276 (1999)
30. J.P. Williams, P.C. Myers, D.J. Wilner, J. Di Francesco: ApJ, 513, L61 (1999)
31. A. Zucconi, C.M. Walmsley, D. Galli: A&A (in press) (2001)

Horsehead Nebula
(Kueyen/UT2 + FORS2)

João Alves

Molecular Cloud Structure: The VLT View

João Alves[1], Charles Lada[2], Elizabeth Lada[3], Marco Lombardi[4], and Edwin A. Bergin[2]

[1] European Southern Observatory, Garching bei München, Germany
[2] Harvard-Smithsonian Center for Astrophysics, Cambridge, USA
[3] University of Florida, Gainsville, USA
[4] Institüt für Astrophysik und Extraterrestrische Forschung, Bonn, Germany

1 The Search for the Initial Conditions to Star Formation

Despite 30 years of molecular spectroscopy of dark clouds little is understood about the internal structure of these objects and consequently about the initial conditions that give rise to star and planet formation. This is largely due to the fact that molecular clouds are primarily composed of molecular hydrogen, which is virtually inaccessible to direct observation. The traditional methods used to derive the basic physical properties of these clouds therefore make use of observations of trace H_2 surrogates, namely those rare molecules with sufficient dipole moments to be easily detected by radio spectroscopic techniques (e.g., Lada 1996, Myers 1999), and interstellar dust, whose thermal emission can be detected by radio continuum techniques (e.g., André et al. 2000). However, as discussed in the previous article in this book by M. Walmsley and collaborators, the interpretation of results derived from these methods is not always straightforward (see also Alves, Lada, & Lada 1999 and Zucconi, Walmsley, & Galli 2001). Several poorly constrained effects inherent in these techniques (e.g., deviations from local thermodynamic equilibrium, opacity variations, chemical evolution, small-scale structure, depletion of molecules, unknown emissivity properties of the dust, unknown dust temperature) make the construction of an unambiguous picture of the physical structure of these objects a very difficult task. There is a clear need for a less complicate and more robust tracer of H_2 to access not only the physical structure of these objects but also to accurately calibrate molecular abundances and dust emissivity inside these clouds. The deployment of sensitive, large format infrared array cameras on large telescopes however, has fulfilled this need by enabling the direct measurement of the dust extinction toward thousands of individual background stars observed through the densest regions of a molecular cloud. Such measurements are free from the complications that plague molecular-line or dust emission data and enable detailed maps of cloud density structure to be constructed.

2 The Method

The most straightforward and reliable way to measure molecular cloud structure is to measure dust extinction of background starlight. We have developed a

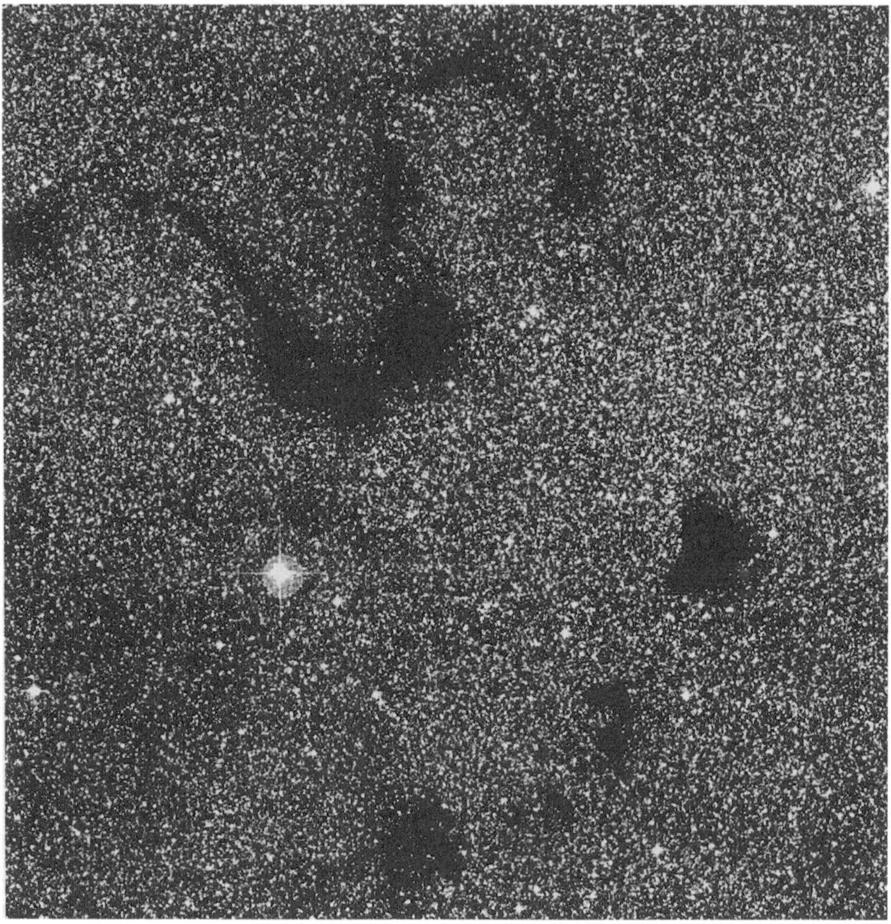

Fig. 1. Barnard 72 (the S-nebula) and Barnard 68 (right of center) (DSS plate)

powerful technique for measuring and mapping the distribution of dust through a molecular cloud using data obtained in large-scale, multi-wavelength, infrared imaging surveys. We have conclusively demonstrated the efficacy of this technique with our study of the dark cloud complex IC 5146 (Lada et al. 1994, Lada et al. 1999), L 977 (Alves et al. 1998), and Barnard 68 (Alves, Lada, & Lada 2001), where we detected nearly 7000 infrared sources background to these clouds and produced detailed maps of the extinction across the cloud to optical depths and spatial resolution an order of magnitude higher than previously possible (Av $\sim$ 40 magnitudes, spatial resolution $\sim$ 10 arcsec).

We have used our extinction observations to measure the masses, density structure, extinction laws, and distances to these objects. We found the radial density profiles of filamentary clouds (IC 5146 and L 977) to be well behaved and smoothly falling with a power-law index of $\alpha = -2$, significantly shallower than

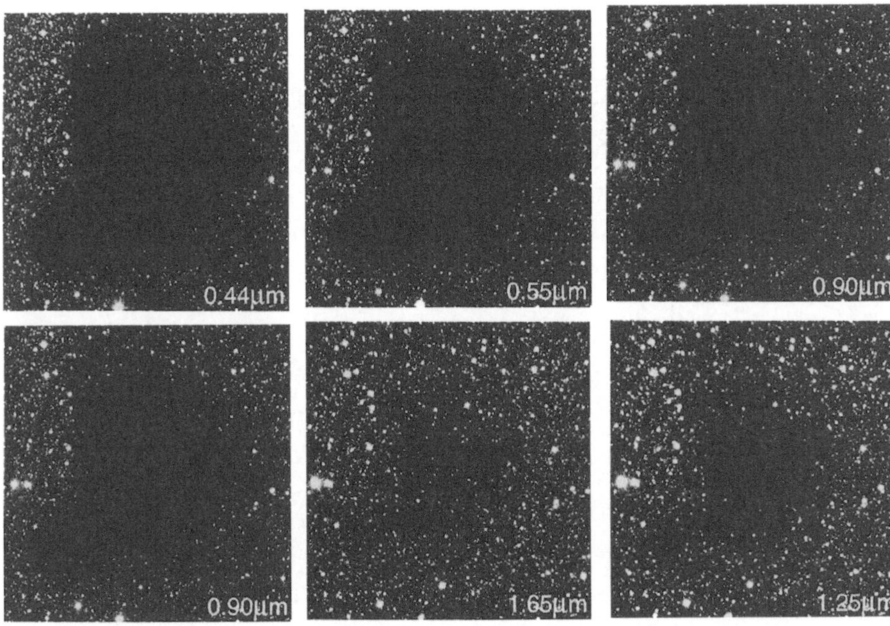

Fig. 2. Deep BVIJHK-band (0.44μm, 0.55μm, 0.9μm, 1.25μm,1.65μm, 2.16μm) images of dark molecular cloud Barnard 68 taken with ESO's Very Large Telescope (VLT) and New Technology Telescope (NTT). The wavelength dependence of interstellar dust extinction in Barnard 68 is clearly depicted in these images. The analysis of the near-infrared colors of the stars seen through the dark cloud allow the construction of a high resolution mass map as traced by dust extinction and the most finely sampled and higher S/N density profile ever obtained for a cold dark cloud

predicted by early theoretical calculations of Ostriker (1964) ($\alpha = -4$), but in good agreement with recent work on helical magnetic fields in filamentary molecular clouds by Fiege & Pudritz (2000). Moreover because we are using pencil beam measurements of dust column density along the line of sight to background stars we were able to demonstrate that the small-scale structure of the clouds is surprisingly smooth with random density fluctuations ($\delta A_V/A_V$) present at very small levels ($<$ 3%!) on size-scales of a few hundred AU. This result is in very good agreement with optical studies of the small scale structure of the diffuse Interstellar Medium (ISM) (Thoraval, Boissé, & Duvert 1999).

When convolved to the appropriate spatial resolution our maps showed structure in the dust distribution which was strikingly well correlated with millimeter-wave CO and CS emission maps of the cloud, although showing crucial differences at high optical depths where these other tracers of column density become unreliable. These comparisons enabled us to directly derive CO, CS, and N_2H^+ abundances, and variations of, over an extinction range of 1–30 magnitudes, a range nearly an order of magnitude greater than achieved previously with optical star counting techniques. In a recent experiment we were able to make a direct

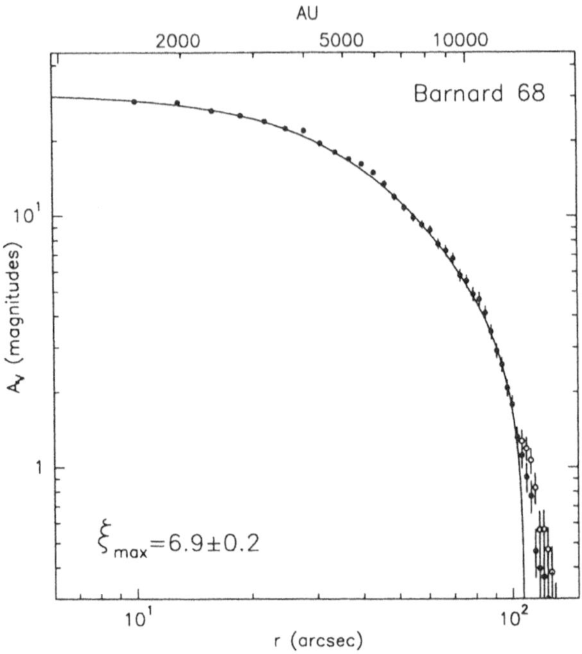

Fig. 3. Azimuthally averaged radial density profile of Barnard 68. The filled circles show the data points for the averaged profile of a sub-sample of the data that do not include the cloud's South-East prominence, seen in Figure 1. The open circles, on the other hand, include this prominence. The error bars were computed as the rms dispersion of the extinction measurements in each averaging annulus and are smaller than the data points for the central regions of the cloud

measurement of molecular depletion in a cold cloud core (Kramer et al. 1999, Bergin et al. 2001). Finally, a comparison between our extinction data and millimeter continuum emission data, allowed a most accurate measurement of the ratio of dust absorption coefficients at millimeter and near-infrared wavelengths (Kramer et al. 1998).

3 Barnard 68 as a Stellar Seed

Recently we have been concentrating efforts on mapping the densest regions of the ISM that are likely places of future star formation. We performed very sensitive near-infrared imaging observations to map the structure of a type of dark cloud known as a Bok globule (Bok & Reilly 1947; Reipurth 1983), one of the least complicated configurations of molecular gas known to form stars. The first target of our study, Barnard 68 (Figure 1 and 2), is itself one of the finest examples of a Bok globule, and was selected because it is a morphologically simple molecular cloud with distinct boundaries, a known distance (125 pc; Launhardt & Henning 1997) and temperature (16 K; Bourke et al. 1995) and no signatures

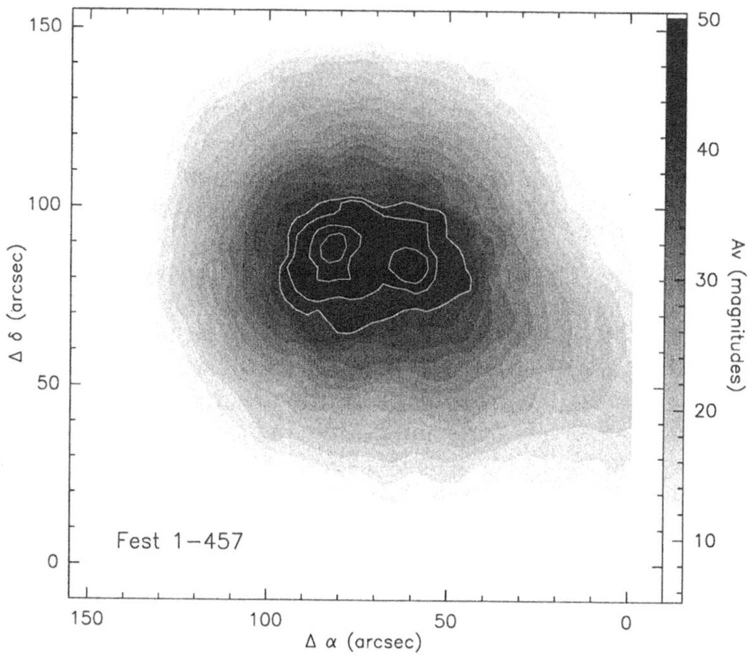

Fig. 4. Preliminary extinction map of Fest 1-457 dense cloud core obtained with ISAAC on VLT Antu. This is the first 5″ resolution extinction map ever constructed for a molecular cloud core. Although a fairly round core at lower column densities, Fest 1-457 overall structure is clearly flattening towards its center where it "breaks" into two extinction maxima

of ongoing star formation, such as IRAS sources, outflows, or mm continuum sources (Reipurth, Nyman, Chini 1996).

We have accurately sampled the dust extinction and column density distribution through the Barnard 68 cloud at more than a thousand positions with extraordinary (pencil beam) angular resolution. We smoothed these data to construct a map of extinction and an azimuthally averaged dust column density profile of the cloud (Figure 3). This is the most finely sampled and highest signal-to-noise radial column density profile ever obtained for a dense and cold molecular cloud. For the first time, the internal structure of a dark cloud has been specified with a detail only exceeded by that characterizing a stellar interior.

3.1 Bonnor-Ebert Spheres and Reverse Engineering: Distance and Gas-to-Dust Ratio to Few Percent

The close correspondence of the observed extinction profile with that predicted for a Bonnor-Ebert sphere strongly suggests that Barnard 68 is indeed an isothermal, pressure confined, and self-gravitating cloud. It is also likely to be in a state near hydrostatic equilibrium with thermal pressure primarily supporting the cloud against gravitational contraction. For Barnard 68, ξ_{max} is very near and slightly in excess of the critical radial parameter and the cloud may be only marginally stable and on the verge of collapse. If this is the case we should expect molecular radio-spectroscopy of this cloud to reveal a quiet, non-turbulent cloud with narrow molecular emission lines. Indeed, preliminary results from our radio-spectroscopy observing campaign of Barnard 68 (with the IRAM 30m Radio Telescope at Pico Veleta, Granada) reveal that the linewidth of the $C^{18}O$ line in this cloud is $\Delta v \sim 0.18$ kms^{-1}, one of the narrowest lines ever observed in molecular clouds, in perfect agreement with the Bonnor-Ebert sphere nature of Barnard 68.

The exact physical state of the Barnard 68 cloud is further constrained by the fact that ξ_{max}, which is solely derived from the shape of the observed column density distribution, uniquely specifies the combination of central density, sound speed and physical size that characterizes the cloud (e.g., Alves et al. 2001). Independent knowledge of any two of these parameters directly determines the value of the third. For example, in Barnard 68 we independently measure the dust extinction (which is directly related to the mass column density, via the gas-to-dust ratio in the cloud) and the angular size of the cloud (which is directly related to its physical size via the cloud's distance). Thus, if we know the temperature of the cloud, our measurement of its extinction and angular size (combined with the constraint that $\xi_{max} = 6.9 \pm 0.2$) independently gives the distance to the cloud, provided the gas-to-dust ratio is assumed. The temperature of the molecular gas in Barnard 68 has been previously measured using observations of emission from the (1,1) and (2,2) metastable transitions of the ammonia molecule and found to be 16 ± 1.5 K (Bourke et al. 1995). For a canonical gas-to-dust ratio (1.9×10^{21} protons/magnitude; Bohlin, Savage, & Drake 1978), we derive a distance to Barnard 68 of 112 ± 3 pc. Here the quoted uncertainty (3%) arises solely from the uncertainty in ξ_{max}. Accounting for the uncertainty in the temperature measurement the overall uncertainty increases only to 8%, or ± 9 pc. From its association with the Ophiuchus complex the distance to the cloud has been estimated to be 125 ± 25 pc (de Geus et al 1989), which within the uncertainties agrees with our derivation. This, in turn, implies that the gas-to-dust ratio in this dense cloud must be close to the canonical interstellar value. Indeed, if we independently know the distance to the cloud, our modeling directly yields the gas-to-dust ratio in the cloud. Assuming a distance of 125 pc our measurements allow a high precision determination (2%) for the gas-to-dust ratio in this cloud of $1.73 \pm 0.04 \times 10^{21}$ protons/magnitude. If we also account for the uncertainties in distance and temperature the overall uncertainty (accuracy) of our determination increases to $\pm 0.4 \times 10^{21}$ protons/magnitude, or 23%. Within the overall error this ratio is the same as the long accepted value characterizing low-density interstellar gas and is the first independent and relatively accurate determination of this important astrophysical parameter in a dense molecular cloud core.

4 The Future

Significant progress in extinction mapping studies will result when the DENIS and 2MASS all sky near-infrared imaging surveys are released (Lombardi & Alves 2001). This surveys will be sufficiently sensitive to produce moderate depth extinction maps ($A_V \leq 25$ mags) of many nearby dark clouds in those directions of the Galaxy where field stars suffer little extraneous extinction. In the immediate future large aperture telescopes, such as the VLT outfitted with ISAAC and NAOS/CONICA, or space telescopes (see Harvey & Wilner's poster in this book and Harvey et al. 2001), will provide the complementary deeper surveys of the higher extinction regions ($25 \leq A_V \leq 60$ mags) in these clouds (see Figure 4 for an example). Very soon one will be able to obtain a high-resolution column density map of a molecular cloud from a few tenths of a magnitude (below the threshold for the detection of ^{12}CO) to over 60 magnitudes of visual extinction.

The authors are thankful to Monika Petr-Gotzens for helpful discussions during the preparation and acquisition of the VLT observations.

References

1. Alves, J., Lada, C.J., Lada, E.A., Kenyon, S., & Phelps, R. 1998, ApJ, 506, 292
2. Alves, J., Lada, C.J., & Lada, E.A. 1999, ApJ, 515, 265
3. Alves, J., Lada, C.J., & Lada, E.A. 2001, Nature, 409, 159
4. André, P., Ward-Thompson, D., & Barsony, M. 2000, *Protostars and Planets IV*, (Tucson: University of Arizona Press), eds. Mannings, V., Boss, A.P., Russell, S.S.
5. Bergin, E., Ciardi, D., Lada, C., Alves, J., Lada, E. 2001, 557, 209
6. Bok, B. & Reilly, E. 1947, ApJ, 105, 255
7. Bourke, T., Hyland, A., Robinson, G., James, S. Wright, C. 1995, MNRAS, 276, 1067
8. de Geus, E., de Zeeuw, P. & Lub, J. 1989, A&A, 216, 44
9. Fiege, J., & Pudritz, R. 2000, MNRAS, 311, 85
10. Harvey, D., Wilner, D., Lada, C., Myers, P., Alves, J., Chen, H. 2001, ApJ, in press
11. Kramer, C., Alves, J., Lada, C., Lada, E., Sievers, A., Ungerechts, H., Walmsley, M. 1997, A&A, 329, 33
12. Kramer, C., Alves, J., Lada, C.J., Lada, E.A., Sievers, A., Ungerechts, H., Walmsley, C.M. 1999, A&A, 342, 257
13. Lada, C.J., Lada, E.A., Clemens, D.P., & Bally, J. 1994, ApJ, 429, 694
14. Lada, C.J. 1996, Proceedings of IAU Symposium *CO: Twenty-Five Years of Millimeter-Wave Spectroscopy*, eds. W.B. Latter et al., (Kluwer-Dordrecht), 387
15. Lada, C.J., Alves, J., Lada, E.A. 1999, ApJ, 512, 250
16. Launhardt, R. & Henning, T. 1997, A&A 326, 329
17. Lombardi, M. & Alves, J. 2001, A&A, in press
18. Ostriker, J. 1964, ApJ, 140, 1056
19. Reipurth, B. 1983, A&A, 117, 183
20. Reipurth, B., Nyman, L. & Chini, R. 1996, A&A, 314, 258
21. Thoraval, S., Boissé, P., Duvert, G. 1999, A&A, 351, 1051
22. Zucconi, A., Walmsley, M., Galli, D. 2001, A&A, in press

Lex Kaper and Carsten Kramer

The Physical Structure of a Prestellar Core in IC 5146

Carsten Kramer[1], João Alves[2], and Charles Lada[3]

[1] Universität zu Köln, KOSMA, Germany
[2] ESO Garching, Germany
[3] Harvard-Smithsonian Center for Astrophysics, USA

Abstract. We observed an evenly sampled map of the $^3P_1$–$^3P_0$ line of neutral atomic carbon in a $0.5 \times 0.5\,\mathrm{pc}^2$ quiescent, dark core region of IC 5146 immersed in the interstellar radiation field. For comparison at the same spatial resolution of 0.067 pc, we mapped in addition the emission of ^{13}CO 2–1, and, previously, of C^{18}O 2–1 and A_V. The evenly sampled map of A_V was created by Lada et al. [14] from NIR pencil beam observations towards background stars. We find that C I and ^{13}CO 2–1 line intensities are well correlated reflecting their similar line shapes. In contrast to C^{18}O, their integrated intensities stay almost constant with extinctions and are poorly correlated. We argue that this is indicative of unresolved small scale structure and use a spherical symmetric model of photon dominated regions (PDRs) to study the emergent line intensities. The observations are consistent with a large number of small clumps having radii of e.g. 760 AU, masses of 10^{-3} M$_\odot$ and surface densities of 10^5 cm^{-3}. The volume filling factor is $\sim 10\%$.

1 Introduction

Cooling due to atomic carbon and carbon monoxide are of the same order of magnitude in most external galaxies [5]. Also, Galactic emission of the two atomic fine structure lines of carbon is widespread and the average cooling power of C I is about one third of that of CO [4,1]. Studies of individual Galactic giant molecular clouds reveal an abundance ratio [C I]/[CO] of typically 0.1–0.2 [8]. There is a strong correlation of C I and ^{13}CO line intensities in molecular clouds with nearby OB stars heating the clouds with their strong FUV radiation, as was found by many authors [15,7].

Extended C I emission is observed also in dark clouds where there are no strong sources of UV radiation: Tatematsu et al. [17] recently observed selected positions in L134N, TMC-1, and IC 5146 at 2.5′ angular resolution. In strong contrast to giant molecular clouds with high impinging UV fields, Tatematsu et al. find only small variations of integrated C I intensities over the surface of the dark clouds, and in addition, only low correlations between C I and ^{13}CO integrated intensities.

Our study aims at mapping at high spatial and spectral resolutions C I and ^{13}CO in one dark cloud core of IC 5146 subject to weak impinging UV fields. NIR observations were used to derive a map of optical extinctions at the same spatial resolution, thus allowing to accurately study the total H_2 column density for comparison.

The core we are studying is part of an elongated filament, the "Northern Streamer", lying at a distance of 460 pc. This filament is delineated by large-scale maps of optical extinctions and by ^{13}CO, C^{18}O, CS at moderate resolutions [14,13]. Only few IRAS point sources and CO outflows were detected along this ridge, none in the core studied here [3,2]. The small IC 5146 star cluster lies at a projected distance of 8.1 pc to the east of the core. Optical extinctions vary between 3 and 27 mag over the core of $0.5 \times 0.5\,\mathrm{pc}^2$ and $40\,\mathrm{M}_\odot$. Observations of low-J C^{18}O and C^{17}O lines [11], at the same angular resolution of 0.067 pc (30″), are consistent with densities between 10^4 and $10^5\,\mathrm{cm}^{-3}$ and a gas kinetic temperature of 10 K. A comparison with optical extinctions indicates depletion of CO onto dust grains in the interior of the core at $A_\mathrm{V} \geq 10$ mag. Mapping of the dust emission at 1.2mm indicates, again in comparison with the A_V map, a dust temperature gradient from 20 K at the core surface to 8 K in the interior [10,9]. Recent models of the temperature and density distribution in prestellar cores (Walmsley et al., this volume) appear to confirm these observational results.

As part of a continuing study of this core, we constructed fully sampled maps of the ground state submillimeter $^3P_1$–$^3P_0$ line of atomic carbon and for comparison of the ^{13}CO 2–1 line. These data were resampled to the same angular resolution of 30″ as the C^{18}O and A_V data previously observed. Figure 1 shows scatterplots of the observed line integrated intensities versus optical extinctions.

2 PDR Model: Towards a Consistent Picture

A self consistent description of the cooling lines of C I and ^{13}CO escaping from the surfaces of dense, cold cores may be possible in the framework of PDR models taking into account heating by external UV fluxes leading to temperature gradients, density gradients, a network of gas phase reactions, and, if needed, the cloud clumpiness.

We have used PDR models of spherical symmetric clumps [16] to model the observed C I and ^{13}CO line ratios and line integrated intensities (Fig. 1). A model clump is characterized by its total Mass M, its surface H_2 density n_s, and the strength of the impinging isotropic UV field χ heating the clump. All clumps have a H_2-density profile following a power law: $n(r) = n_s(r/R)^{-\alpha}$ for $(R \geq r \geq 0.2\,R)$ and constant density $n(r)$ for $(r \leq 0.2R)$, where n_s is the surface density and R is the total radius of the clump. The PDR code calculates the relative density of each species and the gas and dust temperatures within each layer of the clump using a simple escape probability formalism. In a second step, surface integrated intensities of dedicated species are calculated more accurately using the `ONION` radiation transport code for spherically symmetric clouds [6].

Probably the main source of UV photons, the IC 5146 star cluster, lies at a projected distance of 8.1 pc to the south-east of the region mapped in C I. Assuming that BD+46.3474, a B0V star and the most massive member of the cluster, is the only source of FUV photons, that the star is a black body at the effective temperature of 30900 K corresponding to its spectral type, and that its

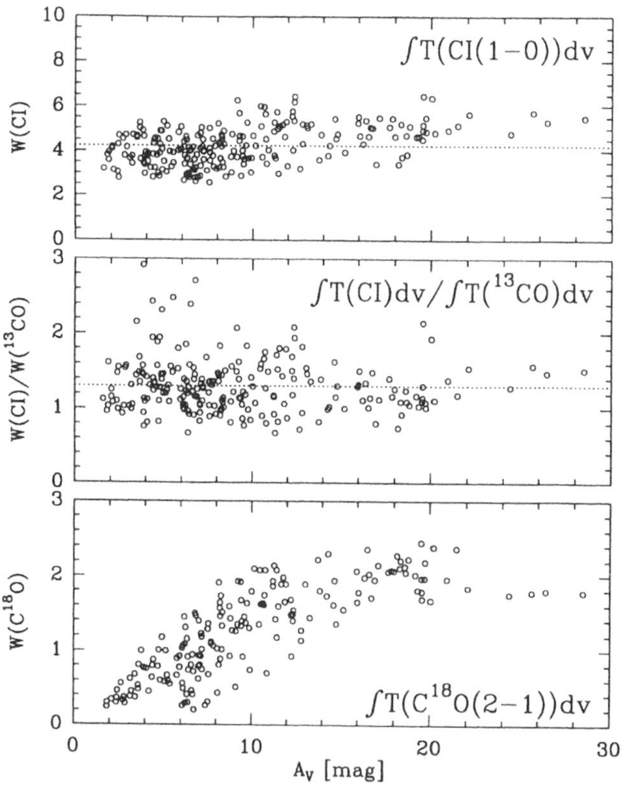

Fig. 1. Velocity integrated intensities on a $T_{\rm mb}$ scale at all positions on a 15″ grid with 30″ common resolution plotted versus optical extinctions. (**a**) Integrated C I intensities stay constant at $4.2 \pm 0.9\,{\rm Kkms}^{-1}$. (**b**) The ratio of CI 1–0 over ^{13}CO 2–1 integrated intensities stays constant as well at 1.3 ± 0.4. Both findings indicate that C I and ^{13}CO only trace the surface of the core and not the cloud interior, in contrast to e.g. C^{18}O (**c**)

radiation is not diluted by material along the way, we derive a flux of $\chi = 3$ in units of the average interstellar radiation field at a distance of 8.1 pc.

The total mass of the IC 5146 core as derived from the NIR extinction data is $40\,{\rm M}_\odot$ [11]. We expect the region to be clumpy and therefore varied the clump mass of the PDR models between $10^{-4}\,{\rm M}_\odot$ and $100\,{\rm M}_\odot$ (Fig. 2). In addition, we varied the average density between $10^3\,{\rm cm}^{-3}$ and $10^6\,{\rm cm}^{-3}$.

Figure 2 shows the resulting ratio of C I over ^{13}CO velocity integrated intensities, averaged over the surface of the clumps, for the modelled range of clump masses and surface densities. The ratio varies independent of the clump radii between 0.3 for massive clumps with high densities and more than 1000 for low mass clumps with low densities. The interior of the former clumps is well shielded from FUV radiation, they thus possess only a thin layer of C I. In contrast,

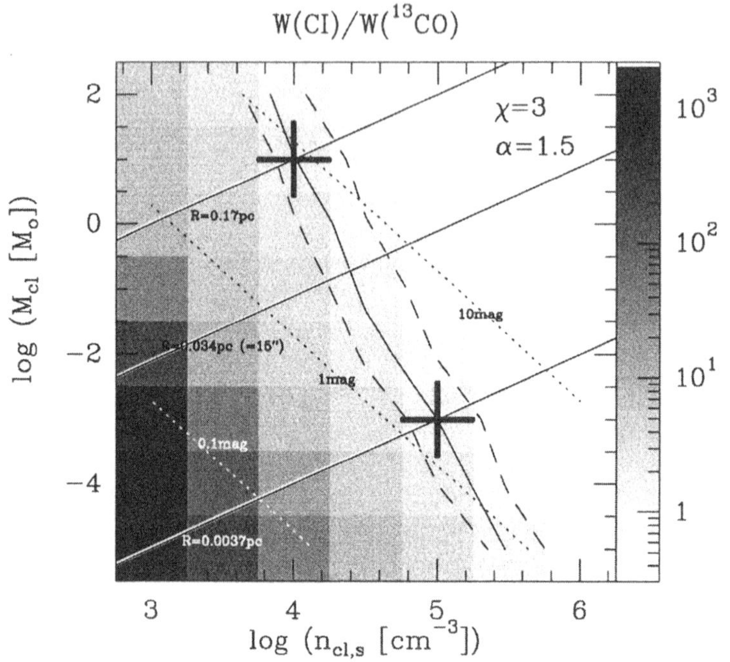

Fig. 2. Results from PDR models for an external isotropic FUV-field of $\chi = 3$ and a density profile of $\alpha = 3/2$. The ratio of line integrated intensity of C I 1–0 over ^{13}CO 2–1, averaged over the clump surface, is shown for a range of clump masses $M_{\rm cl}$ and surface H_2 densities n_s. Drawn lines show the observed ratio 1.3 ± 0.4. Crosses mark the two scenarios discussed in the text: the core could consist of one massive clump of $M = 10\,{\rm M}_\odot$ and $R = 0.17$ pc or many small, dense clumps, each having $M \sim 10^{-4}\,{\rm M}_\odot$, $n_s = 10^5\,{\rm cm}^{-3}$, and $R = 0.0037$ pc. White lines delineate constant radii. Dotted lines delineate constant average optical extinctions: 0.1, 1, and 10 mag. At low A_V most of the molecular gas is dissociated, i.e. the C I/^{13}CO line ratio is high

the latter clumps are easily penetrated, the contribution of C I is therefore much stronger than that of ^{13}CO.

The observed ratio of 1.3 ± 0.4 (Fig. 1) and the total core mass of $40\,{\rm M}_\odot$ are consistent with either one massive clump of relatively low density, $n_s = 10^4\,{\rm cm}^{-3}$, and radius of $R = 0.17$ pc or many small clumps of low mass, high density, and small radii. The following paragraphs will show that the latter model is more appropriate.

2.1 A Single Massive Clump Fails

We have calculated the emission of the C I 1–0 and ^{13}CO 2–1 lines, as seen in pencil beams across the center of a massive, large clump of $10\,{\rm M}_\odot$, $n_s = 10^4\,{\rm cm}^{-3}$, and $\chi = 3$. With varying impact parameter, the ratio C I/^{13}CO varies between 0.5 at the clump center and more than 10^3 at the clump edge, i.e. at

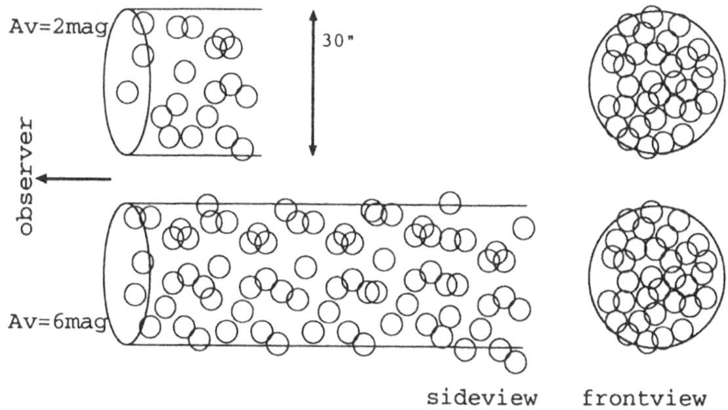

Fig. 3. Sketch of core region consisting of many small, dense PDR clumps

$R = 0.17\,\mathrm{pc}$ ($\alpha = 1.5$). A steeper density profile, $\alpha = 2$, leads to only minor changes: ^{13}CO emission from the clump center is enhanced while emission from the edge is reduced. Due to the fixed mass and density, this clump is slightly smaller. The average C I/^{13}CO ratio is almost the same for both clumps: 1.2–1.3, i.e. it equals the observed ratio (Fig. 2). The strong C I emission at the clump edge is caused by a combination of high abundance near the clump surface, where CO is photodissociated, and limb brightening.

Sampling the clump at 0.067 pc (30″) resolution, would still lead to a marked variation of the ratio while, instead, we observe a nearly constant ratio. This indicates that we do not sample such a single massive clump. Enhanced C I emission at the cloud edges would not show up in smaller, unresolved clumps of $R < 0.034\,\mathrm{pc}$ (Fig. 2). Their mean extinctions must however be above ~ 1 mag, otherwise CO would be completely dissociated.

2.2 A Clumpy Model

Unresolved clumps with average extinctions of more than about 1 mag reproducing the observed line ratio are found for a range of clump masses between about $0.3\,\mathrm{M}_\odot$ and $3\,10^{-5}\,\mathrm{M}_\odot$ and surface densities between $3\,10^4$ and $3\,10^5\,\mathrm{cm}^{-3}$ (Fig. 2). A realistic model would need to take into account clumps of different masses within the beam. Here, we only argue for unresolved small scale structure and therefore select one type of clumps, assuming here that the cloud consists only of those.

A single small clump of mass $M = 10^{-3}\,\mathrm{M}_\odot$, surface density $n_s = 10^5\,\mathrm{cm}^{-3}$, and $\chi = 3$ emits a velocity and surface integrated C I 1–0 intensity of $5\,\mathrm{Kkms}^{-1}$ and exhibits a line ratio of 1.3 ($\alpha = 1.5$). The averaged optical extinction is 1.6 mag and its radius is $R = 0.0037\,\mathrm{pc}$ (760 AU, 1.7″). The optical depths at line and clump center are $\tau(\mathrm{C\,I}) \sim 0.5$ and $\tau(^{13}\mathrm{CO}) = 1.7$. These clumps are only moderately optically thick.

To reproduce an observed extinction of 1.6 mag within a 30″ beam, we assume that the area filling factor is unity, $F_A = 1$, and that the clumps do not shadow each other along the lines of sight (Fig. 3. This is consistent with 82 clumps within a cube of 30″ diameter, leading to a volume filling factor F_V of 11%.

To reproduce an observed extinction of any other observed magnitude within a 30″ beam at the center of the observed region, we assume the cylinder of 30″ diameter along the line of sight to be filled with the same small clumps, with a constant area and volume filling factor (Fig. 3. An observed extinction of e.g. 27 mag is then reproduced by 1400 small clumps within a cylinder of 30″ diameter and 8.5′ depth.

The observed C I over ^{13}CO line ratio and its constancy with extinctions is reproduced by the above simple clumpy model. The observed C I intensities of 4.2 ± 0.9 are slightly smaller than the modelled intensity of 5.4 Kkms^{-1}, which may be caused by the moderate optical thickness of the modelled, small clumps. Deeper lying layers of clumps do slightly add to the observed emission.

In conclusion, we find that our C I and ^{13}CO observations, combined with observations of the optical extinctions at the same resolution, are consistently described by a clumpy PDR model. This result is surprising in view of the smooth density profile found by averaging optical extinction data along the major axis of the IC 5146 Northern Streamer [14]. In addition, the prestellar core B68 presented by Alves et al. elsewhere in this volume, also shows a very smooth radially averaged density profile with no indications of small scale structure. We will discuss these apparent contradictions in a forthcoming paper [12].

References

1. Bennett, C., Fixsen, D., et al., 1994, ApJ, 434, 587
2. Dobashi, K., Onishi, T., Iwata, T., Nagahama, T., Patel, N., Snell, R. L., & Fukui, Y. 1993, AJ, 105, 1487
3. Dobashi, K., Yonekura, Y., Mizuno, A., & Fukui, Y. 1992, AJ, 104, 1525
4. Fixsen, D. J., Bennett, C. L., & Mather, J. C. 1999, ApJ, 526, 207
5. Gerin, M. & Phillips, T. 2000, ApJ, 537, 644
6. Gierens, K., Stutzki, J., & Winnewisser, G. 1992, A&A, 259, 271
7. Ikeda, M., Maezawa, H., et al. 1999, ApJ, 527, L59
8. Keene, J. 1995, in The physics and chemistry of the interstellar medium, ed. G. Winnewisser & G.C.Pelz (Berlin: Springer)
9. Kramer, C. 1999, in The physics and chemistry of the interstellar medium, ed. V. Ossenkopf, J. Stutzki, & G. Winnewisser (Herdecke: GCA Verlag)
10. Kramer, C., Alves, J., Lada, C. J., et al. 1998a, A&A, 329, 33
11. Kramer, C., Alves, J., Lada, C. J., et al. 1999, A&A, 342, 257
12. Kramer, C., Alves, J., Lada, C. J. 2001, A&A, in preparation
13. Lada, C. J., Lada, E. A., Clemens, D. P., & Bally, J. 1994, ApJ, 429, 694, lLCB94
14. Lada, C. J., Alves, J., & Lada, E. A. 1999, ApJ, 512, 250
15. Plume, R., Jaffe, D., Evans, K. T. N., & Keene, J. 1999, ApJ, 512, 768
16. Störzer, H., Stutzki, J., & Sternberg, A. 1996, A&A, 310, 592
17. Tatematsu, K., Jaffe, D., Plume, R., & Evans, N. 1999, ApJ, 526, 295

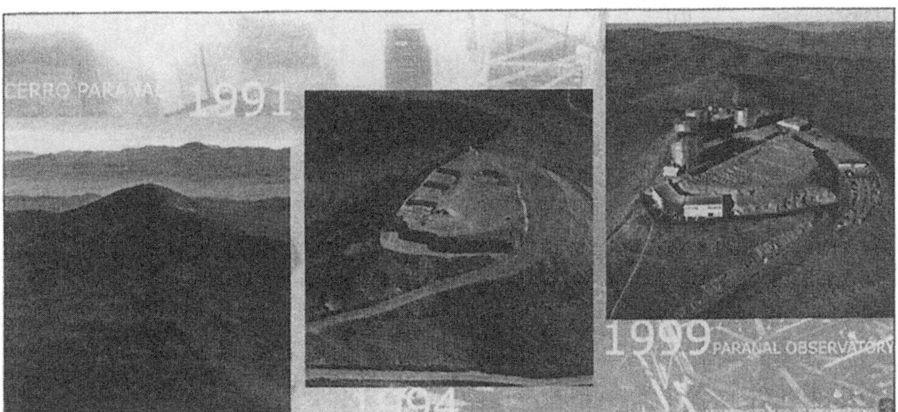

Paranal 1991–1994–1999

Michiel Hogerheijde and Floris van der Tak

From Infall to Rotation Around Young Stellar Objects

Michiel R. Hogerheijde[1], Adwin C. A. Boogert[2], Ewine F. van Dishoeck[3], and Geoffrey A. Blake[4]

[1] Radio Astronomy Laboratory, University of California, 601 Campbell Hall #3411, Berkeley, CA 94720-3411, USA
[2] California Institute of Technology, Downs Laboratory of Physics 320–47, Pasadena, CA 91125, USA
[3] Sterrewacht Leiden, P.O. Box 9513, 2300 RA, Leiden, The Netherlands
[4] California Institute of Technology, Division of Geological and Planetary Sciences 150–21, Pasadena, CA 91125, USA

Abstract. We present evidence of a short-lived, transitional stage between the fully embedded and optically visible stages of low-mass star formation. This stage is characterized by a large (2000 AU radius) disk with close-to Keplerian motions, but with a significant inward component to the velocity field. Millimeter-interferometric observations of HCO^+ J=1–0 and 3–2 first identified this structure around the embedded object L1489 IRS, with inward motions present on scales of several hundred AU. Subsequent $R \approx 25,000$ M-band spectroscopy showed that the inward motions extend to within 0.1 AU from the star. We conclude that angular-momentum transfer is crucial in determining the final density distribution in the disk and the initial conditions for planet formation.

1 Disks Around Young Stars

The cloud cores from which stars form have a certain amount of angular momentum imposed on them by the shear in the Galactic potential. This angular momentum is conserved as the cores collapse, and material accretes onto a rotationally supported disk rather than directly onto the star. Such circumstellar disks are found around the majority of T Tauri stars [1] and are believed to be the progenitors of planetary systems where dust particles coagulate into planetesimals over a time period of several times 10^7 yr [2].

During the earliest T Tauri stages, immediately following the fully embedded phase of star formation, disk masses range between 0.001 and 0.1 $M_\odot$ with disk outer radii of up to several hundred AU [1]. A recent survey [3] has shown that the velocity field in these disks are Keplerian, although it remains to be seen what limits can be placed in radial transport of material at a fraction of the orbital velocities. Such radial motions may be expected based on theoretical models of disk structure [4], in which viscous forces help to transport angular momentum outward and material inward, and based on the observed continued accretion of mass onto the star [5].

Very little is known observationally about the formation of circumstellar disks from collapsing cloud cores, in part because of the difficulty in separating the emission of the disk from that of the surrounding envelope. For example, in one scenario disks are formed early on and keep their size constant by balancing the inflow of material and the accretion of material through the disk onto the star. In another scenario, disks grow gradually in mass and size throughout the embedded phase as material carrying angular momentum flows in.

Here, we present millimeter-interferometer observations of two embedded, low-mass young stellar objects (YSOs), one that shows the effects of rotation on collapse (TMC 1) and another that appears to represent a transitional stage between a fully embedded YSO and a T Tauri star (L1489 IRS). These observations suggest that the disk grows gradually to a size of a few thousand AU in radius, but subsequently contracts as the inflow of material ceases and the redistribution of angular momentum catches up.

2 Collapsing Cores in Continuum and HCO^+ Emission

Recent surveys show that the envelopes around many embedded YSOs have the expected characteristics of collapsing cores. Their radial density profiles, derived from the spatially resolved submillimeter emission [6–8] follow power-laws with indices between -1 and -2 as predicted by theory [9,10]. And a large fraction of objects reveal asymmetric, double-peaked profiles in optically thick molecular emission lines [11,12] that can be understood in terms of radiative transfer in a collapsing cloud where material accelerates as it falls inward (so-called 'inside-out' collapse) [13].

A critical test for collapse models is their ability to reproduce simultaneously the observed density and velocity profiles. Submillimeter-continuum maps obtained with SCUBA on the James Clerk Maxwell Telescope and HCO^+ J=1–0, 3–2, and 4–3 spectra of four embedded YSOs in Taurus [14] showed that the envelopes around three of these objects are indeed collapsing in an inside-out fashion (L1527 IRS, L1535 IRS, and TMC 1). The fourth object, L1489 IRS, did not follow this description, although the SCUBA maps and the HCO^+ line profiles individually did seem to fit inside-out collapse. Within the framework of a particular model [9], however, both sets of data are not described by the same model parameters.

3 A Large Rotating Disk Around L1489 IRS

If not a collapsing envelope, what surrounds the object L1489 IRS? Millimeter-wavelength aperture-synthesis observations of L1489 IRS in HCO^+ J=1–0 and 3–2 (Fig. 1) show an elongated structure of approximately 4000 AU × 2000 AU, with a distinct velocity gradient of several km s^{-1} along the long axis. This gradient strongly suggests Keplerian rotation, and the obvious explanation is that l1489 IRS is surrounded by a rotating disk. Although the mass in this disk, 0.02 $M_\odot$, falls within the range of typical masses of T Tauri disks, with an outer radius of 2000 AU its size is significantly larger than commonly observed. Incidentally, it follows that the densities in the disk are therefore below normal

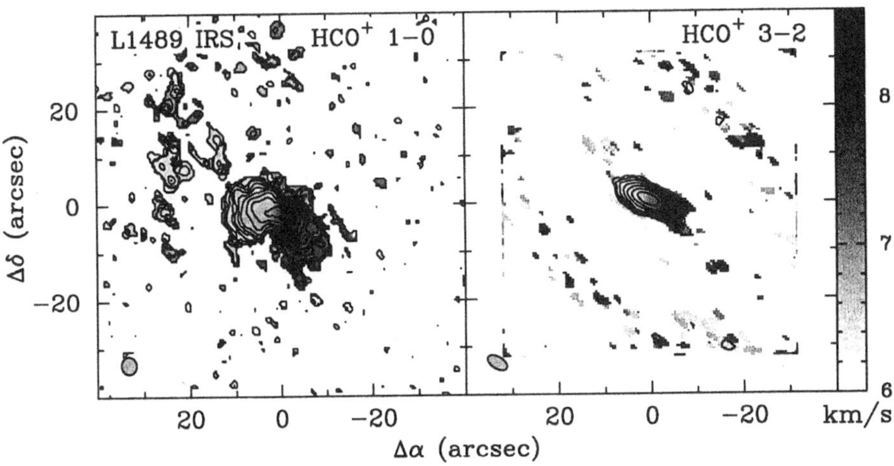

Fig. 1. Aperture synthesis maps of L1489 IRS in lines of HCO^+ J=1–0 (left) and 3–2 (right), obtained with the millimeter arrays of the Berkeley-Illinois-Maryland Association (1–0 and 3–2) and Owens Valley Radio Observatory (1–0). The contours show the integrated intensity; the gray scale shows the velocity centroid of the emission

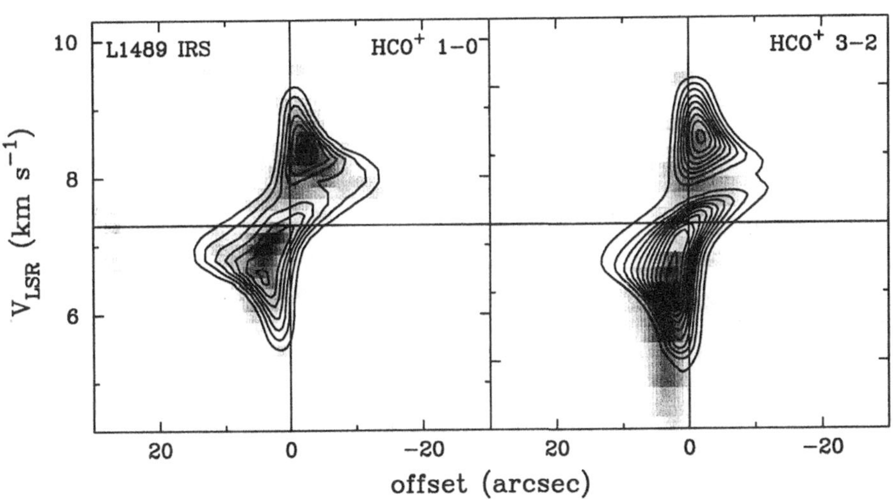

Fig. 2. Position-velocity diagrams of HCO^+ 1–0 (left) and 3–2 (right) along the long axis of L1489 IRS's disk (greyscale), overlaid with best-fit model (contours)

for a circumstellar disk, which may affects its ability to form planets through coagulation.

A closer look at the velocity field in L1489 IRS [15] reveals that rotation alone cannot explain everything. Instead we find that inward motions are also present, as was already traced through the asymmetric double-peaked HCO^+ line profiles. The entire velocity field is described with Keplerian rotation around a

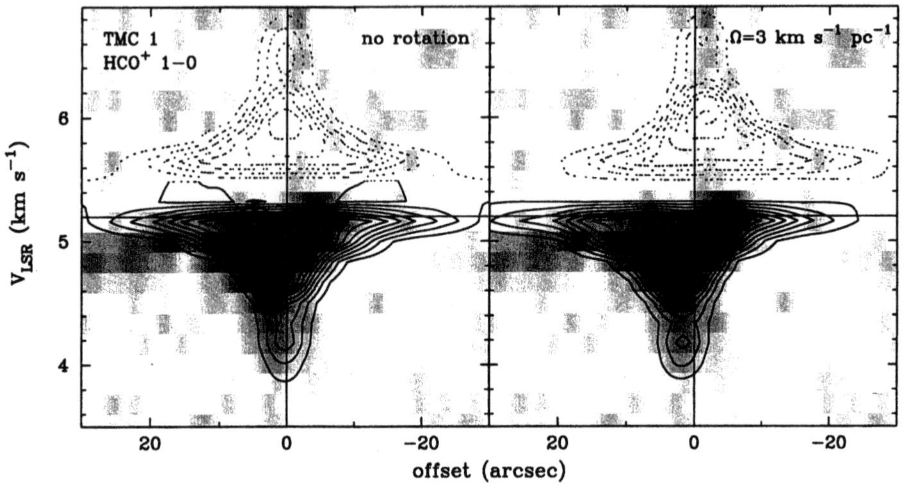

Fig. 3. Position-velocity diagrams of the HCO^+ 1–0 emission along the equator of TMC 1's envelope (greyscale), overlaid with best-fit models (contours) Left: pure infall without rotation. Right: including rotation). The envelope's emission is absorbed by a foreground cloud in the range V_{LSR} = 5.1–6.5 km s^{-1}, represented by dashed contours

0.65 $M_\odot$ star and free-fall–like inward motions with $V_R = 1.3\,(R/100\,\mathrm{AU})^{-0.5}$ km s^{-1} (Fig. 2). The radial component amounts to approximately 10% of the total velocity vector, and one could describe the resulting field as sub-Keplerian. More details about the adopted physical model for L1489 IRS's disk and the modeling procedure can be found elsewhere [15]. Suffice to say here that we use a flared disk [16] with a total mass of 0.02 $M_\odot$. Although only a small fraction of the rotational velocities, the inward component will cause the disk to contract in only 2×10^4 yr if no additional material accretes onto it. This is much shorter than the typical life time of embedded YSOs and T Tauri stars (at several times 10^5 and 10^6 yr, respectively).

4 Comparison to the Embedded YSO TMC 1

Do other embedded YSOs also have such a rotationally supported structures at the center of their envelopes? The velocity field of the material around the embedded YSO TMC 1 closely follows inside-out collapse, especially if we use a model [17] that includes a small amount of rotation in addition to infall (Fig. 3). Compared to L1489 IRS (Fig. 4), rotation plays a much smaller role in TMC 1. In other YSOs for which rotation has been inferred (e.g., L1527 IRS and L1551 IRS 5 [18,19]), rotation is a similarly small fraction of the inward motions.

The adopted model [17], however, breaks down at a 'rotational' radius where rotation balances gravity; inside this radius the model assumes material to flow onto a disk. For TMC 1 this radius is at 360 AU, but the model predicts it to

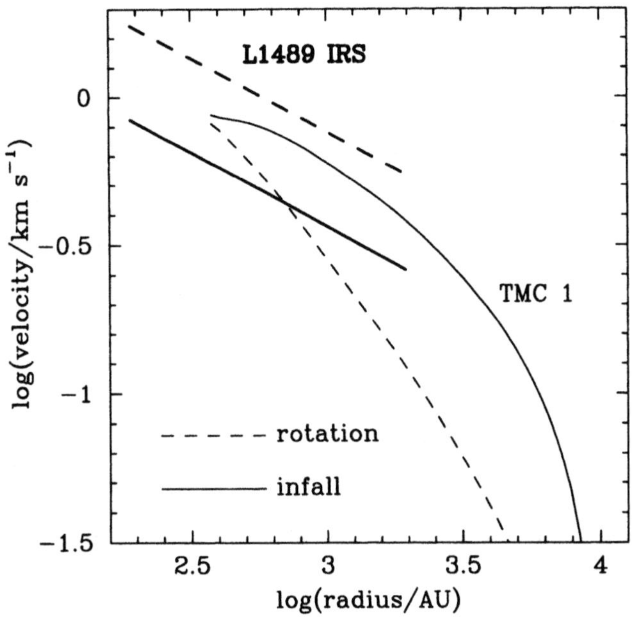

Fig. 4. A comparison of the rotational (dashed line) and radial (solid line) velocity fields in the disk around L1489 IRS and envelope around TMC 1, as inferred from model fits to the HCO^+ interferometer data

increase with time as $R_c \propto t^3$ assuming that the disk cannot efficiently loose angular momentum. In a few times 10^5 yr, comparable to the current age of the object, TMC 1's disk will have grown to 2000 AU, and little envelope material will be left outside this radius at that time: TMC 1 will look very similar to L1489 IRS.

This leads us to propose that different rates of angular-momentum inflow and angular-momentum redistribution result in a gradual growth of the disk throughout the embedded phase, to a maximum size of a few thousand AU. After mass infall ceases, the disk rapidly adjusts to a smaller size of several hundred AU. As a consequence, viscosity or other means for angular-momentum transfer are crucial in determining the final size and density distribution of the disk.

5 Prospects for ALMA and VLT

The presented millimeter-interferometer data only sample the velocity field in L1489 IRS's disk on scales of several hundred AU. It is only because this disk is large to begin with, that the velocity field can be resolved. Disks around T Tauri stars are smaller, and it will take a large array such as ALMA to probe the velocity structure in these disks in detail. This will be an important step in determining if inward motions continue throughout the life time of the disk.

Another way to probe the velocity in smaller scales is through spectroscopy of CO ro-vibrational absorption lines around 4.65 μm. Such M-band spectra, recently obtained of L1489 IRS with NIRSPEC on Keck and with ISAAC on VLT/Antu [20,21], reveal a broad absorption feature due to CO ice along the line-of-sight as well as many narrow absorption lines in the P- and R-branches of gaseous CO. With a spectral resolution of $R = 25,000$, the NIRSPEC data [20] show that the ^{12}CO absorption lines have redshifted wings out to 80 km s^{-1} from the source's velocity. Given the excitation of the lines and the magnitude of the velocities, this suggests that the inward motions as derived above continue to within 0.1 AU from the star. This means that the entire disk is participating in the contraction, as opposed to the outer regions settling onto the smaller, 'real' circumstellar disk.

High-resolution M-band spectroscopy on VLT of transitionary objects like L1489 IRS, embedded objects like TMC 1, and T Tauri stars will offer unique insight into the disk dynamics, especially when combined with ALMA data on the cooler material further from the star. This will lead to a better understanding of the processes that determine the initial conditions of planet formation.

References

1. S. V. W. Beckwith, A. I. Sargent: Nature **383**, 139–144 (1996)
2. S. P. Ruden: 'The Formation of Planets'. In: *The Origin of Stars and Planetary Systems*. ed. by C. J. Lada, N. D. Kylafis (Kluwer: Dordrecht 1999) pp. 643–680
3. M. Simon, A. Dutrey, S. Guilloteau: ApJ **545**, 1034–1043 (2000)
4. D. Lynden-Bell, J. E. Pringle: MNRAS **168**, 603–637 (1974)
5. S. J. Kenyon: 'Accretion Disks and Eruptive Phenomena'. In: *The Origin of Stars and Planetary Systems*. ed. by C. J. Lada, N. D. Kylafis (Kluwer: Dordrecht 1999) pp. 613–640
6. C. J. Chandler, J. S. Richer: ApJ **530**, 851–866 (2000)
7. Y. L. Shirley, N. J. Evans, J. M. Rawlings, E. M. Gregersen: ApJS **131**, 249–271 (2000)
8. F. Motte, Ph. André: A&A **365**, 440–464 (2000)
9. F. H. Shu: ApJ, **214**, 488–497 (1977)
10. A. Whitworth, D. Summers: MNRAS **214**, 1–25 (1985)
11. E. M. Gregersen, N. J. Evans, S. Zhou, M. Choi: ApJ **484**, 256–276 (1997)
12. D. Mardones, P. C. Myers, M. Tafalla, D. J. Wilner, R. Bachiller, G. Garay: ApJ **489**, 719–733 (1997)
13. C. K. Walker, G. Narayanan, and A. P. Boss: ApJ **431**, 767–782 (1994)
14. M. R. Hogerheijde, G. Sandell: ApJ **534**, 880–893 (2000)
15. M. R. Hogerheijde: ApJ **553**, 618–632 (2001)
16. E. I. Chiang, P. Goldreich: ApJ **490**, 368–376 (1997)
17. S. Terebey, F. H. Shu, P. Cassen: ApJ **286**, 529–551 (1984)
18. N. Ohashi, M. Hayashi, P. T. P. Ho, M. Momose: ApJ **475**, 211–223 (1997)
19. M. Saito, R. Kawabe, Y. Kitamura, K. Sunada: ApJ **473**, 464–469 (1996)
20. A. C. A. Boogert, M. R. Hogerheijde, G. A. Blake: In Preparation
21. E. F. van Dishoeck: This Volume

Cerro Paranal and VISTA (NTT) Peak (setting sun from La Montura)

Ralf Klessen

Mass Spectra from Turbulent Fragmentation

Ralf S. Klessen[1,2]

[1] UCO/Lick Observatory, University of California at Santa Cruz, Santa Cruz, CA 95064, USA
[2] Max-Planck-Institut für Astronomie, Königstuhl 17, 69117 Heidelberg, Germany

Abstract. Turbulent fragmentation determines where and when protostellar cores form, and how they contract and grow in mass from the surrounding cloud material. This process is investigated, using numerical models of molecular cloud dynamics. Molecular cloud regions without turbulent driving sources, or where turbulence is driven on large scales, exhibit rapid and efficient star formation in a clustered mode, whereas interstellar turbulence that carries most energy on small scales results in isolated star formation with low efficiency. The clump mass spectrum of shock-generated density fluctuations in pure hydrodynamic, supersonic turbulence is not well fit by a power law, and it is too steep at the high-mass end to be in agreement with the observational data. When gravity is included in the turbulence models, local collapse occurs, and the spectrum extends towards larger masses as clumps merge together, a power-law description $dN/dM \propto M^{\nu}$ becomes possible with slope $\nu \lesssim -2$. In the case of pure gravitational contraction, i.e. in regions without turbulent support, the clump mass spectrum is shallower with $\nu \approx -3/2$. The mass spectrum of protostellar cores in regions without turbulent support and where turbulence is replenished on large-scales, however, is well described by a log-normal or by multiple power laws, similar to the stellar IMF at low and intermediate masses. In the case of small-scale turbulence, the core mass spectrum is too flat compared to the IMF for all masses.

1 Introduction

Stars are born in turbulent interstellar clouds of molecular hydrogen. The location and the mass growth of young stars are hereby intimately coupled to the dynamical cloud environment. Stars form by gravitational collapse of shock-compressed density fluctuations generated from the supersonic turbulence ubiquitously observed in molecular clouds (e.g. Elmegreen 1993, Padoan 1995, Klessen, Heitsch, & Mac Low 2000). Once a gas clump becomes gravitationally unstable, it begins to collapse and the central density increases considerably, giving birth to a protostar. As stars typically form in small aggregates or larger clusters (Lada 1992, Adams & Myers 2001) the interaction of protostellar cores and their competition for mass growth from their surroundings are important processes determining the distribution of stellar masses. These complex phenomena are addressed by analyzing and comparing numerical models of self-gravitating supersonic turbulence. Focus is on the connection between supersonic turbulence and local collapse and on the relation between the mass spectra of transient gas clumps and protostellar cores.

2 Numerical Models of Turbulent Molecular Clouds

To adequately describe turbulent fragmentation and the formation of protostellar cores, it is necessary to resolve the collapse of shock compressed regions over several orders of magnitude in density. To achieve this, I use SPH (*smoothed particle hydrodynamics*) which is a well-known particle-based Lagrangian method to solve the equations of hydrodynamics. Details about the numerical implementation can be found in Klessen & Burkert (2000).

In the current paper I consider different scenarios for interstellar turbulence. Model 1 completely lacks turbulent support (it describes the contraction of Gaussian density distribution, Klessen & Burkert 2000, 2001), model 2 describes freely decaying supersonic turbulence, and in models 3 to 5 turbulence is continuously driven. The non-local driving scheme inserts kinetic energy in a specified range of wavenumbers, $1 \leq k \leq 2$, $3 \leq k \leq 4$, and $7 \leq k \leq 8$, corresponding to sources that act on large, intermediate, and small scales, respectively (Mac Low 1999, Klessen et al. 2000), such that an rms Mach number of $\mathcal{M}_{\rm rms} = 5.5$ is maintained throughout the evolution.

3 Star Formation from Turbulent Fragmentation

Stars form from turbulent fragmentation of molecular cloud material. Supersonic turbulence, even if strong enough to counterbalance gravity on global scales, will usually *provoke* local collapse. Turbulence establishes a complex network of interacting shocks, where converging shock fronts generate clumps of high density. This density enhancement can be large enough for the fluctuations to become gravitationally unstable and collapse. However, the fluctuations in turbulent velocity fields are highly transient and the same random flow that creates local density enhancements can disperse them again. For local collapse to actually result in the formation of stars, Jeans-unstable shock-generated density fluctuations must collapse to sufficiently high densities on time scales shorter than the typical time interval between two successive shock passages. The shorter the time between shock passages, the less likely these fluctuations are to survive. Hence, the timescale and efficiency of protostellar core formation depend strongly on the wavelength and strength of the driving source (Klessen et al. 2000, Heitsch et al. 2001), and accretion histories of individual protostars are strongly time varying (Klessen 2001).

The velocity field of long-wavelength turbulence is dominated by large-scale shocks which are very efficient in sweeping up molecular cloud material, thus creating massive coherent density structures. When a coherent region reaches the critical density for gravitational collapse its mass typically exceeds the local Jeans limit by far. Inside the shock compressed region, the velocity dispersion is much smaller than in the ambient turbulent flow and the situation is similar to localized turbulent decay. Quickly a cluster of protostellar cores forms. Therefore, models 1 to 3 lead to a *clustered* mode of star formation. The efficiency of turbulent fragmentation is reduced if the driving wavelength decreases. When

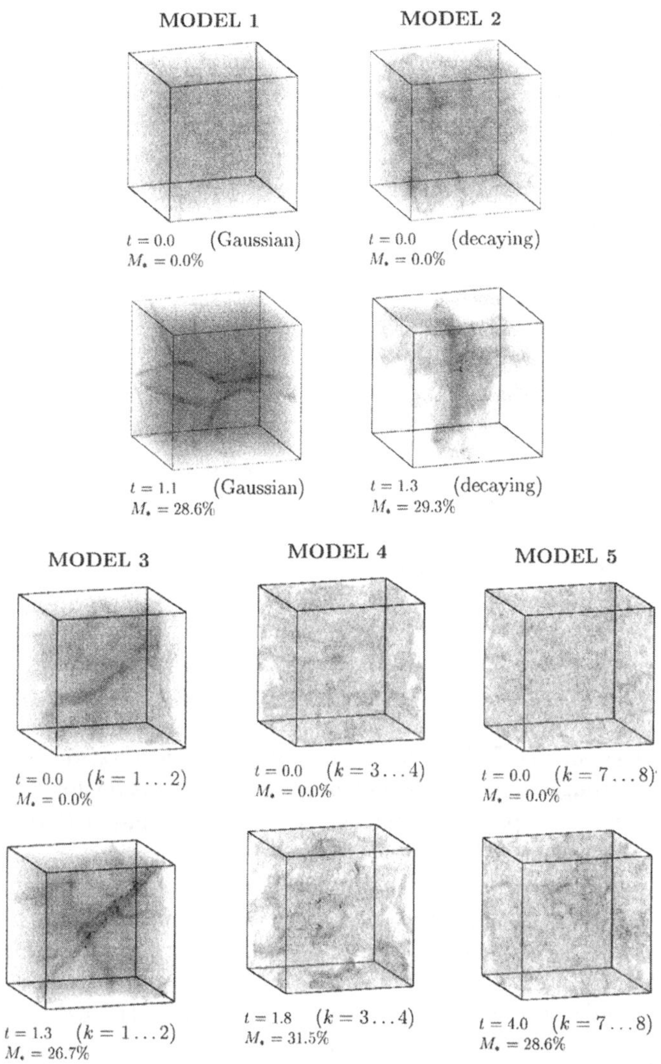

Fig. 1. Comparison of the gas distribution in the five models at two different phases of the dynamical evolution, at $t = 0$ indicating the initial density structure, just before gravity is 'switched on', and after the first cores have formed and accumulated roughly $M_* \approx 30\%$ of the total mass. The high-density (protostellar) cores are indicated by black dots. Note the different time interval needed to reach the same dynamical stage. Time is normalized to the global free-fall timescale of the system, which is $\tau_{\rm ff} = 10^5\,{\rm yr}$ for $T = 11.4\,{\rm K}$ and $n({\rm H_2}) = 10^5\,{\rm cm^{-3}}$. The cubes contain masses of $220\,\langle M_{\rm J}\rangle$ (models 1 and 2) and $120\,\langle M_{\rm J}\rangle$ (models 3 to 5), respectively, where the average thermal Jeans mass is $\langle M_{\rm J}\rangle = 1\,{\rm M_\odot}$ with the above scaling. The considered volumes are $(0.32\,{\rm pc})^3$ and $(0.29\,{\rm pc})^3$, respectively

energy is inserted mainly on small spatial scales, the network of interacting shocks is very tightly knit, and protostellar cores form independently of each other at random locations throughout the cloud and at random times. Individual shock generated clumps have lower mass and the time interval between two shock passages through the same point in space is small. Hence, collapsing cores are easily destroyed again and star formation is inefficient. This scenario corresponds to the *isolated* mode of star formation.

This is visualized in Figure 1, showing the density structure of all five models initially and at a time when the first protostellar cores have formed by turbulent fragmentation and have accreted roughly 30% of the total mass. Note the different time interval needed to reach this dynamical state and the large variations in the resulting density distribution between the various models. Dark dots indicate the location of dense collapsed cores.

4 Mass Spectra from Turbulent Fragmentation

To illustrate the relation between the masses of molecular clumps, protostellar cores and the resulting stars, Figure 2 plots for the five models the mass distribution of all gas clumps, of the subset of Jeans-critical clumps, and of collapsed cores at four different evolutionary stages – clumps are defined as described in Appendix 1 of Klessen & Burkert (2000).

In the initial, completely pre-stellar phase the clump mass spectrum is not well described by a single power law. The distribution has small width and falls off steeply at large masses. Below masses $M \approx 0.3\,\langle M_{\rm J}\rangle$ the distribution becomes shallower, and strongly declines at and beyond the resolution limit (indicated by a vertical line). Clumps are on average considerably smaller than the mean Jeans mass in the system $\langle M_{\rm J}\rangle$.

Gravity strongly modifies the distribution of clump masses during the later evolution. As gas clumps merge and grow bigger, their mass spectrum becomes flatter and extends towards larger masses. Consequently, the number of clumps that exceed the Jeans limit grows, and local collapse sets in leading to the formation of dense condensations. This is most evident in models 1 and 2 where the velocity field is entirely determined by gravitational contraction on all scales. The clump mass spectrum in intermediate phases of the evolution (i.e. when protostellar cores are forming, but the overall gravitational potential is still dominated by non-accreted gas) exhibits a slope -1.5 similar to the observed one. When the velocity field is dominated by strong (driven) turbulence, the effect of gravity on the clump mass spectrum is much weaker. It remains steep, close to or even below the Salpeter value. This is most clearly seen for small-wavelength turbulence. Here, the short interval between shock passages prohibits efficient merging and the build up of a large number of massive clumps. Only few fluctuations become Jeans unstable and collapse to form protostars. These form independent of each other at random locations and times and typically do not interact. Increasing the driving wavelength leads to more coherent and rapid star formation.

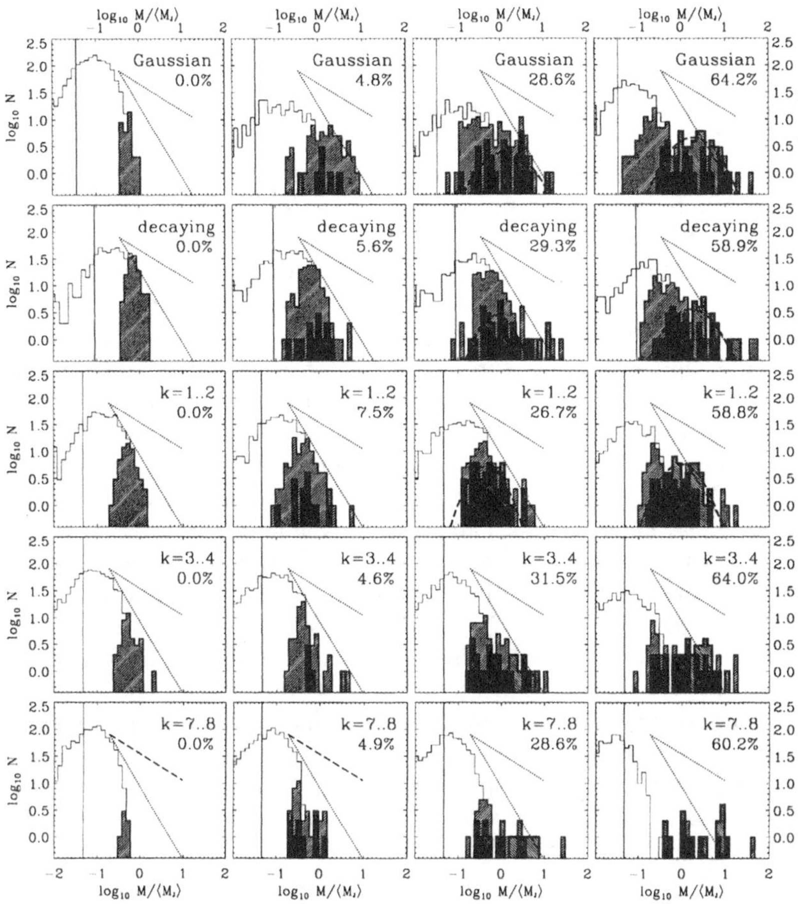

Fig. 2. Mass spectra of gas clumps (thin lines), and of the subset of Jeans unstable clumps (thin lines, hatched distribution), and of dense collapsed cores (hatched thick-lined histograms). Masses are binned logarithmically and normalized to the average thermal Jeans mass $\langle M_J \rangle$. The left column gives the initial state of the system, just when gravity is 'switched on', the second column shows the mass spectra when $M_* \approx$ 5% of the mass is accreted onto dense cores, the third column describes $M_* \approx 30\%$, and the last one $M_* \approx 60\%$. For comparison with power-law spectra ($dN/dM \propto M^{\nu}$), a slope $\nu = -1.5$ typical for the observed clump mass distribution, and the Salpeter slope $\nu = -2.33$ for the IMF at intermediate and large masses, are indicated by the dotted lines in each plot. Note that with the adopted logarithmic mass binning these slopes appear shallower by +1 in the plot. The vertical line shows the SPH resolution limit. In columns 3 and 4, the long dashed curve shows the best log-normal fit to the core mass spectrum. To compare the distribution of core masses with the stellar IMF, an efficiency factor of roughly 1/3 to 1/2 for the conversion of protostellar core material into single stars needs to be taken into account. For $T = 11.4\,\mathrm{K}$ and $n(H_2) = 10^5\,\mathrm{cm}^{-3}$, the average Jeans mass in the system is $\langle M_J \rangle = 1\,\mathrm{M}_\odot$

Long-wavelength turbulence or turbulent decay leads to a core mass spectrum that is well approximated by a *log-normal*. It roughly peaks at the *average thermal Jeans mass* $\langle M_J \rangle$ of the system and is comparable in width with the observed IMF (Klessen & Burkert 2000, 2001). The log-normal shape of the mass distribution may be explained by invoking the central limit theorem (e.g. Zinnecker 1984), as protostellar cores form and evolve through a sequence of highly stochastic events (resulting from supersonic turbulence and/or competitive accretion). To find the mass peak at $\langle M_J \rangle$ may be somewhat surprising given the fact that the local Jeans mass strongly varies between different clumps. In a statistical sense the system retains knowledge of its mean properties. The total width of the core distribution is about two orders of magnitude in mass and is approximately the same for all four models. However, the spectrum for intermediate and short-wavelength turbulence, i.e. for isolated core formation, is too flat (or equivalently too wide) to be comparable to the observed IMF. This is in agreement with the hypothesis that most stars form in aggregates or clusters.

The current findings raise considerable doubts about attempts to explain the stellar IMF from the turbulence-induced clump mass spectrum *only* (e.g. Elmegreen 1993, Padoan 1995, Padoan & Nordlund 2001). Quite typically for star forming turbulence, the collapse timescale of shock-compressed gas clumps often is comparable to their lifetime (molecular cloud clumps appear to be very transient, e.g. Bergin et al. 1997). While collapsing to form or feed protostars, clumps may loose or gain matter from interaction with the ambient turbulent flow. In a dense cluster environment, collapsing clumps may merge to form larger clumps containing multiple protostellar cores, which subsequently compete with each other for accretion form the common gas environment. As a consequence, the resulting distribution of clump masses in star forming regions strongly evolves in time (Figure 2). It is not possible to infer a *one-to-one* relation between the clump masses resulting from turbulent molecular cloud fragmentation and the stellar IMF. It is not appropriate to take a snapshot of the turbulent clump mass spectrum and think it describes the IMF.

References

1. Adams, F. C., Myers, P. C. 2001, ApJ, in press (astro-ph/0102039)
2. Bergin, E. A., Goldsmith, P. F., Snell, R. L., Langer, W. D. 1997, ApJ, 428, 285
3. Elmegreen, B. G. 1993, ApJ, 419, L29
4. Heitsch, F., Mac Low, M.-M., Klessen, R. S. 2001, ApJ, 547, 280
5. Klessen, R. S. 2001, ApJ, 550, L77
6. Klessen, R. S., Burkert, A. 2000, ApJS, 128, 287
7. Klessen, R. S., Burkert, A. 2001, ApJ, 549, 386
8. Klessen, R. S., Heitsch, F., Mac Low, M.-M. 2000, ApJ, 535, 887
9. Lada, E., 1992, ApJ, 393, L25
10. Mac Low, M.-M., 1999, ApJ, 524, 169
11. Mac Low, M.-M., Klessen, R. S., Burkert, A., Smith, M. D. 1998, PRL, 80, 2754
12. Padoan, P. 1995, MNRAS, 277, 337
13. Padoan, P., Nordlund, Å. 2001, ApJ, submitted (astro-ph/0011465)
14. Zinnecker, H. 1984, MNRAS, 210, 43

From Molecular Clouds to Planets: Prospects for ALMA

Ewine F. van Dishoeck

Leiden Observatory, P.O. Box 9513, NL-2300 RA Leiden, The Netherlands

Abstract. Submillimeter observations with ALMA will provide a major step forward in our understanding how stars and planets form. Key projects range from detailed imaging of the collapse of pre-stellar cores and measuring the accretion rate of matter onto deeply embedded protostars, to unravelling the chemistry and dynamics of high-mass star-forming clusters and high-spatial resolution studies of protoplanetary disks down to the 1 AU scale. The synergy with other major (ESO) facilities is emphasized.

1 Introduction

The formation of stars and planets occurs deep inside clouds and disks of gas and dust with hundreds of magnitudes of extinction, and can therefore only be studied at long wavelengths. The Atacama Large Millimeter Array (ALMA) will provide images in the millimeter and submillimeter range with unprecedented sensitivity and angular resolution, improving on existing facilities by up to three orders of magnitude in observing speed. ALMA will be the first telescope capable of resolving the key physical processes taking place during the collapse of molecular clouds, imaging the structure of protostars and of proto-planetary disks, and determining the chemical composition of the material from which future solar systems are made. It will be the long wavelength counterpart of large optical telescopes such as the VLT and HST, with similar angular resolution but unhindered by dust opacity. It is also the perfect complement to the near- and mid-infrared instruments on NGST, which will operate in the same time frame. Although this review focusses on the galactic star-formation science case, ALMA will have a major impact in every branch of astronomy.

ALMA is a joint project of many nations around the globe. In the baseline North American – European proposal, ALMA consists of 64 $\times$ 12m antennas equipped with sensitive submillimeter SIS receivers covering initially 4 of the 10 accessible atmospheric windows from 40–900 GHz (0.35–7 mm). ALMA will have baselines extending up to 12 km, providing a 'zoom-lens' capability with resolutions ranging from $0.01''$ to a few $''$ (given by $0.25''$ λ (mm) / baseline (km)). With the strong possibility of Japan joining ALMA, an 'enhanced' three-way ALMA project is being defined, which may eventually cover all of the atmospheric windows, provide a compact array of smaller antennas for improved wide-field imaging, and give increased flexibility and sensitivity with a next generation correlator. Further details can be found at `http://www.alma.nrao.edu` and `http://www.eso.org/projects/alma`, in the ALMA science case submitted

Table 1. ALMA resolution and sensitivity at 230 GHz in 1 hr

		— Galactic —		— Extragalactic —		Continuum	
Max. baseline	230 GHz beam	resolution at 10 kpc	5σ in 0.5 km/s	resolution at 10 Mpc	5σ in 20 km/s	5σ in $\Delta\nu$=8 GHz	
150 m	2″	0.1 pc	0.05 K	100 pc	7.5 mK	60 μJy	0.4 mK
1.0 km	0.3″	0.015 pc	2 K	15 pc	0.3 K	60 μJy	0.015 K
12.0 km	0.025″	250 AU	290 K	1.2 pc	45 K	60 μJy	2.1 K

to the ESO council in 2000 (`http://iraux2.iram.fr/~guillote/node1.html`) and those submitted to other agencies, and in the proceedings of the conference on *Science with Large Millimeter Arrays* (Wootten 2001).

The dominant radiative processes at (sub-)millimeter wavelengths are continuum emission from cold (<100 K) dust and line emission from molecules. ALMA will have the sensitivity to detect dust masses as small as 10^{-2} M_{Earth} at 150 pc in 2 hr, and image thermal molecular line emission at $< 0.1''$ resolution in 8 hr integration time (see Table 1). The wide frequency coverage of ALMA provides access to molecular lines with a wide range of critical densities and temperatures which can constrain the physical structure of the source. The different receiver bands are also important for astrochemistry, with lines from simple light molecules and atomic carbon found at high frequencies and those from complex heavy organic species at low frequencies. The long wavelength continuum data will allow the contribution of dust emission to be distinguished from free-free and synchrotron emission, whereas the combination with shorter wavelength observations will allow the spectral index of the dust to be measured, providing constraints on the dust optical depth and grain growth.

The spatial resolution of ALMA is well matched to the sizes of galactic star- and planet-forming regions. Table 2 contains a summary of different components at distances ranging from the nearest low-mass star-forming clouds to distant high-mass GMCs. It is clear that ALMA can resolve planet formation out to ~150 pc and high-mass star-formation throughout the Galaxy.

In this brief review, the star- and planet formation process will be followed from the earliest pre-stellar cores to protoplanetary disks around young stars and debris disks around main-sequence stars, emphasizing the areas in which ALMA will make a major impact.

2 Pre-Stellar Cores

The advent of large bolometer arrays on single-dish submillimeter telescopes has revived detailed studies of the structure of molecular clouds and cores just prior to and during the star formation process (see Figure 1). On large pc-size scales, molecular clouds have a highly inhomogeneous or 'clumpy' structure, from which the mass distribution $\Delta N/\Delta M$ can be measured. Interestingly, the mass spectrum of pre-stellar cores in the ρ Oph cloud follows a law $\propto M^{-1.5}$ below 0.5 $M_\odot$ and $\propto M^{-2.5}$ above 0.5 $M_\odot$ (Motte et al. 1998), similar to the stellar initial mass function (IMF) (e.g., Bontemps et al. 2001). This suggests that the IMF

Table 2. Physical sizes and chemical characteristics of star-forming regions

Component	Size (AU)	Nearby star 15 pc (″)	Taurus 140 pc (″)	Orion 450 pc (″)	GMC 3 kpc (″)	Chemical Characteristics
Pre-stellar core	>10,000	...	> 70	> 20	> 3	Ions, Long-chains (HC_5N, DCO^+, ...)
Cold envelope	5000	...	35	10	1.7	Simple species, Heavy depletions (CS, N_2H^+, ...)
Warm inner envelope	500	...	3	1	0.17	Evaporated species, High-T products (CH_3OH, HCN, ...)
Hot core (high-mass only?)	500	...	...	2	0.17	Complex organics (CH_3OCH_3, CH_3CN,... vib. excited mol.)
Outflow: direct impact	<100–500	...	<0.7–4	<0.2–1	<0.03–0.2	Si- and S-species (SiO, SO_2, ...)
Outflow: walls, entrainment	100–1000	...	0.7–7	0.2–2	0.03–0.3	Evaporated ices (CH_3OH, ...)
PDR, compact H II regions (high-mass only)	100–3000	...	...	0.2–7	0.03–1	Ions, Radicals (CN/HCN, CO^+)
Massive Disk	100–500	...	0.7–3.5	0.2–1.1	0.03–0.15	Ions, D-rich species, Photoproducts (HCO^+, DCN, CN, ...)
Debris Disk	100–500	7–35	0.7–3.5	...	...	Dust, CO

may already be determined at the pre-stellar stage during the fragmentation of a molecular cloud. If confirmed by further studies, this is a key result for studying the evolution of molecular clouds in our Galaxy and galaxies as a whole. The sensitivity and spatial resolution of ALMA are needed, however, to link this work with the optical and near-infrared determinations of the low-mass end of the IMF in young clusters. For example, the mass spectrum is still unknown for clump masses below 0.1 $M_\odot$, whereas ALMA can probe the mass function down to planetary masses and study the origin and distribution of brown dwarfs and free-floating Jupiter-mass exo-planets.

Detailed studies of line profiles in selected cores provide evidence for collapse in some objects through asymmetric self-absorption features which are skewed to the blue (e.g., Mardones et al. 1997, Gregersen et al. 1997). But do these cores indeed form stars? ALMA will produce images in (polarized) dust and molecular lines of individual cores with hundreds of spatial pixels at subarcsec resolution, which will allow detailed tests of models of the physical and dynamical state of pre-stellar cores. With current interferometers, the clumps are found to fragment into smaller units, which may eventually lead to binary or multiple systems (Looney et al. 2000, Kamazaki et al. 2001). The high angular resolution of ALMA is needed to study the formation of nearby binaries with a few AU separation as well as small clusters or groups of stars at larger distances.

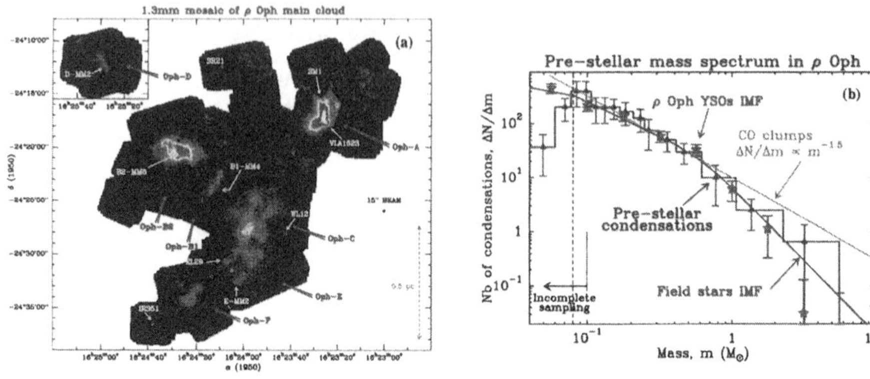

Fig. 1. Left: Millimeter continuum mosaic of the ρ Oph main cloud obtained with the MAMBO 19-channel bolometer array at the IRAM 30-m telescope. Right: Mass spectrum of the 59 pre-stellar condensations identified in the map (Based on: Motte et al. 1998)

3 Embedded Low-Mass Protostars

Deeply embedded young stellar objects (the so-called 'Class 0' and 'Class I' objects) have a complex physical and kinematical structure, with envelopes, disks and outflows all blurred together in current single-dish observations. High spatial and spectral resolution ALMA data will be essential to disentangle the infall, outflow and rotation components of these systems and study their evolution.

Collapsing envelopes. Models of protostellar collapse predict subtle changes in the density profiles of the protostellar envelopes with time. For example, the popular Shu (1977) model has $n \propto r^{-1.5}$ out to the radius to which the collapse has propagated and $\propto n^{-2}$ outside this radius. The inclusion of magnetic fields, turbulence and core rotation can lead to different distributions. Current studies have been restricted to low-excitation lines at low spatial resolution, probing the outer envelope. ALMA will provide a high-resolution view of the innermost, densest regions close to the protostar itself. By using appropriate tracers, the mass accretion can be measured directly by looking for red-shifted absorption against the disk-protostar continuum.

Rotating disks. The study of disk formation in the earliest stages will be a central scientific goal of ALMA. Key questions include whether most of the disk mass is already assembled in the earliest class 0 phase, or whether disk growth continues in the later stages. Also, can asymmetries in the disks, indicative of giant planet formation, be observed? Only ALMA has the spatial resolution and sensitivity to address these questions. ALMA can also study in detail the kinematics of disks around transitional objects between the embedded and T Tauri phase (see Hogerheijde et al., this volume).

Outflows. Violent outflows are a key characteristic of star formation (Richer et al. 2000). These outflows provide not only an efficient mechanism to carry away excess angular momentum, but also limit the mass of the forming star. Although outflows have been studied with single-dish telescopes for more than 20 years, the mechanism of their formation remains poorly understood. An intriguing correlation has been found between the mechanical power of the outflows and the bolometric luminosity of the central protostar (Bontemps et al. 1996). In the earliest Class 0 objects, highly collimated jet-like outflows are commonly observed, but when the protostar evolves from Class 0 to I, both the mechanical power and the collimation seem to decrease, suggesting that the former is due to a decline in the mass accretion rate. ALMA will permit studies of many different kinds of YSOs, and, combined with independent estimates of the mass accretion rate (see above), test magnetohydrodynamical models of their evolution. The actual location and mechanism by which outflows are launched is still a subject of intense debate. ALMA will provide detailed images of the disk/outflow interface down to 10 AU scales where the outflow is accelerated and where the most intense interactions between the outflow and its surroundings take place.

Dispersal of envelopes. Outflows can sweep up and carry away significant amounts of circumstellar material. In $\sim 10^6$ yr, the envelopes are completely dispersed and the young stars become visible as 'Class II' objects. For higher mass stars, other dispersal mechanisms may start to play a role, in particular photodissociation, ionization and evaporation of the molecular material. ALMA will be able to trace $M_{\rm env}/M_*$ and $M_{\rm disk}/M_*$, as well as the kinematics of the dispersal mechanisms, for a wide variety of stellar types as a function of evolution.

4 High-Mass Star Formation

O and B stars are the prime tracers of star formation throughout our Galaxy as well as extragalactic systems. In contrast with the situation for Sun-like stars, however, no detailed theory or scenario yet exists for the formation of massive (> 10 $M_\odot$) stars. Observations suggest that most massive stars do not form in isolation, but in a cluster with many low-mass stars and often together with other high-mass stars. Competing theories include the collapse and fragmentation of a single massive cloud clump to the merging of low-mass stars in the center of the cluster (Stahler et al. 2000).

Most massive star formation occurs at distances of a few kpc, so for this topic the high angular resolution of ALMA, rather than its sensitivity, is of key importance. Take as an example the nearest region of high-mass star formation, the Orion nebula at ~450 pc. Several O stars have formed in a region of a few arcmin-squared, and even close to IRc2 itself, several clumps of hot and dense gas —either 'hot cores' around the protostars themselves or remnant cloud material heated from the outside— are present within 10–20″ (Blake et al. 1996). Thus, a region like Orion ten times further away would be only a couple of arcsec in size. Another example is provided by the case of W 3(OH)/W 3(H_2O), located at a distance of 2.2 kpc (Schilke et al. 2000) (see Fig. 2). Orion and W 3 are puny compared with giant clusters such as NGC 3603, where 20 O- and Wolf-Rayet stars have formed in the last few $\times 10^6$ yr (Brandner et al. 2000) or 30 Doradus

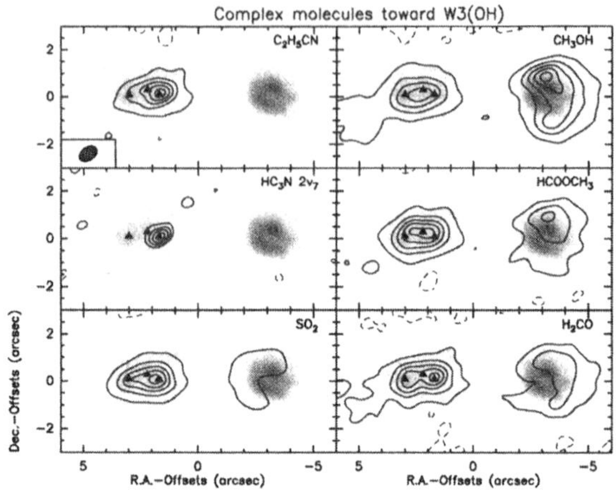

Fig. 2. Maps of different molecular lines in the well-known compact H II region W 3(OH) and hot core W 3(H_2O) ($d \approx 2.2$ kpc). The grey-scale indicates the 3.6 cm continuum. Note that W 3 (H_2O) (left source) breaks up in three millimeter continuum peaks at high spatial resolution. Different molecules have radically different distributions due to a mixture of excitation, chemistry and evolution effects (Based on Wyrowski et al. 1999)

in the LMC. ALMA studies of such massive young clusters at $< 0.1''$ resolution will complement near-infrared imaging on 8-m class telescopes and allow the relation between the optically visible and optically invisible components to be investigated.

In the earliest stages of massive star formation, preceding the ultracompact H II region phase, the star has not yet started to ionize its surroundings but stands out at submillimeter wavelengths because of its strong emission from warm dust. Comparison between high-resolution ALMA and radio images can reveal the progression of star formation within a cloud and test theories of triggered star formation. The resolution of ALMA is essential to trace the outflows back to their sources, and disentangle the complex web of motions.

ALMA will also be able to investigate whether circumstellar disks are a necessary ingredient for the formation of intermediate- and high-mass young stars. Current limits on the disk mass around B stars of $M_{\rm disk}/M_* < 10^{-3}$ (Fuente et al. 2001) are significantly lower than the typical ratio of 0.04 around lower-mass (< 5 $M_\odot$) objects (Natta et al. 2000). ALMA will be able to detect disk masses of $\sim 10^{-3}$ $M_\odot$ out to distances of a few kpc, thus improving significantly on the existing upper limits.

5 Chemistry of Star-Forming Regions

From the collapse of a dense cloud core to the formation of a young star and its circumstellar disk, molecules undergo a series of complex chemical changes. Processes such as depletion of molecules onto the cold icy grains during the collapse phase, evaporation of newly-formed species when the protostar starts to heat its surroundings, and high temperature reactions in shocked zones created by the impact of the outflow, cycle molecules from one compound into another (see van Dishoeck & Blake 1998, Langer et al. 2000 for overviews). These changes are not only of chemical interest, but can also be used as diagnostics of the physical state of evolution of the object. Moreover, knowledge of the chemistry is needed to choose the proper molecular line to trace the physical structure of a particular component (see Table 2). In single-dish observations with 15–30″ beams, all of these different chemical processes are 'blurred' together, whereas current interferometers with a few arcsec resolution suffer from poor spatial sampling. ALMA, with its unprecedented sensitivity, resolution and *uv* coverage will be necessary to zoom in and image these different chemical regimes and quantitatively address the chemical evolution in the initial stages of star formation.

A particularly timely question is the level of chemical complexity in star-forming clouds. Surprisingly large and complex molecules have been found in 'hot cores' associated with massive young stars. Fully saturated species like CH_3OCH_3 (dimethyl-ether) and CH_3CH_2CN (ethylcyanide) are found with abundances that are enhanced by many orders of magnitude with respect to quiescent clouds (see Figure 2). These complex molecules are thought to originate from rapid gas-phase reactions of evaporated molecules (especially CH_3OH) in the warm gas (Charnley et al. 1992). This chemistry only lasts for about 10^4-10^5 yr, after which the normal ion-molecule gas-phase chemistry takes over. Some of these complex species may be the precursors of biogenic molecules such as amino acids, bases and sugars. Searches for the simplest amino acid, glycine, have so far been unsuccessful, but the detection of the first sugar, glycolaldehyde, has just been reported (Hollis et al. 2000). Current instrumentation prevents deep searches for more complex molecules for several reasons: (a) the regions of high chemical complexity are often very small (<0.1″); (b) the crowding of lines is usually so high that the confusion limit is reached; and (c) large molecules have many close-lying energy levels so that the intensity is spread over many different lines, each of them too weak to detect. ALMA will be able to push the searches for prebiotic molecules two orders of magnitude deeper to abundances of $< 10^{-13}$ with respect to H_2, because it will resolve the sources, filter out confusing lines from larger scales and have a higher sensitivity owing to its large collecting area.

6 Protoplanetary Disks

Protoplanetary disks around young stars are prime targets for ALMA. With typical temperatures of order $\sim 30-300$ K at 10–100 AU from the star, the disks are cold, making submillimeter observations a key to their study. Because of their small angular size (see Table 2), current millimeter arrays have only unveiled the largest disks. The orders of magnitude improvement provided by ALMA in both sensitivity and angular resolution will revolutionize the study of protoplanetary disks. Key questions that can be addressed with ALMA are: (i) How do disks evolve from the massive gas-rich disks seen around pre-main sequence T Tauri stars to the tenuous dust debris disks seen around mature stars such as Vega? When are the gas and dust dissipated? (ii) How many disks have formed planets that are revealed by tidal gaps in the disks? What is the associated time-scale for planet-formation? (iii) What is the chemical composition of the gas in disks, providing the building blocks from which future planetary systems are made?

Massive disks. Large gas disks with masses of $\sim 10^{-2}$ $M_{\odot}$ have been revealed around classical T Tauri and Herbig Ae stars with ages of a few Myr by millimeter continuum emission and rotational lines of CO, but these data have only probed the outer disk region (e.g., Dutrey et al. 1996, Mannings & Sargent 1997). Figure 3 compares the brightness distribution of the 1.3 millimeter dust emission and the optically thin ^{13}CO J=2–1 line for a standard disk model with the sensitivity curves of the IRAM and ALMA interferometers (see Dutrey et al. 2000). It is clear that ALMA will detect *all* the dust in the disk down to the 1 AU scale. By using the highest angular resolution and multifrequency observations, ALMA will also be able to measure the change of dust properties within disks, perhaps showing direct evidence for dust settling in the midplane. ALMA will have the sensitivity to map optically thick lines at a few AU resolution, providing information about the gas content and its kinematics. By imaging lines with different excitation conditions, maps of the H_2 density distribution will become possible. Together, such ALMA data can provide complete unbiased surveys of different star-forming regions, down to an equivalent sensitivity of a few Earth masses of dust and gas, and probe the distribution of disk masses, sizes and chemical composition with stellar mass, luminosity, age and environment. By using the kinematics from CO images, direct tests of pre-main-sequence stellar evolution models can be performed (Simon et al. 2000). Finally, ALMA can also reveal the tidal gaps created by protoplanets in circumstellar disks, when pushed towards its highest angular resolution. Thus, surveys for gaps in disks can provide statistics on the frequency and timescale for planet formation.

Debris disks. The IRAS and ISO satellites have detected the presence of disks around mature A-stars such as Vega (e.g., Habing et al. 1999, Spangler et al. 2001). Contrary to the case for pre-main sequence stars, the gas and dust in these so-called 'debris disks' are not the remains of the accreted interstellar material, but they are thought to originate from collisions and evaporation of larger planetary bodies. The amount of dust in these disks is very small, less than the mass of the Moon, and most of it is optically thin even at near-infrared and optical wavelengths.

The first images of cold dust in debris disks at submillimeter wavelengths have been obtained by SCUBA on the JCMT (Holland et al. 1997, Greaves et al. 1998). Because many of the objects are nearby (<30 pc), the dust can be detected and resolved even with single-dish telescopes, at least for the strongest sources. In a few cases, evidence for cleared inner holes suggests the existence of planetary bodies. ALMA will be able to image the dust and CO gas in debris disks with an order of magnitude higher spatial resolution in systems which are

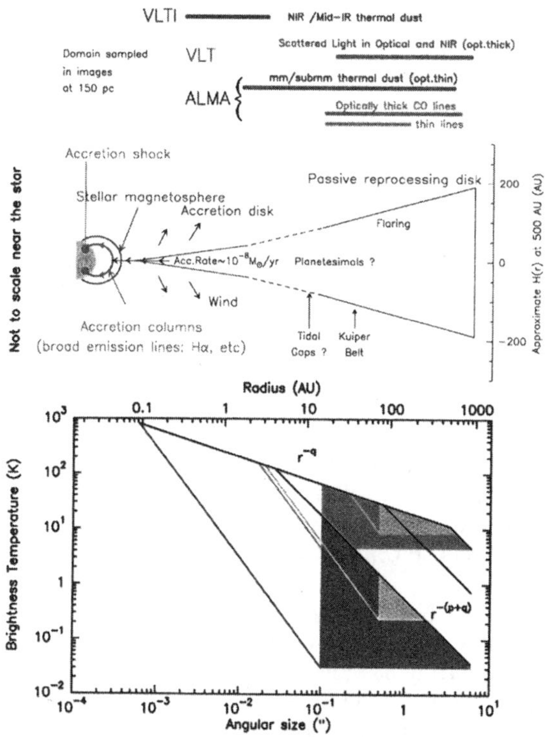

Fig. 3. Top: Schematic drawing of a protoplanetary disk. Bottom: Brightness distribution in K of dust at 1.3 mm (lower lines and lower dark area) and ^{13}CO and $C^{18}O$ 2–1 lines (upper lines and upper dark area) compared with the 3σ sensitivity curves of the IRAM interferometer (light) and ALMA (dark) at resolutions of 0.5″ and 0.1″, respectively, at 150 pc. Current interferometers can only probe the outer edges (> 100 AU) of disks, but ALMA can detect and resolve dust emission into the planet-forming regions (Dutrey et al. 2000)

more than an order of magnitude weaker. It will also be able to search for analogs of the cold zodiacal dust found in our Solar system in the Kuiper Belt.

This review is based on the star formation science case in the ALMA proposal submitted to the ESO council in December 2000. The author is grateful to R. Bachiller, A. Dutrey, S. Guilloteau, K. Menten, P. Shaver and M. Walmsley for using parts of their material.

References

1. Blake, G.A., Mundy, L.G., Carlstrom, J.E., Padin, S., Scott, S.L., Scoville, N.Z., Woody, D.P.: ApJ **472**, L49 (1996)
2. Bontemps, S., André, P., Kaas, A.A. et al.: A&A **372**, 173 (2001)
3. Bontemps, S., André, P., Terebey, S., Cabrit, S.: A&A **311**, 858 (1996)

4. Brandner, W., Grebel, E., Chu, Y.-H., et al.: AJ **119**, 292 (2000)
5. Charnley, S.B., Tielens, A.G.G.M., Millar, T.J.: ApJ **399**, L71 (1992)
6. Dutrey, A., Guilloteau, S., Duvert, G., et al.: A&A **309**, 493 (1996)
7. Dutrey, A., Guilloteau, S., Guélin, M.: in 'Astrochemistry: from molecular clouds to planets', IAU Symposium 197, ed. Y.C. Minh & E.F. van Dishoeck (ASP), p. 415 (2000)
8. Fuente, A., Neri, R., Martín-Pintado, J., Bachiller, R., Rodriguez-Franco, A., Palla, F.: A&A **366**, 873 (2001)
9. Greaves, J.S., Holland, W.S., Moriarty-Schieven, G., et al.: ApJ **506**, L133 (1998)
10. Gregersen, E.M., Evans, N.J., Zhou, S., Choi, M.: ApJ **484**, 256 (1997)
11. Habing, H.J., et al.: Nature **401**, 456 (1999)
12. Holland, W. et al.: Nature **392**, 788 (1997)
13. Hollis, J.M., Lovas, F.J., Jewell, P.R.: ApJ **540**, L107 (2000)
14. Kamazaki, T., Saito, M., Hirano, N., Kawabe, R.: ApJ **548**, 278 (2001)
15. Langer W.D., van Dishoeck E.F., Blake G.A. et al.: In *Protostars & Planets IV*, eds. V. Mannings et al. (Univ. Arizona), p. 29 (2000)
16. Looney, L.W., Mundy, L.G., Welch, W.J.: ApJ **529**, 477 (2000)
17. Mannings, V.G., Sargent, A.I.: ApJ **490**, 792 (1997)
18. Mardones, D., Myers, P.C., Tafalla, M.., Wilner, D.J., Bachiller, R., Garay, G.: ApJ **419**, 789 (1997)
19. Motte, F., André, P., Neri, R.: A&A **136**, 150 (1998)
20. Natta, A., Grinin, V., Mannings, V.: In *Protostars & Planets IV*, eds. V. Mannings et al. (Univ. Arizona), p. 559 (2000)
21. Richer, J., Shepherd, D.S., Cabrit, S., Bachiller, R., Churchwell, E.: In *Protostars & Planets IV*, eds. V. Mannings et al. (Univ. Arizona), p. 867 (2000)
22. Schilke, P., Menten, K.M., Wyrowski, F., Walmsley, C.M.: in 'Astrochemistry: from molecular clouds to planets', IAU Symposium 197, ed. Y.C. Minh & E.F. van Dishoeck (ASP), p. 125 (2000)
23. Shu, F.: ApJ **214**, 488 (1977)
24. Simon, M., Dutrey, A., Guilloteau, S.: ApJ **545**, 1034 (2000)
25. Spangler, C., Sargent, A.I., Silverstone, M.D., Becklin, E.E., Zuckerman, B.: ApJ **555**, 932 (2001)
26. Stahler, S.W., Palla, F., Ho, P.T.P.: In *Protostars & Planets IV*, eds. V. Mannings et al. (Univ. Arizona), p. 327 (2000)
27. van Dishoeck E.F., Blake G.A.: ARA&A **36**, 317 (1998)
28. Wootten, A., ed.: 'Science with Large Millimeter Arrays' (ASP), in press (2001)
29. Wyrowski, F., Schilke, P., Walmsley, C.M., Menten, K.M.: ApJ **514**, L43 (1999)

Four VLT enclosures

Thomas Henning and Sebastian Wolf

Continuum Polarization as a Tool: A Perspective for VLT and ALMA

Thomas Henning[1], Ralf Launhardt[2], Bringfried Stecklum[3], and Sebastian Wolf[3]

[1] Astrophysical Institute and University Observatory, Friedrich Schiller University, Schillergaesschen 3, D-07745 Jena, Germany
[2] Division of Physics, Mathematics, and Astronomy, California Institute of Technology, MS 105-24, Pasadena, CA 91125, USA
[3] Thuringian State Observatory, Sternwarte 5, D-07778 Tautenburg, Germany

Abstract. We will discuss how polarization maps, obtained with near-infrared imaging instruments such as ISAAC and CONICA/NAOS at the VLT, can be used to determine the structure of star-forming regions.

The analysis of submillimetre polarization maps provided by ALMA will allow the study of magnetic field configurations in the environment of protostars. The potential of submillimetre polarimetry will be demonstrated by a SCUBA polarization map of the globule DC 253-1.6.

1 Introduction

The polarization state of dust continuum radiation at near-infrared to submillimetre wavelengths provides more information about star-forming regions, protostars, and circumstellar disks than just the specific intensity. There are two main processes which lead to polarization of continuum radiation: The scattering of light on dust grains and the dichroic emission/absorption by aligned particles. For submicrometre-sized particles, scattering is only important at near-infrared wavelengths because the albedo of the grains becomes very low at far-infrared and submillimetre wavelengths where the emission from aligned particles is the dominant polarization effect.

The properties of polarized light can be described by the Stokes vector (I, Q, U, V), a particularly useful formulation for scattering problems [2]. There are three quantities I (specific intensity), Q^2+U^2, and V which are invariant under rotation of the reference directions. When light passes a medium - the astronomical system or the telescope/detector - the corresponding Stokes vector is transformed linearly by a 4x4 matrix (Mueller matrix) which contains real elements. Different types of the Mueller matrix can be found in [15].

We should also note that the parameters are not all independent, but we have for fully polarized light

$$I^2 = Q^2 + U^2 + V^2. \tag{1}$$

For partially polarized light, the corresponding relation is

$$I^2 > Q^2 + U^2 + V^2. \tag{2}$$

The total degree of polarization is given by $(Q^2+U^2+V^2)^{(1/2)}/I$. The degree of linear polarization is defined by $(Q^2+U^2)^{(1/2)}/I$, and the degree of circular polarization by V/I. Linear polarization will be induced by light scattering on dust grains. The vibration direction of the electric vector of the scattered radiation becomes perpendicular to the scattering plane. The scattering plane contains the incident and scattered rays. Circular polarization can be produced by scattering from spheroidal dust grains and has been observed in several star-forming regions [9].

Optical and infrared imaging polarimeters [15] use mostly a Wollaston prism, a rotatable halfwave plate, or a wire-grid polarizer to determine the polarization degree of light. The Wollaston prism belongs to the birefringent prisms which divide light into two components (an ordinary and extraordinary ray). The two emerging beams are mutually perpendicularly polarized.

2 Polarization Maps at Near-Infrared Wavelengths

Radiative transfer calculations, mostly using the Monte Carlo technique, nicely demonstrated that polarization images at near-infrared wavelengths can be used to determine the geometry and structure of outflow cavities, the circumstellar disk size and orientation, the location of embedded protostars, and the optical properties and size distribution of dust grains [1], [18], [19], [6], [7], [20], [8]. As an example, Fig. 1 summarizes the results of Monte Carlo radiative transfer simulations of a disk around a low-mass young stellar object.

Near-infrared polarization images can be obtained with the La Silla instrument SOFI at the NTT. As part of a detailed investigation of the structure of massive star-forming regions [10], we obtained polarization images of ultracompact HII regions and embedded luminous IRAS sources with this instrument. Fig. 2 shows an image of the ultracompact HII region G 5.89-0.39 obtained with SOFI (see [5] for a detailed description of the source). This image is characterized by a centrosymmetric polarization pattern which allows the determination of the position of the illuminating source. To get this position, the intersection of the lines perpendicular to the polarization vectors has to be found. This image demonstrates how such data can be used to learn more about the structure and location of embedded sources. Near-infrared polarization images are particularly useful to identify "sources" which are just scattering peaks. We also used SOFI to find out if disks around binaries are randomly aligned or co-planar [21]. A statistical analysis of the data led to the conclusion that the disks are preferentially coplanar.

ISAAC at Antu is now offering the capability of imaging polarimetry in its short wavelength arm. CONICA coupled with the adaptive optics system NAOS will provide polarization maps with the appropriate spatial resolution to

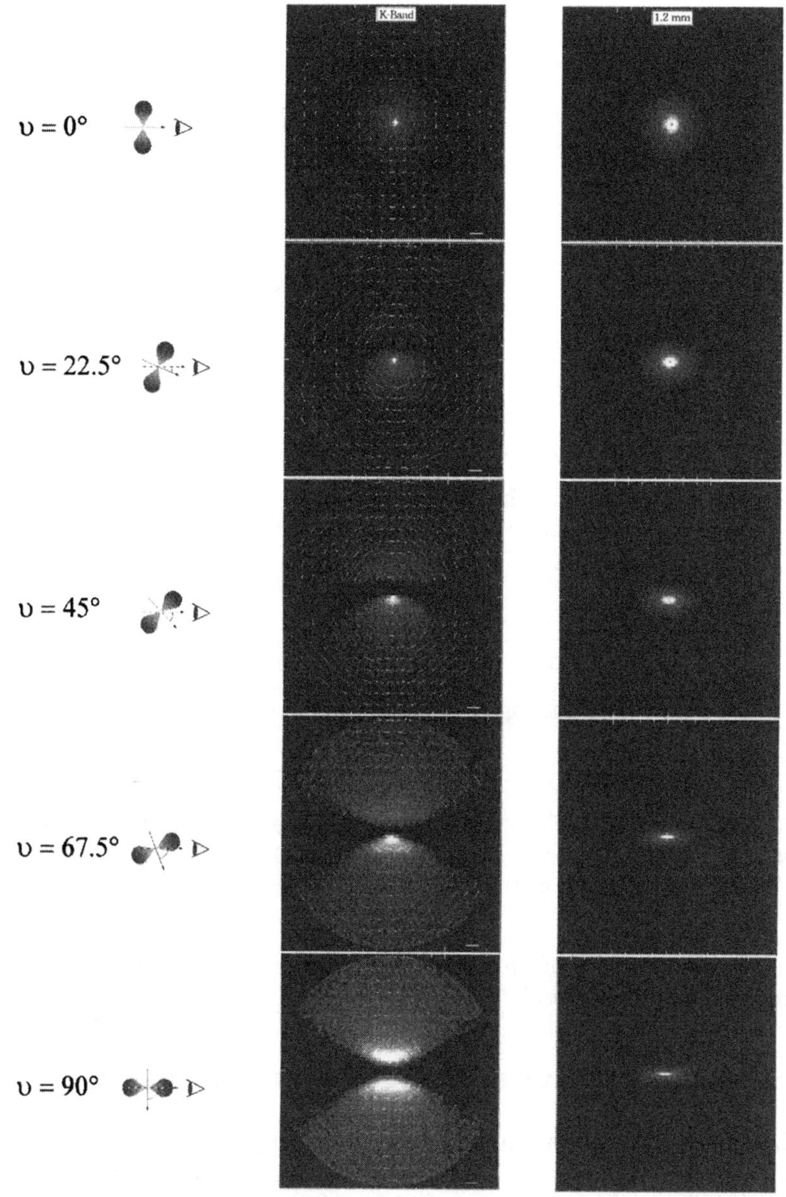

Fig. 1. Intensity and polarization images of a disk with different inclination angles ranging from 0° (pole-on) to 90° (edge-on). Left column: Wavelength of 2.2 μm. Right column: Wavelength of 1.2 mm. The polarization produced at millimetre wavelengths by scattering on grains is practically zero. The bar in the right corner of the K-band images stands for a polarization degree of 100%.
Figure available in colour on the CD-ROM

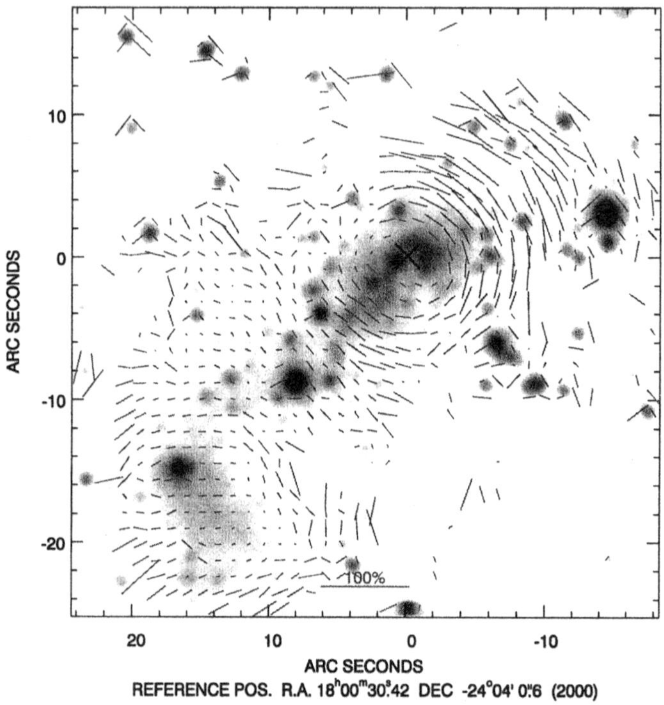

Fig. 2. Combination of Br γ and NB 2.195 μm polarization image of the ultracompact HII region G 5.89-0.39. The image was obtained with SOFI in March 1999. The position of the illuminating source is given by the cross

investigate the structure of protoplanetary disks in nearby regions of low-mass star formation in more detail.

3 Polarization Maps at Submillimetre Wavelengths

Polarized dust radiation at submillimetre wavelengths, produced by spinning and aligned dust grains, is an important signature of magnetic fields [17]. The sensitivity and angular resolution of ALMA will allow a detailed mapping of magnetic field configurations in star-forming regions, protostellar envelopes, and high-mass protostars. Similar information may be provided by mid-infrared polarimetry with TIMMI II at the 3.6 m telescope.

The alignment of dust particles can be caused by a variety of mechanisms (e.g. paramagnetic relaxation, supersonic flows, radiative torques; see [12], [4] for a detailed discussion). Charged dust grains become partially aligned with the magnetic field irrespective of the underlying alignment mechanism, generally with their long axes perpendicular to the field [4]. Thus, the thermal emission from grains at far-infrared and submillimetre wavelengths is partially linearly polarized, with a polarization direction perpendicular to the magnetic field as projected onto the plane of the sky.

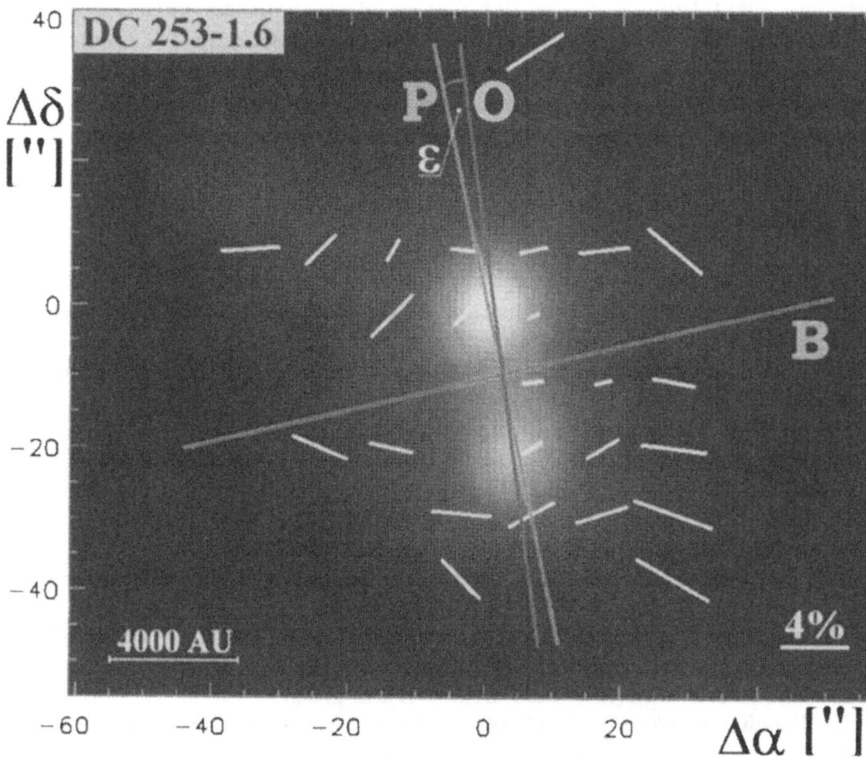

Fig. 3. 850 μm image of DC 253-1.6 with overlayed polarization vectors rotated by 90° representing the spatial distribution of the magnetic field direction. The **P** axis represents the preferential orientation of the linear polarization, the **O** axis connects the centers of both sources, and the **B** axis is oriented perpendicular to the **P** axis. The angle between the **O** and **P** axis amounts to 4°.
Figure available in colour on the CD-ROM

We should note that there is mounting evidence that the emission from the densest regions of low-mass protostellar cores is nearly unpolarized and does no longer trace the magnetic field geometry and strength. Possible mechanisms for this effect are discussed in [11], [14].

Recently, first observations of the magnetic field geometry in three pre-protostellar cores [16] and three protostellar cores in Bok globules [11] have been performed. In another study the polarization of the 850 μm dust continuum emission associated with class 0/I protostars in the Serpens dark cloud core was measured [3]. Polarization measurements of OMC-3 in Orion A led to the conclusion that the magnetic field is predominantly perpendicular to the filament along most of its length [13].

Fig. 3 shows the 850 μm SCUBA polarization map of DC 253-1.6 (CG 30, BHR 12), a low-mass protostellar double core in a nearby Bok globule [11].The projected orientation of this double core is oriented nearly perpendicular to the magnetic field direction derived from the polarization map. This supports the view that the fragmentation process of a collapsing molecular cloud core occurs perpendicular to the magnetic field lines. The image gives a glimpse of the possibilities, we will have with ALMA at much higher spatial resolution.

References

1. P. Bastien, F. Ménard: ApJ **364**, 232 (1990)
2. C.F. Bohren, D.R. Huffman: *Absorption and Scattering of Light by Small Particles.* (John Wiley and Sons, New York 1983)
3. C.J. Davis, A. Chrysostomou, H.E. Mathews, T. Jenness, T.P. Ray: ApJ **530**, L115 (2000)
4. Draine, B.T., Weingartner, J.C.: ApJ **480**, 633 (1997)
5. M. Feldt, B. Stecklum, Th. Henning, R. Launhardt, T.L. Hayward: A&A **346**, 243 (1999)
6. O. Fischer, Th. Henning, H.W. Yorke: A&A **284**, 187 (1994)
7. O. Fischer, Th. Henning, H.W. Yorke: A&A **308**, 863 (1996)
8. K.D. Gordon, K.A. Misselt, A.N. Witt, G.C. Clayton: ApJ **551**, 269 (2001)
9. T.M. Gledhill, A. McCall: MNRAS **314**, 123 (2000)
10. Th. Henning, B. Stecklum: 'The Formation of Massive Stars'. In: *Modes of Star Formation and the Origin of Field Populations, Heidelberg, 2000*, ed. by E.K. Grebel, W. Brandner (ASP Conf. Ser., 2001), in press
11. Th. Henning, S. Wolf, R. Launhardt, R. Waters: ApJ (2001), in press
12. A. Lazarian, A.A. Goodman, P.C. Myers: ApJ **490**, 273 (1997)
13. B.C. Matthews, C.D. Wilson: ApJ **531**, 868 (2000)
14. P. Padoan, A. Goodman, B. Draine et al.: ApJ (2001), in press
15. J. Tinbergen: *Astronomical Polarimetry.* (Cambridge University Press, Cambridge 1996)
16. D. Ward-Thompson, J.M. Kirk, R.M. Crutcher et al.: ApJ **537**, L135 (2000)
17. D.A. Weintraub, A.A. Goodman, R.L. Akeson: 'Polarized Light from Star-Forming Regions'. In *Protostars and Planets IV*, ed. V. Mannings, A.P. Boss, S.S. Russell (Univ. Arizona Press, Tucson 2000), pp. 247–271.
18. B. Whitney, L. Hartmann: ApJ **395**, 529 (1992)
19. B. Whitney, L. Hartmann: ApJ **402**, 605 (1993)
20. S. Wolf, Th. Henning, B. Stecklum: A&A **349**, 839 (1999)
21. S. Wolf, B. Stecklum, Th. Henning: 'Pre-main sequence binaries with aligned disks?'. In: *The Formation of Binary Stars, IAU Symp. 200, Potsdam, 2000*, ed. by B. Mathieu, H. Zinnecker (ASP Conf. Ser., 2001), in press

The Mineralogy and Magnetism of Star and Planet Formation as Revealed by Mid-Infrared Spectropolarimetry

Christopher M. Wright[1], David K. Aitken[2], Craig H. Smith[3], Patrick F. Roche[4], and Rene J. Laureijs[5]

[1] Physics, University College, UNSW, ADFA, Canberra ACT 2600, Australia
[2] Physical Sciences, University of Hertfordshire, Hatfield, Herts AL10 9AB, UK
[3] EOS Technologies Inc., 925a West Grant Rd., Tucson AZ 85705, USA
[4] Dept. of Astrophysics, Oxford University, Keble Road, Oxford OX1 3RH, UK
[5] ISO Data Centre, ESA Astrophysics Division, Villafranca del Castillo, Spain

Abstract. We present mid-infrared polarisation observations which elucidate the magnetic field and dust properties of the target objects, including sources lying behind large columns of the interstellar medium and Young Stellar Objects in molecular clouds. Comparisons between the observed magnetic field direction and disc, outflow and interstellar field axes reveal several correlations, whilst modelling of the 8–13 μm polarisation spectra using laboratory optical constants indicates that in all cases amorphous olivine is the dominant dust component.

1 Introduction

Magnetism is but one of a range of physical effects which may influence star formation. From a theoretical standpoint, magnetic fields can provide support for cloud cores against gravitational collapse ([17]), whilst the most efficient mechanisms for bipolar outflow typically invoke magnetic fields embedded within a circumstellar disk (e.g. [18]). Such fields have the potential to transport angular momentum out of the system, so that magnetism may be regarded as an essential element in the formation of stars. A strong case for observing the magnetic field in star forming regions using polarimetric techniques may then be put.

As well as the magnetic field, mid-infrared polarisation observations provide unique information on the dust grain properties. At these wavelengths many of the proposed components of interstellar dust possess signature resonances, such as amorphous and crystalline silicates, metal oxides, and carbonaceous materials. Since the polarisation cross section depends on a difference between cross sections evaluated along the principal axes of a spheroidal grain, which in turn sensitively depend on the optical constants and grain shape, polarisation is a much more sensitive probe of dust grain physics and chemistry than spectroscopy alone.

The mid-infrared observations were performed at one of either the AAT, UKIRT or the IRTF, using a cooled grating array spectropolarimeter. A spectrum of the massive YSO AFGL 2136 is shown in Fig. 1, whilst the complete set has been published by [9]. Further observations at 25 μm were made using the ISOPHOT photo-polarimeter on board ISO, and have been presented by [24].

2 Astromagnetism

The position angle of mid-infrared (8–13 and 16–22 μm) polarisation spectra can be directly related to the magnetic field direction projected onto the plane of the sky. In standard interstellar dust grain alignment scenarios the position angle of absorptive polarisation is parallel to the magnetic field, whilst that of emissive polarisation is orthogonal to it. Within our sample we observe all three cases, pure absorptive, pure emissive, and a mixture (e.g. Fig. 1).

2.1 Mid-Infrared and Interstellar Polarisation Comparison

Figure 2, which includes all objects in our sample with detected polarisation, shows a comparison between the interstellar and source magnetic fields. Clearly evident are two equally populated groups, with the difference in the two directions of $< 30°$ and $> 30°$ respectively. We use 30° as the boundary as it is larger than both the uncertainty in position angle of our data and the dispersion of the interstellar data, obtained from [5], and so represents a significant and real change. Similar histograms have been presented by [7] and [10] using K-band data. We interpret the $> 30°$ group as implying that the evolution of the source itself has had a substantial perturbing effect on its ambient magnetic field, rather than vice-versa. Such perturbations may even be seen in the mid-infrared data itself, which in many cases shows a twist in the magnetic field direction between the cold, outer region and warmer, inner region implied by absorptive and emissive polarisation components respectively.

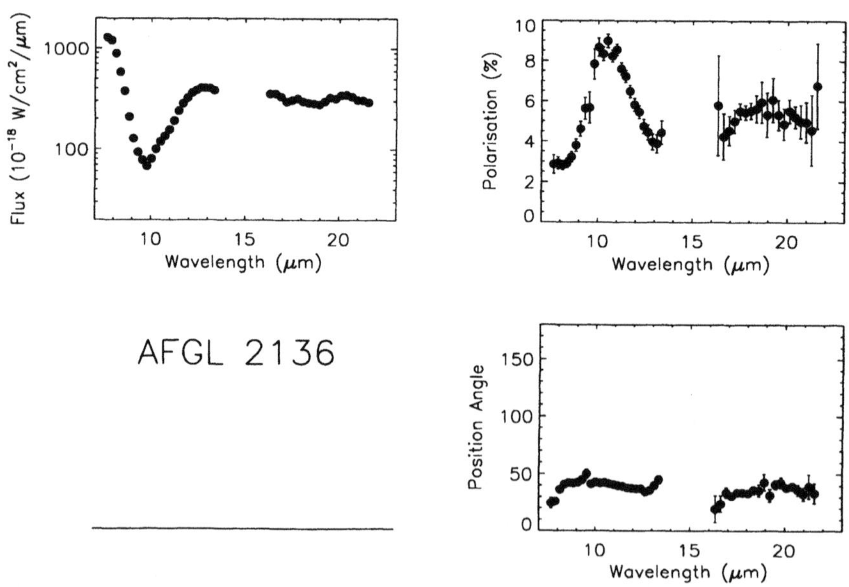

Fig. 1. Example of a mid-infrared polarisation spectrum, in this case of the massive YSO AFGL 2136. The changing position angle through the spectrum indicates both emissive and absorptive polarisation are contributing

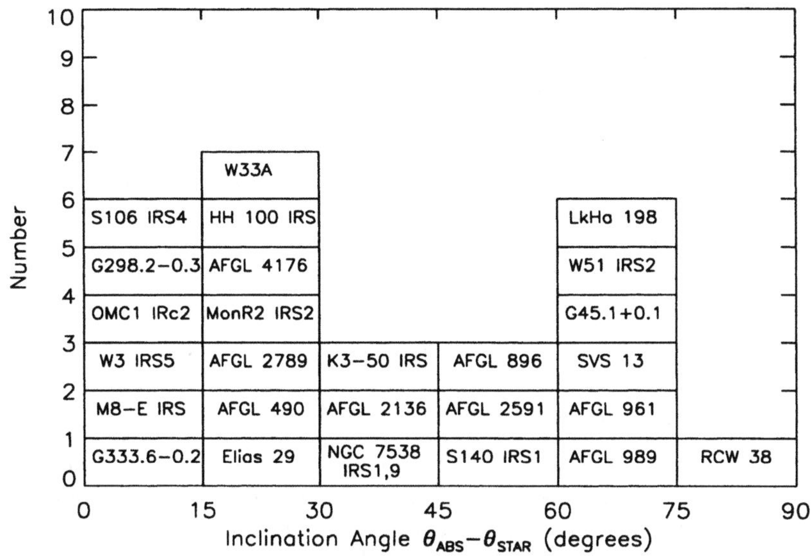

Fig. 2. Histogram of the mid-infrared absorptive polarisation position angle, identified with the magnetic field direction at the IR source, compared with the large scale field direction indicated by the position angle of the polarisation of nearby stars

2.2 Mid-Infrared Polarisation, Outflows and Flattened Structures

Considering only the absorptive polarisation component, the magnetic field tends to lie in the plane of flattened circumstellar envelopes, and orthogonal to the bipolar outflow axis ([3], [22], [24]). This may have implications for the formation of bipolar outflows, favouring a magnetic pressure mechanism (e.g. [21]), where the disk field is predominantly toroidal, rather than a centrifugal mechanism (e.g. [18]), where it is poloidal, although each case should be judged independently.

For instance, in some cases we may probe closer to the source by considering the emissive polarisation, which arises from warm dust, $T \geq 200$ K. Most of the sources with a large change in the magnetic field direction between the absorbing and emitting regions, i.e. $\geq 30°$, are deeply embedded, presumably very young, objects, whilst most of those with a twist of $< 30°$ are more evolved H II regions. These latter objects may be cases where infall and outflow has ceased, and the field has "relaxed" back to its original configuration. In 3 of the "large twist" sources, AFGL 2136, K3-50 IRS and NGC 7538 IRS1, the field twists between lying in the plane of the disk to being closely aligned with the outflow. This may indicate that either a centrifugal mechanism is driving the outflow, or that the flow is being steered, or collimated, by the ambient field.

3 Astromineralogy

The life-cycle of interstellar dust, from its creation and ejection from evolved stars, its transport through the interstellar medium and molecular clouds, to its deposition into planet-forming disks, is a rich area of research. Each environment may have different effects on the nature of the dust, for example cosmic ray hits and/or supernova shocks in the diffuse ISM may induce structural changes, whilst agglomeration of different components (e.g. silicates, oxides, ices, carbonaceous material) may occur in molecular clouds. It is of interest to determine if such changes do indeed occur by looking at the corresponding spectral changes. The ISO satellite provided great impetus by detecting new dust spectral features around old and young stars, such as crystalline silicates, which are similar to those seen in comets (eg. [14]). With the demise of ISO, mid-infrared spectropolarimetry provides the best means of continuing investigations into the mineralogy of cosmic dust (e.g. [1], [2], [23]).

3.1 The Interstellar Medium

Figure 3a shows a polarisation spectrum viewed through the diffuse ISM toward two bright Wolf-Rayet stars, WR 48A and AFGL 2104. A model using optical constants (n, k) of a bare, glassy olivine silicate from [6], $Mg_{0.8}Fe_{1.2}SiO_4$, provides an almost perfect match to the data, including the FWHM and the wavelength of peak polarisation. It may only fail at the very shortest wavelengths, where the laboratory silicate underestimates the observed polarisation.

3.2 Molecular Clouds

Assuming that this laboratory silicate is an adequate description of the ISM material, we can use it as a template to model sources in other environments.

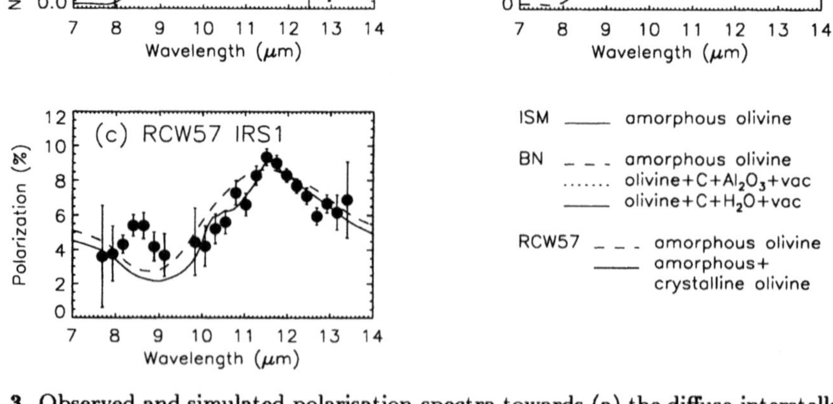

Fig. 3. Observed and simulated polarisation spectra towards (a) the diffuse interstellar medium, (b) the BN Object, and (c) RCW57 IRS1

For instance, the BN Object in Orion lies within a dense molecular cloud, and has the highest known absorptive polarisation ([2]). In comparison to the DISM, the BN profile peaks at the same wavelength, but is significantly broader, with polarisation excess on both the short and long wavelength sides of the peak. There are two possible explanations, firstly that the silicate itself is different, or secondly that additional dust components contribute to the emissivity at these wavelengths. So far, no laboratory silicate that we have tested, olivine or pyroxene, has been able to match the BN width, let alone its detailed profile. On the other hand, it is expected in molecular clouds that grains may grow through either agglomeration or mantle deposition.

Figure 3b shows a model using a composite of amorphous olivine with inclusions of graphitic carbon (n, k from [12]), aluminium oxide (Al_2O_3, n, k from [8]) and vacuum, with volume fractions of 0.1. The carbon and Al_2O_3 provide the emissivity at short and long wavelengths respectively. Aluminium oxide has been used as it is observed in the spectra of evolved stars (e.g. [16]), as well as in meteorites (e.g. [13]), and similarly for graphitic carbon, whilst vacuum is used to simulate the possible porosity of dust grains in molecular clouds. An alternative model uses water ice instead of Al_2O_3, and has the advantage that it also produces 3 μm ice-band polarisation as observed by [11]. The carbon and vacuum inclusions have a volume fraction of 0.1 and 0.25, and the water ice mantle to core volume ratio is 0.25. The models fit well at 8–13 μm, but the 20 μm model polarisation is typically too high, due primarily to the silicate itself.

3.3 Circumstellar Material

We now turn our attention to circumstellar material, where grains may reside in a disk, and be subject to such things as infall and/or outflow shocks and thermal radiation. Fig. 3c shows just such an example, the massive YSO RCW57 IRS1. In this case the polarisation spectrum is indicative of pure emission, but which approximates closely the ratio of P_a/τ ([15]). We thus use the the ratio of the polarisation and extinction cross sections to model such data. Amorphous olivine is not able to match the sharp peak at about 11.5 μm, but inclusion of a small amount of crystalline olivine (15% by volume, n, k from [17]) reproduces it nicely. This would represent the second detection of crystalline olivine in such an object using polarimetric means, the only other being AFGL 2591 ([1], [23]).

4 Opportunities with Large Aperture IR Telescopes

Since polarisation depends on measurement of the small difference between incoming intensities at different analyser orientations, on 4 m class telescopes it has been primarily restricted to bright and massive objects, e.g. $F_{10\mu m} \geq 30$ Jy and $M \geq 10_{\odot}$. Few measurements at 20 μm or of less massive objects ($\leq$ 3 $M_{\odot}$) have been made at adequate signal-to-noise. Further, most objects have only been observed at low spectral resolution, i.e. $\sim$ 50, which is adequate for amorphous silicates, but not for studying detailed profiles of trace constituents, such as the sharper crystalline silicate bands. There is a great need therefore for high resolution spectropolarimeters on 8–10 m class telescopes, such as Michelle on Gemini, which can extend the range of observed objects down to the solar mass regime, such as T Tauri and Herbig Ae/Be stars, in nearby clouds. Further, imaging polarimeters, such as TIMMI2 at ESO and CanariCam on the Spanish GTC, will constrain the role of magnetic fields in star formation, the potential for which has already been demonstrated by [4] and [9].

References

1. D.K. Aitken, P.F. Roche, C.H. Smith, S.D. James, J.H. Hough: Mon. Not. R. Astron. Soc., **230**, 629 (1988)
2. D.K. Aitken, C.H. Smith, P.F. Roche: Mon. Not. R. Astron. Soc., **236**, 919 (1989)
3. D.K. Aitken, C.M. Wright, C.H. Smith, P.F. Roche: Mon. Not. R. Astron. Soc., **262**, 456 (1993)
4. D.K. Aitken, C.H. Smith, T.J.T. Moore, P.F. Roche, T. Fujiyoshi, C.M. Wright: Mon. Not. R. Astron. Soc., **286**, 85 (1997)
5. D.J. Axon, R.S. Ellis: Mon. Not. R. Astron. Soc., **177**, 499 (1976)
6. J. Dorschner, B. Begemann, Th. Henning, C. Jäger, H. Mutschke: Astron. Astrophys., **300**, 503 (1995)
7. H.M. Dyck, C.J. Lonsdale: Astron. J., **84**, 1339 (1979)
8. T.S. Eriksson, A. Hjortsberg, G.A. Niklasson, C.G. Granqvist: Appl. Optics, **20**, 2742 (1981)
9. T. Fujiyoshi, C.H. Smith, C.M. Wright, T.J.T. Moore, D.K. Aitken, P.F. Roche: Mon. Not. R. Astron. Soc., in press (2001)
10. P.A. Heckert, M. Zeilik II: Astron. J., **86**, 1076 (1981)
11. J.H. Hough, A. Chrysostomou, D.W. Messinger, D.C.B. Whittet, D.K. Aitken, P.F. Roche: Astrophys. J., **461**, 902 (1996)
12. C. Jäger, H. Mutschke, Th. Henning: Astron. Astrophys., **332**, 291 (1998)
13. J.F. Kerridge: 'Interstellar Material in Meteorites'. In: *Formation and Evolution of Solids in Space, Italy, March 10–21, 1997*, ed. by J.M. Greenberg, A. Li (Kluwer, Dordrecht 1999), pp. 447–484
14. K. Malfait, C. Waelkens, L.B.F.M. Waters, B. Vandenbussche, E. Huygen, M.S. de Graauw: Astron. Astrophys., **332**, L25 (1998)
15. P.G. Martin: Astrophys. J., **202**, 393 (1975)
16. T. Miyata, H. Kataza, Y. Okamoto, T. Onaka, T. Yamashita: Astrophys. J., **531**, 917 (2000)
17. T. Mukai, C. Koike: Icarus, **87**, 180 (1990)
18. R.E. Pudritz, C.A. Norman: Astrophys. J., **301**, 571 (1986)
19. F.H. Shu, F.C. Adams, S. Lizano: Ann. Rev. Astron. Astrophys., **25**, 23 (1987)
20. C.H. Smith, C.M. Wright, D.K. Aitken, P.F. Roche, J.H. Hough: Mon. Not. R. Astron. Soc., **312**, 327 (2000)
21. Y. Uchida, K. Shibata: Pubs. Astr. Soc. Jap., **37**, 515 (1985)
22. C.M. Wright: Aust. J. Phys., **45**, 581 (1992)
23. C.M. Wright, D.K. Aitken, C.H. Smith, P.F. Roche: 'Mid-infrared Spectropolarimetry and the Composition of Cosmic Dust'. In: *Formation and Evolution of Solids in Space, Italy, March 10–21, 1997*, ed. by J.M. Greenberg, A. Li (Kluwer, Dordrecht 1999), pp. 77–83
24. C.M. Wright, R.J. Laureijs: 'ISOPHOT polarimetry of Young Stellar Objects'. In: *ISO Polarisation Observations, Spain, May 25–28, 1999*, ed. by R. Laureijs, R. Siebenmorgen (ESA SP-435, Noordwijk 1999), pp.49–52

ISAAC mounted at Antu

Thierry Montmerle, Michel Mayor, and Isabelle Baraffe

From the Sun to Jupiter: Evolution of Low Mass Stars and Brown Dwarfs down to Planetary Masses

Isabelle Baraffe[1], Gilles Chabrier[1], France Allard[1], and Peter Hauschildt[2]

[1] Ecole Normale Supérieure, 69364 Lyon Cedex 07, France
[2] Center for Simulational Physics, University of Georgia, Athens, GA

Abstract. We present current models for low mass stars and substellar objects down to planetary masses and discuss the success and remaining uncertainties of the theory. We focus on early evolutionary phases during the first $\sim$ 10 Myr, covering for stars the initial deuterium burning phase to the zero-age Main Sequence. Uncertainties of models at young ages due to the choice of initial conditions are emphasized.

1 Introduction

Within the past years, important efforts have been devoted to the observation and the theory of very-low-mass stars (VLMS) and substellar objects (brown dwarfs BD and giant planets GP). The main theoretical improvements involve the description of the interior of these cool and dense objects (equation of state for dense plasmas, screening factors, etc...; see the review by Chabrier and Baraffe 2000) and the model atmosphere (molecular opacity, formation of dust, etc...; see the review by Allard et al. 1997). A major advance in the field is the development of a new generation of consistent models based on the coupling of interior and atmosphere models, providing direct comparison of evolutionary models with observations in colour-colour and colour-magnitude diagrams (CMD). Several observational tests, mainly provided by relatively old objects (age $\gtrsim$ 100 Myr), now assess the validity of this theory devoted to stellar and substellar objects with masses $\leq 1 M_{\odot}$. General agreement is found with (i) the mass - radius relationship of observed eclipsing binaries, (ii) mass - magnitude relationships in $VJHK$ provided by binary systems, (iii) mass - spectral type relationships for M-dwarfs, (iv) colour - magnitude relationships for intermediate age open clusters (Pleiades, Hyades, etc...), field disk M- and L-dwarfs, halo stars, and globular cluster Main Sequences and (v) spectra of M-dwarfs. Details and references for such confrontations of models with observations can be found in Chabrier and Baraffe (2000). Although some discrepancies between models and observations remain, uncertainties due to the input physics are now significantly reduced.

Numerous surveys devoted to the search for substellar objects have been conducted in young clusters with ages spanning from $\sim$ 1-10 Myrs, providing a wealth of data for pre-Main Sequence (PMS) objects. The reliability of the present theory for VLMS and BD allows now a thorough analysis for such PMS objects. Unlike to older Main Sequence stars and BDs, comparison between observations and models for very young objects presents some difficulties: (i) extinction due to the surrounding dust modifies both the intrinsic magnitude and the colours of the object, (ii) gravity affects both the spectrum and the evolution, (iii) the evolution and spectrum of very young objects ($t \lesssim 1$ Myrs) may still be affected by the presence of an accretion disk or circumstellar material residual from the protostellar stage.

This contribution is devoted to models at early ages for VLMS and BDs down to the planetary mass regime and completes the work of Baraffe et al. (1998, BCAH98) and Chabrier et al. (2000) which was essentially devoted to the comparison of models with observations of older objects ($t \gtrsim 100$ Myrs). We discuss the remaining uncertainties of the models and analyse their comparison with observations.

2 Evolutionary Tracks

The models analyzed in the present paper are based on the input physics already described in BCAH98 and CBAH00. Both sets of models use the same ingredients describing the stellar interior but use different sets of atmosphere models, which provide the outer boundary conditions and the synthetic spectra. The BCAH98 evolutionary tracks are based on the non-grey atmosphere models by Hauschildt, Allard and Baron (1999). These models are dust-free and are appropriate to the description of objects with effective temperatures $T_{\rm eff} \gtrsim 2300$ K. The CBAH00 models are based on atmospheres including the formation and opacity of dust (Allard et al. 2001, hereafter DUSTY models). As illustrated in CBAH00, dust must be taken into account in order to explain the near-IR colors of late M-dwarfs and L-dwarfs. The latter models are thus more appropriate to the description of objects with $T_{\rm eff} \lesssim 2300$ K. As emphasized in CBAH00, the DUSTY models are not appropriate for the description of spectral and photometric properties of methane dwarfs ($T_{\rm eff} < 1600$K), which require a different treatment of dust (the so-called COND models in CBAH00 and Allard et al. 2001).

The BCAH98 grid covers a mass range from 0.02 $M_\odot$ to 1.4 $M_\odot$, for ages $\geq$ 1 Myr up to the Main Sequence for stars. Originally, the CBAH00 grid covers masses from 0.01 $M_\odot$ to 0.1 $M_\odot$ for ages $\geq$ 10 Myr. In the present work, we extend it down to 1 $M_{\rm J}$ ($10^{-3} M_\odot$) and ages $\geq$ 1 Myr. Figure 1 presents the complete grid of models in a Hertzsprung-Russell diagram (HRD) from 0.001 $M_\odot$ to 1.4 $M_\odot$.

Objects below 2 $M_{\rm J}$ evolve essentially with $T_{\rm eff} < 1600$K (see Fig. 1), even at very early ages. Their atmospheric properties are thus better described by the COND models.

As shown in CBAH00, grains bear little effects on the evolution of $L(t)$ and $T_{\rm eff}(t)$, because of the reduced dependence of evolution upon opacity. We verified that the difference between the different TiO and H_2O molecular line lists used in BCAH98 and CBAH00 models (see §3 in CBAH00), respectively, affect essentially the outer atmospheric layers, and thus the synthetic spectra and colors, but not the deeper atmospheric layers, and thus the outer boundary

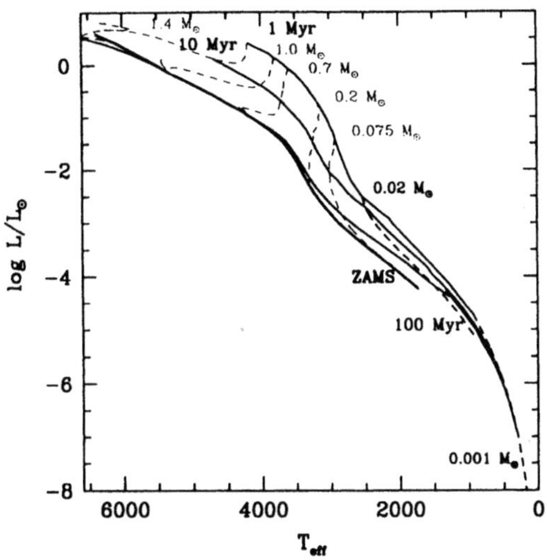

Fig. 1. Evolutionary tracks in the Hertzsprung-Russell diagram for masses from 1.4 $M_\odot$ to 0.001 $M_\odot$ (dashed lines) and ages spanning from 1 Myr to the ZAMS (for stars). Several isochrones for 1, 10 and 100 Myr are indicated by solid lines from right to left. The location of the ZAMS for stars down to 0.075 $M_\odot$ is also indicated (left solid line)

conditions. The effect of these different molecular linelists on the evolution of the effective temperature $T_{\rm eff}(t)$ and the bolometric luminosity $L(t)$ is small, less than 100 K in $T_{\rm eff}$ and 10% in L at a given age. As stressed in CBAH00 and Baraffe et al. 2001a, the computation of more reliable H_2O and, to a lesser extent, TiO linelists is badly needed to solve this shortcoming in the present theory.

3 Main Uncertainties: Initial Conditions

Although shortcomings still remain in current molecular opacities, the resulting uncertainty on the evolution is small. The treatment of convection remains an important source of uncertainty, above all for masses $m \gtrsim 0.6 M_\odot$ (see Chabrier and Baraffe 2000; Baraffe et al. 20001a,b). One of the main source of uncertainty for models at early stages of evolution is the choice of the initial conditions. Most of low mass pre-Main Sequence (PMS) models available in the literature (D'Antona & Mazzitelli 1994, 1997; Burrows et al. 1997; BCAH98; Siess et al. 2000) start from arbitrary initial conditions, totally independent on the outcome of the prior proto-stellar collapse and accretion phases. The initial configuration is that of a fully convective object starting its contraction along the Hayashi line from arbitrary large radii. Evolution starts prior to or at central deuterium ignition, with initial central temperature $\lesssim 10^6$ K. According to studies of low-mass protostellar collapse and accretion phases, such initial conditions are oversimplified, and low mass objects should rather form with relatively small radii (Hartmann et al. 1997, and references therein). Based on spherical accretion protostellar models, Stahler (1983, 1988) defined a birthline in the Hertzsprung-Russell diagram where young objects become visible. Evolutionary tracks should then start from this birthline, which fixes the age $t = 0$. Ages determined from models based on the above-mentioned oversimplified conditions should then be corrected accordingly, with substantial corrections for systems younger than a few Myr (see Palla and Stalher 1999). Furthermore, collapse and accretion are unlikely to proceed spherically. Spherical collapse does not consider angular momentum transport, an important issue of the early phases, which affects the subsequent cooling and formation of the protostar. Hartmann et al. (1997) recently illustrate the sensitivity of the birthline locus assuming that accretion proceeds through a disk rather than spherically. This work stresses again the high uncertainty of assigning ages from HRD positions for the youngest objects. Such analysis demonstrates convincingly that assigning an age to objects younger than a few Myr is totally meaningless when the age is based on models using oversimplified initial conditions.

As shown also in Baraffe et al. (2001a), for $t \lesssim 1$ Myr, the evolutionary tracks themselves are sensitive to the initial conditions, whereas after a few Myr, the models converge toward the same track. To illustrate such effect, Baraffe et al. (2001a) construct a first set of models with initial radii fixed to obtain initial surface gravities $\log g \sim 3 - 3.5$ and initial thermal time-scales $t_{th} \sim$ a few Myr. Such initial conditions are similar to that used in BCAH98 and CBAH00. A second set of models starts with larger radii such that the initial surface gravity $\log g \sim 2.5$. These initial models are more luminous than the previous ones, with central temperatures below 5 10^5K and initial thermal time-scale $t_{th} \sim 10^5$ yr. We also analyze (see Baraffe et al. 2001a for details) the sensitivity of the models to the mixing length $l_{\rm mix}$, characteristic of the mixing length formalism (MLT) used to describe convection.

The effect of different initial radii on evolutionary tracks in a HRD is displayed in Figure 2. Models starting with the lowest gravity are cooler by up to several hundreds K compared to initially denser, less luminous models with the same mass. Note also that for the second set of models, $T_{\rm eff}$ increases during the early evolution, under contraction, in contrast to the first set of models with initial $\log g \gtrsim 3.0$. This is the consequence of the different surface gravities, which strongly affect the atmosphere profiles for $T_{\rm eff}$ = 2200-3500K. Note that evolution along the Hayashi line does not necessarily proceed at constant $T_{\rm eff}$ and the common picture of vertical (constant $T_{\rm eff}$) Hayashi tracks is therefore an oversimplified picture of PMS evolution (see Baraffe et al. 2001a for details).

The two sets of models based on different initial radii follow *the same track* for a given mass and $\alpha_{\rm mix}$, but do not reach *the same position at the same age.* Significant differences appear at ages $\lesssim$ 1 Myr but vanish after a few Myr. We thus consider 1 Myr as the characteristic time required to forget our arbitrary initial conditions and below which models are too sensitive to input physics and thus too uncertain. This is the main reason why we provide confidently evolutionary models for ages $t \geq 1$ Myr. To solve this substantial uncertainty requires the consistent evolution between the 3D collapse of the protostellar phase and the subsequent PMS evolution.

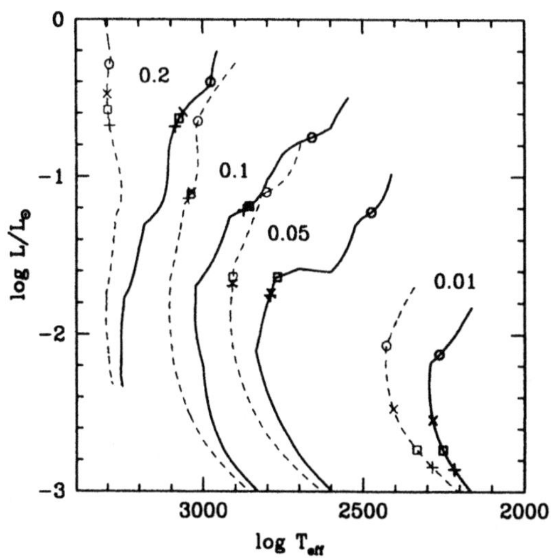

Fig. 2. Evolutionary tracks in the Hertzsprung-Russell diagram for masses from 0.2 $M_\odot$ to 0.01 $M_\odot$, as indicated. The solid curves correspond to $\alpha_{mix} = (l_{mix}/H_P) = 1$ and the dashed curves to $\alpha_{mix} = 2$. Ages of 0.1 and 1 Myr are indicated by respectively crosses and plus for models with initial gravity $\log g = 3.0$-3.5 and by open circles and squares for models with initial $\log g = 2.5$. Note that for the former set of models the initial position (age=0) is essentially the same as the position at 0.1 Myr (cross) and can differ significantly from position of the latter set of models at $t = 0$

4 Observational Tests

A better knowledge of initial conditions may come from the determination of the minimum age below which present models start to depart significantly from observations. Estimation of this age can constrain the characteristic time-scales and accretion rates of the protostellar collapse phase. Unfortunately, direct comparisons of observations with models directly in colour - magnitude diagrams are extremely uncertain due to the large extinction in star formation regions, which affects the observed energy distribution and thus the spectra and the colors. Only very few exceptions, such as σ Orionis, exhibit low extinction. Recently, Béjar et al. (1999) and Zapatero et al. (1999, 2000) obtained optical and near-IR photometry for low mass objects in this cluster. In a $(I-J)$ vs M_I CMD, the data lie between the 1 and 10 Myr isochrones, respectively, for masses down to $\sim 0.01 M_\odot$, using the BCAH98 and CBAH00 models (Zapatero et al. 2000; Béjar et al. 2001). If statistics is improved and if the membership of the objects to the cluster is confirmed, such observations provide an unique opportunity to test directly the validity of young theoretical isochrones. They also offer the best chance to determine the Initial Mass Function (IMF) down to the substellar regime and the minimum mass formed by a collapse process (see Béjar et al. 2001).

Young multiple systems provide also excellent tests for PMS models at young ages, because of the assumed coevality of their different components. In addition, another strong constraint is supplied by the estimate of dynamical masses deduced either from binary systems (Covino et al. 2000; Steffen et al. 2001) or determined from the orbital motion of circumstellar/circumbinary disks (Simon, Dutrey & Guilloteau 2000). An example is provided by the quadruple system GG TAU (White et al. 1999), with components covering the whole mass-range of VLMS and BDs from 1 $M_\odot$ to ~ 0.02 $M_\odot$. Orbital velocity measurements of the circumbinary disk surrounding the two most massive components imply a constraint on their combined stellar mass. This mass constraint and the hypothesis of coevality provides a stringent test for PMS models. The BCAH98 models are the most consistent with GG Tau (for details see White et al 1999; Luhman 1999) and provide the closest agreement with derived masses of other young systems (see Baraffe et al. 2001a).

5 Conclusion

The very good agreement of models based on improved physics with observations for relatively old ($t \gtrsim 100$ Myr) low-mass objects yields confidence in the underlying theory. Such evolutionary models can now be confronted to the complex realm of very young objects, providing important information on star formation processes and initial conditions for PMS models. Although based on extremely simple initial conditions (no accretion phase, no account of protostellar collapse phase and time scale, spherical symmetry), these models provide the most accurate comparison with present observations of very young objects (dynamical masses, tests of coevality in multiple systems, CMDs). Given the combining effects of large observational and theoretical uncertainties at very young ages, however, one must remain cautious. It is probably too premature to conclude on the validity of the present models at early phases of evolution.

Realistic initial conditions can only be provided by multi-dimensional protostar collapse simulations, not by spherically-symmetric models for PMS initial conditions involving many free, ill- or unconstrained parameters. Because of numerical subtleties and complex physical processes (accretion fronts, turbulent time-dependent convection, hydrodynamical radiative transfer, magnetic field etc...), the construction of star formation models is a harsh task, which very likely will necessitate several years of efforts. Besides these theoretical difficulties, observations of very young objects can provide only limited guidance to such simulations, since most phases involved during the collapse are embedded in dusty cocoons. Only the final product can be observationally tested, when the protostar becomes visible. This stage marks essentially the beginning of PMS evolutionary tracks. PMS tracks tested against observations thus provide a precious link to gather insight about star formation models from subsequent evolution.

References

1. Allard, F., Hauschildt, P. H., Alexander, D. R., & Starrfield, S. 1997, ARA&A, 35, 137
2. Allard, F., Hauschildt, P.H., Alexander, D.R., Tamanai, A., Schweitzer, A.. 2001, ApJ, in press
3. Baraffe I., Chabrier G., Allard F., Hauschildt P.H. 1998, A&A, 337, 403 (BCAH98)
4. Baraffe I., Chabrier G., Allard F., Hauschildt P.H. 2001a, A&A, submitted
5. Baraffe I., Chabrier G., Allard F., Hauschildt P.H. 2001b, *From darkness to light: origin and evolution of young stellar clusters*, ASP Conf Series, Vol. 243, Cargese 2000, astro-ph/0007157
6. Béjar, V.J.S., Zapatero Osorio M.R., Rebolo R. 1999, ApJ, 521, 671
7. Béjar, V.J.S., Martín, E.L., Zapatero Osorio, M.R., Rebolo, R., Barrado y Navascués, D., Bailer-Jones, C.A.L., Mundt, R., Baraffe, I., Chabrier, G., Allard, F. 2001, ApJ, in press
8. Burrows A., Marley M., Hubbard W.B., Lunine, J.I., Guillot, T., Saumon, D., Freedman, R., Sudarsky, D., Sharp, C. 1997, ApJ, 491, 856 (B97)
9. Chabrier, G., Baraffe, I. 2000, ARA&A, 38, 337
10. Chabrier, G., Baraffe, I., Allard, F., Hauschildt, P.H. 2000a, ApJ, 542, 464 (CBAH00)
11. Covino, E., Catalano, S., Frasca, A., Marilli, E., Fernandez, M., Alcala, J.M., Melo, C., Paladino, R., Sterzik, M.F., Stelzer, B. 2000, A&A, 361, L49
12. D'Antona, F. and Mazzitelli, I, 1994, ApJS, 90, 467 (DM94)
13. D'Antona, F. and Mazzitelli, I, 1997, in "Cool stars in Clusters and Associations", eds R. Pallavicini and G. Micela, Mem. S. A. It., 68, 807 (DM97)
14. Hartmann, L., Cassen, P., Kenyon, S.J. 1997, ApJ, 475, 770
15. Hauschildt P.H., Allard F., Baron E. 1999, ApJ, 512, 377
16. Luhman KL. 1999, ApJ, 525, 466
17. Palla, F., & Stahler, S.W. 1999, ApJ, 525, 772 (PS99)
18. Siess, L., Dufour, E., & Forestini, M. 2000, A&A, 358, 593
19. Simon, M., Dutrey, A., & Guilloteau, S. 2000, ApJ, 545, 1034
20. Stahler, S.W., 1983, ApJ, 274, 822
21. Stahler, S.W., 1988, ApJ, 332, 804
22. Steffen, A.T., Mathieu, R.D., Lattanzi, M.G., Latham, D.W., Mazeh, T., Prato, L., Simon, M., Zinnecker, H., Loreggia, D. 2001, AJ, in press, astro-ph/0105017
23. White RJ, Ghez AM, Reid IN, Schultz G. 1999, ApJ, 520, 811
24. Zapatero Osorio, M.R., Béjar V.J.S., Rebolo, R., Martín, E.L., Basri, G. 1999, ApJL, 524, 115
25. Zapatero Osorio, M.R., Béjar V.J.S., Martín, E.L., Rebolo, R., Barrado y Navascués, D., Bailer-Jones, C.A.L., Mundt, R. 2000, Sciences, 290, 103

Fabian Heitsch

Core Collapse and the Formation of Binaries

Andreas M. Burkert[1] and Peter Bodenheimer[2]

[1] Max-Planck-Institute for Astronomy, Königstuhl 17, D-69117 Heidelberg, Germany
[2] University of California, Santa Cruz, CA 95064, USA

Abstract. We examine the effect of an initial turbulent velocity field on the line-of-sight properties of cloud cores, their collapse and their fragmentation into multiple stellar systems. The multiplicity m depends strongly on the adopted initial turbulent Mach number M: $m \approx 1 + M$, however with a large spread as a result of different choices of the random initial velocity field. The random initial superposition of modes also leads to a large spread in specific angular momenta of the cores which might explain the large period spread in observed binary systems.

1 Initial Conditions for Binary Formation

Molecular cores are known to be the sites of star formation. They are found preferentially through optically thin tracers like CS inside molecular clouds with clumpy substructures and turbulent velocity fields ([9]). In contrast to clumps that are diffuse and gravitationally unbound with high internal supersonic velocity dispersions, the cores are quiescent, strongly gravitationally bound and dense regions of gas that have decoupled from the turbulent environment and are in a state of gravitational contraction or collapse ([8]). The density distribution of pre-stellar cores resembles Bonnor-Ebert spheres ([1]), indicating that these objects formed close to hydrostatic equilibrium. Whereas the clumps have a power-law mass distribution with no preferred mass scale, recent submillimeter continuum surveys indicate a characteristic mass for cores of order 1 solar mass with a mass spectrum reminiscent of the stellar initial mass function ([7]). Thus the stellar mass function is determined by processes that lead to the formation of cores inside turbulent clouds ([6]). The frequency and properties of multiples is however determined by the physics of the collapse and fragmentation of individual cores.

Previous theoretical models of core collapse adopted very simplified initial conditions of perturbed spherically symmetric initial density distributions and rigid body rotation. In such calculations the main parameters were the ratios, α and β of thermal and rotational energies, respectively, to the gravitational energy. More recently observations have however demonstrated that cores are turbulent with non-thermal gas motions of order $\sigma \approx 0.13$ km/s which is slightly less than the typical sound speed ([2]). Turbulent motions in this case will produce significant non-linear deviations from rigid body rotation and spherical symmetry which have to be taken into account in numerical simulations of core fragmentation.

Burkert and Bodenheimer ([4]) analysed the line-of-sight velocity structure of turbulent cloud cores, adopting Gaussian random velocity fields with power spectra of $P(k) \sim k^{-3}$-k^{-4}. They found that these initial conditions are in good agreement with the observed distribution of line-of-sight velocity gradients of cloud cores ([2]) and their linewidth-size relationship. They concluded that the observed projected velocity gradient of a turbulent core is in general not a good indicator of its intrinsic angular momentum. The cores however still contain a non-zero specific angular momentum which results from the random superposition of the various velocity modes. As a result, statistically identical cores will show a large spread in their internal rotational properties which is comparable to the observed large spread in binary periods. We therefore can conclude, that such initial conditions should lead to more realistic simulations of cloud collapse.

Klein et al. ([5]) were the first who studied the collapse of turbulent cores, using an adaptive mesh refinement method to resolve the condensations. They investigated three cases with different turbulent Mach numbers and found no fragmentation for M=0.58, a binary system for M=1.7 and at least 6 fragments for M=2.9. From this result they concluded that observed Mach numbers $M < 1$ do not lead to binary formation in contrast to the observations that most stars form as members of binaries. It is however not clear whether other random generations of the M=0.58 turbulent core could result in binaries as Klein et al. did not explore the full parameter space in any details.

We report results from a large parameter study of turbulent core collapse starting with different initial Mach numbers and field realisations, adopting a power-spectrum of $P(k) \sim k^{-4}$. The starting point is an isothermal critical Bonnor-Ebert sphere which is supported by the sum of the turbulent and thermal pressure. No intrinsic ordered rotation is added to the turbulent velocity field. To prevent numerical fragmentation, a transition of the isothermal to a polytropic equation of state is assumed at a critical density such that the Jeans mass is always resolved ([3]). The evolution is followed with an SPH code of 200,000 particles. During the early evolution density fluctuations are generated from the initial velocity field while, at the same time, the turbulence decays leading to collapse.

2 Results

Figure 1 shows an example of the evolution of a turbulent core with M = 1.3 and a total mass of 4.4 $M_\odot$. In the (x, z) projection of the top panel of the figure, the core shows a curved, elongated, banana-like shape which resembles recent observations of prestellar cloud cores by Walmsley (this volume). The middle panel shows on a smaller scale the condensation which has not yet fragmented. The maximum density is now 3 orders of magnitude higher than the initial mean density. Shortly afterwards this condensation fragments into a binary (lower panel) with separation 9×10^{14} cm and a total mass about 10% that of the cloud. At this stage the maximum density is about 6 orders of magnitude higher

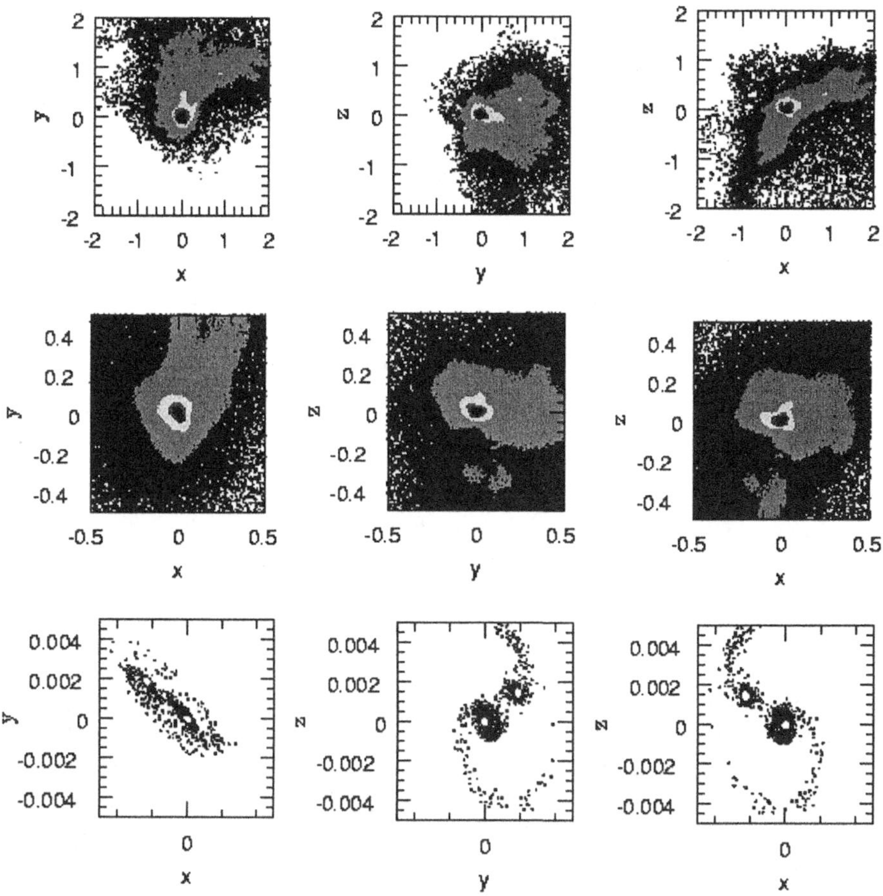

Fig. 1. Projected surface density distribution of a cloud core at three different times during collapse. The unit of length is 10^{17} cm. Top panel corresponds to a time of 5.4 $\times 10^{12}$ s, middle panel to a time of 5.68 $\times 10^{12}$ s, lower panel to a time of 5.75 $\times 10^{12}$ s

than that of the initial cloud. The orbital parameters are likely to change due to future accretion and interaction with the circumbinary disk.

A number of additional simulations were made, in which both M and the seed for the Gaussian random field were varied. The multiplicity of the resulting system is shown as a function of M in Figure 2. The points with the same value of M refer to different initial random velocity fields, all with the same power spectrum $P(k) \propto k^{-4}$. There is a definite trend of increasing multiplicity with increasing M, which is approximately linearly with $(1 + M)$, as shown by the dotted curve. However there is also a large spread about this curve. Observed cloud cores have a typical $M = 0.7$ and typically tend to condense into binaries. One of our cases with M=0.3 did not fragment, in agreement with the conclusion of Klein et al. ([5]) that there might be a problem in forming binaries in cloud

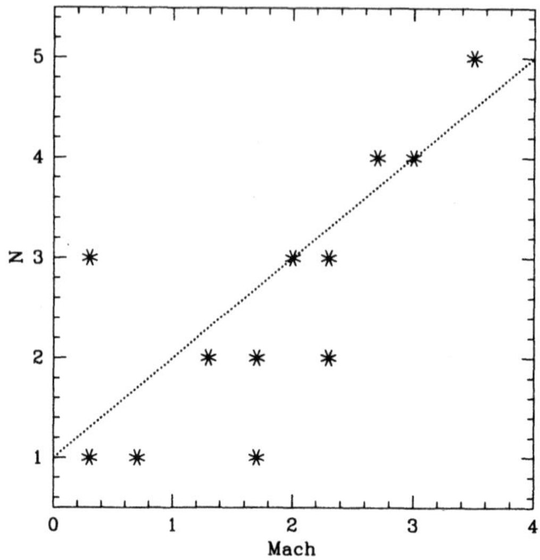

Fig. 2. Multiplicity as function of initial turbulent Mach number

cores with small M. However in another case with $M = 0.3$ a hierarchical triple forms with a separation of 10^{14} cm for the close orbit and 10^{16} cm for the wide orbit. This indicates that it is not only the Mach number but also the details of the initial superposition of the various velocity modes that play a critical role in determining the formation of binaries. In agreement with previous predictions ([4]) the simulations also lead to binaries with a wide spread in separations ranging from 10^{14} cm to 10^{16} cm with a peak around 10^{15} cm.

Although these results look promising more simulations will be needed to determine the likelihood for binary formation at low Mach numbers and their statistical properties.

References

1. J.F. Alves, C.J. Lada, E.A. Lada: Nature, **409**, 159 (2001)
2. J.A. Barranco, A.A. Goodman: ApJ, **504**, 207 (1998)
3. M.R. Bate, A. Burkert: MNRAS, **288**, 1060 (1997)
4. A. Burkert, P. Bodenheimer: ApJ, **543**, 822 (2000)
5. R.I. Klein, R. Fisher, C.F. McKee: In *The Birth and Evolution of Binary Stars: IAU 200*, ed. R. Matthieu and H. Zinnecker, in press (2000)
6. R.S. Klessen, A. Burkert: ApJS, **128**, 287 (2000)
7. F. Motte, Ph. André, R. Neri: Astron. Astrophys., **336**, 150 (1998)
8. M. Tafalla, D. Mardones, P.C. Myers, P. Caselli, R. Bachiller, P.J. Benson: Astron. Astrophys, **339**, L49 (1998)
9. J.P. Williams, L. Blitz, C.F.McKee: In *Protostars and Planets IV*, ed. by V. Mannings, A.P. Boss, S.S.Russell (Univ. of Arizona Press, Tucson, 2000) pp. 97–120

Enclosures in moonlight

Paulo Garcia, Estelle Moraux, Fabien Malbet,
Jérôme Bouvier, and Cathie Clarke

The Formation and Evolution of Low-Mass Binary Systems: How Will VLT(I) Help?

Jérôme Bouvier[1] and Gaspard Duchêne[1,2]

[1] Laboratoire d'Astrophysique, Observatoire de Grenoble, Université Joseph Fourier, B.P. 53, 38041 Grenoble Cedex 9, France
[2] UCLA Division of Astronomy and Astrophysics, Los Angeles, CA 90095-1562, USA

Abstract. We review current estimates of the binary frequency among young low-mass stars belonging to star forming clusters and associations and to young open clusters, and briefly discuss the implications for the binary formation process. We outline the major advances expected in this field by using the VLT/I instrumentation.

1 Introduction

Thanks to the advent of efficient high-angular resolution imaging devices, such as adaptive optics and speckle interferometry, and the development of sensitive spectrographs, a wealth of new results have been obtained in the last decade regarding the frequency of multiple systems among low-mass stars. Systematic searches for spectroscopic and visual binaries have been carried out in various types of low-mass populations ranging from pre-main sequence T Tauri stars in star forming associations and clusters, to low-mass dwarfs in young open clusters, and field G and K dwarfs in the solar neighbourhood. In Section 2, we summarize the present status of binary frequency measurements among low-mass stars at various stages of their evolution and in different types of environments, and discuss in Section 3 the implications of these results for current scenarios of formation and evolution of binary systems. The VLT/I will have a major impact in characterizing the properties of young multiple systems, which is outlined in Section 4.

2 The Binary Frequency Among Low-Mass Stars

Low-mass binary systems have orbital periods ranging from a few days to more than 10^8 days. Spectroscopic studies sample short period binaries, typically with $\log P$ (days) ≤ 3, while, at the distance of star forming regions and young open clusters (100-500 pc), high angular resolution imaging surveys reveal long period systems with $\log P$ (d) $\simeq$ 4-7. Still today, the pioneering study by [11] of binaries among G-type field dwarfs is the only one that covers the full range of orbital periods. It yielded an overall frequency of binary orbits[1] amounting to 61% with

[1] [11] counted the number of binary orbits regardless of the degree of multiplicity of the systems. Their binary frequency is thus obtained as (B+2T+3Q)/(S+B+T+Q),

a log-normal distribution of periods which peaks at $\log P(d) = 4.8$. This result provides a fiducial distribution of orbital periods to which all subsequent studies performed over a limited range of periods can be compared.

Once the detection biases specific to each observational technique are consistently taken into account in order to convert the observed, incomplete binary frequency (BF) to an estimate of the true BF (see [8]), the results obtained in the last decade can be summarized as follows:

- Within uncertainties of a few percent, the BF measured over a typical range of $\log P(d) \simeq 4.5 - 7.5$ for G and K-type stars in young open clusters with an age between 80 and 700 Myr (Alpha Persei [12], Pleiades [5], Praesepe [6], Hyades [28]) is similar to that of G-type field dwarfs (BF $\simeq$ 29%, [11]). The frequency of spectroscopic binaries ($\log P(d) \leq 3$) in the Pleiades [25] and Praesepe [24] clusters also appear to be similar to that derived by [11] for field dwarfs.
- The BF measured for low-mass T Tauri stars in star forming clusters (Orion [30], IC 348 [9]) over ranges of $\log P(d) \simeq 5-8$ or narrower is indistinguishable from that of solar-type field dwarfs.
- The binary content of T Tauri stars in loose associations, such as Taurus [20,14] and Ophiuchus [31,16] appears to be much higher than that measured for field G-dwarfs and T Tauri stars in clusters. This result is particularly significant for the Taurus population, where the binary excess amounts to nearly a factor of 2 compared to field dwarfs over $\log P(d) \simeq 3.5$-7.5 [23].

Beyond the mere binary frequency, other quantities such as the shape of the distributions of orbital periods and mass ratios are relevant to constrain the binary formation process. Unfortunately, these distributions are affected by larger uncertainties than is the BF integrated over the log P range under study and urgently need more precise determinations from larger telescopes (cf. Section 4). Yet, trends already seem to emerge. The mass ratio distributions derived for low-mass visual binaries in young open clusters and star forming regions are usually consistent with their being uniform between $M_2/M_1 = 0.2$ and 1 [10,33], or slightly increasing towards lower mass ratios down to 0.3 [28] as derived by [11] for field G-dwarfs. The distribution of orbital periods over the range $\log P(d) \simeq 4.5 - 7.5$ for low-mass binaries in young open clusters is consistent (albeit with large uncertainties again) with the log-normal orbital period distribution of G-type field dwarfs [11]. However, wide binaries with $\log P(d) \geq 7.5$, appear to be strongly deficient in young clusters compared to the field population [29,13].

where S=single stars, B=binary, T=triple, Q=quadruple systems. This quantity is sometimes referred to as the companion star fraction or, as here, as the binary frequency.

3 Scenarios of Binary Formation and Evolution

Among various plausible mechanisms able to form binaries, the fragmentation of collapsing molecular cloud cores appears today the most promising one (e.g. [4]). We review below the two main scenarios proposed so far to account for the binary frequency measured among low-mass young stars belonging to different environments.

3.1 A Universal Population of Proto-Binaries?

The first scenario assumes that the star formation process is universal in the sense that it yields the same fraction of (proto-)binaries in all types of regions (clusters, associations) at the end of the protostellar collapse. Since observations indicate that at an age of ~1 Myr associations (e.g. Taurus) harbour about twice as many binaries as clusters do (e.g. Orion), some evolution of the proto-binary population must have occurred within the first Myr after their formation.

Stellar densities in associations ($n_\star \sim 10$ pc^{-3}) are much too low to allow the formation of new binaries by capture (or their destruction by gravitational encounters) within such a short timescale. It is therefore likely that the proto-binary population in clouds like Taurus does not evolve over a few Myr. The high BF ($\geq 80\%$) measured today in Taurus is therefore assumed to reflect the pristine population of proto-binaries formed at the end of the gravitational collapse.

Starting from such a high proto-BF, two mechanisms act to disrupt a fraction of proto-binaries in dense clusters ($n_\star \geq 10^4$ pc^{-3}) [19]. As a result of fragmentation some proto-binaries form with a semi-major axis of order or greater than the mean distance between protostellar systems in the center of the cluster and are therefore immediately disrupted. Then, on a timescale of 1 Myr, frequent gravitational encounters between protostellar systems further disrupt the long period binary systems (typically $\log \mathrm{P}(d) \geq 7$), whose binding energy is lower than the cluster's stars average kinetic energy.

This dynamical scenario thus suggests that most stellar systems form as binaries (which may help to solve the initial angular momentum problem), a significant fraction of which are subsequently destroyed by gravitational encounters in the center of dense clusters on a timescale of ~1 Myr. These predictions are consistent with the lower binary fraction observed in clusters compared to associations at an age of a few Myr, and more generally with the qualitative trend of a decreasing BF as the local stellar density increases [27], as well as with the apparent deficit of wide binaries in clusters [29]. After a few Myr, dynamical models [17] predict that the binary population stabilizes, which also agrees with the observations of a similar BF in clusters spanning an age range from 2 to 700 Myr.

One potential difficulty with assuming that gravitational encounters alone are responsible for the observed BF in clusters past 1 Myr is that one would then expect to observe some variations of the BF from clusters to clusters depending on their initial stellar density. As an illustration, numerical modelling [18, see their Fig. 10 and 11] predicts a BF in the $\log \mathrm{P}(d)$ range from 4.5 to 7.5 which differs by nearly a factor of 2 for clusters having a factor of 10 difference in their initial stellar density (BF = 0.34 and 0.20 at 1 Myr, 0.28 and 0.16 at 100 Myr, for $\log n_\star = 4.8$ and 5.8 pc^{-3}, respectively). Such differences in BF are not observed, with all clusters studied so far exhibiting the same BF to within a few percent in this log P range, regardless of their age. Even more puzzling is the fact that the BF in young open clusters is the same as in the field. Yet, no more than 10% of field stars are thought to have formed in dense clusters [1] and the majority probably originated from intermediate-density star forming regions. This tends to indicate that the fraction of visual binaries among low-mass stars is not extremely sensitive to the initial stellar density.

3.2 The Impact of Initial Conditions

As an alternative (or a complement) to the dynamical evolution of clusters, the differing BF in clusters and associations might reflect the impact of initial conditions on the star formation process. Why would fewer binaries originate in cluster forming regions than in regions of distributed star formation is currently unknown. However, various calculations have pointed out the sensitivity of the fragmentation process to local conditions, such as cloud turbulence, density profile, magnetic field, or to external impulses triggering the collapse (e.g. [3,15,2,32]). It is therefore conceivable that the properties of molecular cloud cores and pre-stellar condensations may differ from one region to the other and that their evolution during collapse produces a different output as far as protostellar multiplicity is concerned. Recent surveys of star forming regions at millimeter wavelengths do suggest that specific differences exist between the properties of protostellar objects located in clusters and in associations (e.g. [8]).

The belief that local conditions significantly impact on the product of the star formation process is reinforced by considering other fundamental properties of young low-mass populations that also appear to differ between clusters and associations. This seems to be the case for both the distribution of angular momenta [7] and the distribution of stellar masses [22], with clusters exhibiting a larger fraction of rapid rotators and of substellar objects than associations. It is tempting to ascribe these fundamental properties to a common origin which would then point to an intrinsically different mode of star formation between clusters and associations.

4 Prospects with the VLT/I

At the beginning of the 90's, the most puzzling result has been the finding that binaries are much more frequent in regions of distributed star formation than in cluster forming ones. At the beginning of the 00's, with considerably more observational data, an equally puzzling result is that all clusters appear to exhibit very similar fractions of visual binaries, regardless of their age, and this fraction is the same as that measured for field dwarfs.

Some of the ideas put forward to explain these results will strongly benefit from the advent of the VLT/I. Among the many new capabilities provided by the VLT/I and associated instrumentation, some will be particularly suited to progress in the understanding of binary formation and evolution:

- **The determination of complete distribution of orbital periods** will be feasible using both NAOS/VLT and AMBER/VLTI, complemented by spectroscopic studies for the shortest period systems. Current adaptive optics systems on 4m-class telescopes probe the orbital range $\log P(d) \geq 4.2$ in nearby star forming regions ($\sim$ 140 pc), while spectroscopic studies are usually restricted to $\log P(d) \leq 3$. More than a decade of orbital periods is thus not covered by current instrumentation and it is precisely the range of orbital periods over which the binary excess in associations compared to clusters is expected to start to show up since the BF of spectroscopic binaries appear to be similar in both types of environments, while there is a strong excess of binaries with $\log P(d) \geq 4.5$ in associations. With an angular resolution of 0.030 arcsec at J, corresponding to 4 AU at the distance of the nearest star forming regions, NAOS/VLT will be able to detect solar-mass systems with an orbital period down to about 3000 days (i.e. $\log P(d) = 3.5$). AMBER/VLTI will provide an order of magnitude higher angular resolution, i.e., will probe systems with an orbital period down to about 100 days which, complemented by spectroscopic studies, will additionally yield dynamical mass estimates for the components.
- **A more reliable determination of the mass-ratio distributions** will be provided by NAOS/VLT. The current detection limit of companions using PUEO adaptive optics system at CFHT is shown in Figure 1. It clearly shows that at small separations, faint companions are difficult to detect, which prevents one to obtain a reliable estimate of the mass-ratio distribution below about $M_2/M_1=1/3$. Thanks to the improved Strehl ratio it will deliver, the detection limit of NAOS/VLT is expected to be 3-4 magnitude fainter than that of PUEO, thus providing complete and reliable mass-ratio distributions down to 0.1, a critical constraint to the theories of binary formation.
- Most importantly, a major breakthrough will be the first opportunity to investigate **the multiplicity of protostars at very high angular resolution** thanks to the infrared wavefront sensor associated to NAOS/VLT. So far, observation of embedded young stellar objects with adaptive optics was forbidden by the lack of nearby visible stars required to close the adaptive loop, and most protostellar sources are too faint to be observed by speckle techniques. Hence, little is known today of the degree of multiplicity of protostars, whose direct determination would obviously provide the clearest answer regarding the impact local conditions may have onto the frequency of proto-binaries. Progress has recently been made in this field through radio millimeter interferometry [21] which provides an angular resolution of order of 0.5 arcsec. With more than a tenfold increase in angular resolution, imaging with NAOS/VLT will provide unprecedented insight into the multiplicity of newly born objects.

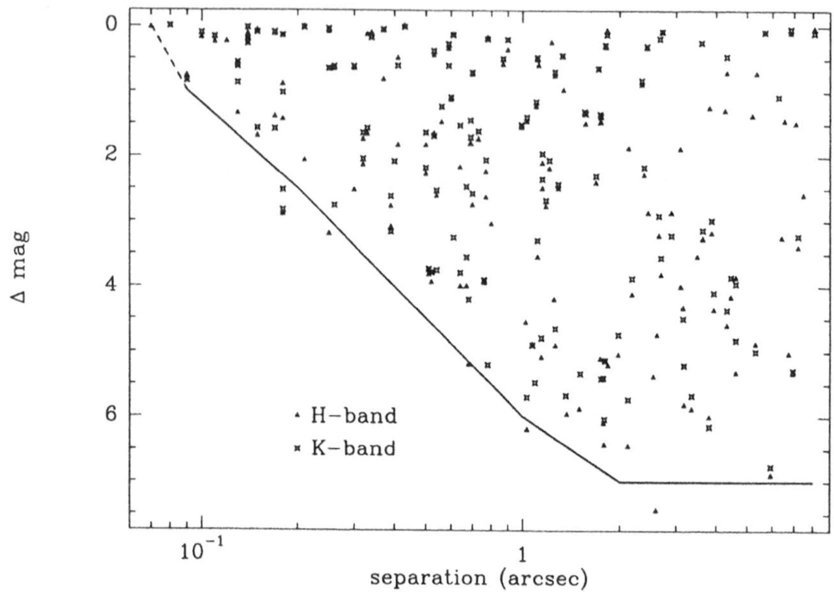

Fig. 1. Detection limit of faint companions with CFHT's PUEO adaptive optics system. Symbols (triangle: H-band, starred: K-band) indicate the magnitude difference between the companion and the primary as a function of angular separation for the 156 binary systems we detected with PUEO in various star forming regions and young open clusters. The solid curve indicates the detectability limit computed as the 5σ (speckle+photon+background) noise level measured on images of single stars obtained in the same conditions

References

1. Adams F.C., Myers P.C.: Astrophys. J. **553**, 744 (2001)
2. Boss A.P.: In: *The Formation of Binary Stars*, ed. by Zinnecker & Mathieu,, ASP Conf. Ser., Vol.200, in press (2001)
3. Burkert A., Bate M.R., Bodenheimer P.: MNRAS **289**, 497 (1997)
4. Bodenheimer P.: In: *The Formation of Binary Stars*, ed. by Zinnecker & Mathieu,, ASP Conf. Ser., Vol.200, in press (2001)
5. Bouvier J., Rigaut F., Nadeau D.: Astron. Astrophys. **323**, 139 (1997)
6. Bouvier J., Duchêne G., Mermilliod J.-C., Simon T.: Astron. Astrophys., in press (2001)
7. Clarke C.J., Bouvier J.: MNRAS **319**, 457 (2000)
8. Duchêne G.: Astron. Astrophys. **341**, 547 (1999)
9. Duchêne G., Bouvier J., Simon T.: Astron. Astrophys. **343**, 831 (1999)
10. Duchêne G., Bouvier J., Eislöffel J., Simon T.: In: *From Darkness to Light*, ed. by T. Montmerle & Ph. André, ASP. Conf. Ser., in press (2001)
11. Duquennoy A., Mayor M.: Astron. Astrophys. **248**, 485 (1991, DM91)
12. Eislöffel J., Simon T., Close L., Bouvier: In: *11th Cambridge workshop on Cool Stars, Stellar Systems and the Sun*, ed. by Garcia Lopez et al., ASP Conf. Ser. Vol. 223, in press (2001)

13. Ghez A.M.: In: *The Formation of Binary Stars*, ed. by Zinnecker & Mathieu,, ASP Conf. Ser., Vol.200, in press (2001)
14. Ghez A., Neugebauer G., Matthews K.: Astron. J. **106**, 2005 (1993)
15. Klein R.I., Fisher R., McKee C.F.: In: *The Formation of Binary Stars*, ed. by Zinnecker & Mathieu,, ASP Conf. Ser., Vol.200, in press (2001)
16. Köhler R., Leinert Ch.: Astron. Astrophys. **331**, 977 (1998)
17. Kroupa P.: In: *Massive Stellar Clusters*, ed. by Lançon & Boily, ASP Conf. Ser. Vol. 211, p.233 (2000)
18. Kroupa P., Aaserth P., Hurley S.J.: MNRAS **321**, 699 (2001)
19. Kroupa P., Petr M.G, McCaughrean M.J.: New Astron. **4**, 495 (1999)
20. Leinert Ch., Zinnecker H., Weitzel N. et al.: Astron. Astrophys. **278**, 129 (1993)
21. Looney L.W., Mundy L.G., Welsch W.J.: Astrophys. J. **529**, 477 (2000)
22. Luhman K.L.: Astrophys. J. **544**, 1044 (2000)
23. Mathieu R.D.: Ann. Rev. Astron. Astrophys. **32**, 465 (1994)
24. Mermilliod J.-C., Mayor M.: Astron. Astrophys. **352**, 479 (1999)
25. Mermilliod J.-C., Rosvick J.-M., Duquennoy A., Mayor M.: Astron. Astrophys. **265**, 513 (1992)
26. Motte F., André P.: Astron. Astrophys. **365**, 440 (2001)
27. Patience J., Duchêne G.: In: *The Formation of Binary Stars*, ed. by Zinnecker & Mathieu,, ASP Conf. Ser., Vol.200, in press (2001)
28. Patience J., Ghez A., Reid I., Weinberger A., Matthews K.: Astron. J. **115**, 1972 (1998)
29. Scally A., Clarke C., McCaughrean M.J.: MNRAS **306**, 253 (1999)
30. Simon M., Close L.M., Beck T.L.: Astron. J. **117**, 1375 (1999)
31. Simon M., Ghez A., Leinert Ch. et al.: Astrophys. J. **443**, 625 (1995)
32. Whitworth A.P.: In: *The Formation of Binary Stars*, ed. by Zinnecker & Mathieu,, ASP Conf. Ser., Vol.200, in press (2001)
33. Woitas J., Leinert C., Köhler R: Astron. Astrophys., in press (2001)

The Formation of Brown Dwarfs

Bo Reipurth

Center for Astrophysics and Space Astronomy
University of Colorado, Boulder, CO 80309, USA

Abstract. Numerous brown dwarfs have been found in recent years, re-igniting debate about their origin. A natural way to stunt the growth of a nascent stellar embryo is to eject it out of the infalling gas envelope. This can be achieved by dynamical interactions in newborn multiple systems or small-N clusters of stellar embryos. In this scenario brown dwarfs form like stars, but are sent into distant orbits or completely ejected, thus depriving them of sufficient growth to ever exceed the hydrogen burning limit of 0.08 $M_{\odot}$. Decay of a multiple system is a stochastic process and, with better luck, brown dwarfs could indeed have become normal stars.

1 The Ubiquity of Brown Dwarfs

The first discoveries of brown dwarfs were made as companions to main sequence or evolved stars (e.g. Nakajima et al. 1996). Subsequent surveys have demonstrated that brown dwarfs are far more common as free floating, isolated bodies, and such objects are now found so frequently that brown dwarfs may be as common as low mass stars (e.g. Kirkpatrick et al. 1999, Martín et al. 2000, Comerón et al. 2000).

Numerous studies have explored the origin of the initial mass function (see e.g. Larson 1999, Kroupa 2001). Increasingly accurate determinations of the mass function reaching into the very low mass and substellar regimes have firmly established that the standard Salpeter power law slope of the IMF at higher masses gives way to a much shallower slope at lower masses or even a turn-over.

The main power law slope of the IMF is probably related to the hierarchical structure of interstellar clouds, and it has been widely assumed that brown dwarfs are formed in the same way as stars, except from very small, dense clumps of gas. The minimum stellar mass may be related to the Jeans mass $M_J \propto T^2 P^{-1/2}$, and stellar objects with much smaller mass are not likely to form because the cloud clumps that should form them will not be strongly self-gravitating (e.g. Larson 1992). One possibility for forming numerous brown dwarfs would be to consider special environments like the ultracold, high pressure molecular clouds found in the inner disk of M31 (Elmegreen 1999).

Because of this difficulty, alternative models have been considered. Numerical simulations of self-gravitating protostellar disks have indicated that gravitational instabilities could lead to the formation of brown dwarf companions. However, Pickett et al. (2000) conclude that isothermal disk calculations cannot demonstrate that disk fragmentation will form such low mass companions.

Taking a very different approach, Lin et al. (1998) suggested that an encounter between two protostars of approximately solar mass and with massive disks can result in the formation and ejection of a tidal filament of the order of 1000 AU long. Out of this gas an unbound brown dwarf might subsequently form. If so, brown dwarfs should be exceedingly rare in loose associations like Taurus, where protostellar encounters are likely to be very infrequent.

In an effort to devise a formation mechanism for brown dwarfs that only relies on well known physical processes without a dependence on a special environment, Cathie Clarke of the Institute of Astronomy, Cambridge, and I have considered the possibility that brown dwarfs are stellar embryos which have been ejected in dynamical interactions in small multiple systems (Reipurth & Clarke 2001). The advantage of this approach is that in this picture brown dwarfs form in exactly the same way as stars, and indeed would have become real stars, if they had not been prematurely evicted because of dynamical processes.

2 Dynamical Interactions in Multiple Systems

The chaotic motions of members of non-hierarchical triple or higher order systems must be studied numerically and analyzed statistically; for a review see Valtonen & Mikkola (1991). The motions can be divided into three categories: *interplay*, in which the members move chaotically among each other; *close triple approach*, when all members at the same time occupy a small volume of space; and *ejection*, in which one member departs the system after exchanging energy and momentum with the two others. A close triple approach is a necessary but not sufficient condition for ejection. Ejections often lead to escape, but can also result in the formation of a hierarchical triple system, with one body in an extended orbit. The remaining two members are bound closer to each other, forming a tighter and highly eccentric binary system. Most often, although not always, it is the lowest mass member that is ejected; the escape probability scales roughly as the inverse third power of the mass (e.g. Anosova 1986). The ejected member acquires a velocity $v_{eject} \sim 15\ D_{ca}^{-1/2}$, where v_{eject} is in km s^{-1} and the closest approach D_{ca} is in AU (Armitage & Clarke 1997). Sterzik & Durisen (1995, 1998) have performed numerical simulations of triple T Tauri systems and, under their assumptions, find ejection velocities of typically 3-4 km s^{-1} but with a higher velocity tail. The decay of a triple system occurs stochastically and can only be described in terms of the half-life of the process. Anosova (1986) finds that within about one hundred crossing times t_{cr} almost all systems have decayed, where $t_{cr} \sim 0.17(R^3/M)^{1/2}$, and R is a characteristic length scale for the system in AU, and M is the total system mass in $M_\odot$.

Three-body dynamical simulations including substellar members have been performed by Sterzik & Durisen (1999) and their dynamical results can be summarized as *brown dwarfs are ejected because they are of low mass*, in agreement with the general theory just outlined. This is in contrast to the evolutionary scenario advocated here which can be summed up as *brown dwarfs are of low mass because they are ejected.*

3 The Ejection Model

How do stars in the process of forming figure out what their ultimate mass should be? For single stars the conventional answer is that stellar masses are determined by the size of the mass reservoir available plus the effect of outflow, which eventually halts the inflow. However, for multiple stars, or stars in dense clusters, dynamical interactions shift individual members in and out of the mass center where gas densities are higher, thus forcing the members to accrete at highly variable rates (Bonnell et al. 2001). Competitive accretion and dynamical processes open the way for forming very low mass objects even in an environment in which stars, if accreting equitably, would have much higher masses. In particular, an object with as little mass as a brown dwarf can form if dynamical interactions either send it into a large orbit away from the main reservoir of infalling gas, or ejects it outright (Reipurth & Clarke 2001).

Obviously, the two main competing processes is the accretion of gas versus displacement or eviction by dynamical processes. The accretion rate is highly dependent on time and location, and the decay of a multiple system is a completely stochastic event, for which no analytic solution exists. Nonetheless, to illustrate the process, consider the following order-of-magnitude estimates. For simplicity assume a constant infall rate $\dot{M}_{infall} \sim 6 \times 10^{-6}(T/10K)^{3/2}$ M$_\odot$ yr^{-1}. A 1 M$_\odot$ cloud at 10 K would then take 1.7×10^5 yr to collapse. Assume that the collapse fragments into a number N_{mul} of stellar embryos and that the infalling mass is distributed among them according to their mass ratios $q_i = M_i/M_{prim}$ (which may be time-dependent), where $1 \leq i \leq N_{mul}$, and M_{prim} is the time-dependent mass of the most massive embryo. Further assume that a part f of the infalling gas is lost in outflow activity, where we adopt $f{\sim}0.3$. The growth rate of the ith embryo is then $\dot{M}_i = \dot{M}_{infall} \times (1-f) \times q_i/N_{mul}$, and for the simplest case where $N_{mul} = 3$ and $q_i = 1$ it will then take a typical embryo about 6×10^4 yr to grow beyond a substellar mass.

If three stellar embryos occupy a volume with diameter 200 AU, and if we consider a total mass in the range 0.06 to 0.24 M$_\odot$, then the characteristic crossing time $t_{cr} \sim 0.17(R^3/M)^{1/2}$ is between 1000 yr and 2000 yr. If, on the other hand, the configuration is tighter, occupying a volume with a diameter of only 20 AU, then T_{cr} is merely 30-60 yr.

The usual decay equation is $n_t/n_o = exp(-0.693t/\tau)$, where τ is the half-life of the decay. In their numerical simulations, Sterzik & Durisen (1995) found that about 95% of all systems had decayed after about 100 crossing times. This allows us to link the crossing time and the half life of the decay, $\tau = 23.1t_{cr}$, so that to first order $n_t/n_o = exp(-0.18\dot{M}_{infall}^{1/2}(1-f)^{1/2}N_{mul}^{-1/2}R^{-3/2}t^{3/2})$.

While the stellar embryos are still very small, the crossing time, and therefore the half life, is extremely long, i.e. the embryos do not effectively start to interact until a time T$_i$ when they have a certain mass. When the embryos are very small, their awareness of each other is limited because they are surrounded by the massive infalling envelope, and this further adds to T_i. But as the envelope material thins out and the embryo masses increase, their dynamical in-

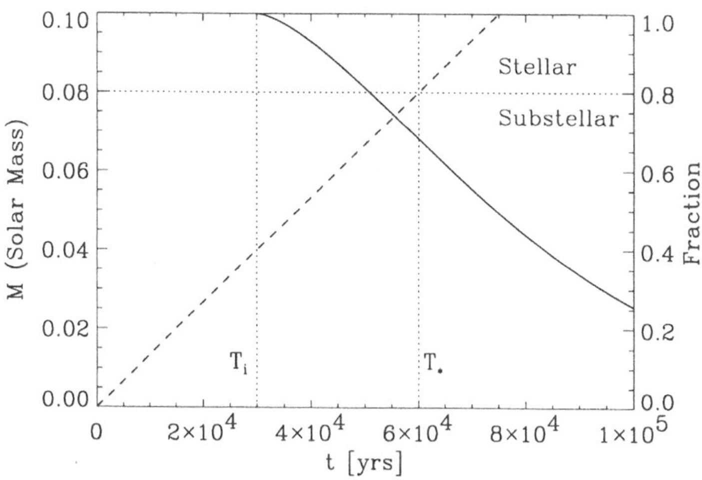

Fig. 1. A growth vs decay diagram for disintegrating multiple systems

teractions become important. A more precise value of T_i can only be determined by numerical experiments.

We will say that there is a reasonable chance that a brown dwarf is ejected if the half-life τ of the decay is less than $T_* - T_i$, where T_* is the free-fall time required to build up a multiple system of objects of average mass 0.08 $M_\odot$, i.e. $T_* = (N_{mul} \times 0.08)/\dot{M}_{infall}$. If the ejection occurs at a time later than T_*, then the ejected object is a star.

Figure 1 shows a simple, schematic presentation of these processes in a growth vs decay diagram, based on the values already discussed. The stapled line represents the growth of an embryo, which reaches the stellar mass threshold after 6×10^4 yr. The solid curve shows n_t/n_o, which is a measure of the probability that the multiple system has not yet decayed. To be conservative, we have assumed the embryos occupy a volume of diameter 200 AU, resulting in a large t_{cr}. The time when interactions begin to take place, T_i, has arbitrarily been set to 3×10^4 yr. In the particular example shown, about one third of a population of triple systems would have decayed before the expulsed member reached the stellar mass limit.

4 Observational Consequences

The ejection model has a number of observational predictions, which can be used to test the concept.

1. Since most ejections will take place during the collapse phase, it follows that the very youngest brown dwarfs (and therefore the most luminous ones) should be found in the near vicinity of Class 0/I sources. The projected separation s in arcsec should be $s = 0.21\ v\ t\ d^{-1}\ sin\alpha$, assuming that the nascent brown dwarf is observed at a time t [yr] after ejection and moving with the space velocity v [km s^{-1}] at an angle α to the line-of-sight at a distance d [pc].

2. A number of binary brown dwarfs are now known (e.g. Martín et al. 2000; Reid et al. 2001). Also, it has recently been found that while close brown dwarf companions to normal stars are rare, this is not the case for more distant brown dwarf companions (Gizis et al. 2001). Brown dwarf binarity and companionship pose very important constraints on any model of brown dwarf formation. Table 1 summarizes in a compact form the predictions of the ejection model. If the primary is a late-type F-G-K star, it may form a wide pair with a brown dwarf companion, because the brown dwarf has been ejected into a distant, but bound orbit. Such a primary, however, would only under special circumstances form a close pair with a brown dwarf. For M type primaries, a brown dwarf can be a distant companion because it has been ejected, but it can also be a close companion, because the whole binary can have been ejected. If the primary is a brown dwarf, it is unlikely to have a distant brown dwarf companion, but may well have a close brown dwarf companion, because the brown dwarf pair was ejected. In those cases where the brown dwarf has been ejected into a distant orbit, the primary itself should normally be a close binary (although dynamical situations could allow exceptions to this).

Table 1. Binaries with BD Companions

	F-G-K	M	BD
Wide Pair	yes	yes	rare
Close Pair	rare	yes	yes

3. An ejected object, whether star or brown dwarf, will have a velocity comparable to the velocity attained at pericenter in the close triple encounter. Because the ejection velocity may well be within the velocity dispersion of the gas in the associated cloud, such excess velocities are likely to be very difficult to detect. In the case of dense clusters, the excess velocity will with time translate into a slight difference in spatial distribution. However, brown dwarfs in very massive and/or very old clusters have velocities that are higher because their kinematics have been dominated by two-body relaxation. Therefore, such clusters should *also* be surrounded by a halo of brown dwarfs, but not because of their additional ejection velocities. A proper test of the ejection scenario is made only by finding a halo of brown dwarfs around clusters small enough or young enough that relaxation has not yet dominated their kinematics.

4. When a stellar embryo is ejected, its circumstellar disk will be pruned during the close interactions that led to its ejection. Armitage & Clarke (1997) showed that very close encounters, which truncate the disks to radii of a few AU, could promote the rapid decline of classical T Tauri star characteristics thereafter. Such truncations would not affect the near-infrared excesses of brown dwarfs, because they are generated well within 1 AU, and indeed observations of young brown dwarfs in the Orion Cluster show that many have such near-infrared excesses (Muench et al. 2001). Spectroscopic signatures of infall and outflow, which we have come to associate with extremely young stellar objects, may well be more shortlived than for normal T Tauri stars. An interesting object recently discovered in the Chamaeleon clouds by Fernández and Comerón (2001) which appears to be at the border between stellar and substellar objects and shows evidence of both accretion and outflow may well be a rarity, if indeed it has originated in an ejection event.

An interesting issue has arisen with the observations by Muench et al. (2001) which suggest that a number of their substellar candidate objects may reside in proplyds (see also the paper by Lada in this volume). If those objects, which are selected photometrically by their colors and luminosities, are indeed confirmed spectroscopically to be bona fide brown dwarfs, we would then have evidence for circumstellar material stretching over many hundreds of AU, which is in apparent contradiction to the expected disk truncation in a triple encounter. However, if a disk is truncated at the moment of ejection, it will not remain forever confined to a few AU once the dynamical interaction is over, but will through scattering processes spread over a much larger volume. Exposed to intense UV radiation, the outer layers of this circumstellar material will also heat and expand. Further theoretical work is required to study if the magnitude and time scales of these effects suffice to fit the observations.

References

1. Anosova, J.P. 1986, Ap&SS, 124, 217
2. Armitage, P.J., Clarke, C.J. 1997, MNRAS, 285, 540
3. Bonnell, I.A., Bate, M.R., Clarke, C.J., Pringle, J.E. 2001, MNRAS, 323, 785
4. Comerón, F., Neuhäuser, R., Kaas, A.A. 2000, A&A, 359, 269
5. Elmegreen, B.G. 1999, ApJ, 522, 915
6. Fernández, M., Comerón, F. 2001, A&A, in press
7. Gizis, J.E., Kirkpatrick, J.D., Burgasser, A. et al. 2001, ApJ, 551, L163
8. Kirkpatrick, J.D., Reid, I.N., Leibert, J. et al. 1999, ApJ, 519, 802
9. Kroupa, P. 2001, MNRAS, 322, 231
10. Larson, R.B. 1992, MNRAS, 256, 641
11. Larson, R.B. 1999, in *Star Formation 1999*, ed. T. Nakamoto, Nobayama, p. 336
12. Lin, D.N.C., Laughlin, G., Bodenheimer, P., Rozyczka, M. 1998, Science, 281, 2025
13. Martín, E.L., Brandner, W., Bouvier, J. et al. 2000. ApJ, 543, 299
14. Muench, A.A., Alves, J.A., Lada, C.J., Lada, E.A. 2001, ApJL, in press
15. Nakajima, I. et al. 1996, Nature, 378, 463
16. Pickett, B.K., Durisen, R.H., Cassen, P., Mejia, A.C. 2000, ApJ, 540, L95
17. Reid, I.N., Gizis, J.E., Kirkpatrick, J.D., Koerner, D.W. 2001, AJ, in press
18. Reipurth, B., Clarke, C.J. 2001, AJ, 122, 432
19. Sterzik, M.F., Durisen, R.H. 1995, A&A, 304, L9
20. Sterzik, M.F., Durisen, R.H. 1998, A&A, 339, 95
21. Sterzik, M.F., Durisen, R.H. 1999, *Star Formation 1999*, ed. T.Nakamoto, p.387
22. Valtonen, M., Mikkola, S. 1991, Ann. Rev. Astr. Ap., 29, 9

Daniel Apai, Roland Vavrek, and Zoltan Balog

Disk Orientations in PMS Binary Systems Determined Through Polarimetric Imaging with UT1/FORS

Jean-Louis Monin[1], François Ménard[2,1], and Nicolas Peretto[1]

[1] Laboratoire d'Astrophysique de Grenoble, Observatoire de Grenoble, BP 53, 38041 Grenoble Cedex, France
[2] Canada-France-Hawaii Telescope Corporation, PO Box 1597, Kamuela, HI 96743, USA

Abstract. The IPOL mode of FORS on UT1 allows to measure high quality polarimetric images on a wide field of view. We have used this instrument during ESO observation period 65 to determine the respective orientation of circumstellar disks in pre Main Sequence (PMS) binaries. In this paper, we present our method and first results obtained with FORS/IPOL. Thanks to the top quality assistance of the VLT and FORS teams, our run has been remarkably successful and the instrument appears extremely well adapted to fine polarimetric measurements on a large field of view. This large field of view is essential to disentangle between intrinsic polarisation and interstellar polarisation. Our first results suggest that circumstellar disks tend to be aligned in PMS binaries.

1 Disks in PMS Binary Systems

Our current understanding of low mass stellar formation has to take into account two very different yet complementary constraints. On one hand, when we consider individual stars, the current model put forward for embedded Young Stellar Objects (YSOs) includes a central stellar core, surrounded by an equatorial accretion disk and a remnant infalling envelope (see e.g. [1]).

On the other hand, we also know that a large fraction of T Tauri stars (TTS) form in binary or multiple (N>2) systems ([2]). This ubiquitous property of the stellar formation process has a potentially enormous influence on the previous one because the circumstellar environment of the individual components of a multiple system can be deeply modified by the presence of a companion. The study of individual disks in PMS binary systems and in particular their relative orientations can provide strong constraints on the star formation process.

Polarimetry can give access to the projected disk orientation in the plane of the sky, provided the disk is far enough from pole-on, a condition met in most of the cases.

2 Determining Disk Orientation Using Polarimetry

2.1 Method

Models of bipolar reflection nebulae by [3] have shown that the position angle of the integrated linear polarization of the scattered starlight is parallel to the equatorial plane of the disk, provided the inclination is sufficiently large. The method is thus likely to give good results when circumstellar disks are present around the two stars in the binary, ie. when both are Classical TTS (CTTS). We have chosen our targets so that at least one of the components can be classified as an active T Tauri star. Previous observations of PMS binary stars have shown indeed that most of the time, if one of the components of a young binary system has an active disk, so has the other ([4]).

2.2 The Case of the Interstellar Polarisation

One of the main problem of polarimetric measurements in young systems, is that such stars are often deeply embedded in molecular clouds and subject to interstellar medium polarization. When we measure two different polarization directions in a binary system, we can be fairly sure that they are actually different, but if they are similar, there is a chance that this equality is due to a common interstellar polarization. This is why a large field polarimetric imaging system like the one available with FORS/IPOL on UT1 is essential in this context. We are now able to measure at the same time the polarization on the central system and on a large number of nearby field stars. These complementary measurements allow to estimate the local interstellar polarisation, hence to subtract its contribution from the central object's. They can also bring very valuable informations on the large scale interstellar polarization pattern.

3 Measurements

The FORS instrument is equipped with a Wollaston prism that splits the beam into two different directions with orthogonal polarization states, the so-called ordinary (o) and extraordinary (e) beams. A stepped half wave plate retarder is placed at the entrance of the incident beam and can be rotated at various angles multiple of 22.5^o (16 positions in a complete rotation, see the FORS/IPOL documentation for details). For each position of the plate, a CCD image is recorded. The separation of the two O and E beams on the CCD is performed via the Wollaston prism, using a focal 9-slit mask, so that a given polarization state (e or o) occupies half of the focal plane image. To get a complete square field of view requires two measurement sets, at two positions on the sky separated by an inter-slit offset. The total field of view in our measurements is $6.8' \times 6.8'$ in the Standard Resolution (SR) mode with a focal scale of $0.2''$/pixel and $20''$ wide slits. The FORS/IPOL observing blocs allow to take 4, 8 or 16 CCD frames on the corresponding positions of the rotating plate. Then a Fourier series is computed to extract the Stokes parameter U & Q, hence P and θ from the data.

4 Data Reduction

4.1 Reduction Pipeline

We have written a dedicated data reduction pipeline using NOAO/IRAF. The first step concerns of course bias, bad pixels and flat-field corrections. Then the images go through a polarization pipeline. Two options are available: P and θ can be estimated on a pixel per pixel basis, a useful possibility to map extended structures like reflection nebulosities, at the cost of a loss of accuracy on point sources when the image quality (FWHM) changes during acquisition of a full data set (i.e. between different rotation positions of the half-wave plate). Another option of our pipeline uses aperture photometry to estimate precise polarization measurements on point like objects. Indeed, the individual CTTS components and their disks in our binaries are most of the time spatially unresolved, and the disk orientation information is obtained via aperture photometry. When using aperture photometry, we can account for any FWHM change if a large enough aperture is used. Both techniques are available in our pipeline on request, to estimate the polarization of point sources and extended features alike.

We estimate the errors using 2 independent methods: first from the statistical photon noise on the *e* & *o* beams separately, and then propagating the errors up to U, Q, P and θ; second by measuring the standard deviation on the 4, 8 or 16 images from the half wave plate rotation. Both estimations are consistent except in some pathological cases (e.g. severe hit by cosmic rays, images in a set too close to dead zones between orthogonal polarisation strips, etc.), and this is extremely useful to check our results. Our conclusion on the observing strategy is that it is strongly recommended to systematically record 16 images using all the available $\lambda/2$ plate positions. The result is without question worth the time investment. Most of the time, the residual error is less than $\Delta P = 0.1\%$ (absolute value) when the binary components are well separated ($\geq$1.3 arcsec). Our program also includes tight binaries for which we have to adjust and subtract a PSF, a third method our pipeline provides. Depending on the contrast and separation, the errors can at worse reach $\Delta P = 0.4\%$, a value well within our goal.

4.2 Instrumental Polarization

Of crucial importance is the determination of the instrumental polarization $P_{\rm inst}$. We have carefully measured it against nearby unpolarized targets. We have observed GJ 781 and GJ 781.1, two high proper motion stars. As the close solar neighborhood is remarkably devoid of dust, the interstellar polarization of near Earth objects can be considered null. The average of our 4 measurements on both GJ objects gives $P_{\rm inst} = 0.02\% \pm 0.03\%$. Even if bad luck could have made the instrumental polarization just cancel the possible intrinsic polarization of these test stars, we independently found some of our low-polarization scientific targets to present linear measurements very close to previously published measurements. We believe that FORS/IPOL (+ incident optics) instrumental polarization is actually very low, well below 0.1%, so we considered it as negligible for practical purposes and we did not remove it from other measurements.

5 Preliminary Results

The complete results of our run, as well as a detailed analysis of their implication on star formation, will be published in a forthcoming paper. The alignment of circumstellar disks in PMS binary systems has been studied previously by [5] and [6]. These previous results indicate that disks are preferentially aligned. However, the question of a precise and simultaneous determination of the local interstellar polarisation in the immediate vicinity of the central binary has not been correctly addressed yet. Our new results with FORS/UT1 should fix this issue.

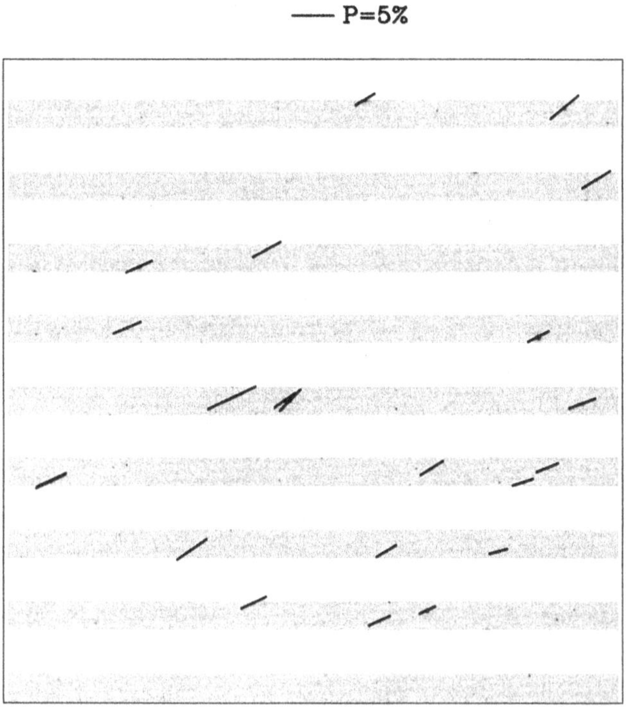

Fig. 1. Resulting polarisation vectors for individual sources superimposed on a 6.8′ × 6.8′ intensity map in the I band (shaded every other 20″), around the binary SZ 60. We can simultaneously trace the local interstellar polarisation field on a large number of surrounding field stars, and from this estimation we can extract the central object intrinsic polarisation

Figure 1 shows that the interstellar polarization can indeed dominate over the central object intrinsic polarization: on this image, all the polarization vectors are almost exactly aligned, with a possible fluctuation on the central object, most probably indicating that it possesses intrinsic polarisation. One of the great interest of FORS/IPOL in this matter is that our wide field polaro-imaging data allow us to precisely measure, and subtract the local interstellar polarization.

We are currently working on a method to remove this ambient polarization in order to study the intrinsic target polarization. The complete presentation and results of this method will be published in a forthcoming paper. In the meantime, we have plotted the respective orientation of both polarisation vectors in 20 binaries (see fig. 2), as obtained from direct unsubtracted measurements. Unless the polarisation is always dominated by the interstellar polarisation in every object, a result that will be thoroughly checked using our VLT measurements, this plot shows that the disks tend to be aligned in PMS binaries. If this result holds when the local interstellar polarisation contribution has been subtracted, this will be the first firm confirmation that circumstellar disks are aligned in newly formed binary systems. As these systems are young (1-3 Myr), it is unlikely that tidal interactions have had the time to realign the disks, and this alignment must be a result of the binary formation process. If disks around the components in young binaries are coplanar, then they may provide stable favorable environments to build large planetary bodies.

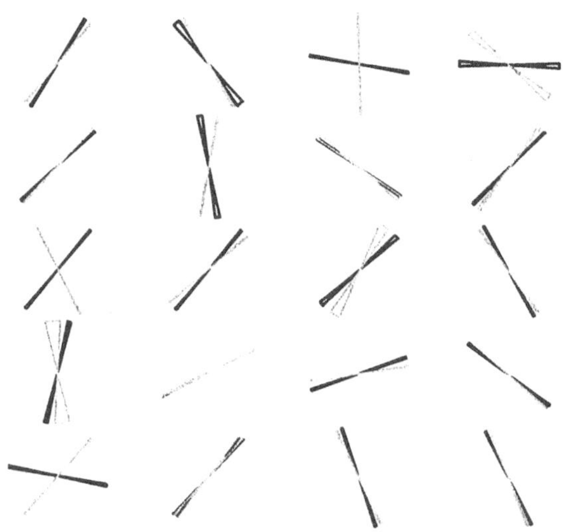

Fig. 2. Sketch of the orientation of the disks in 20 binaries. In every case, the disk is traced as a bow tie with an angular opening of $\pm 1\,\sigma_\theta$. Black: primary ; grey: secondary

Another valuable result from our work is that we have obtained images at different wavelengths (V, R & I bands) so that we will be able to check whether the expected λ dependence of the polarization is recovered. For most of our sources, the result will remain unchanged, with both polarizations remaining parallel, but in a few cases, the changes can be clear, in magnitude and / or orientation.

References

1. F.H. Shu, F.C. Adams, S. Lizano: ARA&A **25**, 23 (1987)
2. G. Duchêne: A&A **341**, 547 (1999)
3. P. Bastien, F. Ménard: ApJ **364**, 232 (1990)
4. G. Duchêne, J.-L. Monin, J. Bouvier, F. Ménard: A&A **351**, 954 (1999)
5. J.-L. Monin, F. Ménard, G. Duchêne: A&A **339**, 113 (1998)
6. E.L.N. Jensen, A.X. Donar, R.D. Mathieu: 'Aligned Disks in Pre–Main Sequence Binaries'. In: *Birth and Evolution of Binary Stars, IAU Symposium 200 at Potsdam, Germany April 10–15, 2000*, poster p85

Multiplicity of Young Brown Dwarfs in Cha I

Viki Joergens[1], Eike Guenther[2], Ralph Neuhäuser[1], Fernando Comerón[3], Nuria Huélamo[1], João Alves[3], and Wolfgang Brandner[3]

[1] Max-Planck-Institut für Extraterrestrische Physik, Giessenbachstr. 1, D-85748 Garching, Germany
[2] Thüringer Landessternwarte Tautenburg, Karl-Schwarzschild-Observatorium, Sternwarte 5, D-07778 Tautenburg, Germany
[3] European Southern Observatory, Karl-Schwarzschild-Str. 2, D-85748 Garching, Germany

Abstract. How frequent are brown dwarf binaries? Do brown dwarfs have planets? Are current theoretical pre-main-sequence evolutionary tracks valid down to the substellar regime? – Any detection of a companion to a brown dwarf takes us one step forward towards answering these basic questions of star formation.

We report here on a search for spectroscopic and visual companions to young brown dwarfs in the Cha I star forming cloud. Based on spectra taken with UVES at the VLT, we found significant radial velocity (RV) variations for five bona-fide and candidate brown dwarfs in Cha I. They can be caused by either a (substellar or planetary) companion or stellar activity. A companion causing the detected RV variations would have about a few Jupiter masses. We are planning further UVES observations in order to explore the nature of the detected RV variations. We also found that the RV dispersion is only ~2 km/s indicating that there is probably no run-away brown dwarf among them.

Additionally a search for companions by direct imaging with the HST and SOFI (NTT) has yielded to the detection of a few companion candidates in larger orbits.

1 Multiplicity of Brown Dwarfs

High-precision radial velocity (RV) surveys have brought up more than 60 planetary candidates in orbit around stars (mostly G- and K-type) but only a few brown dwarf candidates despite the fact that these surveys are more sensitive to higher masses. This is referred to as the 'brown dwarf desert'.

The search for fainter companions by direct imaging yielded to the discovery of seven brown dwarf companions to stars confirmed by both spectroscopy as well as proper motion. These are Gl 229 B (Nakajima et al. 1995, Oppenheimer et al. 1995), G 196-3 B (Rebolo et al. 1998), Gl 570 D (Burgasser et al. 2000), TWA 5 B (Lowrance et al. 1999, Neuhäuser et al. 2000), HR 7329 (Lowrance et al. 2000, Guenther et al. 2001), Gl 417B and Gl 584C (Kirkpatrick et al. 2000, 2001). Furthermore GG Tau Bb (White et al. 1999) and Gl 86 (Els et al. 2001) are good candidates of brown dwarf companions to stars.

These detections show that there are at least a few brown dwarfs orbiting stars but what about brown dwarfs having companions themselves? Up to now there are three brown dwarf binaries known, i.e. brown dwarf – brown dwarf pairs: the brown dwarf spectroscopic binary PPl 15 (Basri & Martín 1999) and two brown dwarf binaries confirmed by both imaging and common proper motion, DENIS-P J1228.2-1547 (Martín et al. 1999) and 2MASSW J1146 (Koerner et al. 1999). Very recently another object, 2MASSs J0850359+105716, has been detected as likely brown dwarf binary (Reid et al. 2001).

There is no planet known orbiting a brown dwarf. The lowest mass star with a RV planet candidate is the M4-dwarf Gl 876 (Delfosse et al. 1998).

Do brown dwarfs have companions at all? Planetary or substellar companions around brown dwarfs may form in a circumstellar disk but it is also possible that they form by fragmentation like low-mass stars. Although circumstellar disks around mid- to late M-dwarfs might be expected to have insufficient mass to form companions around them, recent observational evidence hints at significant reservoirs of gas and dust even around these objects (Persi et al. 2000, Fernández & Comerón, 2001). There are also indications for the presence of significant circumstellar material around a few of the Cha I and ρ Oph bona fide and candidate brown dwarfs, which show IR excess (Comerón et al. 2000, Wilking et al. 1999).

2 Bona Fide and Candidate Brown Dwarfs in Cha I

The Cha I dark cloud is a site of on-going and/or recent low- and intermediate-mass star formation. It is one of the most promising grounds for observational projects on young very low-mass objects, since it is nearby (160 pc) and the extinction is low compared to other star forming regions.

By means of two Hα objective-prism surveys 12 new low-mass late-type objects (M6–M8) have been detected in the center of Cha I, named Cha Hα 1 to 12 (Comerón et al. 1999, 2000). Their masses are below or near the border line separating brown dwarfs and very low-mass stars according to comparison with evolutionary tracks by Baraffe et al. (1998) and Burrows et al. (1997). Four of them have been confirmed as bona fide brown dwarfs (Neuhäuser & Comerón 1998, 1999 and Comerón et al. 2000).

3 The Radial Velocity Survey

We used UVES, the high-resolution Echelle spectrograph at the VLT in order to search for companions to the bona fide and candidate brown dwarfs in Cha I. We took at least two spectra separated by a few weeks of each of the nine brightest objects in the red part of the wavelength range, since the objects are extremely red (Fig. 1).

The determination of precise RVs requires the superposition of a wavelength reference on the stellar spectrum so that both light beams follow exactly the same path in the spectrograph. We used the telluric O_2 lines as wavelength reference, which are produced by molecular oxygen in the Earth atmosphere and show up in the red part of the optical spectral range (Fig. 1). It has been shown that they are stable up to about 20 m/s (Balthasar et al. 1982, Caccin et

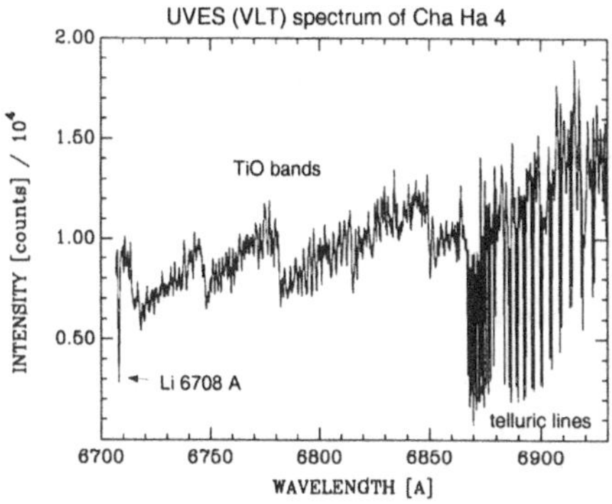

Fig. 1. UVES Echelle spectrum. For clarity only a small part of the total observed spectrum (6700 Å to 10400 Å) is displayed. Late M-dwarfs exhibit a wealth of spectral features in the red part of the wavelength range, which are used for the RV determination by means of cross correlation. The telluric lines served as wavelength reference

al. 1985). The iodine cell, often used for extrasolar planet searches, produces no lines in this wavelength region.

RVs are determined by cross-correlating plenty of stellar lines of the object spectra against a template spectrum and locating the correlation maximum. As a template we used a mean spectrum of a young, cool star also obtained with UVES. The spectral resolution of the UVES spectra is about 40 000. We achieved a velocity resolution of about 200 m/s for a S/N of 20 in agreement with expectations (Hatzes & Cochran 1992).

3.1 Small Radial Velocity Dispersion

Neuhäuser & Comerón (1999) constrained the RV dispersion of the Cha I bona fide and candidate brown dwarfs from medium resolution spectra to be 11 km/s. Based on the high-resolution UVES spectra of nine of the Cha I bona fide and

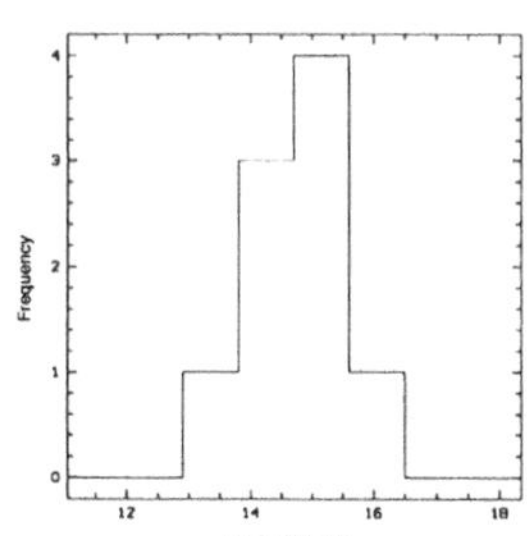

Fig. 2. A histogram of mean RVs of nine bona fide and candidate brown dwarfs in Cha I clearly depicts the very small RV dispersion of the studied sample of only ~2 km/s. This detection indicates that there is no run-away brown dwarf among the sample

candidate brown dwarfs in Cha I we find that the RV dispersion is even smaller, namely ∼2 km/s (Fig. 2). This finding gives suggestive evidence that there is no run-away brown dwarf among them and does not support the formation scenario that brown dwarfs are ejected stellar 'embryos' proposed by Reipurth & Clarke (2001).

3.2 Jupiter Mass Companions Around Brown Dwarfs?

The analysis of UVES spectra taken at different times yielded to the detection of significant RV variations for five bona fide and candidate brown dwarfs in Cha I. They could be caused by reflex motion due to orbiting objects or by shifting of the spectral line center due to surface features (stellar activity).

The detected RV variations are of the order of 1 km/s. If they are caused by companions they would have masses of a few Jupiter masses depending on the orbital parameters as shown in Fig. 3. We found a *preliminary* RV orbit for one of the studied objects (Fig. 4) with an approximate M sin i for a hypothetical companion of 4.8 M_{Jup}. This shows that the detection of companions with masses of a few times the mass of Jupiter in orbit around brown dwarfs or very low-mass stars is clearly feasible with these data. An orbiting planet around a brown dwarf has a much larger effect on its parent object than a planet in orbit around a star and is therefore easier to detect. If an absorption cell for the red part of the optical wavelength range would be available for UVES a RV precision of 3 m/s might be feasible and with it the detection of planets with a few Earth masses in orbit around brown dwarfs.

It would be an interesting finding if the existence of planets around the studied bona fide and candidate brown dwarfs in Cha I can be confirmed since they would be the objects (very low-mass stars or even brown dwarfs) with the latest spectral types, i.e. the lowest masses, harboring planets. Furthermore the confirmation of planets around these extremely young objects would show that fully formed giant planets can already exist around low-mass objects that are only a few million years old. All the up to date known brown dwarf binaries (cp. Sec. 1) are considerably older than the ones in Cha I.

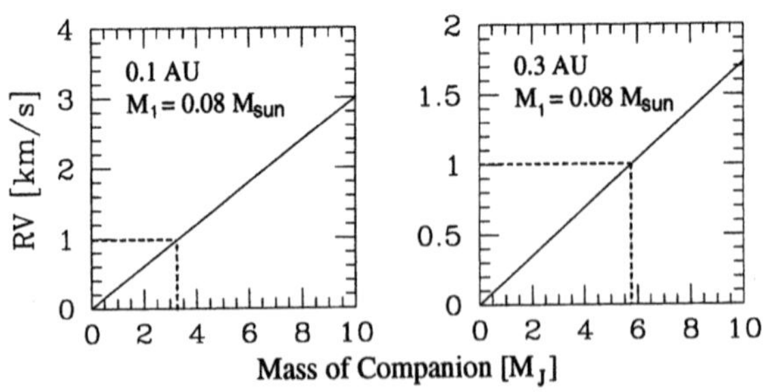

Fig. 3. The detected RV variations of about 1 km/s for five bona fide and candidate brown dwarfs in Cha I could be caused by planetary companions: A RV variation of 1 km/s corresponds to a few Jupiter masses depending on the orbital parameters

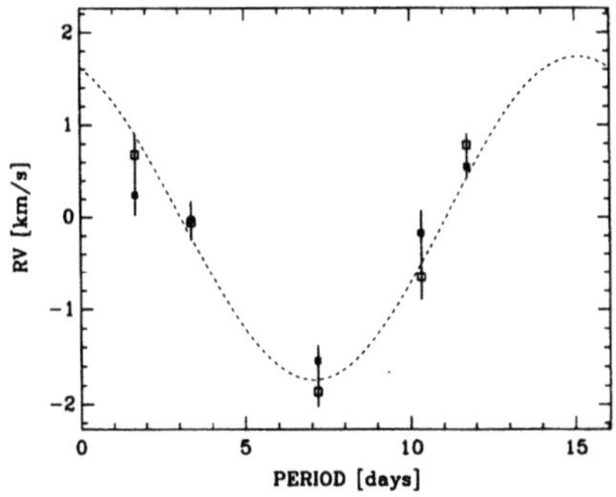

Fig. 4. Preliminary RV orbit of one of the bona fide and candidate brown dwarfs. While five observations are not enough to derive a perfect fit we can nevertheless determine an approximate M sin i for the hypothetical companion of 4.8 M_{Jup}

3.3 Stellar Spots on Brown Dwarfs?

We know from young T Tauri stars that they exhibit prominent surface features causing RV variations of the order of 2 km/s (Guenther et al. 2000) due to a high level of magnetic activity of these stars.

It is an outstanding question if brown dwarfs have magnetically active photospheres and consequently prominent surface features. Recent detections of X-ray emission from young brown dwarfs may be explained by magnetic activity (Neuhäuser & Comerón 1998, Neuhäuser et al. 1999, Comerón et al. 2000). Furthermore a few brown dwarfs have shown indications for periodic photometric variabilities with periods less than one day, which may be caused by spots or weather (Bailer-Jones & Mundt 2001, Eislöffel & Scholz, this conference).

We investigated the UVES spectra for hints of magnetic activity. We detected Ca II emission for some of the bona fide and candidate brown dwarfs in Cha I, but did not find a correlation between the presence of this emission and the RV variations. Some of the targets have been detected as X-ray emitters (Neuhäuser et al. 1999, Comerón et al. 2000) but there is no correlation between the X-ray luminosity and the amplitude of the detected RV variations.

Photometric data of the objects will be used to search for rotational periods. If the RV variations are caused by spots the brown dwarfs should exhibit photometric variations with the same period. Furthermore RV monitoring may yield useful complementary information on the appearance and evolution of cool spots in brown dwarfs and very low-mass stars.

4 Direct Imaging Campaign

In a complementary project to the RV survey we are searching for visual companions to the Cha I bona fide and candidate brown dwarfs with larger separations by means of direct imaging. Images obtained with the HST and with SOFI at the NTT yielded to the detection of a few companion candidates and are the subject of further analysis.

References

1. C.A.L. Bailer-Jones, R, Mundt: A&A **367**, 218 (2001)
2. H. Balthasar, U. Thiele, H. Wöhl: A&A **114**, 357 (1982)
3. I. Baraffe, G. Chabrier, F. Allard, P.H. Hauschildt: A&A **337**, 403 (1998)
4. G. Basri, E.L. Martín: ApJ **118**, 2460 (1999)
5. A.J. Burgasser, J.D. Kirkpatrick, R.M. Cutri et al.: ApJ **531**, L57 (2000)
6. A. Burrows, M. Marley, W.B. Hubbard et al. ApJ **491**, 856 (1997)
7. B. Caccin, F. Cavallini, G. Ceppatelli, A. Righini, A.M. Sambuco: A&A **149**, 357 (1985)
8. F. Comerón, G.H. Rieke, R. Neuhäuser: A&A **343**, 477 (1999)
9. F. Comerón, R. Neuhäuser, A.A. Kaas: A&A **359**, 269 (2000)
10. X. Delfosse, T. Forveille, M. Mayor et al. A&A **338**, L67 (1998)
11. S.G. Els, M.F. Sterzik, F. Marchis et al.: A&A **370**, L1 (2001)
12. M. Fernández & F. Comerón: A&A, submitted (2001)
13. E.W. Guenther, V. Joergens, R. Neuhäuser et al.: "A spectroscopic and photometric survey for pre-main sequence binaries". In: *Birth and Evolution of Binary Stars, IAU Symposium No. 200, Potsdam, Germany, April 10–15, 2000*, ed. by B. Reipurth, H. Zinnecker (ASP Conference Series, in press)
14. E.W. Guenther, R. Neuhäuser, N. Huélamo, W. Brandner, J. Alves: A&A **365**, 514 (2001)
15. A.P. Hatzes, W.D. Cochran: "Spectrograph Requirements for Precise Radial Velocity Measurements". In: *High Resolution Spectroscopy with the VLT, ESO workshop, Garching, Germany, February 11–13, 1992*, ed. by M.-H. Ulrich
16. J.D. Kirkpatrick, I.N. Reid, J.E. Gizis et al.: AJ **120** 447 (2000)
17. J.D. Kirkpatrick, C.C. Dahn, D.G. Monet et al.: AJ in press (2001)
18. D.W. Koerner, J.D. Kirkpatrick, M.W. McElwain, N.R. Bonaventura: ApJ **526**, L25 (1999)
19. P.J. Lowrance, C. McCarthy, E.E. Becklin et al.: ApJ **512**, L69 (1999)
20. P.J. Lowrance, G. Schneider, J.P. Kirkpatrick et al.: ApJ **541**, 390 (2000)
21. P.J. Lowrance, E.E. Becklin, G. Schneider, AAS **197**, 5203 (2000)
22. E.L. Martín, W. Brandner, G. Basri: Science **283**, 1718 (1999)
23. T. Nakajima, B.R. Oppenheimer, S.R. Kulkarni et al.: Nature **378**, 463 (1995)
24. R. Neuhäuser, F. Comerón: Science **282**, 83 (1998)
25. R. Neuhäuser, F. Comerón: A&A **350**, 612 (1999)
26. R. Neuhäuser, C. Briceño, F. Comerón et al.: A&A **343**, 883 (1999)
27. R. Neuhäuser, E.W. Guenther, W. Brandner et al.: A&A **360**, L39 (2000)
28. B.R. Oppenheimer, S.R. Kulkarni, K. Matthews, M.H. van Kerkwijk: Science **270**, 1478 (1995)
29. Persi P., Marenzi A.R., Olofsson G. et al.: A&A, 357, 219 (2000)
30. R. Rebolo, M.R. Zapatero-Osorio, S. Madruga et al.: Science **282**, 1309 (1998)
31. I.N. Reid, J.E. Gizis, J.D. Kirkpatrick, D.W. Koerner: ApJ **121**, 489 (2001)
32. B. Reipurth, C. Clarke: ApJ in press (2001), astro-ph/0103019
33. R.J. White, A.M. Ghez, I.N. Reid, G. Schulz: ApJ **520**, 811 (1999)
34. B.A. Wilking, T.P. Greene, M.R. Meyer: AJ **117**, 469 (1999)

Tom Greene, Alan Moorwood, and Timo Prusti

Imaging with the VLT

Ralf Siebenmorgen

European Southern Observatory, Karl-Schwarzschild-Str. 2,
D-85748 Garching b. München, Germany

1 Introduction

For each of the four VLT unit telescopes there are at least three instruments. Already in operation for a couple of years are FORS1, FORS2 and ISAAC. By the end of next year imaging will be possible with VIMOS and NIRMOS, NAOS/CONICA and VISIR. In Fig. 1 a schematic overview of the VLT instrumentation as of April 2001 is given together with first light dates.

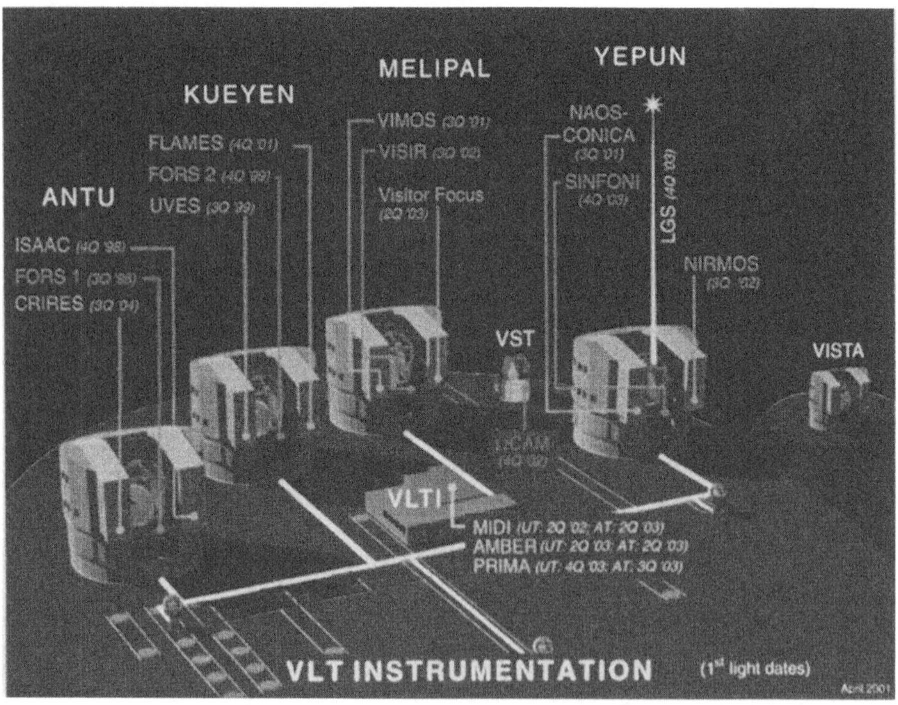

Fig. 1. Schematic view of the VLT instrumentation.

Figure available in colour on the CD-ROM

2 Instruments

2.1 VIMOS and NIRMOS

The visible multi–object spectrograph (VIMOS) and the near infrared multi–object spectrograph (NIRMOS) are devices with large field imaging capabilities. Both instruments have similar outline. One main characteristic is that there are four quadrants, each with a 2k × 2k detector, giving a total field of $4 \times 7' \times 8'$. The quadrants are separated by a cross of $2'$. There is some slight vignetting on 2 corners along the $8'$ direction. The pixel scale is $0.205''$ per pixel, corresponding to 2048 pixels for $7'$. A schematic view of the detector unit is shown in Fig. 2.

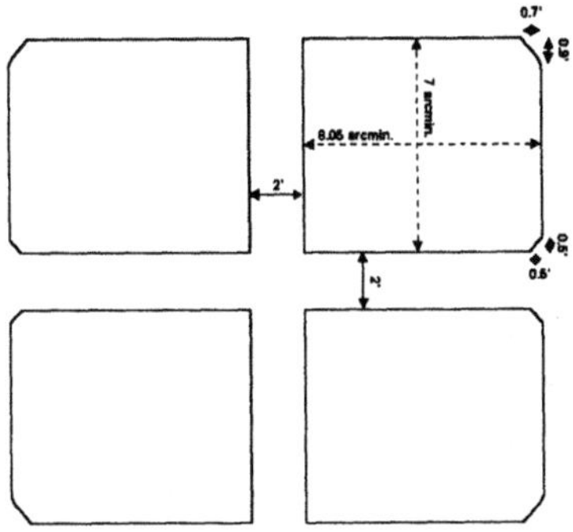

Fig. 2. Layout of the VIMOS imaging field of view

2.2 FORS1 and FORS2

There are two focal reducer and low dispersion spectrographs (FORS). They have imaging as well as polarimetric capabilities in the 0.3 –1 μm range. They feature a low and high resolution mode with a $6.8' \times 6.8'$ field at an image quality of 80% in $0.2''$ within the central $4.0'$ and a $3.4' \times 3.4'$ field at an image quality of 80% in $0.1''$ within the central $2.0'$, respectively. A large selection of filters and grisms is available.

FORS at the VLT has made already several discoveries over the last years. One example of the high image quality is the protostellar system HH34 (ESO PR Photo 40b/99). Regarding search for stellar companions I mention the detection of the young brown dwarf in the nearby TWA-5 constellation (see ESO PR Photo 17a/00by (Neuhaeuser et al. 2000).

2.3 NAOS / CONICA

For each of the four 8m class telescopes there is an *active optics* system making high dynamic range observations feasible. Slight inaccuracies caused by polishing remnants, tube flexures or gravitational forces are corrected by an active computer controlled system of 150 axial supports at a typical update interval of 30s.

Nevertheless, in order to derive diffraction limited images the atmospheric turbulences need to be overcome. While the uncorrected beam is limited by the seeing, a "clever" deformation of the wavefront by means of a flexible mirror in front of the detector (tip/tilt secondary mirror) gives the diffraction pattern. Such an *adaptive optics* system for the VLT is provided by NAOS. NAOS is specified to have a Strehl ratio of 70% in K at V = 12 mag.

NAOS will come together with the coronographic near infrared camera (CONICA). Beside a number of filters for direct and coronographic imaging CONICA will also have polarisation analysing capabilities. Its 1k × 1k detector will operate at different scales from 109.2 milli–arcsec (mas) per pixel and a 73″ diameter field down to 13.6 mas per pixel and a 14″ × 14″ field.

2.4 ISAAC

The infrared spectrometer and array camera (ISAAC) is the workhorse of the VLT in the near infrared. It has two 1k × 1k detectors and a wide field of 2.5′ × 2.5′ for imaging polarimetry. ISAAC operates in its short (1–2.5μm) and long (2.5–5μm) wavelength channel.

Some of the many highlights of imaging with ISAAC are: first light image of the star forming region RCW38 (ESO PR Photo 46b/98), direct observations of low mass stars in the massive HII region NGC3603 (Brandl et al. 1999) and the detection of edge-on circumstellar discs (Brandner et al., this volume).

2.5 VISIR

The VLT mid infrared spectrometer and imager will provide the following baseline observing modes: diffraction limited imaging with variable magnification utilizing broad and narrow-band filters up to a maximum field of 80″ × 80″ between 8 – 13 μm and 16.5 – 24.0 μm; long-slit spectroscopy between 7.9 – 14 μm with $R \sim 250$ and $R \geq 30000$. Additionally there will be long-slit spectroscopy between 7.9 – 14 μm with $R \sim 10000$ and between 16 – 24 μm with $R \sim 3500$ and $R \geq 15000$. For more information see Lagage (this volume).

3 Protostellar Imaging

The mid infrared should be of particular interest for obtaining direct images of the very earliest phase of stellar evolution. Likely the youngest protostar known today is HH108MMS; it is detected with ISOCAM (Fig. 3, see Siebenmorgen & Krügel 2000). Objects of this class are so cold and dense that they are seen in absorption, still at 15 μm, against the diffuse background light. This is remarkable since the dust extinction is already a factor 20 lower than in the visible (Krügel & Siebenmorgen 1994). The VLT goal would be to image these protostars but at a spatial resolution an order of magnitude higher than was possible with ISO.

However, if they are below the detection limit of VISIR we need to wait for the next generation of even larger telescopes.

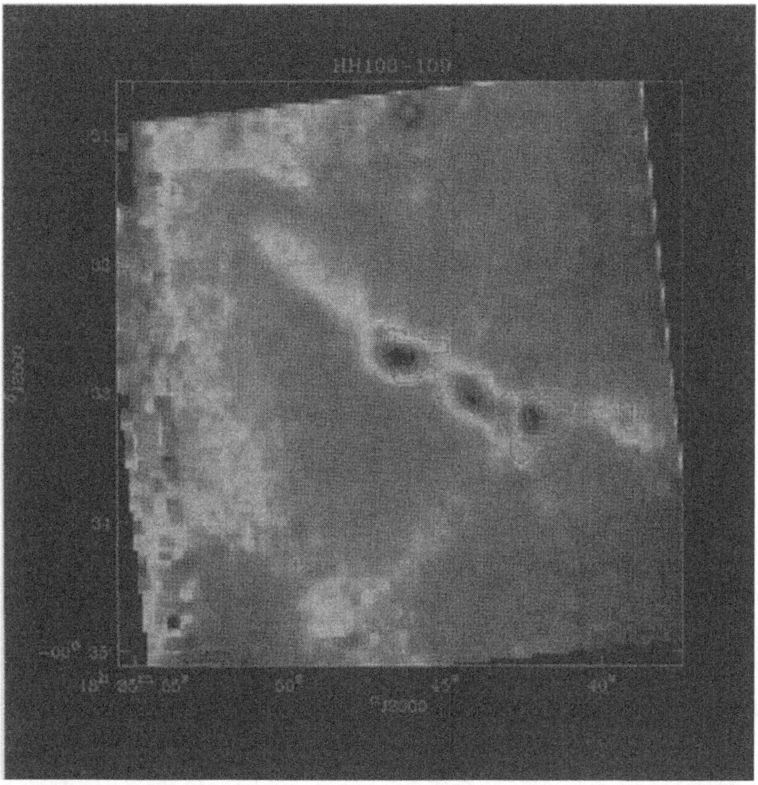

Fig. 3. Possibly the youngest protostar known today is HH108MMS, shown here in grey scale at 14.3μm (Siebenmorgen & Krügel 2000; 30″ ~ 9300AU). The contour lines are an overlay from Chini et al. (1997) and show the 1.3mm dust emission. In the mid infrared, the right 1.3mm source, IRAS18331–0035, is in emission, the left, HH108MMS, is seen in absorption.
Figure available in colour on the CD-ROM

References

1. Brandl B. et al., 1999, A&A 352, 69
2. Brandner W. et al., this volume
3. Chini R. et al., 1997, A&A 325, 542
4. Krügel E. & Siebenmorgen R., 1994, A&A 288, 929
5. Lagage P.O., this volume
6. Neuhaeuser R., et al., ESO PR Photo 17a/00, available at: `http://www.eso.org/outreach/info-events/ut1fl/astroimages`
7. Siebenmorgen R. & Krügel E., 2000, A&A 364, 625

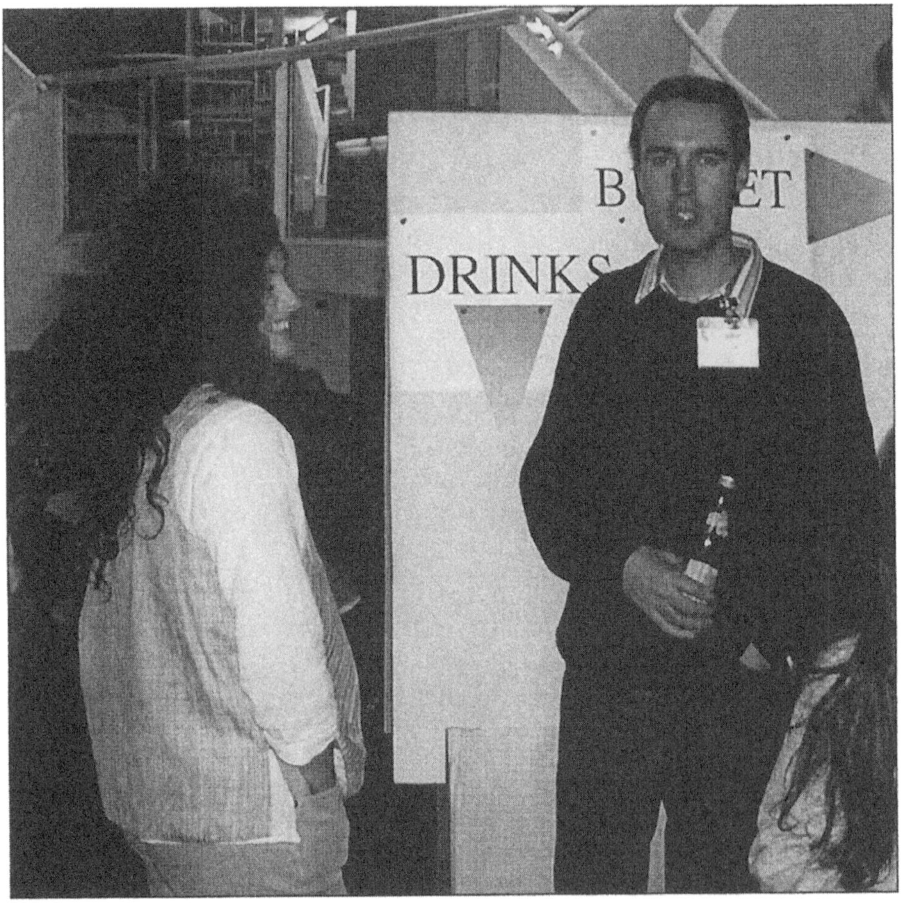

Cathie Clarke and Matthew Bate

The Formation of a Cluster of Stars and Brown Dwarfs in a Turbulent Molecular Cloud

Matthew R. Bate[1,3], Ian A. Bonnell[2], and Volker Bromm[3]

[1] School of Physics, University of Exeter, Stocker Road, Exeter EX4 4QL, United Kingdom
[2] School of Physics and Astronomy, University of St Andrews, North Haugh, St Andrews, Fife, KY16 9SS, United Kingdom
[3] Institute of Astronomy, Madingley Road, Cambridge CB3 0HA, United Kingdom

Abstract. We present preliminary results from the first hydrodynamic calculation to follow the collapse and fragmentation of a large-scale turbulent molecular cloud into a stellar cluster while resolving beyond the opacity limit for fragmentation. The calculation produces a mixture of single and binary stars with comparable numbers of stars and brown dwarfs. Without exception, the brown dwarfs are formed by ejection from unstable multiple systems. The calculation produces a stellar initial mass function in agreement with observations. However, at this stage in the calculation, no high-mass ($> 0.03\ M_\odot$) brown dwarfs have been formed because they are ejected too soon after formation to have accreted much mass.

1 Introduction

The collapse and fragmentation of molecular cloud cores to form single stars and bound multiple stellar systems has been the subject of many numerical studies, e.g. [1–7]. These calculations have resulted in the adoption of fragmentation as the favoured mechanism for the formation of the majority of binary stars, since it can produce a wide range of binary properties (as observed) through simple variations of the pre-collapse initial conditions.

However, while individual binary systems can be reproduced by such fragmentation calculations, it is extremely difficult to predict the statistical properties of stellar systems that should result from the fragmentation model. There are two main reasons for this. First, most fragmentation calculations have only considered the formation of binaries or small multiple systems and not the large numbers of stars for which statistical properties can be determined. Second, the results of these calculations depend primarily on the initial conditions chosen. For example, in forming a binary system, the system's final mass is determined by the mass of the initial cloud core, its separation is determined by the initial angular momentum, and so forth. Thus, the statistical properties of stellar systems such as the frequencies of single, binary and higher-order multiple systems, the initial mass function (IMF), and the distribution of binary separations cannot be determined from such calculations. In order to obtain theoretical predictions of the statistical properties of stellar systems, we must both form a greater number of stars and begin with more realistic initial conditions.

The first such calculation [8] modelled the collapse of a large-scale clumpy molecular cloud to form $\sim$ 60 protostellar cores. It was found that the mass function of the cores could be fit by a lognormal mass function that has a similar width to the observed stellar initial mass function. The core masses were determined by a combination of the initial density structure, competitive accretion (see [9,10] for in depth studies of how competitive accretion leads to an initial mass function), and dynamical interactions. Further calculations [11,12] confirmed the lognormal mass function and showed that the mean mass of the protostellar cores was similar to the mean initial Jeans mass in the cloud.

Unfortunately, while these calculations help us to understand the processes that may determine the initial mass function, they only allow the very widest binaries to be resolved. We do not know which of the unresolved protostellar cores will fragment to form binaries or what their properties will be. Thus, we are also unable to determine the true stellar initial mass function (only the 'system' mass function) or what role multiple systems play in the evolution of a young cluster.

This proceedings presents results from the first calculation to follow the collapse and fragmentation of a large-scale turbulent molecular cloud to form a stellar cluster, while resolving down to the opacity limit for fragmentation [13]. The opacity limit occurs approximately at the point when the centre of a collapsing cloud of molecular gas becomes optically-thick to infrared radiation and begins to heat up. This heating stops the dynamic collapse near the cloud's centre and forms a quasistatic core with a size of $\sim$ 5 AU [14]. The quasistatic core cannot collapse further to form a star until it has accreted more gas and its central temperature exceeds that required for molecular hydrogen to dissociate. This allows a 'second collapse' to form the star. Fragmentation during this second phase of collapse is thought to be inhibited by the high thermal energy [15,16]. Thus, resolving the calculation up to and beyond the opacity limit should allow us to model all potential fragmentation including that resulting in binary and multiple systems.

2 Calculations

2.1 The Code and Initial Conditions

The calculation is being performed using a parallelised version of the three-dimensional smoothed particle hydrodynamics (SPH) code described in [17] on the United Kingdom Astrophysical Fluids Facility (UKAFF), a 128-processor SGI Origin 3000. The code uses a tree structure to calculate gravity and neighbouring particles. High-density gravitationally-bound objects consisting of many SPH gas particles are replaced during the calculation by 'sink' particles. These massive particles interact with the rest of the calculation only via gravity and accrete any SPH gas particles that come within a predefined accretion radius.

The initial conditions consist of an initially uniform-density, spherical cloud of molecular gas with mass of 50 $M_\odot$ and a radius of $\approx$ 0.2 pc. At a temperature of 10 K, this results in a mean thermal Jeans mass of 1 $M_\odot$ (i.e. the cloud contains 50 initial Jeans masses). Although the cloud is uniform in density, we impose an initial supersonic turbulent velocity field on it in the same manner as [18]. We generate a divergence-free random Gaussian velocity field with a power spectrum $P(k) \propto k^{-4}$. In three-dimensions, this results in the velocity dispersion varying with distance, λ, as $\sigma(\lambda) \propto \lambda^{1/2}$ in agreement with the observed Larson scaling relations for molecular clouds [19]. Burger's supersonic turbulence also produces this power spectrum.

2.2 Resolution

One of the goals of this calculation is to resolve the collapse beyond the density at which the gas becomes non-isothermal so that we reach the opacity limit for fragmentation. To model the thermal behaviour of the gas, we use an equation of state that changes from isothermal to barotropic with an adiabatic index of $\gamma = 7/5$ at $\rho = 10^{-13}$ g cm^{-3} (see [15]). Then, once the density in an optically-thick quasistatic core passes $\rho \approx 10^{-11}$ g cm^{-3}, we replace the quasistatic core with a 'sink' particle having an accretion radius of 5 AU. We cannot follow the collapse of quasistatic cores all the way to the actual formation of the stars (as done in [15]) while simultaneously following the evolution of the large-scale cloud: the range of dynamical timescales would be too large.

We must ensure that we have sufficient resolution to resolve the local Jeans mass throughout the calculation [6,7]. The minimum Jeans mass occurs at the maximum density during the isothermal phase of the collapse, $\rho = 10^{-13}$ g cm^{-3}, and is 0.0014 $M_\odot$ (1.4 Jupiter masses). The Jeans mass must be resolved by a minimum of $\approx$ 100 SPH particles [6]. Thus, we use 3.5×10^6 particles to model the 50 $M_\odot$ cloud.

3 Results and Discussion

The divergence-free turbulent velocity field quickly evolves to produce converging flows that result in shocks. Initially, the dense structures formed by the shocks are transient, but as the turbulence decays, dense, self-gravitating cores begin to form. The largest of these cores ($\approx$ 0.05 pc across) undergoes collapse first and forms filamentary structures in which 'stars' form (Figure 1). The stars begin as very low-mass fragments ($\sim$ 0.01 $M_\odot$, set by the opacity limit) that accrete up to their final masses. First to form is a binary system with a separation of $\approx$ 30 AU surrounded by a circumbinary disc that later fragments to form 3 objects. Two of these are quickly ejected in a chaotic interplay which terminates their accretion and fixes their masses to be in the brown dwarf regime. In other parts of the core, four single stars – each surrounded by a disc of radius $\sim$ 100 AU – and another multiple system form. Such a mixture of single and multiple systems and the sizes of the circumstellar discs are consistent with observed pre-main-sequence systems. Eventually, 16 objects form. Most fall together into a small cluster which subsequently breaks up by ejecting the low-mass members.

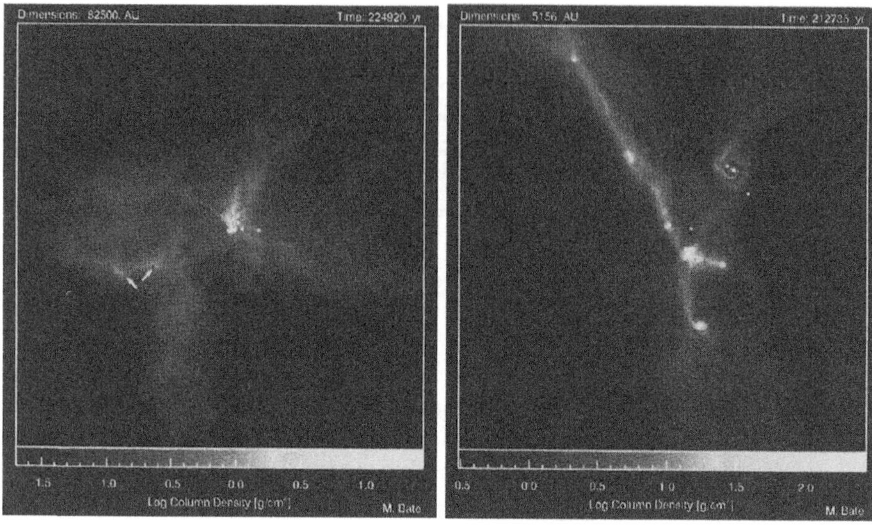

Fig. 1. Column density plots of the molecular cloud during the calculation. *Left:* A view of the entire cloud. The most massive dense core (*near the centre*) forms 10 stars and 6 brown dwarfs. The two smaller cores (*arrows*) are about to form more stars. *Right:* A snapshot of the fragmentation inside the most massive core. A multiple system with circumstellar disc is clearly visible with several single stars and another multiple system forming in the 'filament'.
Figure available in colour on the CD-ROM

Of the 16 objects formed in the collapse of the most massive core, 10 have stellar masses and 6 are brown dwarfs. The brown dwarfs contain only 3% of the total mass of these objects. The these numbers are consistent with current observations of stellar clusters [20] (i.e. a large number of brown dwarfs that do not contain a significant fraction of the mass).

An unsolved problem in the field of star formation is how brown dwarfs form. The typical Jeans mass in molecular clouds is thought to be $\sim 1\ M_\odot$ or greater. Thus, a typical collapsing molecular cloud core should have roughly this mass. Although the cloud core collapses inhomologously, forming one or more low-mass quasistatic cores at the opacity limit for fragmentation, the majority of the envelope should subsequently accrete onto these objects, increasing their masses to those of stars. Thus, there are two possible ways to form field brown dwarfs. Either they must form from particularly dense, low-mass cloud cores (since to have a low total mass while remaining gravitationally unstable they must be very dense), or they must form in typical cores but somehow avoid accreting much of the envelope. The latter is possible if a brown dwarf forms in a multiple system and is quickly ejected (e.g. Reipurth, this volume). In the calculation presented here, without exception, the brown dwarfs are formed in dynamically-unstable multiple systems and are ejected before they can accrete much gas.

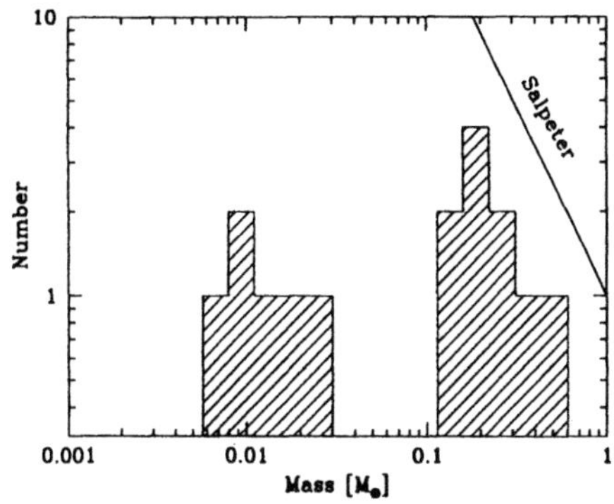

Fig. 2. The mass distribution of the 10 stars and 6 brown dwarfs produced in the calculation thus far. The stellar mass function peaks at ~ 0.2 M$_\odot$, in agreement with observations, but all of the brown dwarfs have very low masses resulting in a gap in the range $0.03 - 0.10$ M$_\odot$

The calculation produces both single stars and multiple systems with masses ranging from 0.6 to 0.008 M$_\odot$. However, there are some potential problems which bear watching as we continue the calculation. Figure 2 gives the mass function of the 16 objects. While the stellar mass function peaks at ~ 0.2 M$_\odot$ in reasonable agreement with observations, all of the brown dwarfs have very low masses. Indeed, there are no objects with masses between $0.03 - 0.10$ M$_\odot$. The brown dwarfs seem to be ejected too quickly, before they have accreted much mass at all. Furthermore, all of the brown dwarfs are single, whereas several binary brown dwarf systems have been found [20]. This is perhaps the most serious problem with the ejection scenario for brown dwarf formation – how are binary brown dwarfs ejected? It is essential that further surveys of field brown dwarfs are performed in order to determine their binarity accurately.

Another potential problem with the calculation is that, although the initial fragmentation of the largest core produces 4 single stars, 2 multiple systems, and discs with radii ~ 100 AU, the stars fall together to form a small bound cluster. This cluster subsequently breaks up into 12 single objects and 2 binary systems and all but two of the large circumstellar discs are destroyed by dynamical interactions (note that we cannot resolve discs smaller than ≈ 10 AU). The high number of single objects is worrying as most stars are observed to be members of binary systems [21]. On the other hand, if only the stellar-mass objects are considered, we have 6 single stars and 2 binaries which, given the small numbers, is reasonable. Again we return to the question of the true frequency of binary brown dwarf systems.

We stress that the results given above are preliminary in that only 6% of the total cloud mass has been converted to stars and brown dwarfs thus far. In particular, there are two lower-mass cores that will shortly form stars. Since each of these cores is likely to produce fewer objects than the largest core, we expect that they may also produce stars with a higher binary fraction. We are continuing the calculation to clarify this.

4 Conclusions

We have presented preliminary results from the first hydrodynamical calculation to follow the collapse and fragmentation of a large-scale turbulent molecular cloud core to form a stellar cluster while resolving beyond the opacity limit for fragmentation. We find

- the initial fragmentation produces a mixture of single stars and multiple systems with typical circumstellar disc sizes of ~ 100 AU
- stars and brown dwarfs are formed in roughly equal numbers, but the stars contain the vast majority of the mass
- all of the brown dwarfs are formed via ejections from multiple systems before they have been able to accrete much mass, but they all have very low masses
- the stellar mass function is in agreement with the observed stellar IMF.

This calculation and similar calculations in the future will allow us to obtain theoretical predictions of the statistical properties of stellar systems for the first time. It is essential that these properties are determined observationally so that the models can be tested and refined. The VLT and other large telescopes will play an important role in obtaining these observations.

Acknowledgments

The computations reported here were performed using the UK Astrophysical Fluids Facility (UKAFF).

References

1. A.P. Boss, P. Bodenheimer: ApJ, **234**, 289 (1979)
2. A.P. Boss: ApJS, **62**, 519 (1986)
3. I. Bonnell, H. Martel, P. Bastien, J.-P. Arcoragi, W. Benz: ApJ, **377**, 553 (1991)
4. R. Nelson, J.C. Papaloizou: MNRAS, **265**, 905 (1993)
5. A. Burkert, P. Bodenheimer: MNRAS, **264**, 798 (1993)
6. M.R. Bate, A. Burkert: MNRAS, **508**, L95 (1997)
7. J.K. Truelove, R.I. Klein, C.F. McKee, J.H.II Holliman, L.H. Howell, J.A. Greenough, D.T. Woods: ApJ, **495**, 821 (1998)
8. R.S. Klessen, A. Burkert, M.R. Bate: ApJ, **501**, L205, (1998)
9. I.A. Bonnell, M.R. Bate, C.J. Clarke, J.E. Pringle: MNRAS, **285**, 201 (1997)
10. I.A. Bonnell, C.J. Clarke, M.R. Bate, J.E. Pringle: MNRAS, **324**, 573 (2001)
11. R.S. Klessen, A. Burkert: ApJS, **128**, 287 (2000)
12. R.S. Klessen, A. Burkert: ApJ, **549**, 386 (2001)
13. C. Low, D. Lynden-Bell: MNRAS, **176**, 367 (1976)
14. R.B. Larson: MNRAS, **145**, 271 (1969)

15. M.R. Bate: ApJ, **508**, L95 (1998)
16. M.R. Bate: in preparation (2002)
17. M.R. Bate, I.A. Bonnell, N.M. Price: MNRAS, **277**, 362 (1995)
18. E.C. Ostriker, J.M. Stone, C.F. Gammie: ApJ, **546**, 980 (2001)
19. R.B. Larson: MNRAS, **194**, 809 (1981)
20. G. Basri: ARA&A, **38**, 485 (2000)
21. A. Duquennoy, M. Mayor: A&A, **248**, 485 (1991)

Rolf Chini and Achim Tieftrunk

Submm and MIR Imaging of Protostellar Clusters

Rolf Chini[1], Markus Nielbock[1], and Ralf Siebenmorgen[2]

[1] Astronomisches Institut, Ruhr–Universität Bochum, Germany
[2] European Southern Observatory, Garching, Germany

Abstract. The detection of protostars is achieved most efficiently at submm wavelengths. With the advent of large bolometer arrays, groups and clusters of protostellar sources become more and more frequent. The physical state of the new sources, however, remains often uncertain due to the absence of data shortward of 350 μm. Recent 10 and 20 μm imaging seem to be able to close this gap and gives more insight into the properties, the morphology and the age of these youngest stellar objects. The present paper gives some examples of newly discovered multiple submm sources, shows the potential of complementary MIR imaging and discusses how future VLT data may contribute to the study of earliest stellar evolution.

1 Introduction

Although searched for decades at various wavelengths, the first detection of protostars had to await the advent of submm techniques (e.g. Chini et al. [6]). Since then, a couple of bona fide low–mass protostellar candidates have been found, whereas the search for the first high–mass protostar is still going on. High–mass protostars are much more difficult to detect because of their short evolutionary time. In this respect, there is no common way of how to define a high–mass protostellar object, while there is fair consensus on low–mass protostars. The latter ones are objects whose submm luminosity exceeds their bolometric luminosity by at least a factor of 200. This requirement is equivalent to the condition that half of the mass is still contained in an envelope and accretes onto a central object (André, Ward-Thompson & Barsony [1]); the luminosity emerging from the protostar is entirely due to accretion. In addition, low–mass protostars are expected (and observed) to be cold with temperatures between 10 and 20 K.

Having found potential protostellar candidates, one faces the fact that further information to study their physical properties is almost impossible to obtain. Usually, FIR data suffer from low spatial resolution and thus cannot be attributed to an individual compact submm source which very often are embedded in confused areas of ongoing star formation. The cold temperatures exclude the emission at optical or near infrared wavelengths. Only radio continuum emission at cm wavelengths which is due to free–free emission from the accretion shock and molecular lines at submm wavelengths which are caused by infall and/or outflow of material help to understand these early stages a little bit better. However, submm and radio data are still lacking the spatial resolution to explore the morphology of the circumstellar environment in terms of accreting core, circumstellar disk and extended envelope. Even the presence of multiple sources cannot be established by the generally available spatial resolution of about 10″.

With the development of sensitive ground-based MIR array detectors, the field has got a new potential tool to exploit both the spectral energy distribution and the morphology of protostellar sources with unprecedented sensitivity and resolution. This paper gives first examples, where MIR imaging has been applied to regions of low and high-mass star formation and thus supplement existing submm and radio data. The data were obtained with TIMMI 2 at the ESO 3.6 m telescope and show the potential of what the MIR regime will contribute to this field of research in an even better way as soon as the sensitivity and the spatial resolution of the VLT will be available.

2 Stars Do Form in Groups

Throughout many years of hunting for protostars, we have never come across isolated objects embedded in a quiescent molecular cloud. Whenever we were lucky to detect protostellar candidates, they were embedded in regions of ongoing star formation where typical signposts like compact H II regions, T Tauri stars, HH objects or molecular outflows indicated the presence of existing (and maybe more evolved) YSOs. In the following, a couple of examples are given to illustrate the occurrence of young stars of different evolutionary stages in the same volume of space.

2.1 Low Mass Star Formation in NGC 1333

Fig. 1 shows the 850 μm image of the HH 7–11 region – a star forming area in the NGC 1333 complex (Chini et al. [9]). The image shows a ridge running roughly from NE to SW, containing three barely resolved sources; they have been detected previously by Chini et al. [7] at 1300 μm. MMS 1 appears to be coincident with a cm radio source VLA 4 which is a suggested driving source of the HH 7–11 bipolar outflow (Rodríguez et al. [18]). MMS 2 has also been found to have a cm counterpart, namely VLA 17 (Rodríguez et al. [19]), while MMS 3 is coincident with VLA 2. The presence of cm radio continuum emission and the values of $L_{\rm FIR}/L_{\rm smm} < 200$ indicate that we deal with protostellar sources.

In addition, Fig. 1 shows four additional sources. The strongest of them (MMS 4) lies east of the ridge just described. This source is also visible in the 1300 μm image of Chini et al. [7]. Its position coincides with the Herbig–Haro object HH 8 (see e.g. Herbig & Jones [13]) and with faint CO emission observed by Grossman et al. [12]. A further source (MMS 7) appears at the eastern edge of the 850 μm map which coincides with the position of HH 7. A stellar origin of MMS 4 and MMS 7 can be ruled out, as Aspin, Sandell & Russell [2] classified them as nebulous near-infrared sources ASR 6 and ASR 22. Additionally, neither source corresponds to any of the VLA sources found by Rodríguez et al. [19]. This indicates that we are seeing – for the first time – HH objects directly at

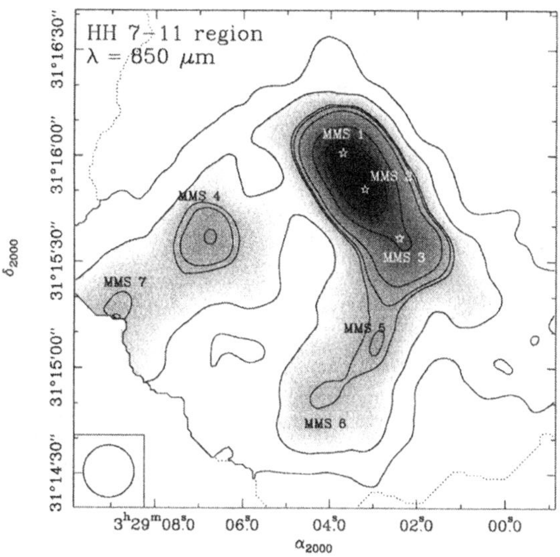

Fig. 1. The HH 7–11 region at 850 μm, containing various sources of different physical nature including three protostars (MMS 1–3), two HH objects (MMS 4, 7) and two T Tauri stars (MMS 5, 6)

mm/submm wavelengths, in which case the emission would probably originate from density enhancements of the dusty environment produced by the interacting HH jet.

Two other new submm sources appear south of the main ridge. Within the positional uncertainties, the stronger one (MMS 5) is coincident with VLA 16, and the fainter one (MMS 6) with VLA 19. Both of them are identified as YSOs by Rodríguez et al. [19]. VLA 16 is a time variable source, exhibiting large circular polarization and coincides with the NIR source ASR 7 (Aspin, Sandell & Russell [2]) which is believed to be a T Tauri star. VLA 19 exhibits a spectral index at cm wavelengths which also suggests an association with a young star. Thus, it is very likely that we have detected the mm/submm counterparts of two further young objects in the HH 7–11 region.

In summary, the region contains a group of young stellar objects, including three protostars (MMS 1–3), two HH objects (MMS 4, 7) and two T Tauri stars (MMS 5, 6). It is thus a good example of how different evolutionary stages populate the same cloud core.

2.2 Intermediate Mass Star Formation in OMC 2/3

OMC 2/3 is a region of intermediate mass star formation. Chini et al. [8] have discovered a filament of new protostellar candidates at 1300 μm which was subsequently observed by Lis et al. [15] at 350 μm; the number of submm condensations is about 30. Reipurth et al. [17] have performed a VLA survey of the region and found 11 sources with radio emission at 3.6 cm which are associated with submm clumps. We have recently imaged the area with TIMMI 2 and discovered 21 MIR sources, most of which coincide with the submm sources. Fig. 2 shows examples, where two submm condensations contain flattened 10 μm nuclei and where extended ridge-like submm contours harbour multiple unresolved 10 μm sources.

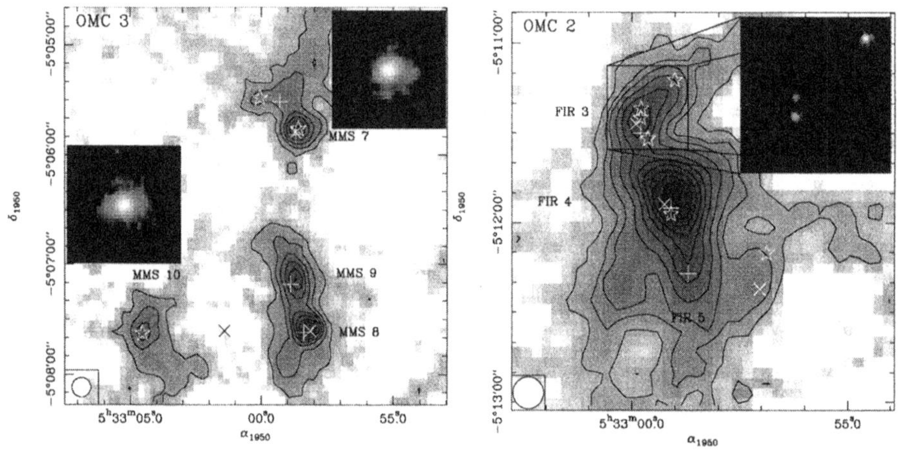

Fig. 2. Detailed views of the 1300 μm emission from the OMC2/3 filaments (grey scale and contour lines); the + denote the positions of the 350 μm sources, × those of the VLA sources and the ☆ represent the positions of the N–band detections. The inserts are the corresponding N–band frames which show that e.g. the spherical submm sources MMS 10 and 7 contain flattened 10 μm nuclei (left), while the extended submm ridge FIR 3 harbours three unresolved 10 μm sources (right)

2.3 High Mass Star Formation

The search for high-mass protostars at submm wavelengths has not been as successful as for the low-mass stars. Either one finds dense massive pre-stellar clumps or objects which have already created a compact H II region. MIR observations are developing into a potential tool to study early stages of massive star formation in more detail. Apart from the better spatial resolution compared to submm imaging, MIR data trace warmer dust and thus probe the circumstellar environment close to the central object.

The BN/KL complex

Fig. 3 shows as an example the famous BN/KL complex in Orion, harbouring a cluster of massive stars with a total luminosity of $\sim 10^6 L_\odot$. The region has been studied before by Gezari et al. [11] at various MIR wavelengths concluding that the number of embedded MIR sources is sufficient to power the IR emission of the area. Our TIMMI image shows unprecedented details of the central region

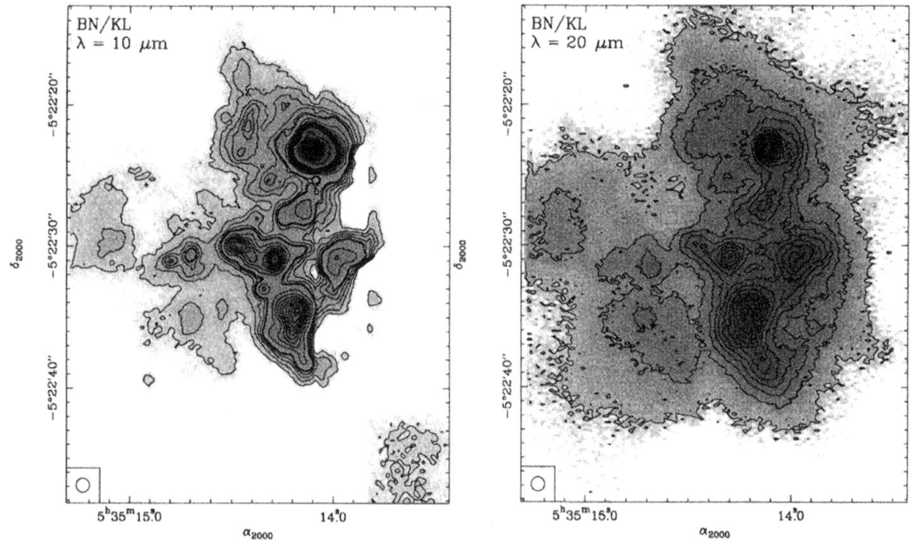

Fig. 3. The BN/KL complex at 10 and 20 μm. Apart from a number of new MIR sources the disturbed morphology of hot circumstellar dust becomes clearly visible

and demonstrates in particular that circumstellar dust may be heavily disturbed by gravitational interaction and/or radiation pressure.

M 17

The Omega Nebula (M 17, NGC 6618) is a site of recent massive star formation (Chini, Elsässer & Neckel [3]). Its stellar content – as observed so far – consists of heavily reddened early type stars. The ongoing process of star formation in this region is witnessed by a large number of embedded IR excess sources, several cocoon stars (Chini [4], Chini & Krügel [5]) and the ultra-compact HII region M 17–UC1 (Felli et al. [10]).

We have imaged the region at 10.5 and 20.0 μm in order to reveal the nature of the IR excess sources (Nielbock et al. [16]). Besides extended emission of warm dust, 22 compact sources have been detected at both wavelengths; one of them is the KW object another one is M 17–UC1. Despite of the diffraction limited images with resolution values of 1.2 and 2.3 arcsec in N and Q, respectively, most of the sources are unresolved. However, there are 6 objects whose 10.5 and 20 μm emission is clearly asymmetric with axes ratios of up to ≈ 1.5. It is very likely that in theses cases we are dealing with flattened dust cocoons and/or circumstellar disks (c.f. Fig. 4).

In order to investigate the evolutionary stage of the MIR sources spectral indices α_{KN} and α_{KQ} have been calculated according to the classification scheme introduced by Lada [14]. All sources qualify as Class I YSOs with formal colour temperatures of 140 K to 440 K for their circumstellar environment.

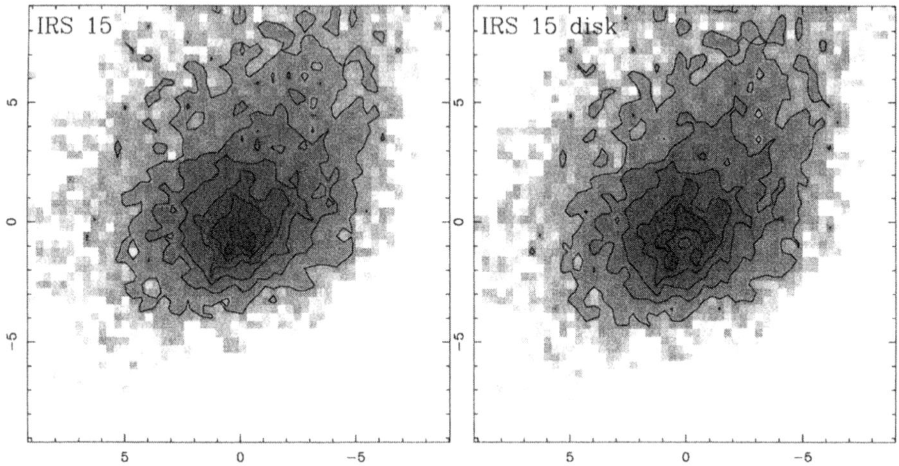

Fig. 4. *N*-band images of IRS 15. After subtracting the point-like compact core from the original image (left), a large disk-like envelope remains (right) which dominates the MIR emission

References

1. P. André, D. Ward-Thompson, M. Barsony: ApJ **406**, 122 (1993)
2. C. Aspin, G. Sandell, A.P.G. Russell: A&AS **106**, 165 (1994)
3. R. Chini, H. Elsässer, Th. Neckel: A&A **91**, 186 (1980)
4. R. Chini: A&A **110**, 332 (1982)
5. R. Chini, E. Krügel: A&A **164**, 175 (1985)
6. R. Chini, E. Krügel, C.G.T. Haslam, E. Kreysa, R. Lemke, B. Reipurth, A. Sievers, D. Ward-Thompson: A&A **272**, L5 (1993)
7. R. Chini, B. Reipurth, A. Sievers, D. Ward-Thompson, C.G.T. Haslam, E. Kreysa, R. Lemke: A&A **325**, 542 (1997)
8. R. Chini, B. Reipurth, D. Ward-Thompson, J. Bally, L.-Å. Nyman, A. Sievers, Y. Billawala: ApJ **474**, L135 (1997)
9. R. Chini, D. Ward-Thompson, J. Kirk, M. Nielbock, B. Reipurth, A. Sievers: A&A **369**, 155 (2001)
10. M. Felli, K.J. Johnston, E., Churchwell: ApJ **242**, L157 (1980)
11. D.Y. Gezari, D.E. Backman, M.W. Werner: ApJ **509**, 283 (1998)
12. E.N. Grossman, C.R. Masson, A.I. Sargent, N.Z. Scoville, S. Scott, D.P. Woody: ApJ **320**, 356 (1987)
13. G.H. Herbig, B.F. Jones: AJ **88**, 1040 (1983)
14. Lada, C.J. 1987, Star Formation: From OB Associations to Protostars. In: Star Forming Regions, ed. M. Peimbert & J. Jugaku, IAU Symp. 115, 1
15. D.C. Lis, E. Serabyn, J. Keene, C.D. Dowell, D.J. Benford, T.G. Phillips, T.R. Hunter, N. Wang: ApJ **509**, 299 (1998)
16. M. Nielbock, R. Chini, M. Jütte, E. Manthey (submitted)
17. B. Reipurth, L.F. Rodríguez, R. Chini: AJ **118**, 983 (1999)
18. L.F. Rodríguez, G. Anglada, S. Curiel: ApJ **480**, L125 (1997)
19. L.F. Rodríguez, G. Anglada, S. Curiel: ApJS **125**, 427 (1999)

Antu enclosure (under the Milky Way)

Charlie Lada

Infrared Imaging of Embedded Clusters: Constraints for Star and Planet Formation

Charles J. Lada[1], Elizabeth A. Lada[2], August A. Muench[2], Karl E. Haisch[3], and João Alves[4]

[1] Smithsonian Astrophysical Observatory, Cambridge, MA 02138, USA
[2] Department of Astronomy, University of Florida, Gainesville, FL 32611, USA
[3] NASA Ames Research Center, Moffett Field, CA 94035, USA
[4] European Southern Observatory, Garching, Germany

Abstract. Since its implementation slightly more than a decade ago, astronomical imaging with near-infrared array detectors has matured into an important tool for star and planet formation studies. In this paper we briefly describe three areas of investigation where infrared imaging plays a major role in advancing the understanding of star and planet formation. We highlight recent findings derived from imaging studies of embedded clusters in local star formation regions which concern: 1) the spatial distributions of young stellar objects, 2) the nature of the IMF, and 3) the frequency and evolution of circumstellar disks. The significance of embedded clusters as laboratories for star formation research is illustrated by each of these examples.

1 Introduction

Infrared array detectors were first deployed on telescopes for astronomical observation less than 15 years ago. Prior to that time, astronomical observations in the infrared were obtained with single channel (pixel) photometric devices. A major drawback of these devices was that they were characterized by relatively large apertures, typically 3-10 arc seconds in size when projected on the sky. The introduction of imaging array detectors revolutionized astronomical data acquisition at near-infrared wavelengths. These detectors simultaneously provided pixel sizes on the order of a typical stellar seeing disk ($\sim$ 1 arc sec), and initial fields of view of a few arc minutes in size that were covered by 10^4 - 10^5 such pixels. This had a profound impact on star formation studies. This is because young stellar objects, the primary specimens for observation and investigation, tend to be buried in molecular clouds and heavily obscured. Consequently these sources are either brightest, or often only visible at infrared wavelengths. With imaging devices, the statistical study of such objects became possible for the first time.

It is in this area that infrared imaging has had, and continues to have, its greatest impact for star formation research. There are at least four fundamental problems which can be addressed primarily by statistical information provided by infrared imaging surveys. These are: 1) the spatial distributions and interrelations of young objects in molecular clouds, 2) the origin and universality of the initial mass function, 3) the frequency and evolution of circumstellar disks and 4) the temporal variability of young stellar objects. In this contribution we will briefly review selected observations which illustrate the impact of infrared imaging studies on the first three of these problems. These examples are drawn largely from studies carried out by various permutations of the authors of this article and are not intended to represent a comprehensive review of the subject. The use of infrared imaging surveys to investigate the fourth problem, that of the variability of young stellar objects, is amply illustrated in the recent paper by Carpenter, Hillenbrand and Skrutskie 2001 and will not be discussed further here.

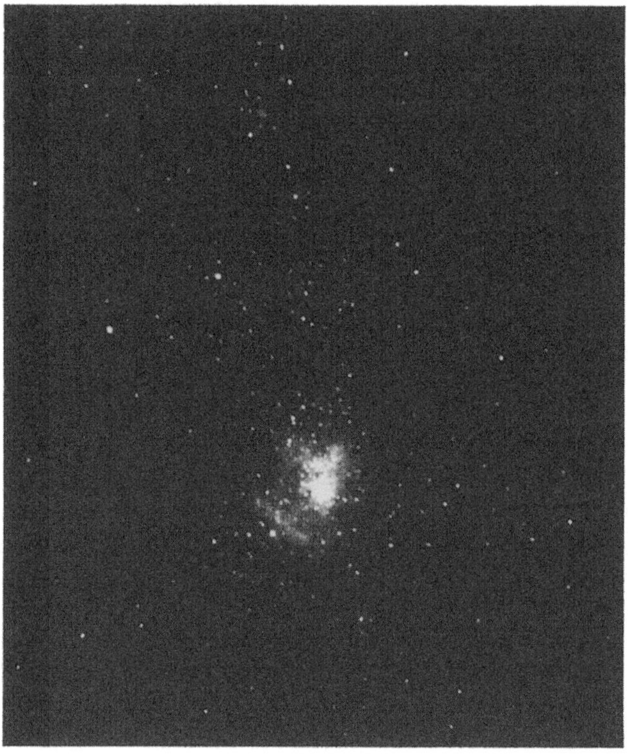

Fig. 1. A single wide field infrared image of the Orion Nebula region obtained with the Flamingos imager/spectrometer of the University of Florida. Such images permit studies of the distributions and interrelations of the young stellar objects in star forming molecular clouds. Courtesy of Richard Elston

2 Spatial Distributions: The Significance of Embedded Clusters

Stellar positions and magnitudes are the basic data that are provided by digital images of the sky at a given wavelength. The angular distribution of infrared sources in a star forming region is likely the most robust result that can be extracted from an imaging survey since it does not require observations of exacting photometric quality to be determined. Therefore, it is not surprising that one of the first new and fundamental advances in star formation research to be derived directly from infrared imaging observations concerned the spatial distributions of young stellar objects in molecular clouds.

Early studies of the two nearest star forming regions provided examples of two different modes of star formation which could occur in molecular clouds. The Ophiuchus cloud was found to produce stars in a clustered mode. That is, the bulk of the young stars in the cloud were found to have been produced within the small spatial confines of an embedded cluster (Grasdalen, Strom and Strom 1973, Wilking and Lada 1981). The Taurus clouds, on the other hand, appeared to form stars in a more isolated, non-clustered mode. Here stars were found to be formed loosely distributed over a relatively large area, comprising an extended association (Ambartsumian 1949, Elias 1979, Myers, Ho and Benson 1979). Because open clusters in the field were known to be responsible for only 10% of all stars formed in a given epoch (Roberts 1957), the isolated mode was naturally assumed to be the dominant mode of star formation. This notion seemed consistent even with the first far-infrared surveys of molecular clouds provided by IRAS.

As is often the case in astronomy, data provided by new technology (in this case infrared imaging surveys) contradicted expectations. The first systematic near-infrared imaging survey of a giant molecular cloud (GMC) was that of the L 1630 cloud in Orion (Lada et al. 1991a). This survey demonstrated that the vast majority ($\sim$ 95%) of stars formed by the star formation process in the cloud were produced in embedded clusters. Indeed, all these newly formed stars were found to be contained in only three rich clusters which together occupy only a very small fraction ($\sim$ 1-2 %) of the cloud's area. Subsequent surveys of this cloud that were either deeper (Li, Evans and Lada 1997) or covered larger areas (Carpenter 2000) confirmed that the cluster mode of star formation is by far the dominant mode of star formation in this cloud. Surveys of additional clouds such as the Rosette GMC (Phelps and Lada 1997), W3 (Carpenter, Heyer and Snell 2000), Perseus, Orion A (L1641), and Monoceros R2 (Carpenter 2001) have produced similar findings suggesting that the clustered mode of star formation is indeed the dominant mode of star formation in GMCs, and consequently in the Galaxy as a whole, since GMCs account for the vast majority of molecular gas contained in the Galaxy (e.g., Lada and Lada 1991). The discovery of numerous embedded clusters in many other star forming regions has provided additional support for the concept that the vast majority (75-95%) of all stars produced by the star formation process form in embedded clusters containing at least a few hundred members (e.g., Hodapp 1994, Lada et al. 1991b, Casali and Eiroa 1992, Barsony, Schombert and Kris-Halas 1991, Persi et al. 1994, Megeath et al. 1996, Horner, Lada and Lada 1997, and others).

Two fundamental consequences for star formation follow from these observations. First, the infant mortality rate for embedded clusters must be very high in order to account for the fact that only 10% of all stars end up in open clusters like the Pleiades, which typically live to ages of a few hundred million years (Lada and Lada 1991; Adams and Myers 2001). This, in turn, likely means that cluster forming cloud cores are characterized by relatively low star formation efficiencies and rapid gas removal or destruction timescales (e.g., Elmegreen 1983, Lada, Margulis and Dearborn 1984). Second, and more significant, any complete theory of star formation must be able to account for star formation within the crowded environments of embedded clusters and their massive cloud cores. Physical processes such as the dynamical interaction of protostellar cores, competitive accretion, and fragmentation probably all operate under such conditions (e.g., Bonnell 1998). Moreover, fundamental properties of the galactic stellar population, such as the initial stellar mass function, stellar multiplicity (e.g., Kroupa 1995), and the frequency of planetary systems around stars must trace their origins to the complex environments of rich embedded clusters.

3 From Luminosity Functions to the Initial Mass Function

Once formed, the life history of a star is entirely pre-determined by its birth mass. Consequently, knowledge of the initial mass function (IMF), that is, the frequency distribution of stellar masses at birth, is critical to understanding the evolution of stellar systems such as clusters and galaxies. However, the most fundamental properties of the IMF, such as its functional form, its universality with respect to space and time, and even the range of stellar masses it encompasses, are not yet well determined. Young stellar clusters offer uniquely important laboratories for investigating each of these properties of the IMF. Such clusters provide the smallest spatial size scale over which a meaningful determination of the IMF can be made. The IMF is a statistical quantity and clusters contain statistically significant numbers of stars which both cover a wide range of mass and share a common heritage, having formed from the same molecular cloud at the same epoch of time. Extremely young embedded clusters are particularly useful because they are not old enough to have lost significant numbers of members due to stellar evolution or dynamical effects, and their lowest mass members are brighter than at any other time in their subsequent evolution (except, of course, when they evolve off the main sequence to become red giants). Thus, the present day mass function of an embedded cluster is its IMF.

However, embedded clusters suffer significant extinction, making optical observations of them very difficult, if not impossible. Thus, until the advent of infrared imaging cameras, little information about the nature of the IMF could be derived from these objects. With infrared cameras an embedded cluster can be detected and imaged in relatively small amounts of observing time. Indeed, even objects with masses significantly below the hydrogen burning limit (i.e., brown dwarfs) can be readily detected in nearby clusters with only modest sized telescopes. The brightness of an astronomical source at a given wavelength is about the most basic astronomical observation one can obtain. Infrared images of an embedded cluster can provide a complete catalog of the (monochromatic) brightnesses of all its members, and directly result in the production of a monochromatic stellar luminosity function for the cluster.

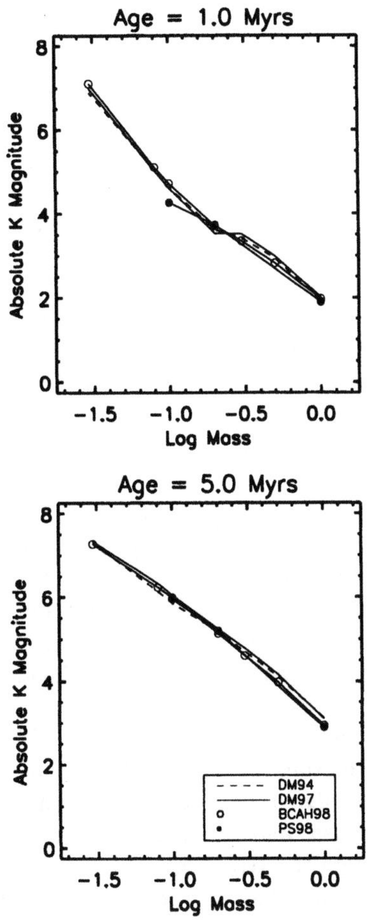

Fig. 2. The predicted absolute K magnitudes of low mass PMS stars as a function of mass at ages of 10^6 (top) and 5 $\times 10^6$ (bottom) years. Calculations of D'Antona & Mazzitelli (1994, 1998), Baraffe et al. (1998) and Palla and Stahler (2000) are compared. The relations are essentially degenerate indicating that the predicted K band fluxes of young PMS stars are not particularly sensitive to the differences in the various models

In order to investigate the IMF of an embedded cluster we need knowledge of the masses of its members. But, luminosity, not mass, is the observable quantity. To derive stellar masses from stellar luminosities requires knowledge of a mass-luminosity relation and bolometric corrections. Although, converting luminosity functions into mass functions would be a relatively straightforward exercise for a cluster of main sequence stars, embedded clusters contain mostly pre-main sequence stars for which there is no unique mass-to-luminosity relationship as there is for main sequence stars. As a result, models for pre-main sequence luminosity evolution must be employed to derive stellar masses from stellar luminosities. The situation is further complicated since star formation within clusters is probably not coeval. Therefore, both a predictive theory of PMS evolution and some knowledge of the ages of cluster members is required to convert observed luminosity functions into mass functions.

Despite these complexities, recent Monte Carlo modeling of the infrared luminosity functions of young clusters (Muench, Lada & Lada 2000) has demonstrated that the functional form of an embedded cluster's luminosity function is considerably more sensitive to the form of the underlying cluster mass function than to any other significant parameter (i.e., stellar age distribution, PMS models, etc.). In fact, despite the significant differences between the parameters that characterize the various PMS calculations (e.g., adopted convection model, opacities, etc.), model luminosity functions are essentially insensitive to the choice of the PMS mass-to-luminosity relations predicted by existing PMS calculations. This is illustrated in Figure 2, where the luminosities predicted by four different PMS models are plotted as a function of stellar mass at two cluster ages, 1 and 5 million years.

The predicted luminosities are essentially degenerate with respect to the PMS models used. Although this is perhaps surprising at first glance, this result can be understood by considering the fact that the luminosity of a PMS star is determined by very basic physics, simply the conversion of gravitational potential energy to radiant luminosity during Kelvin-Helmholtz contraction. This primarily depends on the general physical conditions in the stellar interior (e.g., whether the interior is radiative or partially or fully convective). The close agreement of the model predictions reflects the robust nature of PMS luminosity evolution. On the other hand, predicting the effective temperature of such stars, which depends on detailed knowledge of uncertain conditions in the stellar atmospheres, is a more difficult exercise, and these same models can predict very different locations for such stars on the HR diagram.

Muench, Lada & Lada (2000) further showed that, with knowledge of the mean age of an embedded cluster and appropriate bolometric corrections, the underlying mass function of its members can be derived from the observed infrared luminosity function with minimal ambiguity. This is illustrated in Figure 3a which shows the K luminosity function (KLF) for the Trapezium cluster derived by Muench et al. (2001b). This KLF is corrected for background contamination and is an extinction limited luminosity function. Specifically, this KLF is complete to a depth of 17 magnitudes of visual extinction for all sources in the cluster whose unreddened apparent magnitudes would be brighter than about $m_k = 16$. For an age of one million years this limit corresponds to all sources with masses greater than approximately 20 Jupiter masses (m_J) or 0.02 $M_\odot$ (Baraffe et al. 1998). This is well below the hydrogen burning limit (80 m_J), as indicated in the figure, and encompasses almost the entire range (10-80

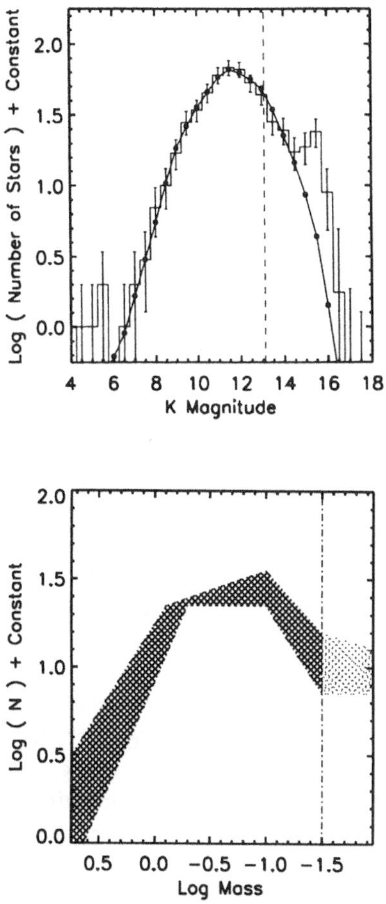

Fig. 3. a(top).- The complete extinction limited KLF of the Trapezium Cluster derived by Muench et al. 2001 is illustrated by the histogram. A vertical dashed line marks the location of the hydrogen burning limit for a million year old population. The smooth line traces the predicted KLF for a best fit IMF of the cluster. b(bottom).- The range of best fit IMFs for the above KLF determined from a chi-squared minimization. The vertical line at logm ~ -1.5 marks the mass that corresponds to the departure of the observed and predicted KLFs at about 15.5 magnitudes seen in a. Although this indicates a relative excess of objects at these low masses, overall, no more than ~ 20% of cluster members are substellar

m_J) of the expected brown dwarf mass spectrum. Also plotted is a best fit model KLF which closely corresponds to the most likely underlying IMF for the cluster. This synthetic KLF has been statistically adjusted for both variable extinction and infrared excess emission of the cluster members. A chi-squared minimization found a set of such "best fit" models with slightly differing IMFs which were all compatible with the observed KLF for member masses greater than about 30 m_J. The range of acceptable IMFs derived by Muench et al. (2001b) is shown in Figure 3b.

For masses m > 0.03 $M_\odot$, the derived IMF is well described by three simple power-law relations. Overall, it is characterized by a broad maximum or peak between about 0.01 - 0.6 $M_\odot$. The derived mass function falls off sharply above 0.6 $M_\odot$ and below the hydrogen burning limit. The overall form of the derived IMF agrees closely with those recently derived for this cluster from other infrared imaging experiments (i.e., Lucas & Roche 2000, Luhman et al. 2000, Hillenbrand and Carpenter 2000, McCaughrean et al. 1995) supporting the notion that infrared luminosity functions can be effective and relatively reliable tools for determining the IMF of a young cluster. Moreover, the overall shape of the Trapezium IMF is consistent with that derived for field stars nearby the sun, at least above the hydrogen burning limit where the field star IMF is known (e.g., Kroupa, Tout and Gilmore 1993, Kroupa 2001). It is quite interesting that the mass function of a group of stars born only 10^6 years ago in a region much less than a parsec in size is even remotely similar to that derived for field stars, formed over billions of years of galactic history and over an immense volume of galactic space. This suggests that a very robust mechanism must responsible for creating the IMF, a mechanism that is independent of time over the history of the Galaxy.

One significant conclusion to draw from all this is that *the primary products of the star formation process are stars with masses between 0.1-0.6 $M_\odot$*. In particular, no more than ~ 20% of all cluster members are found to be substellar (i.e., brown dwarfs). In this regard infrared imaging observations are providing qualitatively new information and insights concerning the extent and form of the IMF. Such information cannot be presently ascertained from studies of the field star population.

The observed KLF of the Trapezium cluster does, however, depart significantly from the predicted KLF at about 15.5 magnitudes, corresponding to a mass of about 30 m_J. This is either due to an unknown and unaccounted for feature in the mass-luminosity relation of the PMS brown dwarf models, or represents a real feature in the mass function which, in turn, would correspond to an "excess" of objects above the extrapolation of a single power-law function in the substellar mass range (Muench et al. 2001). The existence of such an "excess" or feature, or break in the substellar IMF would be interesting since it might provide support for the idea that at least some freely floating brown dwarfs are produced by the disintegration of multiple protostellar systems (Reipurth & Clark 2001, see also Reipurth, this volume).

Another interesting aspect of the infrared luminosity function and its corresponding IMF is that both seem to be continuously populated to the lowest detectable luminosities/masses, all the way down to near the deuterium burning limit. The question that naturally arises therefore is whether there is a low mass limit to the IMF. The present data are not sufficient to answer this question. Not only is there an issue of sensitivity in this regime, but there are also very few objects observed even at the lowest luminosities above the completeness limits, making any statistical investigation of this question very difficult. Resolving this question will require both deeper imaging surveys and sensitive spectroscopic examination of the few extremely faint objects in the cluster. Although the present data do not allow us to determine whether there is a lower limit to the IMF, the scarcity of sources near the completeness limits does suggest that the number of very low mass objects (i.e., m $< 20\ m_J$) produced by the star formation process is very small.

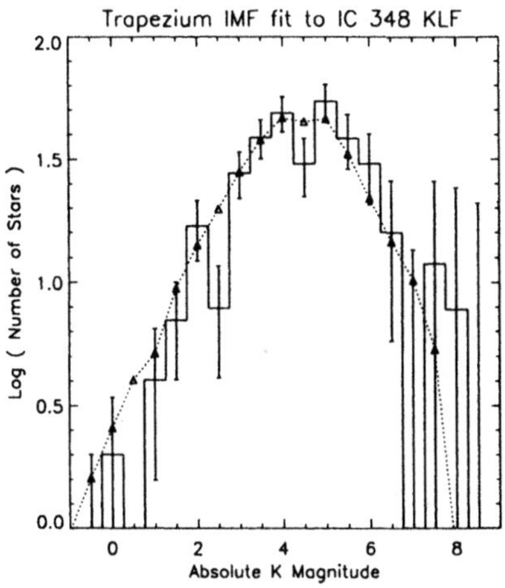

Fig. 4. The KLF observed for the IC 348 cluster (histogram) compared to that (line) predicted for a synthetic cluster of the same age and distance with an underlying IMF identical to that derived for the Trapezium cluster (adapted from Muench, Lada & Lada 2002)

Infrared imaging studies also enable ready comparison of the KLFs and IMFs of embedded clusters in different regions and at different distances from the sun. Thus the universality of the IMF with respect to location and environment can be explored. For example, Figure 4 shows the KLF of the 2-3 million year cluster IC 348 derived from infrared images obtained with a 1.2 meter telescope (Muench, Lada & Lada 2002). Also plotted in the figure is the corresponding synthetic KLF for a cluster of the same size, distance and age as IC 348 but which has an underlying IMF identical to that derived for the Trapezium (i.e., Muench et al. 2000). The synthetic KLF, without any other adjustments, matches the observations quite well. This suggests that the IMFs of the two clusters are very similar. In particular, the decline in the luminosity function below the hydrogen burning limit (i.e., $M_K > 5.5$ magnitudes) corresponds to a decline in the mass function, which is consistent with recent findings derived from infrared HST observations of this cluster (Najita, Tiede, & Carr 2000). This example nicely illustrates the capability for infrared imaging studies to test the universality of the IMF in the Galaxy. With large aperture telescopes such as the VLT, the IMF can be derived for young clusters out to 1-2 kpc from the sun resulting in a significant sample of IMF determinations complete over a relatively large volume of galactic space.

4 Frequency and Evolution of Circumstellar Disks

4.1 Disk Lifetimes and Planet Formation

Multi-wavelength infrared imaging surveys yield colors of detected stars and can produce vital information concerning such parameters as the extinction toward the imaged stars and the presence or lack of infrared excess around them. In particular, infrared color-color diagrams, constructed from multi-wavelength imaging surveys of embedded stellar populations, have been shown to be a very useful tool for identifying infrared excesses caused by circumstellar disks around young stellar objects (Lada & Adams 1992, Meyer, Calvet & Hillenbrand 1997). Circumstellar disks are the likely sites of planet formation. The discovery of numerous extra-solar planets from the detection of small amplitude Doppler periodicities in the spectra of nearby sunlike stars (e.g., Marcy & Butler 1999) has stimulated considerable interest in the question of planet formation, and has raised the issue of exactly how common are planetary systems in the Galaxy. Because a protoplanetary disk is considerably easier to detect than a planetary system with a similar mass of solid material (Beckwith and Sargent 1996), surveys for circumstellar disks in young stellar populations can potentially provide important, albeit indirect, estimates of the frequency of planetary systems in our Galaxy.

The fraction of stars being formed in the Galaxy that will end up with planetary systems depends on two factors, the initial disk frequency and the probability that a circumstellar disk will evolve to form planets. The initial disk frequency can be directly measured with infrared surveys of the youngest stellar populations. The latter factor depends on the physics of circumstellar disk evolution, but can also be addressed by infrared imaging surveys of young stellar populations of differing age and environment. The measurement of disk fraction as a function of age yields the disk lifetime, which sets an upper limit to the timescale for planet formation within the disks. Embedded clusters provide excellent laboratories for such statistical investigations of circumstellar disks. Not only do they contain statistically significant numbers of target stars of widely varying mass, but they also can be reliably characterized by the mean age of their members. Moreover, a sample of clusters can be observed that span a significantly wider range of age than is typical of stars formed in any individual star forming region.

The magnitude of the infrared excess produced by a circumstellar disk rapidly increases with wavelength. Recent studies have shown that measurements at L band (3.5 μm) and the JHKL color-color diagram are the most optimum tools for detecting excesses from protoplanetary disks using data acquired from ground-based telescopes (e.g., Haisch et al. 2000, Lada et al. 2000). In fact, almost all stars with disks produce sufficiently strong L band excess to be clearly identified on JHKL (and HKL) color-color diagrams as disk sources, independent of detailed properties of the disks (e.g., inclination, accretion rate, disk mass, etc.). Indeed, disks with dust masses as small as $\sim 10^{20}$ grams ($\sim 10^{-5}$ Moon masses!) can be identified with JHKL observations! Not to be overlooked is the fact that infrared imaging detectors are also sensitive enough to simultaneously detect JHK and L band emission from the photospheres of relatively low mass young stars in nearby star forming regions. Such observations are essential for the meaningful measurements of disk fractions in these regions.

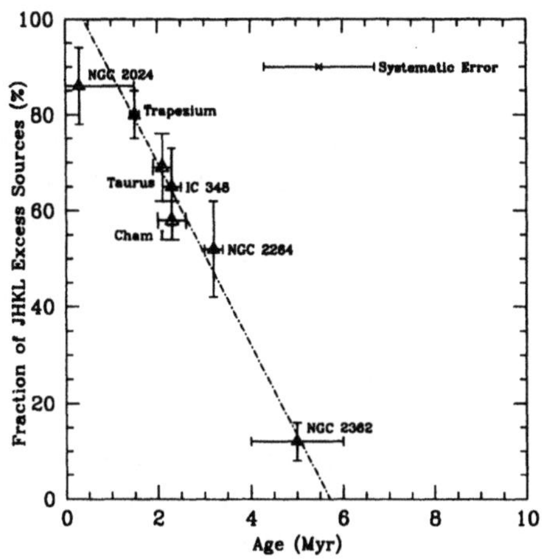

Fig. 5. Circumstellar disk fraction as a function of mean cluster age. Error bars represent statistical errors in the measured disk fraction and mean cluster ages. The systematic uncertainty in cluster age due to uncertainties in PMS evolution models is also plotted. The initially high disk fraction declines rapidly with age, and half of all stars in clusters lose their disks in $\sim 3 \times 10^6$ years. Essentially all the disks are lost in about $5 - 6 \times 10^6$ years

Recently Haisch, Lada & Lada (2001b) summarized the results of a systematic and relatively homogeneous set of JHKL observations of eight nearby young clusters and star forming regions. The data from these studies provided robust determinations of disk frequencies in a set of clusters ranging from approximately 1-30 million years in age. Observations of the two youngest clusters, NGC 2024 (Haisch et al. 2000) and Orion (Lada et al. 2000), revealed a very high disk frequency of 80-85%. This strongly suggests that most stars are born with circumstellar disks and the potential to form planetary systems as a natural consequence of the star formation process. For all but the highest mass (OBA) stars in the Trapezium cluster, the disk fraction was also found to be independent of stellar mass down to the hydrogen burning limit. The OBA stars in the Trapezium displayed somewhat lower disk fractions (~40%).

Clusters of progressively older age were found to have progressively lower disk fractions. This is illustrated in Figure 5 which displays the relation between disk fraction derived from JHKL excess measurements and cluster age for five clusters (including IC348, NGC 2264 and NGC 2362, Haisch et al. 2001a,b) and two non-clustered star forming regions, Taurus (Kenyon & Hartmann 1995) and Chamaeleon (Kenyon & Gomez 2001). The decline in disk fraction with age is both systematic and rapid. After only about 3×10^6 years, half the stars in a cluster have lost their disks! After the passage of another three million years, essentially the entire stellar population of a cluster is devoid of circumstellar disks. In addition, L band observations of the 30 Myr cluster NGC 1960, showed no evidence for disk bearing stars, consistent with the trend found for the younger clusters. The fact that both Taurus and Chamaeleon fall closely on the relation for clusters indicates that the evolutionary timescale for circumstellar disks may be relatively robust or universal in that it is independent of star forming environment. This seems further supported by the recent observations of the poor association of young stars in the MBM 12 cloud, which has an age of about 2 Myr and an L band disk fraction of about 70% (Luhman 2001) and the TW Hydra association, which has an age of about 8 Myr and a disk fraction of only 5-10% (Webb 2001). Both measurements are in good agreement with the overall trend in Figure 5.

Haisch et al. (2001a) also found that the timescale for the decline in disk fraction appears to be a function of stellar mass, with the rate of decline for higher mass stars much greater than that for low mass stars. For example, in IC 348 all stars with spectral types F and earlier were found to be diskless, indicating that the timescale for such massive stars to *all* lose their disks is only $\sim$ 2-3 million years or less.

Although JHKL observations directly trace the inner regions (R $\sim$ 0.1 AU), of circumstellar disks, comparison of JHKL observations with millimeter-wave continuum observations (which trace outer disks, R $\sim$ 10 AU) suggests that inner disks and outer disks evolve on the same timescale (Haisch et al. 2001a,b). Therefore these observations have significant implications for planet formation. In particular, the disk evolution timescale sets the maximum time available for giant planet formation around these stars. For example, it is thought that the gas giant planets were formed in a sequence in which first a solid core of about 10 Earth masses is accumulated from the build up of planetesimals in the disk. This is then followed by the accretion of a gaseous atmosphere (e.g., Lissauer 2001). The timescale for all this to occur is believed to be on the order of ten million years, which at face value seems incompatible with the observed timescales. In addition, the discovery of numerous giant planets very close to their parent stars (e.g., Marcy & Butler 1999) indicates that in many cases such planets migrate inward from the location of their formation. A substantial disk must be present to drive this migration, so the observed disk evolution timescales place an even more severe constraint on the time allowed for construction of giant planets from their parental disks. If planetary systems with giant planets are common, then our theoretical picture of planet formation may need revision. Perhaps models which produce such planets by gravitational instability may need to be reconsidered. Recent versions of such models (e.g., Boss 2000) appear to be able to form Jovian planets on relatively short timescales (10^3 yrs) which would be entirely compatible with the observed disk lifetimes.

4.2 Disks Around Substellar Objects

Another interesting result from the L band disk survey is the finding that the initial frequency of circumstellar disks is high for very low mass stars and does not seem to diminish even at the hydrogen burning limit (Lada et al. 2000, Haisch et al. 2001a). Following these results Muench et al. (2001a) examined deep NTT images of the Trapezium cluster and constructed the JHK_s color-color diagram for the candidate brown dwarf population of the cluster. Figure 6 shows the HLF for the Trapezium cluster. The approximate locations of the hydrogen and deuterium burning limits for a 1 Myr old population of unreddened objects are indicated by vertical lines. A substantial population of substellar candidates is clearly observed in this brightness range (as discussed earlier). However, since this HLF has not been corrected for foreground/background contamination nor adjusted for extinction, some of the faintest objects may be reddened background stars or very deeply embedded stellar mass sources. Also shown is the JHK_s color-color diagram from Muench et al. (2001a) for all brown dwarf/substellar candidate objects with masses between 0.02 - 0.08 $M_\odot$. All faint sources that could possibly be reddened objects of higher mass have been filtered out of this sample by use of infrared color-magnitude diagrams. Plotted on the JHK_s color-color diagram are the loci for main sequence dwarf and giant stars and the appropriate reddening vectors. Approximately 65% of the candidate substellar objects are found in the infrared excess region to the right of the rightmost reddening vector. The presence of infrared excess indicates the extreme youth of these objects and confirms both their membership in the cluster and (consequently) their nature as bona fide substellar objects.

The infrared excess around these substellar objects likely originates in surrounding disks. This interpretation is strongly supported by the fact that 20% of the brown dwarf candidates are spatially coincident with optically identified "proplyds" which are known to be photo-evaporating circumstellar disks (Muench et al. 2001a). Moreover, 70% of the brown dwarf proplyds display JHK_s excess, similar to the proportion of stellar proplyds in the cluster that show JHK excess emission (Lada et al. 2000). Indeed, overall the JHK_s excess fraction for substellar objects is the same as that for stellar objects in the cluster. This, together with the fact that the luminosity function is smooth and continuous across the stellar/substellar boundary, suggests that brown dwarfs and stars are formed by the same physical process. That is, from the collapse of individual protostellar cores and subsequent build up of mass via accretion through a circumstellar disk. In this sense freely floating brown dwarfs are more similar in nature to stars than to planets, which form as a secondary product of star formation within a disk around the central star. Moreover, being surrounded by disks,

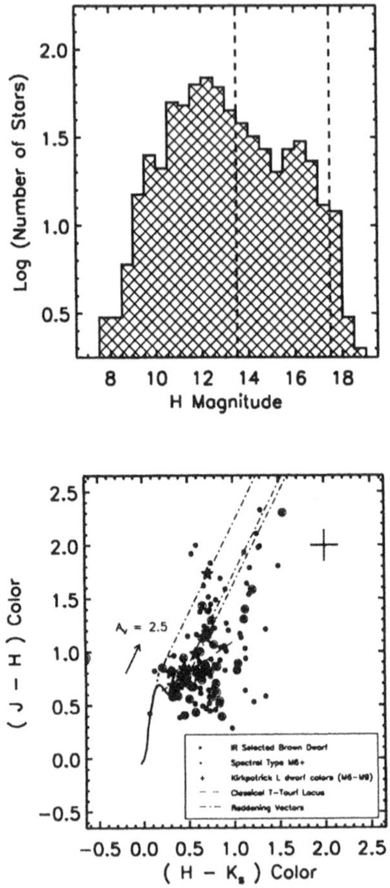

Fig. 6. a(top) - The H band luminosity function of the Trapezium cluster. The vertical lines mark the approximate locations of the hydrogen and deuterium burning limits for a 1 Myr old stellar population, i.e., 0.08 and 0.01 $M_\odot$, respectively. b(bottom) - The JHK_s color-color diagram for the substellar population in the Trapezium. More than half the sources display infrared excess and evidence for circumstellar disks. (Muench et al. 2001a)

substellar objects may be just as likely as stars to form with an accompanying system of planets.

A modified scenario for the formation of brown dwarfs has been proposed by Reipurth and Clark (2001). In this picture freely floating brown dwarfs initially originate in a triple or multiple protostellar system formed from a single protostellar core. The dynamical evolution of this small body system then results in the ejection and escape of its lowest mass member. As mentioned earlier, such a formation mechanism could account for the possible excess of brown dwarfs observed in the IMF. However, the existence of disks around brown dwarfs places strong constraints on this model. This is because during the ejection process the outer parts of a circumstellar disk are likely disrupted and stripped. Very young brown dwarfs should still exhibit near-infrared excess, however it is not clear that substellar objects as old as those in the Trapezium cluster can sustain inner disks on this timescale. This is because material in the disk accretes onto the central object in a viscous accretion time. For typical parameters, this timescale is on the order of 10^5 years for a truncated disk 10 AU in size (Kenyon 1999). Disk surveys of substellar populations in older (> 2 Myr) clusters would provide a strong test of this model. If disks around substellar objects are as long lived (3-5 Myr) as those around stars, then the ejection hypothesis may have to be reconsidered.

5 Future Prospects

Infrared imaging should continue to play a major role in star formation studies as cameras and detectors are deployed on the new generation of large telescopes. Investigations of the stellar distributions, the luminosity and mass functions, and the disk properties of young stellar populations are now becoming possible in more distant reaches of the galaxy. For example, VLT observations are capable of detecting million year old substellar objects in clusters as far away as 2 kpc or more from the sun. Questions about the universality of the IMF, the size of the substellar population in clusters, and the evolution of circumstellar disks will be placed on a more secure footing with infrared observations collected by the VLT and its large cousins in Arizona, Chile, Hawaii and Spain.

References

1. Adams, F.C. & Myers, P.C. 2001, ApJ, in press.
2. Ambartsumian, V.A. 1949 Dokl.Akad.Naus SSR, 68, 22.
3. Baraffe, I., Chabrier, G. Allard, F. & Hauschildt, P. H. 1998, A&A, 337, 403.
4. Barsony, M., Schombert, J. & Kris-Halas, K. 1991, ApJ. 379, 221.
5. Boss, A.P. 2000, ApJL, 536, 101.
6. Beckwith, S.V.W. & Sargent, A.I. 1996, Nature, 383, 139.
7. Bonnell, I.A. 1999 in The Origin of Stars and Planetary Systems, eds. C.J. Lada & N.D. Kylafis, (NATO ASI SER.540) (Kluwer: Dordrecht) p. 479.
8. Carpenter, J. 2000, AJ, 120, 3139.
9. Carpenter, J., Heyer, M.H., & Snell, R.L. 2000, ApJS, 130, 381.
10. Carpenter, J., Meyer, M., Dougados, C., Hillenbrand, L. & Strom S. 1997 AJ, 114, 198.
11. Carpenter, J., Hillenbrand, L. & Skrutskie, M. 2001, AJ, 212, 3160.
12. Casali, M. & Eiroa, C. 1992, A&A, 262, 468.
13. D'Antona, F. & Mazzitelli, I. 1994, ApJS, 41, 467.
14. D'Antona, F. & Mazzitelli, I. 1997, in Cool Stars in Clusters and Associations, ed G. Micela & R. Pallavicini (Mem. S.A.It.), 68, 807.
15. Elias, J. 1979, ApJ, 224, 857.

16. Elmegreen, B.G. 1983, MNRAS, 201, 1011.
17. Grasdalen, G., Strom, S.E. & Strom K.M. 1973, ApJ, 184, L53.
18. Haisch, K.E., Lada, E.A. & Lada, C.J. 2000, AJ, 120, 1396.
19. Haisch, K.E., Lada, E.A. & Lada, C.J. 2001a, AJ, 121, 2065.
20. Haisch, K.E., Lada, E.A. & Lada, C.J. 2001b, ApJL, 553, 153.
21. Hillenbrand, L.A. & Carpenter, J.M. 2000, ApJ., 540, 236.
22. Hodapp, K-W. 1994, ApJS, 94, 615.
23. Horner, D.J., Lada, E.A., & Lada, C.J. 1997, AJ, 113, 1788.
24. Kenyon, S.J. 1999 in The Origin of Stars and Planetary Systems, eds. C.J. Lada & N.D. Kylafis, (Kluwer, Dordrecht), p. 613.
25. Kenyon, S.J. & Gomez, M. 2001, AJ, 121, 2673.
26. Kenyon, S.J., & Hartmann, L. 1995, ApJS, 101, 117.
27. Kroupa, P. 2001, MNRAS, 322, 231.
28. Kroupa, P. 1995, MNRAS, 277, 1491.
29. Kroupa, P., Tout, C. & Gilmore, G. 1993 MNRAS, 262, 545.
30. Lada, C.J. & Adams, F.C. 1992, ApJ, 393, 278.
31. Lada, C.J. & Lada, E.A. 1991 in The Formation and Evolution of Star Clusters, ed K. Janes, (ASP: San Francisco) p 3.
32. Lada, C.J., Alves, J. & Lada, E.A. 1996, AJ, 111, 1964.
33. Lada, C.J., Margulis, M. & Dearborn, D. 1984, ApJ, 285, 610.
34. Lada, C.J., Depoy, D., Merrill, M., & Gatley, I. 1991b, ApJ, 374, 533.
35. Lada, C.J., Muench, A.A., Haisch, K.E., Lada, E.A., Tollestrup, E.V., & Willner, S.P., 2000, AJ, 120, 3162.
36. Lada, E.A. & Lada, C.J. 1995, AJ, 109, 1682.
37. Lada, E.A., DePoy, D., Evans, N. & Gatley, I. 1991a, ApJ, 371, 171.
38. Li, W., Evans, N.J. & Lada, E.A. 1997, ApJ, 488, 277.
39. Lucas, P.W. & Roche, P.F. 2000 MNRAS, 314, 858.
40. Luhman, K.L. et al. 2000, ApJ, 510, 1016.
41. Luhman, K.L. 2001, ApJ, in press.
42. Marcy, G. & Butler, P. 1999, in The Origin of Stars and Planetary Systems, eds. C.J. Lada and N.D. Kylafis,(NATO ASI Ser. 540) (Kluwer: Dordrecht), 681.
43. McCaughrean, M.J., Zinnecker, H., Rayner, J. & Stauffer, J. 1995 in Proc. ESO Workshop, The Bottom Edge of the Main Sequence and Beyond, ed, C.G. Tinney (Heidelberg: Springer), 207.
44. Megeath, S.T., Herter, T., Beichman, C., Gautier, N., Hester, J.J., Rayner, J., & Shupe, D. 1996, A&A, 307, 775.
45. Meyer, M.R., Calvet, N., & Hillenbrand, L. 1997, AJ, 114, 288.
46. Muench, A.A., Lada, E.A. & Lada, C.J. 2000, ApJ, 533, 358.
47. Muench, A.A., Lada, E.A. & Lada, C.J. 2002, in preparation.
48. Muench, A.A., Alves, J.F., Lada, C.J. & Lada, E.A. 2001a, ApJL, in press.
49. Muench, A.A., Lada, E.A., Lada, C.J. & Alves, J.F. 2001b, in preparation.
50. Myers, P.C., Ho, P.T.P., & Benson, P. 1979, ApJ, 233, L141.
51. Najita, J., Teide, G.P. & Carr, J. 2000, ApJ, 541, 977.
52. Palla, F. & Stahler S. 2000, ApJ. 540, 255.
53. Persi, P., Ferrari-Toniolo, M., Marenzi, A.R., Anglada, G., Chini, R., Kruegel, E. & Sepulveda, I. 1994, A&A, 282, 233.
54. Phelps R.L. & Lada, E.A. 1997, ApJ, 477, 176.
55. Reipurth, B. & Clarke, C. 2001, AJ, 122, 432.
56. Roberts, M. 1957, PSAP, 69, 59.
57. Webb, R. 2001, in Young Stars Near Earth: Progress and Prospects, eds., R. Jayawardhana & T.P. Greene, ASP Conf.Series, v 244, in press.
58. Wilking, B.A. & Lada, C.J. 1983, ApJ, 274, 698.

Modes of Star Formation and Constraints on the Birth Aggregate of our Solar System

Fred C. Adams

Physics Department, Univ. Michigan, Ann Arbor, MI 48109, USA

1 Introduction

This work suggests that star forming environments should be classified into finer divisions than the traditional isolated and clustered modes. Using the observed open cluster system, we estimate the fraction of star formation that takes place within cluster environments: Only about 10% of the stellar population originates from star forming regions destined to become open clusters. The smallest clusters included in the observational surveys (containing $N \sim 100 - 300$ members) roughly coincide with the smallest stellar systems that evolve as clusters in a dynamical sense. Specifically, we show that stellar systems with too few members $N < N_\star$ have dynamical relaxation times that are shorter than their formation times ($\sim 1 - 2$ Myr), where the critical number of stars $N_\star \approx 100$. Our results thus suggest that star formation can be characterized by (at least) three principal modes: [**I**] isolated singles and binaries, [**II**] groups ($N < N_\star$), and [**III**] clusters ($N > N_\star$). Many – if not most – stars form through the intermediate mode in stellar groups with $10 < N < N_\star \sim 100$. The groups evolve and disperse much more rapidly than do open clusters and thus represent a different type of astronomical entity. Furthermore, groups and clusters affect star formation in different ways. Groups have a low probability of containing massive stars and are thus unaffected by supernovae and intense ultraviolet radiation fields. In addition, interactions between the young stellar objects are relatively rare in groups because of their short lifetimes and small stellar membership. Finally, we can apply these considerations to the formation of our own solar system and thereby find tight constraints on the solar birth aggregate.

2 Fraction of Star Formation in Clusters

To start, we use the galactic open cluster system to estimate the fraction of stars that are born within clusters [4]. Many authors have studied the distribution of open clusters in our galaxy ([5], [20], [8], [15], [11]). These observational surveys provide the cluster formation rate $\mathcal{R} \approx 0.45 \pm 0.04$ Myr^{-1} kpc^{-2}. Given this rate, the total number of clusters, per square kpc, produced over the age of the galactic disk is given by

$$N_C = \int_0^{t_{\rm disk}} \mathcal{R} e^{q(t_{\rm disk}-t)} \, dt = \frac{\mathcal{R}}{q} \left\{ e^{q t_{\rm disk}} - 1 \right\} , \qquad (1)$$

where $t_{\rm disk} \approx 10$ Gyr. We have used an exponentially decreasing star formation rate, where q can be estimated from chemical evolution models of the galaxy [16] or from the white dwarf luminosity function [21]. If all clusters are slated for destruction, the cluster contribution to the disk surface density is $\Delta\Sigma = N_C \langle M \rangle$, where $\langle M \rangle$ is the average cluster mass. For comparison, the observed surface density of the galactic disk is 26.4 $M_\odot$ pc^{-2} in visible stars, with an additional 18.2 $M_\odot$ pc^{-2} in stellar remnants [6]. If we correct for mass loss in the transformation between progenitor stars and remnants, the total stellar surface density $\Sigma_* \approx 63$ $M_\odot$ pc^{-2}. The fraction f_C of the stellar disk contributed by open clusters is then given by

$$f_C = \frac{\Delta\Sigma}{\Sigma_*} \approx 7 \times 10^{-5} \frac{\langle M \rangle}{1 M_\odot} \frac{1}{q t_{\rm disk}} \left\{ e^{q t_{\rm disk}} - 1 \right\} \approx 0.08 \,, \tag{2}$$

where the typical cluster mass $\langle M \rangle \approx 500$ $M_\odot$. Open clusters thus contribute about 8 percent of the field stars. Although this fraction is substantial, open clusters are *not* the birth places for the vast majority of stars.

The above analysis integrates over the age of the galactic disk. We can obtain a consistency check on this argument using present day values. The cluster formation rate $\mathcal{R}$ and $\langle M \rangle \approx 500 M_\odot$ jointly imply a star formation rate in clusters $(SFR)_C \approx 225$ $M_\odot$ kpc^{-2} Myr^{-1}, whereas the current star formation rate in the solar neighborhood is substantially larger, $(SFR)_T \approx 3000 - 5000$ $M_\odot$ kpc^{-2} Myr^{-1} [16]. These present day values thus indicate that the fraction $f_C \approx 0.045 - 0.075$, consistent with the time integrated estimate given above. We stress that this result for the fraction of star formation in clusters is consistent with previous results (e.g., [17], [8]).

In this discussion, our implied definition of a cluster requires that the system is big and bright enough to be accounted for in the observational surveys. Such systems have $N > 100$ and may require $N > 300$. This estimate of f_C should be considered as the fraction of stars forming in "large clusters" and may not include all clusters with only a few hundred members.

3 Minimum Number of Cluster Members

Next, we estimate the minimum number $N_\star$ of stars required for a stellar system to behave as a cluster [4]. Specifically, we require the dynamical relaxation time $t_{\rm relax}$ to be longer than the cluster formation time $t_{\rm form}$, i.e., $t_{\rm relax} \geq t_{\rm form}$. We want to find the minimum number $N_\star$ of stars necessary to satisfy this inequality. The dynamical relaxation time of a cluster can be written in the form $t_{\rm relax} = Q_{\rm relax} t_{\rm cross}$, where the crossing time $t_{\rm cross} = R/v$ and where $Q_{\rm relax}$ is the number of crossings required to make the velocity of a star change by a relative amount comparable to unity [6]. The velocity v is related to the cluster size R through the depth of the cluster potential well, i.e., $v^2 = GM/R$, where M is the cluster mass (including stars and gas). For a given star formation efficiency $\epsilon = N\langle m_* \rangle / M$, $Q_{\rm relax} \approx 0.1\, N\epsilon^{-2} / \ln[N/\epsilon]$.

We assume that the cluster formation time can be written in terms of the sound crossing time, i.e., $t_{\rm form} \geq \beta\, r_\infty/a$, where a is the effective sound speed, r_∞ is the size of the cloud that forms the cluster, and β is a dimensionless number of order unity. As a simple model, we use an isothermal sphere so that $r_\infty = GM/2a^2$ [18]. Departures from the isothermal model can be incorporated into the parameter β. The sound crossing time of the cluster is $\sim 10^6$ yr, much longer than the time scale for individual star formation events ($\sim 10^5$ yr, [13], [2]). Because $r_\infty > R$, we also define the parameter $\gamma \equiv r_\infty/R$. The above results jointly imply the constraint

$$\frac{N/\epsilon}{\ln[N/\epsilon]} \geq 10\,\sqrt{2}\,\beta\,\gamma^{3/2}\,\epsilon\,. \tag{3}$$

This limit grows weaker as the star formation efficiency ϵ decreases. For the cluster to remain gravitationally bound, however, the star formation efficiency cannot become too small. To incorporate this effect, we require that the final bound cluster (which contains N_f stars after gas removal) have its relaxation time longer than its crossing time, i.e., $N_f \geq N_1$, where $N_1 \approx 36$. We thus impose the additional constraint

$$\mathcal{F}(\epsilon) N \geq N_1\,, \tag{4}$$

where $\mathcal{F}$ is the fraction of stars remaining in the cluster after gas removal (see [1]). For a given function $\mathcal{F}(\epsilon)$, the coupled constraints of eqs. (3 – 4) define a well posed optimization problem. The solution is straightforward [4]: For typical values of parameters, we find the lower bound $N_\star > 60$, independent of star formation efficiency. Even for extreme cases, $N_\star > 100$.

4 Effects of Groups and Clusters on Star Formation

In the previous sections, we made a dynamical distinction between clusters (large systems with $N > N_\star$) and smaller groups ($N < N_\star$). We now discuss different ways that groups and clusters affect the star formation process. In particular, we consider scattering interactions, supernova explosions, and the background ultraviolet radiation field.

4.1 Scattering Interactions

An important channel for clusters/groups to affect star formation is through stellar encounters, which could lead to binary capture, disk disruption, or changes in planetary orbits (see [9], [14], [3]). These effects require young solar systems to experience disruptive close encounters (all solar systems experience distant encounters that lead to dispersal of the cluster [6]). Let σ_{200} denote the cross section for a close encounter in units of $(200\ \mathrm{AU})^2$, a typical cross section required for an encounter to force binary capture or to strongly disrupt a young solar system. In our solar system, e.g., a cross section of $(200\ \mathrm{AU})^2$ is the value required to eject Neptune, give Uranus an orbital eccentricity $e > 0.75$, and/or randomize the orbital inclination angles of the giant planets [3].

To illustrate the transition from groups to clusters, we derive a rough scaling law for the probability P_D for a close encounter, given by $P_D \approx \langle n \rangle \sigma_{200} v (\Delta t)$. Using basic results from cluster dynamics [6], the mean stellar density $\langle n \rangle \propto N/R^3$ and the total lifetime $\Delta t \propto (R/v) N / \ln N$. These results give the probability for disruptive encounters $P_D \propto N^\mu$, where the index $\mu \approx 2$. For systems with $N = 300$, the probability (per star) of disruptive close encounters is $P_D \approx 10^{-2}$ (1%), so about 3 systems are disrupted per cluster. For larger systems with $N = 1000$, the probability of a disruptive encounter becomes significant, $P_D \approx 0.1$, i.e., perhaps 100 out of the 1000 cluster members could experience significant disruption through a close encounter (for further detail, see [3], [4], and references therein).

The probability of a group or cluster instigating disruptive scattering interactions is thus a smooth function of N. The system size N required for a stellar aggregate to support some disruptive encounters is N = *few hundred*. In order for all of the members to experience disruptive encounters with significant probability, however, the system requires N = *few thousand*.

4.2 Probability of a Supernova

A star forming environment can affect its constituent stars through supernova explosions. These energetic events can disrupt star forming regions, remove gas from young clusters, and could possibly trigger the star formation process. We argue that groups have little chance of experiencing a supernova, whereas clusters will often be subjected to their destructive effects. We thus need to find the probability P_{SN} that a stellar system will be subjected to a supernova explosion, as a function of the number N of stars in the system. Only stars more massive than $M_{SN} \approx 8\ M_\odot$ explode at the end of their nuclear burning lives. The fraction f_{SN} of stars that are massive enough to explode ($M_* > M_{SN}$) depends on the stellar IMF; typically $f_{SN} \approx 0.004$. If the stellar IMF is independent of the system size N, the probability that a stellar aggregate (with N members) contains a progenitor star massive enough to explode is given by

$$P_{SN}(N) = 1 - \left[1 - f_{SN}\right]^N . \tag{5}$$

A natural break-even point between systems with supernovae and those without occurs where $P_{SN}(N_{SN}) = 0.5$, where the critical number of stars $N_{SN} \approx 170$. Larger clusters thus have an appreciable chance of containing stars massive enough to explode as supernovae. Keep in mind that the boundary is not perfectly sharp – stellar aggregates follow a smooth probability distribution (given here by eq. [5]).

4.3 UV Radiation Fields

External radiation fields from the background group or cluster have a substantial impact on the star formation process. For example, radiation can remove gas from circumstellar disks and suppress disk accretion and planet formation [19]; radiation fields can also play a role in ending the protostellar infall phase [10]. These processes are driven mostly by the UV portion of the radiation field, which is dominated by the most massive stars in the system. The shape of the stellar IMF dictates that massive stars are rare except in sufficiently large systems. As a result, solar systems forming within large stellar aggregates (clusters) receive an appreciable contribution of UV radiation from their background cluster; solar systems forming within small stellar groups receive relatively little UV radiation from the background.

To substantiate this claim, we estimate the UV radiation field provided by a stellar aggregate, as a function of the number N of stars in the system. This calculation [3] [4] finds the expectation value for the ionizing ultraviolet flux from a background stellar system, i.e.,

$$\langle \mathcal{F}_{uv} \rangle = 1.6 \times 10^{12} \mathrm{cm}^{-2} \mathrm{sec}^{-1} \left(\frac{N}{2000}\right) \left(\frac{R}{1\mathrm{pc}}\right)^{-2} , \tag{6}$$

where we have integrated over stellar orbits in the cluster and (separately) over the IMF. A solar system receives UV photons from the cluster at a rate

$$\Phi_{uv} = 2\pi R_{\mathrm{d}}^2 \langle \mathcal{F}_{uv} \rangle \approx 10^{41} \mathrm{sec}^{-1} \left(\frac{N}{100}\right) \left(\frac{R}{1\mathrm{pc}}\right)^{-2} , \tag{7}$$

where $R_{\mathrm{d}} \approx 30$ AU is the disk size. The nominal value of Φ_{uv} (scaled to $N = 100$) is equal to the total luminosity of ionizing UV photons from a 1 $M_\odot$ star ($\sim 10^{41}$ sec^{-1}). Sufficiently large clusters (those with $N > 100$) produce enough ionizing UV radiation to dominate the UV flux intercepted by circumstellar disks. Smaller groups ($N < 100$) have smaller UV backgrounds and circumstellar disks are primarily irradiated by their central stars.

5 The Birth Aggregate of our Solar System

Next, we turn attention to our own solar system [3]. Here, observations of meteorites indicate that unexpectedly large quantities of short-lived radioactive nuclei were present at the epoch of planet formation. One traditional explanation for this set of abundance anomalies is that the solar nebula was enriched in radioactive species by a nearby massive star [7] through a supernova explosion. For this scenario to work, the exploding star must have a mass of $M_* \approx 25 M_\odot$. Because massive stars are rare, this requirement implies that the solar system formed within a relatively large cluster.

However, we also know that our solar system could not have formed within a cluster containing too many stellar members. In a sufficiently crowded environment, the solar system would be disrupted by gravitational scattering effects from passing stars, binary systems, and other solar systems. The observed orbital elements of the outer planets exhibit low eccentricities and are (almost) confined to the same orbital plane. This relatively well-ordered configuration places tight restrictions on the characteristics of the solar birth aggregate.

Since our solar system must form in a large cluster to provide the massive star and a small cluster to avoid planet disruption, we can constrain birth aggregate of the Sun within this scenario of (external) radioactive enrichment. This calculation [3] yields a probability distribution for the number of stars in the solar birth aggregate. The Sun is most likely to have formed within a cluster containing $N = \langle N \rangle \approx 2000 \pm 1100$ members. The *a priori* probability of a star forming in this type of environment is $\mathcal{P} \approx 0.0085$, i.e., only about 1 out of 120 solar systems are expected to form under similar conditions. In such a birth cluster, the UV radiation fields provided by the putative environment are intense, supplying the early solar nebula with about 40 times more UV photons than the Sun itself. These considerations thus place tight constraints on the solar birth aggregate for the scenario of external enrichment by a massive star. Alternately, these constraints provide as basis for favoring a self-enrichment scenario for the short-lived radioactive species [12].

6 Summary

This work suggests that we should consider star formation to take place in more modes than has been historically recognized. In particular, we must move beyond the traditional dichotomy between single stars and clusters. We thus make the distinction between clusters (robust clusters that live for long times, i.e., those systems that can be observationally identified as open clusters) and groups (smaller systems with intermediate numbers of stars, say, $10 < N < N_\star \sim 100$). In this paper, we have made the distinction between groups and clusters in five different ways. All of these determinations are roughly coincident and imply $N_\star \approx 100 - 300$:
[1] Clusters are big and bright enough, and live long enough, to be included in observational surveys: $N_\star \sim 100 - 300$
[2] Clusters have dynamical relaxation times longer than their formation times: $N_\star = 60 - 100$
[3] Clusters are sufficiently dense and long-lived so disruptive scattering encounters can affect circumstellar disks and planets: $N_\star \sim 100$
[4] Clusters have enough stellar members and live long enough so that supernovae can affect forming stars: $N_\star \approx 170$
[5] In clusters, the ionizing UV radiation field is dominated by the background environment, and not the central star: $N_\star \approx 100$

Some ambiguity remains in the relative portions of stars that form in groups vs clusters. The observational cluster surveys show that only 10% of star formation takes place within *the clusters included in the sample.* The lower boundary of completeness must be defined more clearly in the future. One explanation for why stars are not seen in open clusters is that they form within cluster-sized units, which then disperse quickly after gas removal. If this scenario is true, then the cluster-sized units must be relatively small to avoid detection in the surveys (which include young clusters with ages $t \sim 3$ Myr and all bright clusters). It nonetheless remains possible for a substantial fraction of star formation to take place in systems with N in the intermediate size range $100 \le N \le 300$. However, these stellar systems are not true clusters because they disperse rapidly and stars forming within them would be unlikely to suffer any feedback effects. A crucial challenge for the future is to more clearly determine the fractions of stars that form in each of the classes, and to determine the distribution of star forming environments within the classes.

References

1. Adams, F. C. 2000, *ApJ*, **542**, 964
2. Adams, F. C., & Fatuzzo, M. 1996, *ApJ*, **464**, 256
3. Adams, F. C., & Laughlin, G. 2001, *Icarus*, **150**, 151
4. Adams, F. C., & Myers, P. C. 2001, *ApJ*, in press
5. Battinelli, P., & Capuzzo-Dolcetta, R. 1991, *MNRAS*, **249**, 76
6. Binney, J., & Tremaine, S. 1987, *Galactic Dynamics* (Princeton: Univ. Press)
7. Cameron, A.G.W., Hoeflich, P., Myers, P. C., & Clayton, D. D. 1995, *ApJ*, **447**, L53
8. Elmegreen, B. G., & Clemens, C. 1985, *ApJ*, **294**, 523
9. Heller, C. 1993, *ApJ*, **408**, 337
10. Hester, J. J. et al. 1999, *Bull. AAS*, 194.6810
11. Janes, K. A., Tilley, C., & Lynga, G. 1988, *AJ*, **95**, 771
12. Lee, T., Shu, F. H., Shang, H., Glassgold, A. E., & Rehm, K. E. 1998, *ApJ* **506**, 898
13. Myers, P. C., & Fuller, G. A. 1993, *ApJ*, **402**, 635
14. Ostriker, E. C. 1994, *ApJ*, **424**, 292
15. Pandey, A. K., & Mahra, H. S. 1986, *Astr. Space Sci.*, **126**, 167
16. Rana, N. C. 1991, *ARA&A*, **29**, 129
17. Roberts, M. 1957, *PASP*, **69**, 406
18. Shu, F. H. 1977, *ApJ*, **214**, 488
19. Störzer, H. & Hollenbach D. 1999, *ApJ*, **515**, 669
20. van den Bergh, S. 1981, *PASP*, **93**, 712
21. Wood, M. A. 1992, *ApJ*, **386**, 539

Hans Zinnecker, David Barrado, Ralph Neuhäuser,
Nuria Huélamo, Jochen Eislöffel, and Brigitte König

The VLT and the Powers of 10: Young Clusters Home and Away

Hans Zinnecker

Astrophysikalisches Institut, Potsdam, Germany

Abstract. The purpose of this short paper is to remind the European star formation community, and more specifically the European young star clusters community, of the great potential of the VLT and to encourage the young European astronomers to make more and better use of it. Three classical examples of very young star clusters at 500 pc, 6.5 kpc, and 55 kpc (the Orion Nebula Cluster, NGC 3603 in the Carina arm, and R136 in the LMC) are chosen to illustrate the resolving power of the VLT in direct imaging mode, adaptive optics mode, and interferometric mode. The VLT with its high spatial resolution modes can be used as an astronomical microscope, as it were, with a zoom factor of 10 to 100.

1 Introduction

The VLT with its four 8.2 m unit telescopes and three 1.8 m auxiliary telescopes is about to become fully operative as a powerful high spatial resolution machine. The time is near when it will beat the Hubble Space Telescope in terms of spatial resolution in the near-infrared (J, H, and K) bands. The adaptive optics system CONICA/NAOS is about to be delivered, resulting in 25 mas spatial resolution at 1 micron. As for interferometry, first fringes have been obtained and commissioning of the interferometric mode is in sight, both in the near-infrared (AMBER, 2 mas resolution at 2 microns) and in the mid-infrared (MIDI, 20 mas at 10 microns). Up to now, capitalising on the good seeing at the VLT on Paranal, direct images with 0.4 arcsec resolution have been taken rather regularly, occasionally even 0.3 arcsec, and exceptionally 0.2 arcsec for short exposures.

Here we briefly summarise some VLT results that have been obtained so far concerning imaging studies of young star clusters. Starting with the Orion Nebula Cluster as a reference object, we move further away from the solar neigbourhood to the nearest giant Galactic HII region (NGC 3603 in the Carina arm) and its associated cluster, and on to the nearest extragalactic giant HII region (30 Doradus in the LMC) and its associated exciting dense cluster core (R136). Clearly, by climbing up the distance ladder we need correspondingly higher and higher spatial resolution, in order to resolve the stellar population in these template starburst clusters. On the other hand, we also need better spatial resolution if we want to resolve binaries or even the circumstellar environment (disks, giant planets) in the nearest cluster. Overall we can study scales as small as 5 AU in the Orion Trapezium cluster to 500 AU in the R136 cluster with the VLT(I).

2 Critical Spatial Resolution (and Sensitivity)

There is a critical spatial resolution for each cluster to resolve its members, related to the mean stellar separation in the clusters (10 arcsec for Orion, 1 arcsec for NGC 3603, and 0.1 arcsec for R136, respectively). The critical resolution is likely to be of the order 1/10 of the mean stellar separation, given that a fair number of stars will be projected on top of each other. Unlike old globular clusters, young star clusters exhibit a huge dynamical range in stellar brightness, and often also irregular nebulosity, which requires extra spatial resolution if convincing results for star counts (luminosity function) and photometry (colour-magnitude diagram) are to be obtained. In addition, the better the spatial resolution the better will be the achievable point source sensitivity. From experience with direct imaging, adaptive optics, and HST data, it seems that the critical resolution for the Orion Trapezium Cluster core is 0.5 arcsec, 0.1 arcsec for the NGC 3603 cluster core, and perhaps 0.02 arcsec for R136. With such resolution, it will be possible to detect faint free-floating objects ($K = 20$ mag in 1 hour) in the Orion cluster with masses of order 3 Jupiter masses, substellar brown dwarfs in NGC 3603, and objects near the hydrogen burning limit in R136 in the LMC.

3 VLT Results and Expectations on Young Star Clusters

The VLT has been used to obtain very deep JHK images of the Orion Nebula Cluster [10] in an attempt to discover faint free-floating planets. Near-infrared observations with UKIRT [6] suggested the presence of such objects, but most of the objects appear to be close to the detection limit and therefore controversial. Independent deeper observations with the VLT are needed to confirm or disprove the UKIRT observations. An incomplete list of nearby (distance less than 1 kpc) young clusters (age less than 10 Myr) other than the Orion Nebula Cluster is given in Table 1. These might be good southern targets for direct imaging VLT observations to constrain the shape of the pre-Main Sequence IMF into the brown dwarf regime, as well as the presence of infrared excess, i.e. disks, as a function of stellar mass and age (cf. Lada, this volume).

The VLT has also been used to take deep J, K, and L images of the NGC 3603 giant Galactic HII region [25]. These images have been widely shown as an example what the VLT can achieve in direct imaging mode (seeing 0.4 arcsec). One of the results of the J and K observations has been that the IMF appears to be populated down towards the hydrogen burning limit ($0.1\,M_\odot$), i.e. there is no evidence for a truncation of the IMF in this starburst cluster. Before the VLT became available, Eisenhauer et al. [26] had studied this cluster using adaptive optics observations at the ESO 3.6 m telescope. They, too, concluded that the IMF is not truncated to their completeness limit (around $1\,M_\odot$); furthermore they found that the slope of the IMF for the massive stars is significantly shallower (0.8) than the Salpeter value (1.3), a result that differs from the conclusions on the R136 starburst cluster where the slope of the IMF for massive stars down to $3\,M_\odot$ is indistinguishable from the Salpeter slope [10]; for a discussion of IMF slopes vs stellar number density see [7].

Table 1. Some Galactic young southern clusters

Name	d	m - M
IC 2602	160	6.2
Lupus III	160	6.2
NGC 2264	750	9.4
RCW 38		10.9
Tr 14, 16		~ 14
Arches		~ 15

The low-mass stellar population in the R136 cluster in the LMC has been studied by Zinnecker et al. [17], [18], using HST NICMOS H-band data. The IMF keeps rising down to 1 $M_{\odot}$ according to a Salpeter function, contrary to the results found by Sirianni et al. [12] based on optical HST data. However, in both studies major incompleteness corrections are involved, and it would be most desirable to study the IMF of R136 with the help of the VLT adaptive optics system NAOS/CONICA in the JHK bands. There is a twin cluster to R136 (the core of NGC 2070) in the LMC which is somewhat older (10 Myr) and without nebulosity. This is NGC 2100, a cluster less dense and more relaxed than R136, suitable for direct imaging with the VLT in excellent seeing. Optical and near-infrared VLT observations are in progress for this cluster (PI Meylan), and it will be interesting to compare the stellar content of these two populous clusters.

4 Orion and NGC 3603 Displaced to the LMC

In Fig. 5 we have placed the Orion Nebula Cluster as well as the NGC 3603 cluster at the distance of R136 in the LMC. This shows that the Orion cluster (with more than 500 stars) is a mere 4 arcsec across at the LMC distance, while the NGC 3603 cluster subtends a somewhat larger angular field. In reality, clusters of the kind as in Orion and NGC 3603, if located in the LMC, would look fuzzy and blurred in even the best seeing limited images. They would sharpen up and be partially resolved only if viewed with increased resolution, e.g. through adaptive optics (field of view about 30 arcsec). They would look fully resolved into stars if viewed with VLT interferometric resolution, but it must be noticed that the field of view for interferometry is very small, of the order of 1 arcsec, making it impractical to observe anything but a small subarea, such as the very center of the clusters.

Fig. 1. HST optical/IR image of the R136 cluster (FOV ~30 arcsec × 30 arcsec or 7.5 pc × 7.5 pc). In the bottom left corner a VLT image of NGC3603 [25] and in the top left corner a VLT image of the Trapezium Cluster in Orion [12] are shown, as these two galactic clusters would be seen if they were located at the distance of R136 in the LMC (55 kpc)

5 Further from Home

The VLT will be capable of resolving at least the massive stars in even more distant young star clusters, such as NGC 604 or NGC 595 in M33 at 800 kpc (see [7] for optical Hubble images), using adaptive optics. Using interferometry, another interesting target could be the embedded protoglobular cluster in the center of NGC 5253 at 4 Mpc which has recently been resolved by VLA radio observations [13], [6]. While AMBER might detect its most luminous massive stars, MIDI observations at 10 microns should be able to resolve the bright (2 Jy) warm dust cocoon which completely obscures the supergiant HII region. Other spectacular objects include the chain of HII regions in the blue compact dwarf galaxy He 2-10 [3] at 9 Mpc, and the super star clusters in the Antennae colliding galaxies at 20 Mpc [14], [11], [15], [16]. 20 Mpc is also the distance of the Virgo and Fornax clusters of galaxies, where young star forming regions in individual galaxies (such as M100) can be resolved. Table 2 lists the major young clusters of the distance ladder.

Table 2. Major southern clusters (home and away)

VLT targets	distance	m - M
Rho Oph	125 pc	5.5
Orion TC	500 pc	8.5
NGC 3603	6.5 kpc	14.0
R136	55 kpc	18.5
NGC 5253	4 Mpc	28.0
Antennae	20 Mpc	31.5

6 Zooming in onto the Orion Trapezium Cluster

Rather than using the VLT to step away in distance by factors of 10, we can concentrate on the most famous nearby cluster, the Trapezium Cluster, and study it in greater and greater detail. If nothing else, the resolution achievable with NAOS/CONICA, will yield a new convincing estimate of the binary frequency in the Trapezium Cluster [10]. The field star binary separation distribution peaks at 30 AU, corresponding to 66 mas at a distance of 500 pc. This will allow us to compare the Trapezium binary properties (including the question of infrared excess on the primaries and the companions) with those of the young stars in nearby associations, such as Taurus and Sco-Cen. As an aside, the latter region, due to its ideal location in the southern sky at only 150 pc, will a particularly favourable region for the VLT and the VLTI, e.g. for resolving double-lined spectroscopic binaries to obtain dynamical pre-Main Sequence masses and orbital parallaxes. One of the exciting uses of the VLTI in the Trapezium will be astrometric interferometric imaging with the dual feed system PRIMA, which combines high spatial resolution, sensitivity, and field-of-view (analogous to the LBT). This should enable us, for example, to measure the proper motions of stars and outflow velocities in jets, on short timescales. Individual proper motions on the order of 1 km/s become measurable on the timescale of a year in Orion. Thus we may be able to study the dynamical state of young clusters like the Trapezium, e.g. issues like: is the cluster likely to remain bound; is there evidence for the ejection of very low-mass brown dwarfs and planetary mass objects through dynamical interactions? (cf. [19]).

7 Conclusion

The VLT(I) and young star clusters is all about

RESOLUTION, RESOLUTION, RESOLUTION!!!

8 Appendix

Fig. 2.

Fig. 3.

The formation and evolution of young stellar clusters is the topic of a Research Training Network (RTN1-1999-00436) under the European Commission Fifth Framework Programme. This network involves over 30 scientists from the following seven institutes (see Fig. 2, 3):

- Astrophysikalisches Institut Potsdam, Germany (network coordinator)
- Osservatorio Astrofisico di Arcetri, Firenze, Italy
- Institute of Astronomy, University of Cambridge, England
- Department of Physics and Astronomy, University of Cardiff, Wales
- Laboratoire d'Astrophysique de l'Observatoire de Grenoble, France
- Observatório Astronómico de Lisboa, Portugal
- Direction des Sciences de la Matiére, Commissariat á l'Energie Atomique, Saclay, France

The majority of the young cluster network scientists attended the star and planet formation ESO workshop, thanks to the organisational skills of the network coordinator Mark McCaughrean (`http://www.aip.de/~mjm/ecrtn_clusters.html`).

Acknowledgements

I am grateful to Mark McCaughrean for suggesting the title of this presentation and to Ulfert Hanschur for technical assistance.

References

1. B. Brandl, W. Brandner et al.: A&A **352**, L69 (1999)
2. P. S. Conti & W. D. Vacca: ApJ **423**, L97 (1994)
3. F. Eisenhauer, A. Quirrenbach, H. Zinnecker, & R. Genzel: ApJ **498**, 278 (1998)
4. V. Gorjian, J. L. Turner, & S. C. Beck: ApJ **554**, L29 (2001)
5. D. A. Hunter: 'Star Formation and Molecular Clouds; the Range in Scales of Star Formation (Invited paper)'. In: *Revista Mexicana de Astronomia y Astrofisica Serie de Conferencias, Vol. 3, The Fifth Mexico-Texas Conference on Astrophysics: Gaseous Nebulae and Star Formation, Tequesquitengo, Mor., Mexico, April 3-5, 1995,* p. 1
6. P. W. Lucas & P. F. Roche: MNRAS **314**, 858 (2000)
7. E. M. Malumuth, W. H. Waller, & J. W. Parker: AJ **111**, 1128 (1996)
8. P. Massey & D. A. Hunter: ApJ **493**, 180 (1998)
9. M. J., McCaughrean, J. F. Alves, H. Zinnecker, & F. Palla: ESO Photo Release 03a–d/01 (2001) (`http://www.eso.org/outreach/press-rel/pr-2001/phot-03-01.html`)
10. M. J. McCaughrean: 'Binarity in the Orion Trapezium Cluster'. In: *The Formation of Binary Stars, IAU Symposium 200, Germany, April 10-15, 2000,* ed. by H. Zinnecker & R.,D. Mathieu (ASP, San Francisco, 2001) p. 169
11. I. F. Mirabel, L. Vigroux et al.: A&A **333**, L1 (1998)
12. M. Sirianni, A. Nota et al.: ApJ **533**, 203 (2000)
13. J. L. Turner, S. C. Beck, & P. T. P. Ho: ApJ **532**, 109 (2000)
14. B. C. Whitmore & F. Schweizer: AAS **185**, 104.04 (1994)
15. B. C. Whitmore: In: *A Decade of HST Science,* ed. by M. Livio, K. Noll, & M. Stiavelli (Cambridge University Press, Cambridge 2001) in press
16. Q. Zhang, S. M. Fall, & B. C. Whitmore: ApJ, in press (2001)
17. H. Zinnecker, A. Krabbe et al.: '30 Doradus: The Low-mass Stars'. In: *New Views of the Magellanic Clouds, IAU Symposium 190, Canada, July 12-17, 1998,* ed. by Y.-H. Chu et al. (ASP, San Francisco, 1999) p. 222
18. H. Zinnecker, M. Andersen et al.: 'The infrared luminosity function of the 30 Dor cluster'. In: *Extragalactic Star Clusters, IAU Symposium 207, Chile, March 12-16, 2001,* ed. by E. Grebel, D. Geisler, & D. Minniti, in press
19. H. Zinnecker & M. J. McCaughrean: 'Star Formation Studies with the LBT'. In *Science with the Large Binocular Telescope, Proceedings of a Workshop held at Ringberg Castle, Germany, July 24-29, 2000,* ed. by T. Herbst (ISBN 3-00-00871-6, 2001) p. 137

Leonardo Testi and Daniele Galli

NIR Low-Resolution Spectroscopy of L-Dwarfs: An Efficient Classification Scheme for Faint Dwarfs

Leonardo Testi[1], Francesca D'Antona[2], Francesca Ghinassi[3], Javier Licandro[3], Antonio Magazzù[3,4], Antonella Natta[1], and Ernesto Oliva[1,3]

[1] Osservatorio Astrofisico di Arcetri
[2] Osservatorio Astronomico di Roma
[3] Centro Galileo Galilei and Telescopio Nazionale Galileo
[4] Osservatorio Astrofisico di Catania

Abstract. We present complete near infrared (0.85–2.45 μm), low-resolution (~100) spectra of a sample of 26 disk L-dwarfs with reliable optical spectral type classification. The observations have been obtained with a prism-based optical element (the Amici device) that provides a complete spectrum of the source on the detector. Our observations show that low-resolution near-infrared spectroscopy can be used to determine the spectral classification of L-dwarfs in a fast but accurate way. We present a library of spectra that can be used as templates for spectral classification of faint dwarfs. We also discuss a set of near infrared spectral indices well correlated with the optical spectral types that can be used to accurately classify L-dwarfs earlier than L6. Our results show that with the VLT/NIRMOS instrument we will be able to extend the study of objects below the deuterium burning limit to all young clusters and associations within ~1 kpc.

1 Introduction

The latest years have witnessed the discovery of numerous brown dwarfs close to the Sun, in nearby clusters and associations, and in binaries. The strategy of the optical and near–IR imaging surveys (2MASS [1], [2], the Sloan Digital Sky Survey [3], and DENIS [4], [5]) has been so successful that two new spectral classes (L and T) have been added to the previous types, to help to classify very cool stellar objects.

For L dwarfs, in spite of the remaining uncertainties in model atmospheres for such cool objects, (e.g. [6]), it has been possible to derive a detailed spectral classification system in 9 subclasses from the systematic changes observed in selected spectral features ([1], [7]) . This spectral classification has been developed in the red part of the optical spectrum: the beginning of the L type is set by the weakening of the TiO and VO bands, while the appearance of the CH_4 bands signals the transition to the T type. However, the optical spectral confirmation and classification of a candidate DENIS or 2MASS L-dwarf requires up to ~1 hr of integration time with a low resolution (~1000) optical spectrograph at a large (10m-class) telescope, depending on the spectral type and magnitude of the candidate. This prevents the applicability of the optical classification to deeper surveys.

Given that L-dwarfs emit most of their radiation in the near infrared bands from 1 to 2.5 μm, the advantage of longer wavelengths is obvious. At present, near IR spectra are available for a handful of objects (14) and very recently it has been attempted to establish a near-infrared classification scheme from full (0.9 to 2.5 μm) UKIRT spectra with resolution ~ 500-1000 [8]; each spectrum required integration times between 1 and 4 hours, depending on the spectral type of the star. With a 4-m class telescope the time demand is comparable (or higher) to that required for the optical classification, and thus prohibitive for large surveys. It is clear that intermediate- and high-resolution spectroscopy, while necessary for investigating photospheric properties of selected objects, is not suitable for candidate confirmation and classification of large, deep surveys.

In this contribution we discuss a different approach, based on very low-resolution near infrared spectroscopy using a high throughput device. In [9] we presented complete near-infrared low-resolution (~100) spectra of a sample of 26 L dwarfs with reliable optical spectral classification ([2]). All spectra have been obtained with a prism-based optical element (the Amici device), which provided a complete near-infrared spectrum of each star in less than 15 min on source at the Italian Telescopio Nazionale Galileo (TNG), a 3.56-m telescope. In Sect. 2 and 3 we present our new observations and the proposed low-resolution near infrared classification scheme and in Sect. 4 we discuss the great promises of the next generation VLT low resolution near infrared spectrograph (NIRMOS) for this type of studies.

2 Observations and Results

The observational data were collected at the 3.56m TNG with the Near Infrared Camera and Spectrograph (NICS), a cryogenic focal reducer designed as near-infrared common-user instrument for that telescope. The instrument is equipped with a Rockwell 1024^2 HAWAII near infrared array detector. Among the many imaging and spectroscopic observing modes ([10]), NICS offers a unique, high throughput, very low resolution mode with an approximately constant resolving power of ~50, when the 1″ wide slit is used. In this mode a prism-based optical element, the Amici device, is used to obtain on the detector a complete 0.85–2.45 μm long slit spectrum of the astronomical source ([11]).

The 26 L-dwarfs in our sample cover in an approximately uniform way the optically defined spectral types ranging from L0 to L8. All the selected sources are brighter than $m_{K_s} \sim 14.4$, with 3 exceptions with m_{K_s}=14.5–14.8. The sources were observed during the commissioning of NICS in several observing runs in the December 2000 to February 2001 period. We used the 0.″5 wide slit and the resulting spectra have an effective resolution of ~100 across the entire spectral range. Integration times on source varied from 2 to 15 minutes depending on the source brightness. Wavelength calibration was performed using an Argon lamp and the deep telluric absorption features. The telluric absorption was was

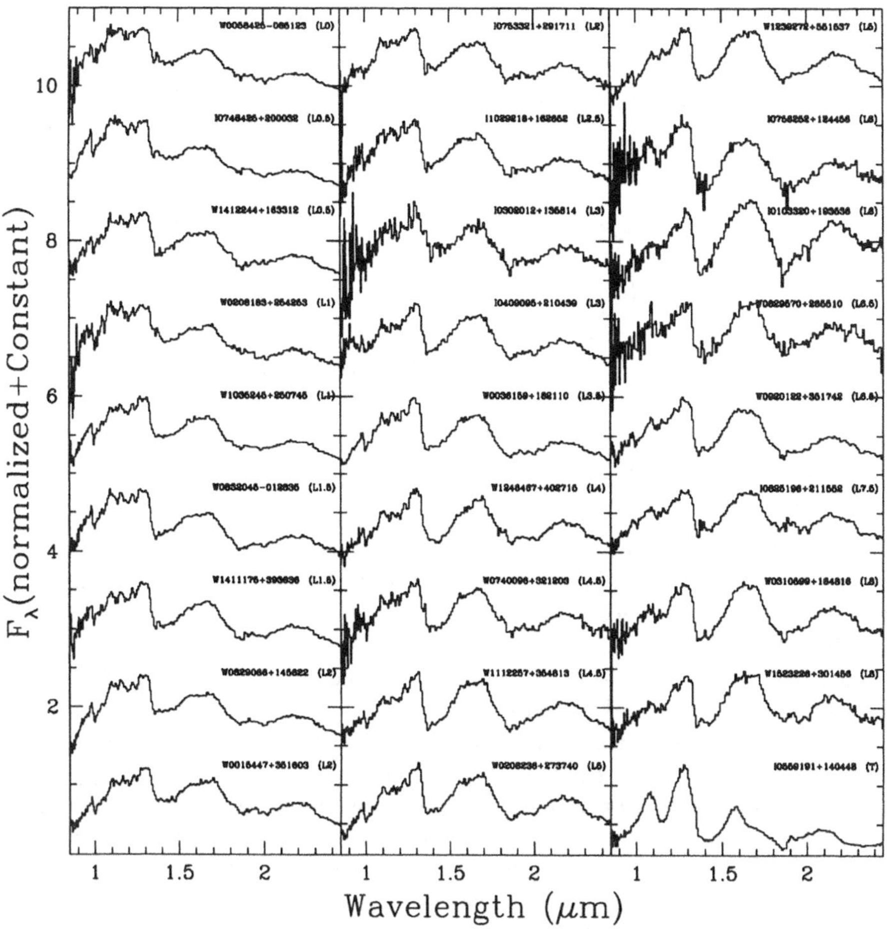

Fig. 1. 0.85–2.45μm low-resolution near-infrared spectra for all the L-Dwarfs in [9]. All spectra have been normalized by the average flux between 1.235 and 1.305 μm and a constant shift has been added to each to separate them vertically. Each spectrum is labeled with the 2MASS name (the 2MASSJ prefix has been omitted) and the optical spectral type ([2]). For comparison, the spectrum of the T-dwarf 2MASSI J0559191+140448 ([12]) at R=25, also obtained with the Amici device, is shown at the bottom right

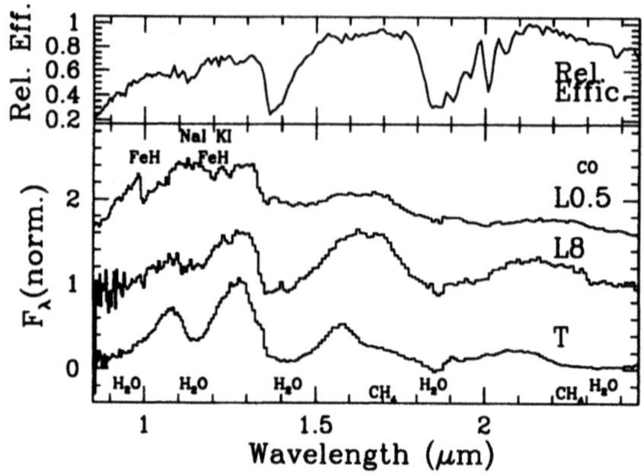

Fig. 2. Top panel: system relative efficiency (including atmosphere). Bottom panel: Amici spectra of three of the dwarfs shown in Figure 1

then removed by dividing each of the object spectra by an A0 reference star spectrum observed at similar airmass. Finally, flux normalization was done using a theoretical A0 star spectrum smoothed to the appropriate resolution. Four of the targets, which are also among the fainter in our sample, were observed in unfavorable weather conditions resulting in a poor compensation of the deep atmospheric features and noisier spectra.

The final spectra are shown in Figure 1. The objects are shown from top to bottom and from left to right in order of increasing optical spectral type from L0 to L8. The spectra have been normalized by the average flux in the 1.235–1.305 μm region, a constant offset has been added to each one to avoid overlap. The spectra show the same general features described in [6] and [8]. In our low resolution spectra the atomic lines of Na I and K I and the FeH lines in the J-band are not resolved, although their blended absorption features are clearly seen in the early type dwarfs. The spectra are dominated by the H_2O features at ~0.95, ~1.15, ~1.40, ~1.85, and ~2.4 μm. TiO, near 0.85 μm, FeH, at 1.00 μm, and CO, longward of 2.3 μm, are visible in some of the spectra, depending on spectral type and signal to noise.

3 The NIR Classification Scheme

Despite their low-resolution, the spectra of Figure 1 allow us to identify a set of spectral indices that can be used to define a near-infrared spectral classification scheme which is well correlated with the widely used optical classification scheme of [1] and [7]. A first attempt in this direction has already been taken by [8]. The main conclusion of their study is that while the J-band atomic lines are only weakly correlated with the optical spectral types, it is possible to define indices based on the H_2O wings which are well correlated with the optical types (at least up to L6). For all the stars in our sample we computed the three indices that [8]

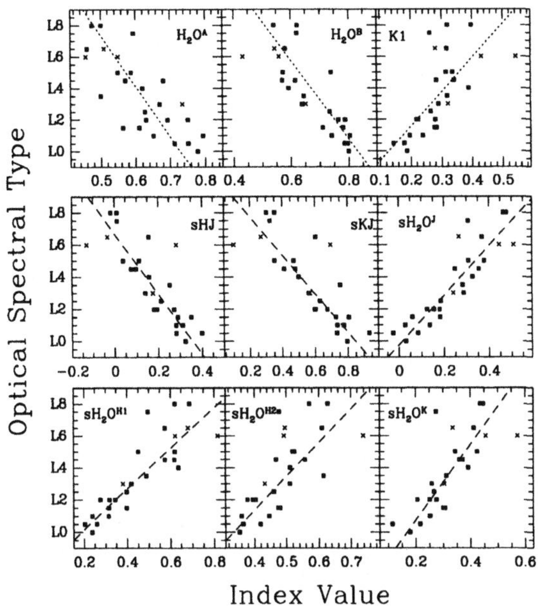

Fig. 3. Correlations between optical spectral types and the near-infrared spectral indices. The top three panels show the H_2O^A, H_2O^B, and K1 indices calculated for the stars in our sample, the dotted lines show the linear fits of [8]. The bottom six panels show the new indices defined in [9]. The four sources with poor telluric correction spectra are shown as crosses

found to be the most correlated with the optical spectral type: K1 (see also [13]), H_2O^A, and H_2O^B, all related to the strength or slope of the water absorption features. In the top panels of Fig. 3 we show the datapoints from our sample compared with the fits reported by [8]; our spectra are generally consistent with their fits. Note that, as in [8], the K1 index can be used only for types earlier than L5; moreover our data indicate saturation at late spectral types also for H_2O^B, while H_2O^A shows a very large scatter. It is possible that this behaviour of the H_2O^A and H_2O^B indices may be caused by the lower resolution of our spectra.

In [9] we also defined six additional indices which are best suited for low-resolution, complete spectra. Two of the new indices (sHJ and sKJ) are based on the slope of the continuum, and can be reliably defined using our spectra because the entire spectral range is observed simultaneously in the same atmospheric conditions, without the need of a problematic intercalibration of various spectral segments. All the other indices measure the slope of the water line wings. They have been defined trying to avoid as much as possible the spectral regions affected by the worse telluric absorption. In Fig. 2 we show the relative system efficiency (including atmosphere), three of the spectra of Fig. 1, representative of the extreme classes (L0.5 and L8) plus the low-resolution (R~25) spectrum of a

T-dwarf. In Figure 3 the value of all the six indices are plotted against the optical spectral type of each star. The sH_2O^J index is a measure of the strength of the water absorption feature at 1.1 μm and, although it shows a very nice correlation with the optical spectral type in our data, it should be used with care as it may be seriously affected by a poor correction of the telluric absorption. With only few exceptions, all stars with good spectra show a tight correlation between the newly defined spectral indices and the optical spectral type.

4 The VLT Promise

Our work has shown that complete 0.85–2.45 μm low-resolution (~100) spectra are a powerful tool for identification and and spectral classification of L dwarfs from large, deep surveys, where the number and magnitudes of potential candidates make other techniques prohibitive, even on large telescopes. The VLT/NIRMOS instrument will allow to perform large area near infrared imaging surveys and low resolution multi-object near infrared followup spectroscopy. We estimate that the NIRMOS will allow to measure in 1 hr the $0.9 - 1.7$ μm spectrum of faint ($m_J > 21$) dwarfs, and classify them using our sHJ, sH_2O^J and sH_2O^{H1} indices. A new era of survey for cool dwarfs in clusters and associations will be open by this VLT/NIRMOS as it will allow to study substellar objects as those presented in the contributions by Roche et al. and Barrado y Navascués et al. in clusters and associations as distant as 1.2 kpc.

References

1. Kirkpatrick J.D., Reid I.N., Liebert J., et al. 1999, ApJ, 519, 802
2. Kirkpatrick J.D., Reid I.N., Liebert J., et al. 2000, AJ, 120, 447
3. Fan, X. *et al.* 2000, AJ, 119, 928
4. Delfosse, X., Tinney, C.G., Forveille, T. et al. 1997, A&A, 327, L25
5. Tinney, C.G., Delfosse, X., Forveille, T. & Allard, F. 1998, A&A, 338, 1066
6. Leggett S.K., Allard F., Geballe T.R., Hauschildt P.H., Schweitzer A. 2001, ApJ, 548, 908
7. Martín E.L., Delfosse X., Basri G., Goldman B., Forveille Th., Zapaterio Osorio M.R. 1999, AJ, 118, 2466
8. Reid I.N., Burgasser A.J., Cruz K.L., Kirkpatrick J.D., Gizis J.E. 2001, AJ, 121, 1710
9. Testi, L., D'Antona, F., Ghinassi, F. et al. 2001, ApJ, 552, L147
10. Baffa C., Comoretto G., Gennari S., et al. 2001, A&A, in press
11. Oliva E. 2001, Mem. Sc. Astr. It., in press (astro-ph/9909108)
12. Burgasser A., Wilson J.C., Kirkpatrick J.D. 2000, AJ, 120, 1100
13. Tokunaga A. & Kobayashi N. 1999, AJ, 117, 1010

VLT Control Center

Artie Hatzes, David Barrado, Nuria Huélamo, and Brigitte Koenig

VLT/FORS Spectroscopy in σ Orionis: Isolated Planetary Mass Candidate Members

David Barrado y Navascués[1,2], María Rosa Zapatero Osorio[3], Víctor Béjar[4], Rafael Rebolo[4], Eduardo L. Martín[5], Reinhard Mundt[2], and Coryn A.L. Bailer-Jones[2]

[1] Universidad Autónoma de Madrid, E-28049 Madrid, Spain
[2] Max-Planck-Institut für Astronomie, D-69117 Heidelberg, Germany
[3] California Institute of Technology, Pasadena, CA 91125, USA
[4] Instituto de Astrofísica de Canarias, E-38205 La Laguna, Spain
[5] Institute of Astronomy, University of Hawaii, HI 96822, USA

Abstract. We present low resolution optical spectra of isolated planetary candidate members of the young cluster σ Orionis. Using several key spectral features, as well as pseudocontinuum indices, we have derived spectral types for this sample and confirmed the nature and the membership of all of them but one. Additionally, the results suggest that one, and possibly other two, could be binary systems. Finally, most of the objects have Hα in emission.

1 Introduction

The cluster associated with the massive multiple star σ Orionis is an excellent "hunting ground" to find very low mass members, since it is near (352 pc), have low reddening, and it is very young, about 1-5 Myr. In the past years, we have conducted several optical-infrared searches, which allowed us not only the identification of a population of brown dwarfs (objects unable to fuse hydrogen, with masses below ~75 jovian masses –1 $M_{\odot}$=1047 jovian masses–), but the discovery of objects even fainter and redder, the so called isolated planetary mass objects. They are characterized by the absence of any nuclear reaction, in particular the deuterium burning, and have masses below 13 jovian masses, as estimated by theoretical models. Additional details can be found in Béjar et al. (1999) and Zapatero-Osorio et al. (2000). Similar objects have been found in IC348 (Najita et al. 2000) and Lucas & Roche (2000, 2001). In Zapatero Osorio et al., we confirmed the nature of a handful of the σ Ori candidates using low resolution spectroscopy. Here, we present a comprehensive study.

2 Observations

We used the FORS1 spectrograph at VLT/UT1 to observed several isolated planetary mass candidates in the σ Orionis cluster. The observations took place during Dec 23-27, 2000. FORS1 has a 0.2 arcsec/pixel scale in the standard resolution, yielding a field of view of 6.8×6.8 arcmin. For our spectroscopic data,

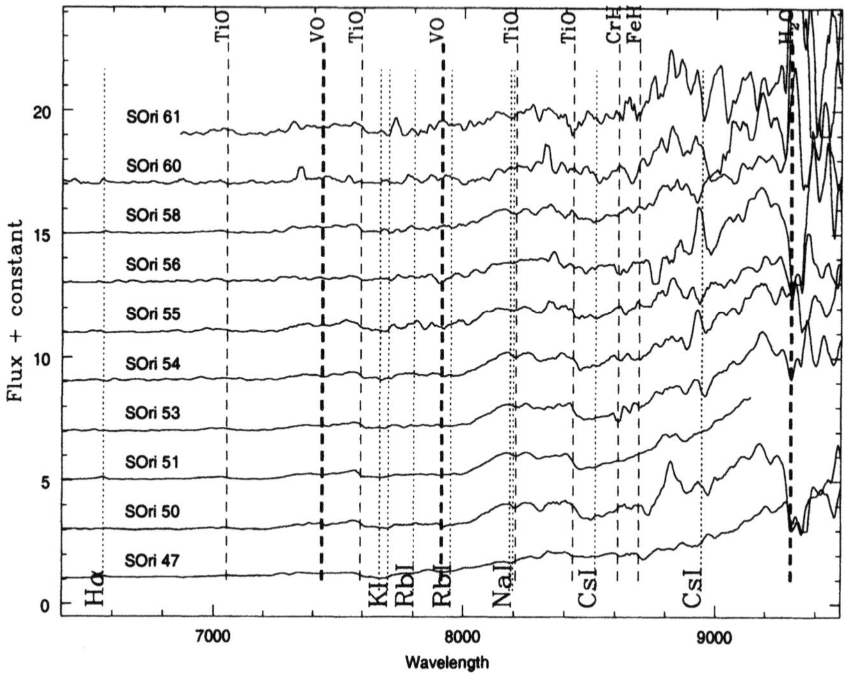

Fig. 1. VLT/FORS spectra for some of our σ Orionis isolated planetary candidate members

we used the grism 150I and the filter OG590, with a nominal spectral resolution of R~280 or 5.52 Å/pixel. With a slit width of 1.4 arcsec, our resolution is ~30 Å, R~250 as measured in the comparison arcs. The data were reduced using standard procedures within the IRAF environment. Individual exposures were added together. Then, the spectra were extracted using the apall package within IRAF, fitting the sky to remove the its emission lines and the background. The wavelength calibration was performed using HeArHgCd comparison arcs taken with the same configuration. Then data were flux calibrated using the photospectroscopic standards. Finally, we improved the signal to noise ratio by smoothing the spectra with a boxcar of 5 pixels. For the faintest objects in the sample, a convolution with Gaussian functions of 12 sigma kernel was performed. Therefore, the final resolutions, as measured in the comparison arcs, are R~210 and R~55, respectively. We also observed several nearby field objects of M and L spectral type. Some spectra are shown in Figure 1.

3 Spectral Types

The Figure 1 depicts some of our VLT/FORS1 spectra, together with the position of several relevant spectral features, including lines from alkali elements (dotted vertical lines), the heads of TiO, FeH and CrH bands (vertical thin dashed lines), and the central position of broad molecular bands of VO and H_2O

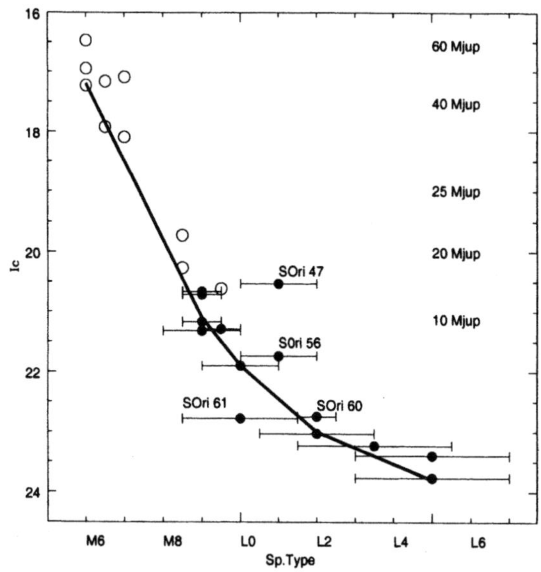

Fig. 2. Magnitudes in the I band versus the spectral types. Solid circles represent the new data, open circles show the location of previous data by Béjar et al. (2001). Masses were derived with a 5 Myr isochrone by Baraffe et al. (1998)

(vertical thick dashed lines). The differences can be appreciated, both in the slope of the pseudo-continuum as well as the change in the strength and width of KI doublet, VO and TiO bands, etc. Detailed studies of characteristics of late M and field L very low mass stars and brown dwarfs can be found in Kirkpatrick et al. (1999) and Martín et al. (1999). Figure 1 shows that, with these resolutions, it is possible to distinguish key spectral features and attempt a spectral classification of our isolated planetary mass σ Orionis candidate members.

We have estimated the spectral type of our targets following the scheme described in Martín et al. (1999). We have measured the ratio between several bands (pseudo-continuum, VO, TiO, CrH, FeH) and compared the results with data previously published in the literature and with the data of our comparison M-L nearby field objects. Most of the objects in our list define a good spectral type sequence (the fainter the magnitude and redder the color, the later the spectral type). This fact is illustrated in Figure 2, where we show the magnitude in the Cousins I band versus the derived spectral type. The solid line represents an empirical isochrone for the cluster. However, there is one exception. SOri 61 is too faint for its spectral type (M8.5), indicating that it is a likely spurious member of the cluster. Therefore, our spectroscopy indicates that all objects in the sample but this one are probable real members of the cluster.

On the other hand, SOri 47 stands out among the sample displayed in Figure 2. The quality of our data and previously published data by Zapatero Osorio et al. (1999, 2000) confirm that this object, with a magnitude above the border line between bona fide brown dwarfs and isolated planetary mass members of

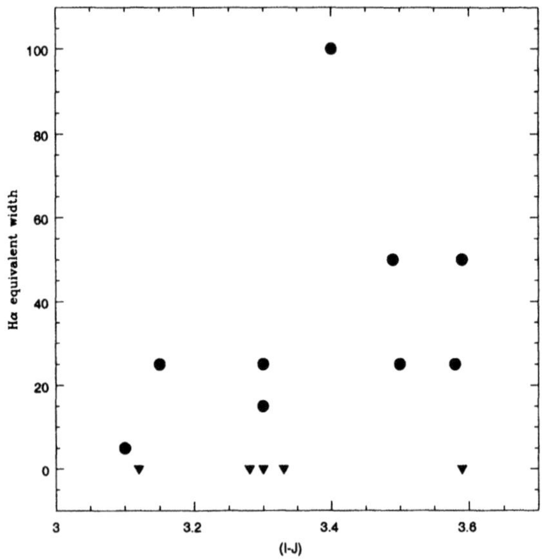

Fig. 3. Hα equivalent widths versus the (I-J) color index. Upper limits are represented as triangles, whereas detections appear as circles

the cluster (the deuterium burning limit, at 13 $M_{\rm jupiter}$), has a spectral type of L1.0–1.5. Actually, SOri 47 could be a photometric binary, containing two objects of similar masses. If this is true, SOri 47 might be composed by the first isolated planetary mass objects discovered, and it would be the first binary planet –two planets of similar masses orbiting around the common center of gravity–, with two components of 9–13 $M_{\rm jupiter}$ each for the age range 1–5 Myr. Analogous cases are SOri 56 and SOri 60, too bright for their spectral types and optical-infrared colors (L1.0, I=21.29, and L2.0, I=22.75, respectively). Moreover, SOri 60 also appears overluminous in the IR color-magnitude diagram. Additionally, a search for visual companions using K imaging can be found in Martín et al. (2001).

4 Hα, Membership and the Initial Mass Function

The equivalent width of the Hα line at 6563 Å is an age indicator for M dwarfs, and it is associated to stellar activity. In general, the stronger the emission line, the younger the object, for a given spectral type. We have identified this feature in emission for the first time in isolated planetary mass objects, and measured the equivalent width. About two thirds have a significant emission in Hα, a strong evidence that they are young and probable real members of the stellar association. We note that very few Hα emissions have been detected in similar spectral type field objects (Kirkpatrick et al. 1999; Gizis et al. 2000), and the EWs of the lines are typically below 10 Å. The origin of this feature is not clear for isolated planetary mass objects. Actually, the emission could due either to stellar-like activity or to the presence of mass accretion from a gas-dust disk

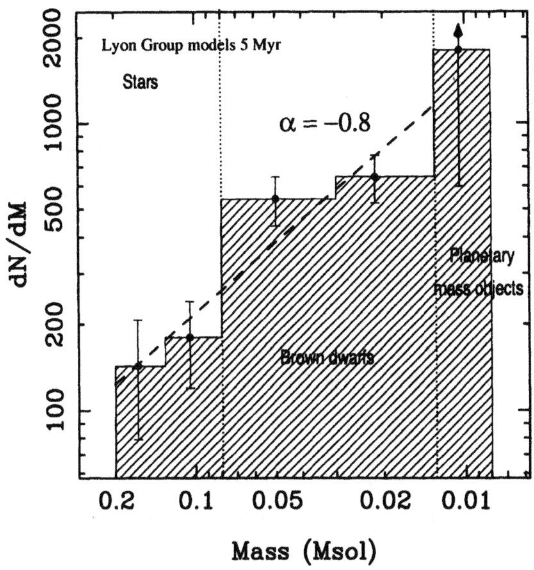

Fig. 4. A complete substellar Mass Function in σ Orionis

(Muzerolle et al. 2000). If this is true this fact might be indicating that these objects have formed in isolation by direct collapse and cloud fragmentation, and they are not runaways from embryonic multiple systems (Boss 2001; Reipurth 2001; Bate 2001). A diagram illustrating the Hα behaviour as a function of (I-J) is presented in Figure 3. One of them, SOri 55, may have experienced a flare-like episode, with variability in the Hα equivalent width (Zapatero Osorio et al. 2001). Moreover, SOri 47 seems to have a variable Hα in emission, since the VLT spectrum shows a equivalent width of 25 Å and Zapatero Osorio et al. (1999) found an upper limit of 6 Å.

All the available information, both spectroscopic and photometric, indicates that most of the objects in our sample are bona-fide members, with masses in the planetary domain or slightly above it. Recently, Béjar et al. (2001) have derived a initial mass function for the substellar domain in σ Orionis cluster, with α=0.8±0.4 (where $dN/dM = kM^{-\alpha}$, see Figure 4). Our spectroscopy confirms that the last bin of that IMF, corresponding to the planetary mass members of the cluster, is essentially unaffected by interlopers or other contaminants. Therefore, the IMF of the cluster, in the form dN/dM, keeps rising below the deuterium burning limit, implying that the cluster contains a large number of brown dwarfs and isolated planetary mass objects.

5 Conclusions

Using VLT/FORS1 low resolution spectrograph, we have derived spectral types for 15 isolated planetary mass candidates discovered by Zapatero Osorio et al. (2000). All of these objects, except one (SOri 61), appear to be bona fide members of the cluster, since they follow a well defined sequence in the diagram magnitude versus spectral type. A significant fraction of the isolated planetary mass objects presents Hα in emission, and at least two of them show variability in this activity indicator (SOri 47 and SOri 55). Since their membership in the cluster is confirmed with our optical spectroscopic data, the masses of these planetary objects must be in the range 18–8 $M_{\rm Jupiter}$ assuming an age of 5 Myr.

The combined information provided by the color-magnitude and the spectral type-magnitude diagrams indicates that SOri 47 probably is a photometric binary. Other two objects (SOri 56 and SOri 60) might also be photometric binaries.

The spectra presented here prove that membership is essentially correct in Zapatero Osorio et al. (2000). Therefore, the substellar IMF derived by Béjar et al. (2001), is not biased by spurious members in the planetary mass domain.

Acknowledgements. We thank the ESO staff at Paranal Observatory. Partial financial support was provided by the Spanish DGES project PB98–0531–C02–02 and CICYT grant ESP98–1339-CO2. DByN thanks the organizing committee for the invitation and for the financial support to attend the workshop.

References

1. Baraffe, I., Chabrier, G., Allard, F., & Hauschildt, P., A&A **337**, 403 (1998)
2. Bate M., In: *'The Origins of stars and planets: the VLT view' Conference at Garching, Germany, April 24–27, 2001*, ed. J. Alves and M. McCaughrean (Springer Verlag, 2001), this volume
3. Béjar, V. J. S., Zapatero Osorio, M. R., & Rebolo, R. ApJ **521**, 671 (1999)
4. Béjar, V., Martín, E.L., Zapatero Osorio, M.R., Rebolo, R., Barrado y Navascués, D., Bailer-Jones, C.A.L., Mundt, R., Baraffe, I., Chabrier, C., Allard, F., ApJ in press. (2001)
5. Boss, A. P. ApJ **551**, L167 (2001)
6. Gizis, J. E., Monet, D. G., Reid, I. N., Kirkpatrick, J. D., Liebert, J., & Williams, R. J. AJ **120**, 1085 (2000)
7. Kirkpatrick, D., et al. ApJ **519**, 802 (1999)
8. Lucas, P. W., & Roche, P. F. MNRAS **314**, 858 (2000)
9. Lucas, P. W., Roche, P. F., Allard, F., & Hauschildt, P. H. MNRAS in press (2001)
10. Martín, E. L., Delfosse, X., Basri, G., Goldman, B., Forveille, F., and Zapatero Osorio, M. R. AJ **118**, 2466 (1999)
11. Martín, E. L., et al. ApJ Letters, submitted (2001)
12. Muzerolle J., et al. ApJ Letters **545**, L141 (2001)
13. Najita, J., Tiede, G.P., Carr., J.S., ApJ **541**, 977 (2000)
14. Reipurth B., In: *'The Origins of stars and planets: the VLT view' Conference at Garching, Germany, April 24–27, 2001*, ed. J. Alves and M. McCaughrean (Springer Verlag, 2001), this volume
15. Zapatero Osorio, M. R., Béjar, V. J. S., Rebolo, R., Martín, E. L., & Basri, G. ApJ **524**, L115, (1999)
16. Zapatero Osorio, M.R., Béjar, V., Martín, E.L., Rebolo, R., Barrado y Navascués, D., Bailer-Jones, C.A.L., Mundt, R., Science **290**, 103 (2000)
17. Zapatero Osorio, M.R., et al., in prep. (2001)

Yepun mirror

Nancy Ageorges and Patrick Roche

Infrared Spectroscopy of Sub-Stellar Objects in Orion

Patrick Roche[1], Philip Lucas[2], F. Allard[3], and P. Hauschild[4]

[1] Astrophysics, Oxford University, NAPL, Keble Road, Oxford, OX1 3RH, UK
[2] Dept. of Physical Sciences, University of Hertfordshire, College Lane, Hatfield, AL10 9AB, UK
[3] CRAL, École Normale Supérieure, 46 Allée d'Italie, Lyon 69364, Cedex 07, France
[4] Dept. of Physics and Astronomy & Center for Simulational Physics, University of Georgia, Athens, GA 30602-2451, USA

Abstract. Near-infrared spectroscopy of a sample of 21 substellar objects in the Orion Trapezium cluster, with masses in the range 0.008-0.10 $M_\odot$ is presented. Most of the targets are located away from the region of brightest nebulosity and have photometric colours that indicate low extinction (A(v) < 5 mag). All but one, of the low-mass candidates display deep water vapour absorption bands, confirming the low effective temperatures indicated by the (I-J) colours. Interestingly, the profiles of the water bands near 1.6 microns are quite distinct from those of field brown dwarf stars, providing convincing evidence that the Trapezium objects are not foreground or background objects. Furthermore, the 2.3 micron CO absorption bands and the 2.21 micron [NaI] absorption line are significantly weaker than field brown dwarfs with similar water absorption band depths. These spectral characteristics are quite well reproduced by the low surface gravity Ames-Dusty models of Allard et al, providing additional evidence that the Trapezium substellar candidates are young and recently- formed.

1 Introduction

The Orion nebula is a particularly favourable target for investigating the low mass end of the initial mass function, as the dense backdrop of the Orion Molecular Cloud limits contamination from background stars, while its relative proximity to the Earth (450pc) means that few foreground objects will be included in the sample. There is good evidence that the great majority of the stars have formed within the last 2-million years, leading to a relatively well-understood population (e.g. [4]). These young objects are much more luminous than old field stars of comparable mass identified through the SLOAN, DENIS and 2MASS surveys (by factors of 10^3 or more [3]). A photometric survey of the central 25 arcmin^2 has revealed >600 point sources down to J ~20 mag. If these objects are indeed cluster members, about one third lie below the hydrogen-burning limit, and a couple of percent appear to be below the deuterium burning limit [6]. The form of the mass function at these low masses is currently not well defined, but there is an indication that it probably is falling slowly at M ~10 Mjup. This suggests that there could be a significant population of very low mass brown dwarfs and objects below the deuterium burning limit (~0.012 $M_\odot$).

Model tracks for young (1-2 Myr) low-mass objects are in broad agreement down to the bottom of the brown dwarf regime, but while there are significant differences in the model predictions below $0.02M_{\odot}$, the uncertainty in the true ages of the low mass candidates probably provides the main source of error in assigning masses.

In this report we present low resolution spectra of a sample of low mass objects, including brown dwarf and planetary mass candidates. The spectra are compared to those of published field brown dwarfs and model predictions for low gravity objects.

2 Observations

We have undertaken near infrared spectroscopy with CGS4 on UKIRT of a small sample of substellar candidates in Orion to search for the expected signatures of low mass, low surface gravity cool stellar atmospheres. In the centre of the cluster, the bright nebular background, which has structure on all spatial scales, limits the sensitivity for detection of faint point sources, so our targets are generally located in the lower surface brightness regions to the west of the Trapezium. The targets selected also have relatively low extinction estimates from the photometric measurements, to reduce the chances of selecting heavily reddened background sources. The 60 arcsec long CGS4 slit meant that it was usually possible to observe 2 targets at once; the high density of objects occasionally led to a third object lying within the slit by chance. 21 candidates with indicative masses in the range 0.008 to 0.10 $M_{\odot}$ were measured in the H band with 11 of these also measured at K. The spectral resolving powers were 330 at H and 440 at K while the integration times ranged up to 90 minutes; spectra of objects measured at both H and K are shown in Fig. 1 .

Spectra of two field brown dwarfs with published spectra [5] were also observed to provide higher surface gravity comparisons with the Orion objects, and to ensure the reliability of the spectral differences between the young Trapezium objects and the field brown dwarfs.

3 Results

All but one of the low mass candidates measured at H have significant water vapour absorption bands, consistent with late spectral types. The exception is 047-436 (where the designation follows the scheme suggested by [8]) which instead shows a strong red continuum with prominent H_2 emission lines, suggesting that this object is very young and still embedded in its molecular envelope. Two other objects, 019-108 and 066-433 suffer from high extinction and have weak water absorption bands consistent with earlier (late K or early M) spectral types than the rest of the sample. These two objects may well be reddened background stars, lying just behind the Trapezium cluster, though they could also be younger and therefore hotter cluster members.

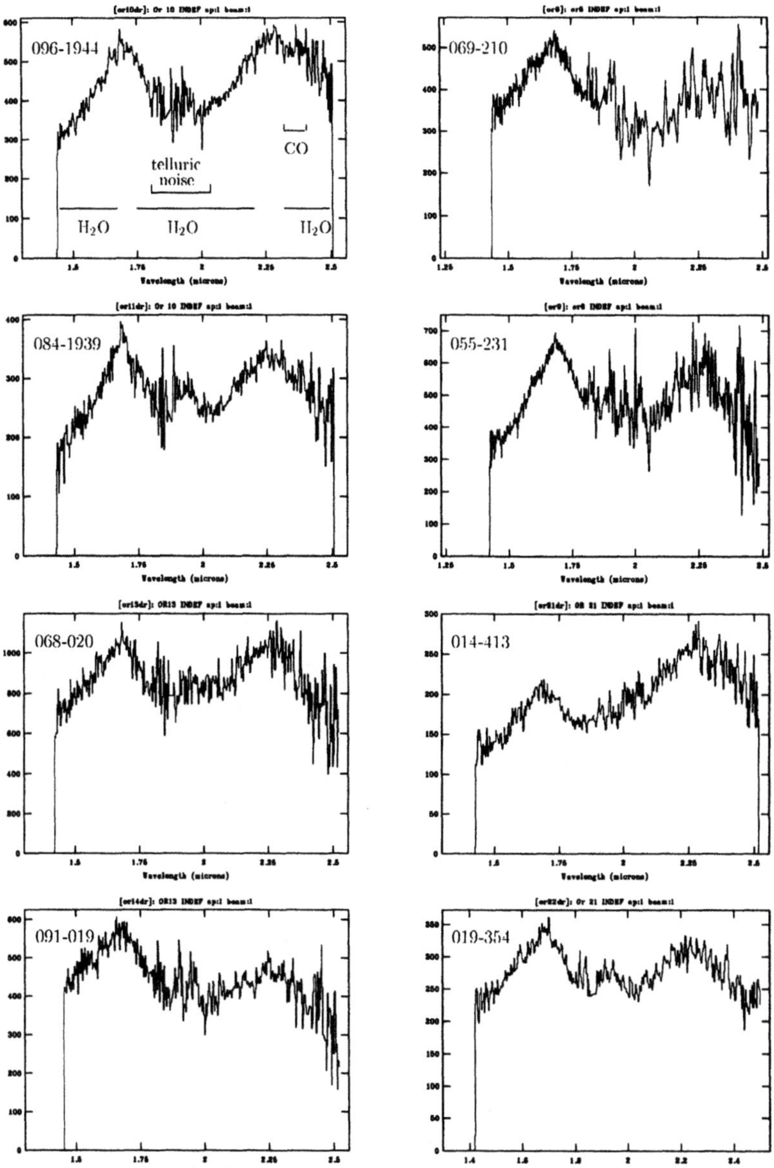

Fig. 1. Dereddened spectra of Trapezium objects observed in both the H and K bands. The excess noise at 1.8-2.0 μm results from low and variable atmospheric transmission there

The remaining objects in the sample have qualitatively similar spectra, with the deep water vapour bands at the edges of the H window combining to give a sharp peak near 1.68 microns. The peak has a triangular profile which is quite different from those of late M- and L-type dwarfs in the local neighbourhood with strong water vapour bands. All 18 of the Trapezium objects measured have similar triangular profiles suggesting that this is a common property of these young low mass objects, and distinct from the older cool field brown dwarfs. While the spectra of the lowest mass candidates have low signal-to-noise ratio, they too appear to resemble the Trapezium band shapes more closely than those of the field brown dwarfs.

A composite spectrum has been produced by averaging the H-band spectra of 12 objects with substantial water absorption bands and reasonably high quality spectra. This gives a higher signal-to-noise ratio than the individual spectra for comparison with the field dwarfs. As the composite is weighted towards the brighter, more luminous objects, it has slightly weaker water absorption than the average, but it clearly shows the distinct triangular profile of the 1.68 μm flux peak which is clearly different from the flatter profiles in the field dwarfs. Apart from the water bands, no other absorption features are evident in the composite spectrum, which is not surprising given the low spectral resolution and modest signal-to-noise ratio; the composite spectrum is compared to the field dwarf spectra in Fig. 2 .

Of the 11 objects measured in the K band, only 4 clearly show the V=2-0 rovibrational bands of CO at 2.3 μm, while the 2.21 μm [NaI] line is not detected in any of the spectra. This indicates that the atomic and molecular absorption lines are relatively weak in the Trapezium objects, consistent with predictions of low gravity models, e.g. [2].

4 Interpretation

The spectra are interpreted with reference to models produced by the Lyon group [1,2]. The low-gravity Ames-Dusty models provide quite a good fit to the H-band profiles in the Trapezium objects. The overall shape is well-matched (fig. 3), but the model spectra fall more sharply than the observations on the long wavelength edge of the 1.68 μm flux peak, unless unreasonably high temperatures are used (T >3300 K). The discrepancy may lie in incomplete or inaccurate water line lists. Emission from hot dust does not appear to contribute significantly to this effect in most objects as the K-band peak is not generally significantly higher than the H-band peak in most objects (014-413 is the best counterexample).

The water absorption depths are quantified by defining water band indices. Comparison with field dwarfs then allows spectroscopic temperatures to be assigned. Comparison of these spectroscopic temperatures with those derived form the photometric fluxes and colours is encouraging with agreement in the trends, but with the spectroscopic temperatures higher by about 250 K.

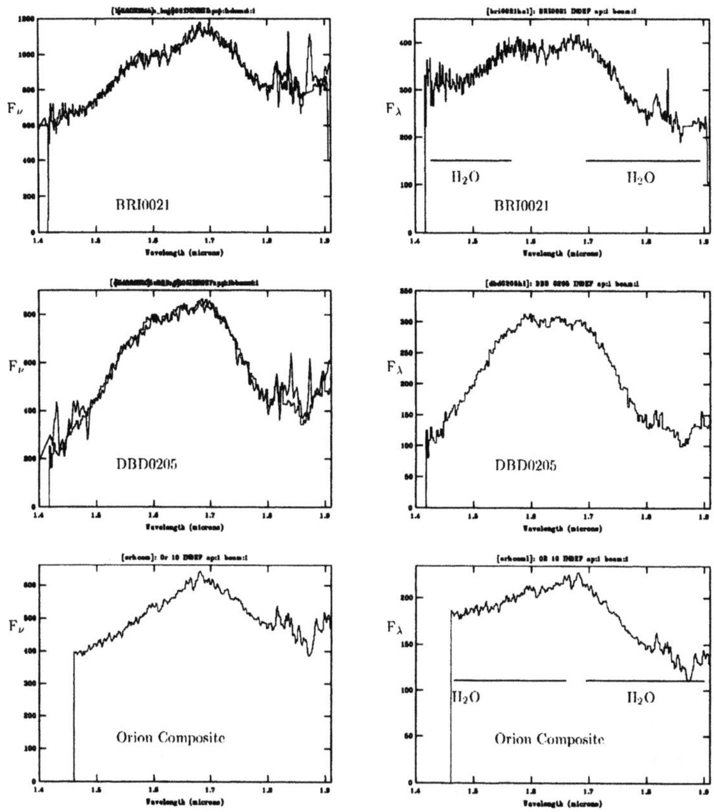

Fig. 2. Composite spectrum of Trapezium objects compared to those of field brown dwarfs where our UKIRT observations are overplotted on the published data [5]. Note the distinct triangular profile of the flux peak near 1.68 μm in the Orion composite spectrum

5 Conclusions

H- and K-band spectroscopy of candidate low mass objects in the Trapezium cluster has revealed deep water absorption bands with distinct profiles. The profiles can be qualitatively matched by low surface gravity models, providing further confirmation that these objects are indeed young, low mass cluster members.

Deep near-infrared, multi-object spectroscopy with an 8-m telescope such as the VLT would provide good quality, higher-resolution observations of very low mass candidates, and yield diagnostics of surface gravity, temperature and possibly abundances and dust content. Adaptive optics would improve the contrast of these faint point sources against the nebular background, allowing observations further into the core.

The full version of this paper is in press in MNRAS [7]

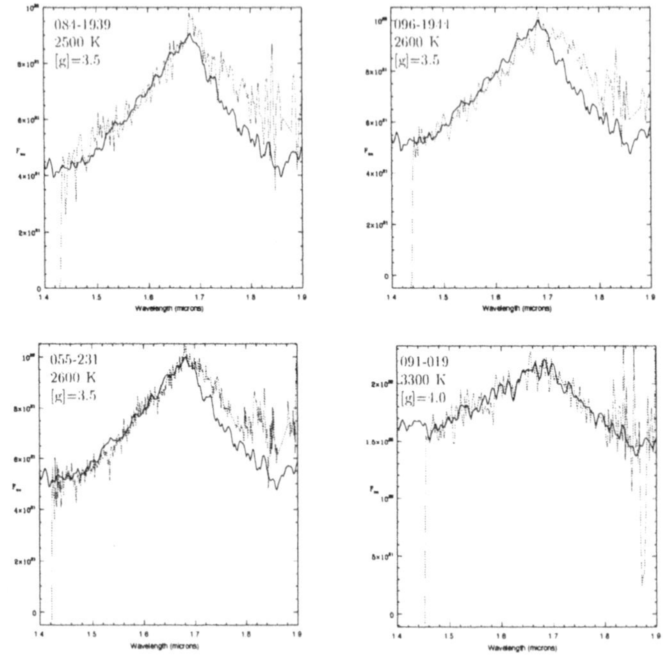

Fig. 3. Ames-Dusty-1999 model fits (solid lines) to four H-band Trapezium object spectra (dotted lines). The 1.4–1.7 μm region and the peak are well matched, but the 1.7-1.9 μm region can be fitted simultaneously, except in the warmest objects; the discrepancy increases for higher gravities

References

1. F. Allard, P.H. Hauschildt, D. Schwenke: ApJ 540, 1005 (2000)
2. F. Allard, et al: ApJ submitted (2001)
3. A. Burrows, et al.:ApJ 491, 856 (1997)
4. L.A. Hillenbrand: AJ 113, 1733 (1997)
5. S.K. Leggett, F. Allard, T.R. Geballe, P.H. Hauschildt, A. Schweitzer: ApJ 548, 908 (2001)
6. P.W. Lucas, P.F. Roche: MNRAS 314, 858 (2000)
7. P.W. Lucas, P.F. Roche, F. Allard, P.H. Hauschildt: MNRAS in press (2001)
8. C.R. O'Dell, K. Wong: AJ 111, 846 (1996)

Kueyen M2 beryllium mirror

Nicola Schneider and Sylvain Bontemps

Young Stars in ρ Ophiuchi: Toward the Completeness down to Substellar Masses

Sylvain Bontemps

Observatoire de Bordeaux, BP 89, 33270 Floirac, France

Abstract. The ρ Ophiuchi molecular cloud hosts the most nearby embedded cluster with more than 150 young stars. Intensively studied since roughly twenty years, a significant fraction of the very low mass cluster members is still to be identified. The high level of dust extinction (A_V up to 40 mag), affecting at most wavelengths the emission arising from the newly formed stars, is responsible for the slow progress in the stellar and substellar census in this cluster. From the results of the recent surveys in the infrared (IR) and especially with the mid-IR camera ISOCAM onboard ISO, I outline the present situation and describe when and how one can expect to reach the ultimate goal of recognizing all young stars and protostars down to the lowest stellar and substellar masses. The required sensitivities at most IR wavelengths are only available from the ground with the new generation of 10-metre class telescopes. I briefly explain how the VLT and its present and future instrumentation can play a major role to reach this goal by using deep photometry in the mid-IR, spectroscopy in the near-IR but also thanks to alternative techniques like the detection of proper motions in the IR bands.

1 Introduction

The ρ Ophiuchi molecular complex is closely associated with the Upper-Sco OB association. While several thousands of stars have been formed (a few Myr ago; e.g. de Geus 1992) in the OB associations, a few hundreds is still presently forming in the two most massive and densest clouds of the complex: L1688 and L1689. These dark clouds have been intensively surveyed in order to study the emergent population of stars. This forming cluster is an important laboratory to constrain the mass distribution of stars at birth and then to question the origin of the initial mass function (IMF). The ρ Ophiuchi embedded cluster is a prime target for this issue because it is fairly close by ($\sim$ 140 pc), it exhibits a clustered mode star formation, and the genetic link to the parental cloud is well readable (a complete population of protostars and pre-collapse condensations is recognized). In contrast, the young cluster is deeply embedded in the parental cloud and the large amount of dust extinction (A_V up to more than 40 mag) complicates the detection of the weakest members, and the derivation of accurate measurements such as luminosities, ages, and masses of young stars and young stellar objects (YSOs).

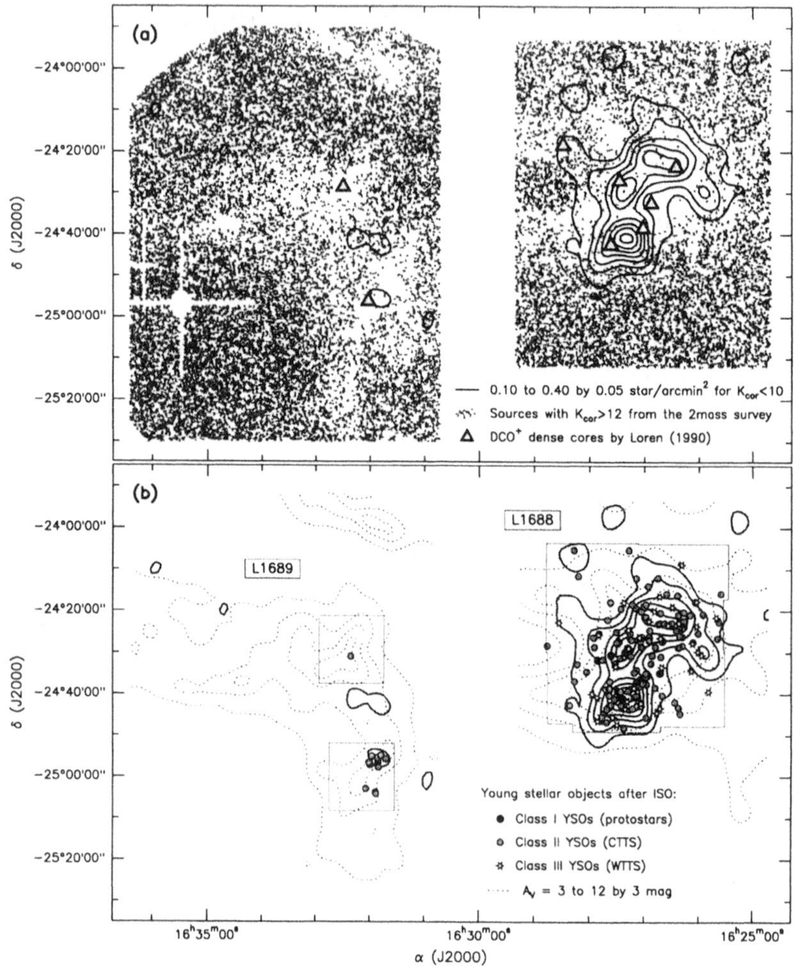

Fig. 1. (a) Contour map of the distribution of the 2 μm sources with $K_s^{cor} < 10$ from the 2MASS survey. All sources with $K_s^{cor} > 12$ which are mostly background stars are plotted as grey dots. $K_s^{cor} = K_s - 0.1 \times 9.09 \times ((J - H) - 0.6)$ is a rough extinction corrected K_s magnitude assuming a typical $(J - H)$ stellar color.
(b) Distribution of the 177 identified YSOs after the ISOCAM survey (the boxes show the ISOCAM covered areas). The (dotted) contour map of A_V is derived from near-IR colors of all 2MASS sources with $K_s^{cor} > 11$ as proposed, for instance, in Lada et al. (1994)

2 An Infrared Cluster of Embedded, Young Stars

Since the ρ Ophiuchi region is close by, the molecular cloud is dark in the optical but also in the near-IR. This can be seen in the distribution of the 2MASS[1]

[1] The Two Micron All Sky Survey is a joint project of the University of Massachusetts and the Infrared Processing and Analysis Center.

sources shown in Fig. 1. The bulk of weak sources which is dominated by background galactic stars clearly draws the contour of the clouds, and only a few near-IR sources are detected inside the cloud. On the other hand, the distribution of the brightest 2 μm sources (with $K_s^{cor} < 10$ in Fig. 1) shows exactly the opposite behavior, being clustered inside the dark clouds. These bright sources are expected to be mostly young stars so that the map in Fig. 1a should actually trace very well the contours of the embedded cluster. Most of these bright near-IR stars actually correspond to the population of young stars studied since long in the IR (e.g. Elias 1978, Lada & Wilking 1984, Wilking et al. 1989, Comerón et al. 1993, Greene et al. 1994, Strom et al. 1995, Luhman & Rieke 1999).

3 The Embedded Cluster after the ISOCAM Survey

The deep mid-IR survey operated by ISOCAM (Cesarsky et al. 1996; Nordh et al. 1998) has revealed a population of 139 YSOs with IR excesses which is constituted of 16 Class I YSOs and 123 Class II YSOs (see Bontemps et al. 2001). The distribution of these confirmed members is shown in Fig. 1b. It perfectly coincides with the distribution of bright 2 μm sources in Fig. 1a. Three sub-clusters are well identified in both the ISOCAM population and the 2MASS map in L1688. A fourth sub-cluster in L1689 has been discovered by ISOCAM (Fig. 1b in L1689S, the southern dense core of L1689). Altogether with 38 confirmed Class III YSOs, a cluster of 177 members is identified up to now.

4 Luminosity and Mass Functions of the Cluster

A detailed description of the ISOCAM survey results for ρ Ophiuchi are given in Bontemps et al. (2001) including the luminosity functions of all Class I, Class II and Class III YSOs, and the mass function of the 123 Class II YSOs (see Fig. 2). Interestingly enough this mass function does not show any low-mass cutoff above the completeness level estimated to be $\sim 0.055\,M_\odot$. If the population of young stars with IR excess is well representative of the global cluster, it shows that the expected turnover in the IMF should therefore be smaller, and well inside the substellar regime. The issue of the IMF turnover is of high importance for constraining the theories for the origin of stellar masses. There are for instance some debates about the value of the turnover mass depending on stellar densities of embedded clusters (and star-forming regions). Luhman (2000) argues that the turnover in the loosely aggregated region of Taurus is close to $0.1\,M_\odot$ while it would be at much lower mass in the Trapezium (high density cluster). Obviously ρ Ophiuchi, which is of intermediate stellar density, is very important to investigate as deep as possible in order to get the most complete view and to compare with these other regions.

Finally we note that the ISOCAM results have been important for identifying members with IR excess: Class I/II YSOs (protostars and CTTS) but they are inappropriate for surveying Class IIIs (embedded WTTS) which might be numerous in the cloud (e.g. Grosso et al. 2000). Deep investigations with the

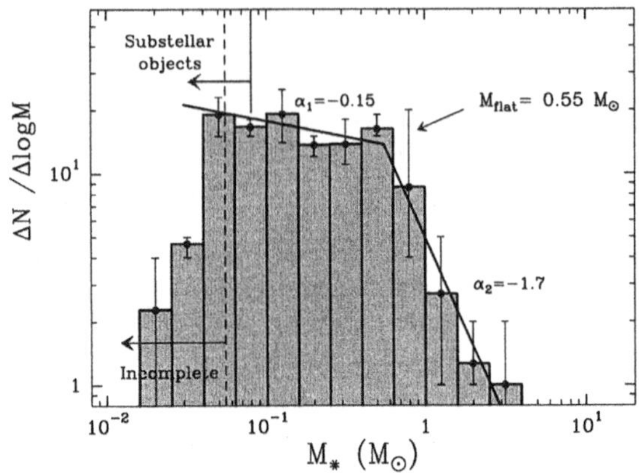

Fig. 2. Mass function of the 123 Class II YSOs (embedded CTTS) identified after the ISOCAM survey. The error bars represent the range of uncertainty on the age distribution of the population (see Bontemps et al. 2001 for details). The best two power-law fit is shown as a heavy solid curve. No low-mass cutoff is observed above the completeness level

new generation of X-ray satellites (Chandra, XMM) are expected to improve the census of this population (e.g. Montmerle et al.; these proceedings).

5 Objectives of Ground-Based Infrared Surveys

The lowest mass substellar objects embedded in ρ Ophiuchi are particularly bright since they are young. They are however strongly affected by the interstellar extinction. For instance, a 0.02 $M_\odot$ object at 1 Myr is expected to have J=14.5, H=13.9, and K=13.5 (from Baraffe et al. 1998) at 140 pc outside the cloud ($A_V \sim 0$) but should have J=25.1, H=20.1, and K=17.1 for $A_V = 40$ (neglecting IR excess in these bands). While K-band imaging with a 1 to 2-metre class telescope can easily reveal all young sources, the detection in the H-band already needs a large telescope, and in the J-band even with the 10-metre class telescopes the weakest most embedded substellar objects are out of reach.

More than the detection, in order to reveal all the weakest members of the cluster, the critical point is to firmly establish the membership of all sources. The near-IR photometry alone does not provide systematic criterium to select all members, even not all "red" members (stars with IR excess). In contrast, photometry in the mid-IR range has proven to be efficient to recognize "red" members using the mid-IR colors (see Sect. 6). Near-IR spectroscopy also provides a way to distinguish weak members from background sources shining through the cloud (see Sect. 7). Finally I argue in Sect. 8 that proper motion measurements in the near-IR might be a very interesting and efficient way to recognize all members in an almost unbiased way, regardless they have IR excess or not.

6 Mid-Infrared Photometric Surveys

Since IRAS the mid-IR and far-IR colors are known to be efficient indicators of the nature of IR sources. They provide a youth criterium and are used to order YSOs along the classical classification Class I to Class III which is, at first order, an evolutionary sequence (Lada 1987). It is actually only at longer wavelengths than $\sim 2\,\mu$m that the IR excesses of embedded YSOs can be systematically detected (e.g. Kenyon & Hartmann 1995).

For ρ Ophiuchi, ISOCAM in the mid-IR has proven to provide efficient mid-IR colors to recognize all IR excess YSOs (Bontemps et al. 2001). This survey was however limited to a detection level of the order of $\sim$ 10 mJy mainly dictated by the difficulties of detecting weak sources in the vicinity of bright YSOs. With SIRTF, the situation might improve but the saturation restrictions on bright sources will also greatly affect the reachable sensitivity. As a consequence ground-based observations on large telescopes can also significantly contribute to deep surveys. The L-band at $3.5\,\mu$m provides a good enough sensitivity, the fast-reading modern cameras do not saturate on astrophysical sources (the bright sky emission always dominates the flux) and can map reasonably large areas, and finally the mid-IR color $K - L$ appears to be efficient to recognize all "red" young stars (see, for instance, Haisch et al.; these proceedings). A $0.02\,M_\odot$ object at 1 Myr with $A_V = 40$ is expected to have $L = 14.9$ to 13.9 depending of the amount of IR excess at $3.5\,\mu$m. With the VLT/ISAAC, under average seeing conditions, L=14.9 is reached in 10 minute integration time. The other ground-based mid-IR bands (M, N or Q) do not provide sensitivities to detect the weakest substellar objects.

7 Spectroscopic Surveys

Near-IR spectroscopy is an efficient technique to derive approximate spectral types of IR sources. Together with an estimate of the stellar luminosity, assuming the source is at the distance of the cluster, it is possible to place the source in a HR diagram and to determine whether it is a PMS or a background star. Only the (few) foreground stars can be mis-interpreted in this scheme (e.g. Luhman & Rieke 1999 for a large spectroscopic survey in ρ Ophiuchi).

From the pure point of view of detectability, the 10-metre telescopes can reach a $0.02\,M_\odot$ object at 1 Myr with $A_V = 40$ (with the VLT/ISAAC a good spectrum for K=17 requires of the order of 1 hour integration time). However it is several hundreds of sources (mostly background) which are expected at K=17 to shine through the cloud. With one hour per source, it then represents several hundreds of hours on a 10-metre telescope which is obviously more than a few night observing time. Such program needs multi-object spectrometer to improve observing efficiency. On the VLT, NIRMOS is a multi-object spectrometer with 180 slits in a relatively large field-of-view. It will however operate only in J and H bands which are less suitable than the K-band for our concern. In the H-band, the most embedded and lowest mass objects in ρ Ophiuchi might be out of reach with NIRMOS.

8 Proper Motion Surveys

Proper motion measurements give an unbiased way to recognize the members of a cluster. While it is often used in optical wavelengths to survey open clusters for instance, this technique has been barely used in the IR. Large IR arrays on large telescopes provide opportunities to actually reach the required precision to investigate the young embedded clusters.

The ρ Ophiuchi cluster is a very good target for such investigations since it has a large proper motion (the dark cloud is tightly associated to the Upper-Sco OB association which has $\mu = 26.1 \pm 0.1$ mas/yr; de Zeeuw et al. 1999). With the ESO/NTT, from 100 × repeated images, we recently reached an astrometric precision of the order of 1 mas (rms) for all IR sources down to K=15.5 (ongoing project together with Ducourant, Daigne, et al.). With the VLT/ISAAC, a slightly better precision is expected (due to the large primary mirror) and it is possible to reach this precision for all weak sources down to K=17. The whole cluster can be systematically covered in a reasonable amount of time, and only two or three epochs (one or two years) can then be enough to get a complete census down to $0.02\,M_{\odot}$. Moreover all members (with or without IR excess) can be recognized directly in an almost unbiased way (only the "red" members are expected to be slightly overbright in K). This technique is therefore very promising to reach the most complete census in ρ Ophiuchi.

References

1. I. Baraffe, G. Chabrier, F. Allard, P.H. Hauschildt: A&A **337**, 403 (1998)
2. S. Bontemps, P. André, A. Kaas, L. Nordh, et al.: A&A **372**, 173 (2001)
3. S. Casanova, T. Montmerle, E.D. Feigelson, P. André: ApJ **439**, 752 (1995)
4. C. Cesarsky, A. Abergel, P. Agnèse, et al.: A&A **315**, L32 (1996)
5. F. Comerón, F., G.H. Rieke, A. Burrows, M.J. Rieke: ApJ **416**, 185 (1993)
6. E.J. de Geus: A&A **262**, 258 (1992)
7. P.T. de Zeeuw, R. Hoogerwerf, J.H.J. de Bruijne, et al.: AJ **117**, 354 (1999)
8. J.H. Elias: ApJ **224**, 453 (1978)
9. T.P. Greene, B.A. Wilking, P. André, et al.: ApJ **434**, 614 (1994)
10. N. Grosso, T. Montmerle, S. Bontemps, P. André, et al.: A&A **359**, 113 (2000)
11. S.J. Kenyon, L. Hartmann: ApJS **101**, 117 (1995)
12. C.J. Lada, B.A. Wilking: ApJ **287**, 610 (1984)
13. C.J. Lada In: *Star Forming Regions*, (IAU Symposium No. 115, 1987), p. 1
14. C.J. Lada, E.A. Lada, et al.: ApJ **429**, 694 (1994)
15. K.L. Luhman, G.H. Rieke: ApJ **525**, 440 (1999)
16. K.L. Luhman: ApJ **544**, 1044 (2000)
17. M.R. Meyer, N. Calvet, L.A. Hillenbrand: AJ **114**, 288 (1997)
18. L. Nordh, G. Olofsson, S. Bontemps, et al.: ASP Conf. Ser. **132**, 127 (1998)
19. K.M. Strom, J. Kepner, S.E. Strom: ApJ **438**, 813 (1995)
20. B.A. Wilking, C.J. Lada, E.T. Young: ApJ **340**, 823 (1989)

The Basecamp

Jochen Eislöffel, Floris van der Tak, João Alves,
Jérôme Bouvier, and Jean-Philippe Beaulieu

Variability and Rotation in Low Mass Stars and Brown Dwarfs

Jochen Eislöffel and Alexander Scholz

Thüringer Landessternwarte Tautenburg,
Sternwarte 5, D-07778 Tautenburg, Germany

Abstract. The evolution of angular momentum in stars is not really understood today. In particular, very little is known about rotation of substellar objects. Therefore, we started a systematic search for periodic variability in very low mass (VLM) stars and Brown Dwarfs. From a deep multi-filter survey of the young open cluster IC4665, accompanied by follow-up spectroscopy with FORS at the VLT, we selected a field with 190 candidate VLM cluster members. This field was then observed in an extended time series to study photometric variability. We find photometric periods for 13 stars with masses $< 0.3\,M_{\odot}$, and for five Brown Dwarfs with masses down to $38\,M_{Jupiter}$. Furthermore, we find no obvious flares in the time series. These results lead to the conclusion, that VLM objects show less activity and faster rotation than solar mass stars of similar age. For the lowest mass objects the surface features leading to the observed variability may be a kind of clouds from condensated dust in the atmospheres.

1 Introduction – The Evolution of Rotation

The investigation of angular momentum regulation in young stars is a crucial step towards a deeper understanding of the early phases of stellar evolution. Young stars have an angular momentum per mass which is by magnitudes lower than that of the molecular clouds from which they formed (Bodenheimer 1995). If stars would preserve all the angular momentum from the molecular cloud, they could not form because the centrifugal forces would surpass the gravitational attraction by far. There must exist mechanisms which allow a star to get rid of the excess angular momentum. How these processes work in detail is still not understood.

The rotation of solar mass stars has been studied by many groups in the past. As an example we refer to Fig. 1 (Bouvier et al. 1997), which shows the evolution of the rotational velocity with stellar age for solar mass stars. At any age objects cover a wide range of velocities, but there is an upper limit which varies with age. During the pre-main sequence (PMS) phase no very fast rotators are found, although the objects still accrete matter and with it angular momentum. It is generally believed, that the rotation of these objects is braked either by magnetic coupling between the star and its disk or by mass loss in highly collimated jets. Once the disk has disappeared, the rotational velocities rise because the objects contract quasi-hydrostatically and reach the Zero Age Main Sequence at their highest rotation rates. From then on, they slow down again following a Skumanich law (Skumanich 1972) through angular momentum loss in stellar winds. Many details of all these processes are not yet understood.

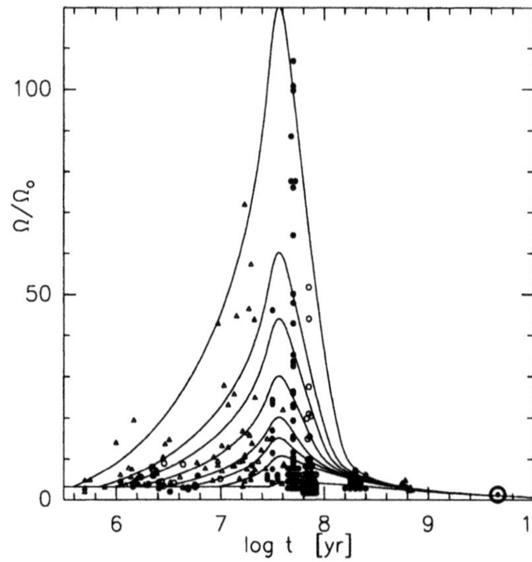

Fig. 1. Evolution of rotational velocities $v \sin i$ for solar mass stars (Bouvier et al. 1997)

It is unknown how the evolution of rotational velocities (Fig. 1) looks like for objects with masses $< 0.3\, M_\odot$. Probably it looks different from that of solar mass stars, because low mass objects are different from their solar mass siblings in several ways:

- Their internal structure is different because they have degenerate cores and are fully convective (Chabrier & Baraffe 1997).
- Their magnetic field structure and the generation of the field may be different. The standard model to explain magnetic activity of solar mass stars is the so-called $\alpha\omega$-dynamo (see Reid & Hawley 2000 and references herein). This dynamo works in a shell below the convection zone of the star. When objects are fully convective, a different mechanism must come into play.
- The radii of objects with masses of $0.1 \ldots 0.001\, M_\odot$ are nearly constant, namely ≈ 1 radius of Jupiter, while for the higher mass stars the radii are increasing steadily with mass (Oppenheimer et al. 2000).

Therefore, we have set out to study the rotation rates of VLM objects by measuring photometric rotation periods.

2 Observations

We concentrated our efforts on the young open cluster IC4665, which has a distance of ≈ 350 pc and an age of ≈ 36 Myr (Mermilliod 1981a,b). We surveyed this cluster in a VLT-preparatory-program with the Danish 1.54m-telescope on La Silla and with the Tautenburg Schmidt telescope. This survey was originally done in the I- and Z-band in 1998/99 and has been extended in the meantime to R, I, Z and J with additional data from the WFI at the ESO/MPI 2.2m-telescope, OmegaPrime at the 3.5m-telescope on Calar Alto and UFTI at UKIRT on Mauna Kea. Finally, we obtained a catalogue of four filter photometry for about 60000 objects in an area of about 1 sqdeg. A mass-luminosity relation for the cluster was calculated using the Baraffe et al. models (1998). Objects which lie close to this isochrone were selected as potential cluster members. This selection of cluster members is still contaminated by other objects in the fore- and background of the cluster, mostly by red dwarf stars. To confirm cluster membership, we took spectra of $\approx$ 350 of our candidates with FORS at the VLT in multi-object mode. A preliminary reduction of these spectra showed that the candidates are indeed mostly late M-type objects.

As a next step, we selected a field in IC4665 covered by WFI at the ESO/MPI 2.2m-telescope which contains 190 of our cluster member candidates to measure photometric rotation periods. This field was observed with the WFI in an observing run in May/June 1999. To obtain the optimum time sampling and to get as many as possible data points for every object, we concentrated just on the one selected field during two consecutive nights. In addition, we took several frames per night in five nights distributed over ten days around these two nights. The resulting time series consists of 96 I-band images.

Image reduction included bias/overscan-subtraction, flatfield correction as well as the removal of the very distinctive fringe structures. Subsequently, differential photometry was done for all objects in the field. We used `daophot`-routines within IRAF to fit the point spread function of every object. A mean lightcurve was constructed by averaging the time series of 140 non-variable reference stars. By subtraction of this mean lightcurve, the time series of all objects were freed from atmospheric extinction influences.

3 Time Series Analysis

We focussed our time series analysis on the search for photometric rotation periods. In addition, the lightcurves were inspected visually to detect obvious signs of non-periodic variability, e.g. flare-like phenomena, but no such events were found. This constrains the flare-rate for VLM objects to be $< 0.0004\,h^{-1}$, i.e. at least a factor of 15 lower than for solar mass stars ($0.006\,h^{-1}$, Guenther & Ball 1999).

For our period search we worked out a catalogue of criteria, which must be fulfilled to unambiguously detect periodic variability.

- Firstly, the commonly used Scargle-periodogram (Scargle 1982) was calculated. A peak was classified as significant, if its False Alarm Probability ($FAP_{Scargle}$, calculated following Horne & Baliunas 1986 is below 1%.
- A sinewave with the frequency of the period in question was fit to the lightcurve and subtracted. Then, the variance of the residual lightcurve was compared with the original in a statistical F-test. The resulting FAP_F gives the significance that the two variances are different. In conformity with Koen

(1990) we learned, that the Scargle-FAP systematically overestimates the significance of a peak. We accepted a period only if its FAP_F is below 10%.
- To make sure, that the periods are really intrinsic properties of the objects, we compared the lightcurves with those of at least three nearby comparison stars. Only if these comparison lightcurves neither in the periodogram nor in the lightcurve itself show a sign of a similar periodicity, the period was accepted.
- We checked, if the period could be detected as well in both single nights, which were completely used for time series observations.
- The lightcurves were checked by eye: The measured period should be clearly visible in the original lightcurve as well as in the phased time series.
- The CLEAN-algorithm (Roberts et al. 1987) was used to free the periodograms from sidelobes and to avoid misinterpretations of such features.

Only 18 out of the 190 objects passed all these tests, so that we consider their derived rotation periods as reliable. Five of these objects were probable Brown Dwarfs. Figure 2 shows examples of the phased lightcurves.

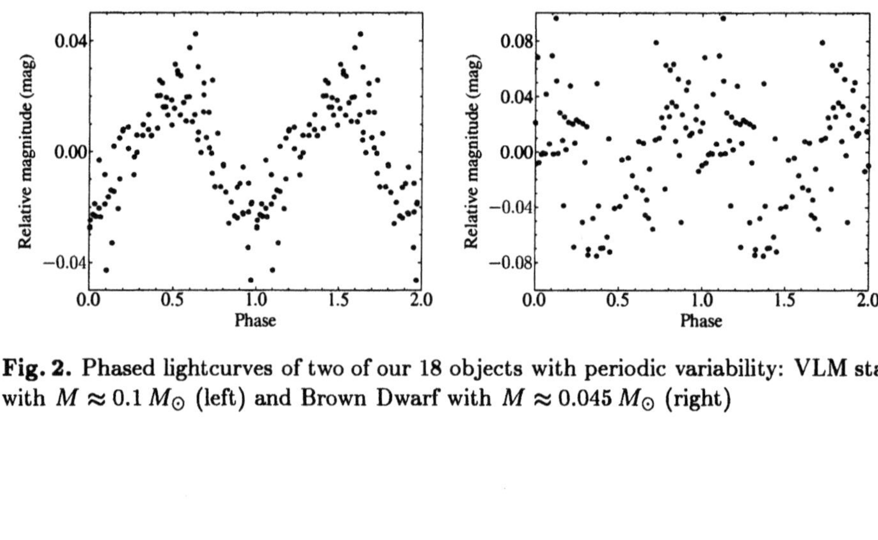

Fig. 2. Phased lightcurves of two of our 18 objects with periodic variability: VLM star with $M \approx 0.1\,M_\odot$ (left) and Brown Dwarf with $M \approx 0.045\,M_\odot$ (right)

4 Discussion – Rotation and the Surface Features of VLM Objects

The $\alpha\omega$-dynamo model which works very well in explaining the activity of solar mass stars, predicts much activity, strongly asymmetrically distributed starspots and a strong correlation between rotation and activity (see Reid & Hawley 2000 and references herein). As we have outlined in Sect. 1, this model may not be applicable for fully convective objects. On the other hand, an alternative turbulent dynamo model for these objects (Durney et al. 1993) produces lower magnetic field strengths. It predicts little activity, more uniformly distributed

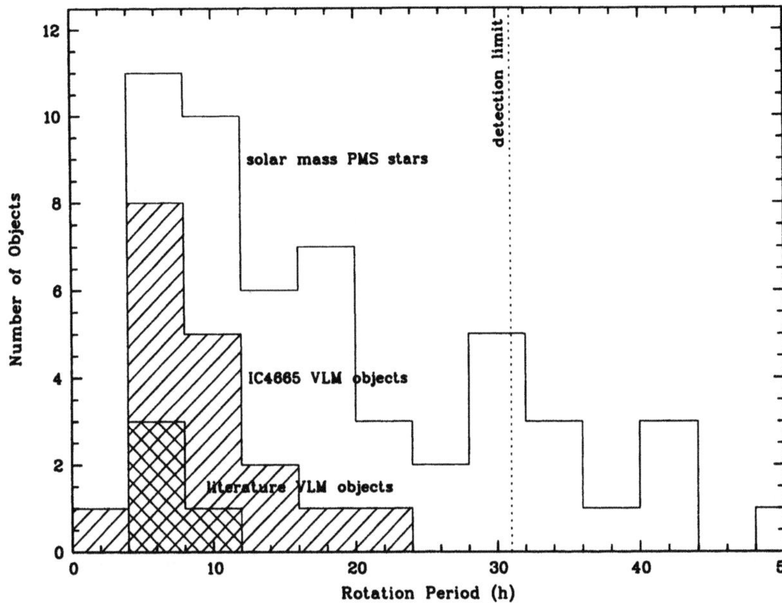

Fig. 3. Rotation period distribution: not hatched - solar mass stars in young open clusters with ages comparable with IC4665 (data taken from Allain et al. 1996, Barnes et al. 1999, Krishnamurthi et al. 1998, O'Dell et al. 1997, Patten & Simon 1996, Prosser et al. 1993, Prosser et al. 1995); single hatched - our VLM objects in IC4665; double hatched - VLM stars in the literature (Bailer-Jones & Mundt 1999, Martin & Zapatero-Osorio 1997, Terndrup et al. 1999)

starspots and at most a weak correlation between activity and rotation. These predictions are consistent with our results for the observed VLM objects.

In Fig. 3 we compare the rotation periods of VLM objects with those of solar mass stars of similar age. A Kolmogoroff-Smirnoff test reveals that the distribution for VLM objects is different from that for solar mass stars with a False Alarm Probability of 2%. It is especially remarkable that up to now no VLM object is known with a rotation period > 21 h, although our period search was sensitive to periods up to 31 h. This leads to the conclusion, that VLM objects rotate significantly faster than solar mass stars. Considering that a main mechanism to brake the rotation of stars probably is magnetic coupling with a circumstellar disk, this result can be understood with the above discussed implications about the magnetic activity of VLM objects. If magnetic field strengths on VLM objects are lower than on solar mass stars, the coupling mechanism works less efficient and consequently the objects are braked less strongly. On the other hand, these objects may have lost their disks at an early stage of their evolution (e.g. by evaporation through the strong UV-light of a nearby hot star), so they cannot be braked through the disk any more. At the same time, their matter reservoir is lost with the disk, so that mass accretion stops at this point. This would give a natural explanation why the lowest mass objects in a cluster rotate so fast.

In young and active stars, rotational variability is attributed to large magnetically induced star spots. This may be the case for the low mass stars in our sample as well. But what is the origin of the observed variability for our five Brown Dwarfs? As we mentioned in Sect. 3, they did not show evidence for magnetic activity in the form of flares. On the other hand, atmosphere models (Baraffe et al. 1998) tell us, that these objects have temperatures of $\approx$ 2650 K and that below 2800 K dust condensation will set in. Therefore, maybe the surface features that we see in the rotation periods of the lowest temperatures Brown Dwarfs are a kind of clouds, i.e. eddies of condensated dust in the atmospheres. Following this interpretation, our observations would be the first clear evidence for weather in extra-solar substellar objects.

References

1. S. Allain, J. Bouvier, C.F. Prosser, L.A. Marschall, B.D. Laaksonen: A&A **305**, 498 (1996)
2. C.A.L. Bailer-Jones, R. Mundt: A&A **348**, 800 (1999)
3. I. Baraffe, G. Chabrier, F. Allard, P.H. Hauschildt: A&A **3**, 403 (1998)
4. S.A. Barnes, S. Sofia, C.F. Prosser, J.R. Stauffer: ApJ **516**, 263 (1999)
5. P. Bodenheimer: ARA&A **33**, 199 (1995)
6. J. Bouvier, M. Forestini, S. Allain: A&A **326**, 1023 (1997)
7. G. Chabrier, I. Baraffe: A&A **327**, 1039 (1997)
8. B.R. Durney, D.S. DeYoung, I.W. Roxburgh: SoPh **145**, 207 (1993)
9. E.W. Guenther, M. Ball: A&A **347**, 508 (1999)
10. J.H. Horne, S.L. Baliunas: ApJ **302**, 757 (1986)
11. C. Koen: ApJ **348**, 700 (1990)
12. A. Krishnamurthi, et al.: ApJ **493**, 914 (1998)
13. E.L. Martin, M.R. Zapatero-Osorio: MNRAS **286**, 17 (1997)
14. J.C. Mermilliod: A&AS **44**, 467 (1981a)
15. J.C. Mermilliod: A&A **97**, 235 (1981b)
16. M.A. O'Dell, R.W. Hilditch, A. Collier Cameron, S.A. Bell: MNRAS **284**, 874 (1997)
17. B.R. Oppenheimer, S.R. Kulkarni, J.R. Stauffer: 'Brown Dwarfs'. In: *Protostars and Planets IV*. ed. by V.Mannings, A.P. Boss, S.S. Russell (Tucson: University of Arizona Press, 2000) pp. 1313
18. B.M. Patten, T. Simon: ApJS **106**, 489 (1996)
19. C.F. Prosser, et al.: PASP **105**, 1407 (1993)
20. C.F. Prosser, et al.: PASP **107**, 211 (1995)
21. I.N. Reid, S.L. Hawley: *New Light on Dark Stars* (Springer, Berlin, Heidelberg, New York 2000), pp. 163-208
22. D.H. Roberts, J. Lehar, J.W. Dreher: AJ **93**, 968 (1987)
23. J.D. Scargle: ApJ **263**, 835 (1982)
24. A. Skumanich: ApJ **171**, 565 (1972)
25. D.M. Terndrup, A. Krishnamurthi, M.H. Pinsonneault, J.R. Stauffer: AJ **118**, 1814 (1999)

Infrared Observation of Hot Cores *

Bringfried Stecklum[1], Bernhard Brandl[2], Markus Feldt[3], Thomas Henning[4], Hendrik Linz[1], and Ilaria Pascucci[4]

[1] Thüringer Landessternwarte Tautenburg, Sternwarte 5, D–07778 Tautenburg
[2] Center for Radiophysics & Space Research, Cornell University, Ithaca, NY 14853
[3] Max-Planck-Institut für Astronomie, Königstuhl 17, D–69117 Heidelberg
[4] Astrophysikalisches Institut und Universitäts–Sternwarte, Friedrich-Schiller-Universität Jena, Schillergäßchen 2–3, D–07745 Jena

Abstract. We report on mid-infrared imaging of hot cores performed with SpectroCam–10 and TIMMI2. The observations aimed at the detection of thermal emission presumably associated with the hot cores. Mid-infrared flux measurements are required to improve the luminosity and optical depth estimates for these sources. Results are presented for W3(H_2O), G9.62+0.19, G10.47+0.03, and the possible hot core candidate G232.620+0.996. They illustrate that the morphology of these sources cannot be described by simple geometries. Therefore, line-of-sight effects and considerable extinction even at mid-infrared wavelengths must not be neglected.

1 Introduction

Hot cores (HCs) are the suspected birthplaces of high-mass stars ($M \gtrsim 8\,M_\odot$) [9]. Based on mm/submm interferometric results which provide sufficient angular resolution to separate HCs from neighbouring ultracompact HII regions (UCHIIs), first attempts were made to derive properties of the embedded high-mass stars as well as the surrounding envelope using infall models in combination with 1D radiative codes [11]. However, these results relied on the mm/submm part of the spectral energy distribution (SED) only. We performed infrared (IR) observations of HCs to provide flux densities (or at least upper limits) for the atmospheric 10 and 20 μm spectral windows in order to complete the coverage of their SEDs with sub-arcsecond beam sizes and to investigate their morphology.

2 Observations and Data Reduction

The observations were carried out with SpectroCam–10 (SC 10) on the 5-m Hale[1] and TIMMI2 on the ESO 3.6-m telescopes. SC 10 is the Cornell-built 8–13 μm spectrograph/camera [6] which utilizes a Rockwell 128×128 Si:As BIB array, providing a circular field of view (FOV) in imaging mode of 16″ (pixel size

* Based on observations collected at the European Southern Observatory, La Silla, Chile, Proposal IDs 62.I-0530,67.C-0359

[1] Observations at the Palomar Observatory were made as part of a continuing collaborative agreement between the California Institute of Technology and Cornell University.

0.″25). The SC 10 imaging was performed in December 1998 and June 1999. The images were filtered using a wavelet algorithm [12] to enhance the signal-to-noise ratio (SNR). TIMMI2 is the new ESO thermal imaging multi-mode instrument covering the wavelength range from 5–20 μm [14]. Its 320×240 Si:As BIB array manufactured by Raytheon provides an unprecedented FOV in imaging mode of 64″ × 48″ (pixel scale 0.″2) at 10 and 20 μm and 96″ × 72″ (pixel scale 0.″3) at 4.7 μm. The observations were carried out in March 2001.

3 Results

3.1 $W3(H_2O)/W3(OH)$

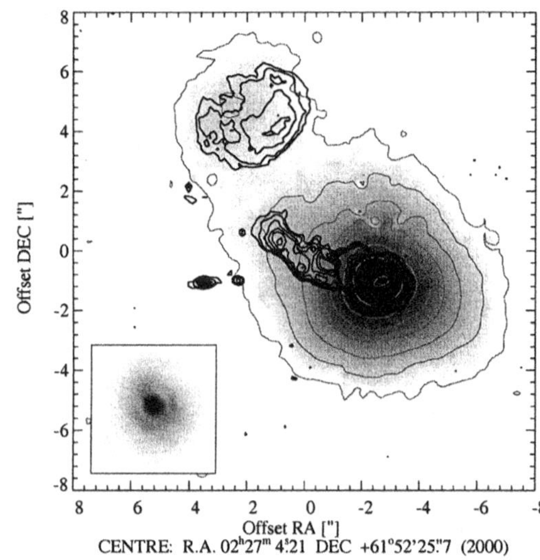

Fig. 1. SC 10 11.7 μm image of W3(OH) with contours of the 11.7 μm (grey) emission and of the 3.6 cm radio continuum (black, from [19]). $W3(H_2O)$ is located at the offset position [+3.″6,-1.″0]. The inset image of α Tau indicates the beam size

The HC $W3(H_2O)$ is located ~ 6″ east of the UCHII W3(OH) at a distance of 2.2 kpc. The proper motion of its H_2O masers [1] and the associated radio continuum jet [13] suggest recent outflow activity. Interferometric radio observations of $W3(H_2O)$ [21] yielded H_2 column densities of up to $1.5\times10^{24}\,cm^{-2}$ and rotation temperatures in the range of 160–200 K. The detection of the HC at mid-infrared (MIR) wavelengths was claimed by [8]. Fig. 1 shows the SC 10 image of $W3(H_2O)$/W3(OH) with black contours delineating the 3.6 cm radio continuum [19]. $W3(H_2O)$ is at the offset position [+3.″6,-1.″0] and traced by its radio continuum jet, but is not detected at 11.7 μm. The feature identified as $W3(H_2O)$ by [8] is presumably the weak cometary UCHII northeast of W3(OH). The radio position of the UCHII W3(OH) served as as astrometric reference.

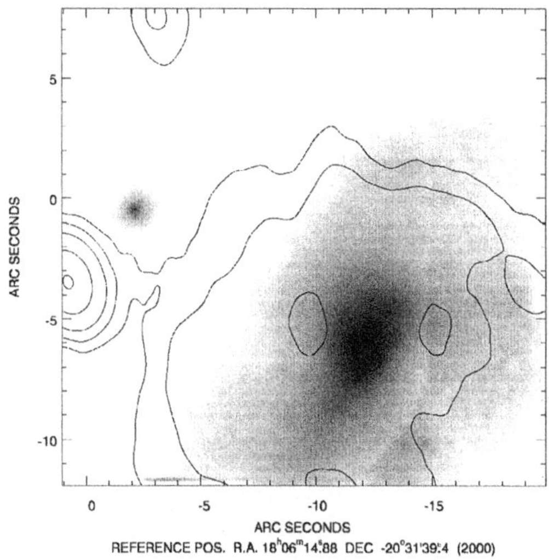

Fig. 2. SC 10 11.7μm image of G9.62+0.19 with contours of the 1.3 cm radio continuum (from [3])

The comparison of the spatial extent of W3(OH) with the standard star α Tau (inset of Fig. 1) shows that it is clearly resolved.

The temperature and size derived from molecular line interferometry for W3(H_2O) suggest that it should be a bright object in the MIR, with a predicted 11.7 μm flux density of $\sim$ 2000 Jy. Our failure to detect it despite a much better sensitivity (3σ point source detection limit of 6 mJy) than [8] implies a large amount of cold dust in front of the hot core. A detailed analysis of the SEDs of W3(H_2O)/W3(OH) is given in [15].

3.2 G9.62+0.19

The object G9.62+0.19 (IRAS18032-2032) comprises several UCHIIs at a distance of 5.7 kpc. Recently, weak radio continuum emission has been detected [17] at the location of the HC (component F) as well as outflow activity [7]. Remarkably, this object seems to be associated with 2.2 μm emission [16], a fact which can be hardly reconciled with spherical models of HCs. Such models imply extinction values which should prevent the detection of the HC at near-infrared (NIR) wavelengths. Fig. 2 shows the 11.7 μm SC 10 image of G9.62+0.19 with contours of the 1.3 cm radio continuum [3]. The bulk of the 11.7 μm emission stems from the extended radio component B which was used as astrometric reference. Surprisingly, no thermal emission was seen from the UCHII D which again implies a considerable optical depth even at MIR wavelengths for this source. This result concerning component D contradicts that of [5] due to differing astrometry. A pointlike source is obvious close to the location of the HC (offset position [-2.''0,-1.''0]). This emission is almost coincident with the 2.2 μm

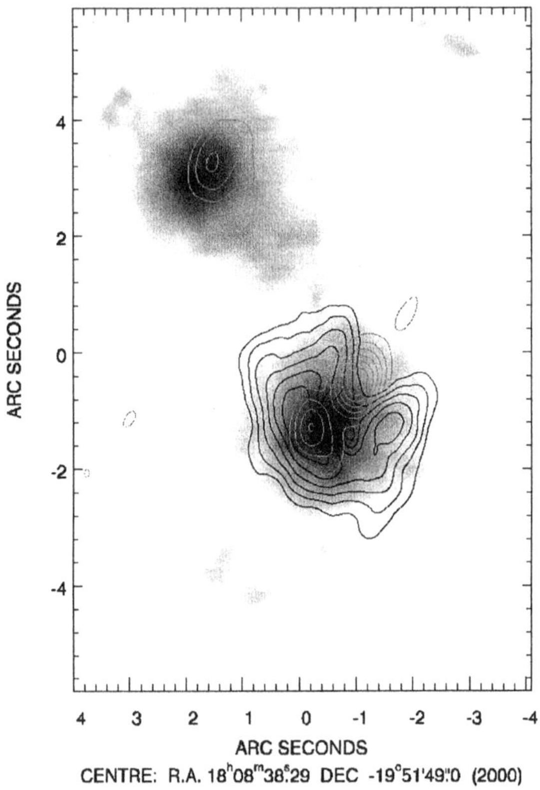

Fig. 3. SC 10 11.7μm image of G10.47+0.03 with contours of the 6 cm radio continuum (grey, from [20]) and of the $NH_3(4,4)$ line (black, from [4])

feature. The poster contribution of Linz et al. (this volume) discusses this source and the relation of the observed IR emission to the recently detected molecular outflow [7] in more detail. Possibly, IR radiation arising from the HC escapes in the outflow lobe inclined towards the observer where the optical depth is lower.

3.3 G10.47+0.03

According to the radial brightness temperature profile based on interferometric maps of the NH_3(4,4) line, the HC G10.47+0.03 (IRAS18056-1952) seemed to be one of the rare cases with *established* evidence for internal heating by embedded OB stars contrary to external heating by adjacent UCHIIs [4]. Two stars from the corresponding 2MASS images of the region were also detected with SC 10 at 11.7 μm. They provide astrometric reference with sub-arcsecond precision. MIR emission from two sites in the G10.47+0.03 region was detected. These sources are not seen on the 2.2 μm 2MASS image (10σ detection limit of 1.2 mJy). Fig. 3 displays the SC 10 11.7 μm image with contours of the 6 cm radio continuum (grey, from [20]) and of the $NH_3(4,4)$ emission (black, from [4]). The HC is located at the offset position $[0\farcs0,-1\farcs6]$. To the northwest of the HC, the NH_3 emission is diminished due to absorption by the UCHIIs in the foreground. The centroid of the 11.7 μm radiation almost coincides with the center of the NH_3 emission. A possible configuration which explains this morphology might be a cometary UCHII, similar to G29.96–0.02 and its adjacent HC (which are seen side-on), with the line of sight slightly inclined with respect to the symmetry axes of the HII region. The fact that there is only one MIR peak associated with the HC suggests that the multiple structure in the radio continuum (3 components are present in the map of [4]) is presumably due to locally enhanced plasma density and not caused by a few widely distributed high-mass stars. The lack of NH_3 emission for the second MIR source (offset position $[+1\farcs5,+3\farcs0]$) indicates that this is presumably a more evolved UCHII.

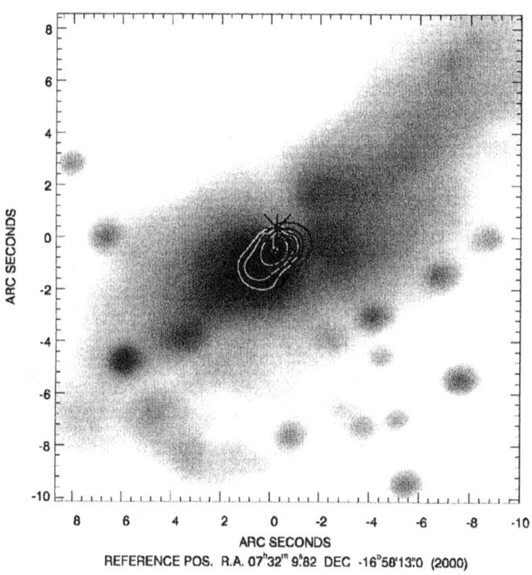

Fig. 4. 2.2μm SOFI image of G232.620+0.996 with contours of the 4.7 μm (white) and the 11.9 μm (black) emission. The reference position is defined by the location of the masers. The black asterisk marks the location of the illuminating source derived from our 2.2 μm polarization map

3.4 G232.620+0.996

The UCHII G232.620+0.996 (IRAS07299-1651) is associated with OH and CH_3OH masers [2],[18] which are signs of newly formed massive stars. The UCHII is $1\farcs5$ northwest of the maser position [18]. These objects are located at the southern border of a dense core seen in the 1.3 mm map taken with the SEST [10]. Our 2.2 μm polarimetric map obtained with SOFI at the ESO-NTT revealed that the bipolar-like extended emission (see Fig. 4) at the southern rim of the dense core is dominated by scattered light. The location of the illuminating source was derived by minimizing the sum of the scalar products between polarisation and radius vector as a function of source position. It is marked in Fig. 4 by the black asterisk, and situated very close to the masers and the UCHII. The large FOV of TIMMI2 covered the target and HD 60068 (situated 47″ northwest of it) simultaneously which allowed to establish accurate astrometry. A resolved IR source was found close to the maser position. Fig. 4 also displays contours of the 4.7 μm and 11.9 μm image taken with TIMMI2. It can be noticed that the emission peaks of both wavelengths are spatially offset, indicating a strong gradient in the optical depth in the immediate neighbourhood of the illuminating source.

Acknowledgements

This work was supported by DFG grants STE 605/17-1 and STE 605/18-1.

References

1. J. Alcolea, K.M. Menten, J.M. Moran, M.J. Reid: In: *Astrophysical Masers*, ed. by A.W. Clegg & G.E. Nedoluha (Berlin: Springer), 225 (1993)
2. J.L. Caswell: MNRAS **297**, 215 (1998)
3. R. Cesaroni, E. Churchwell, P. Hofner, C.M. Walmsley: A&A **288**, 903 (1994)
4. R. Cesaroni, P. Hofner, C.M. Walmsley, E. Churchwell: A&A **331**, 709 (1998)
5. J.M. DeBuizer, R.K. Piña, C.M. Telesco: ApJS **130**, 437 (2000)
6. T.L. Hayward, J.W. Miles, J.R. Houck, G.E. Gull, J. Schoenwald J.: In: *Infrared Detectors and Instrumentation*, ed. by A.W. Fowler, Proc. SPIE, 1946, 334 (1993)
7. P. Hofner, H. Wiesemeyer, Th. Henning: ApJ **549**, 425 (2001)
8. E. Keto, D. Proctor, R. Ball, J. Arens, G. Jernigan: ApJ **401**, L113 (1992)
9. S. Kurtz, R. Cesaroni, E. Churchwell, P. Hofner, C.M. Walmsley: In: *Protostars and Planets IV*, ed. by V. Mannings, A.P. Boss, S.S. Russell, University of Arizona Press, Tucson, 299 (2000)
10. R. Klein, priv. comm. (2001)
11. M. Osorio, S. Lizano, P. D'Alessio: ApJ **525**, 808 (1999)
12. E. Pantin, J.-L. Starck: A&AS **118**, 575 (1996)
13. M.J. Reid, A.L. Argon, C.R. Masson, K.M. Menten, J.M. Moran: ApJ **443**, 238 (1995)
14. H.G. Reimann, H. Linz, R. Wagner, H. Relke, H.U. Käufl, E. Dietzsch, M. Sperl, J. Hron: In: *Optical and IR Telescope Instrumentation and Detectors*, ed. by M. Iye & A.F. Moorwood, Proc. SPIE **4008**, 1132 (2000)
15. B. Stecklum, B. Brandl, T.L. Hayward, Th. Henning, I. Pascucci, J. Wilson: A&A submitted, (2001)
16. L. Testi, M. Felli, P. Persi, M. Roth: A&A **329**, 233 (1998)
17. L. Testi, P. Hofner, S. Kurtz, M. Rupen: A&A **359**, L5 (2000)
18. A.J. Walsh, M.G. Burton, A.R. Hyland, G. Robinson: MNRAS **301**, 640 (1998)
19. D.J. Wilner D.J., W.J. Welch, J.R. Forster: ApJ **449**, L73 (1995)
20. D.O.S. Wood, E. Churchwell: ApJS **69**, 831 (1989)
21. F. Wyrowski, P. Schilke, C.M. Walmsley, K.M. Menten: ApJ **514**, L43 (1999)

Achim Tieftrunk, Tyler Bourke, and Miguel Moreira

An Overview of Induced Star Formation Near the Surfaces of Molecular Clouds

Glenn J. White[1], R.P. Nelson[2], Monica Huldtgren White[3], C.V.M. Fridlund[4], René Liseau[3], Jingqi Miao[1], and Mark A. Thompson[1]

[1] School for Physical Sciences, University of Kent, Canterbury CT2 7NR, England
[2] Astronomy Unit, Queen Mary & Westfield College, University of London, England
[3] Stockholm Observatory, SE–133 36 Saltsjöbaden, Sweden
[4] Astrophysics Division, ESTEC, Noordwijk, The Netherlands

Abstract. The stability of molecular cloud surfaces against radiatively induced star formation is examined in Bright Rim Globules, and the Eagle Nebula.

1 Introduction

Material at the edges of dense molecular clouds is often irradiated by the ultraviolet light from nearby OB stars. We examine how excitation of the gas and dust in these Photon Dominated Regions (PDRs) influences their ionisation, chemistry, thermal, kinematic and chemical structures and determines the future star forming history.

Detailed theoretical models of their structure and physics have been formulated for the cases of radiatively driven implosion (RDI) [1], [5], and cometary globule models [9], [2]. Illumination by an external O or B star causes an *R*-type ionisation front to propagate rapidly through the gas to the Strömgren radius.

There are two distinct types of ionisation fronts, characterised by their propagation speed. *R*-type ('Rarified') ionisation fronts travel through a low density medium with a velocity of $u_{IF}>2c_i$, where c_i = sound speed of the ionised medium $\sim$ 11.4 km s^{-1}. *D*-type ('Dense') fronts travel through denser gas with a velocity of $u_{IF} < (c_i{}^2/2c_I)$, where c_I = the sound speed in the pre-ionised material $\sim \ll c_i$. An intermediate front is called *M*–type, and results in a shock being driven into the clump (RDI stage), compressing the gas until pressure equilibrium between the ionisation front and the cloud interior is attained. Under these circumstances, the gas just ahead of the ionisation front is compressed initially by the preceding shock wave so that an approximately *D*-critical ionisation front is able to form (i.e. $u_{IF} = c_I{}^2/2c_i$) behind the shock front which continues to compress the cloud ahead of it. The structure of the cloud during this implosion stage, moving from the cloud surface towards its interior, consists of a hot, photoevaporating, ionised region; an approximately *D*–critical ionisation front; a dense, neutral post-shock region; a shock front; and then a pre-shocked neutral region composed of the undisturbed gas in its original state. Provided that the implosion does not induce the cloud to collapse, the post-implosion cloud evolves slowly as the ionisation front slowly propagates into the cloud to form the familiar cometary nebula phase. In the objects we are studying (see Fig. 1),

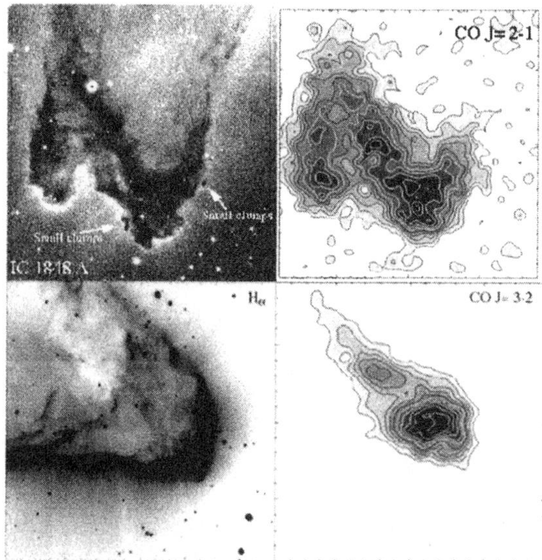

Fig. 1. Bright rim globules from the survey. The boxes are 3′ x 3′ in size

the ionised gas temperature $T_i \sim 10^4$ K, and sound speed $c_i \sim 11.4$ km s^{-1}. An *R*-type front encountering a dense clump, will stall, driving a shock into the clump (RDI) until it becomes sufficiently compressed for a steady *D*-type ionisation front to be maintained.

2 The Data

We have recently obtained JCMT and Onsala molecular line observations and JCMT SCUBA data to measure the column densities, masses, thermal and velocity structures of the gas and dust; optical CCD imaging and high resolution echelle spectroscopy from the Nordic Optical Telescope to estimate the excitation conditions in the ionised gas at the surfaces of the photo-ionised clumps; VLA continuum observations to study the conditions in the externally located ionised gas; ISOCAM CVF data [6] to measure the properties of the warm dust - for a sample of bright rim clouds selected from the catalogue of [7].

3 The Eagle Nebula

The Eagle Nebula, M16, is a prominent HII region lying $\sim$ 2 kpc from the Earth. Crossed by several opaque 'Elephant Trunks' or dense molecular 'fingers' [4], the cloud surface is illuminated by several O5 stars $\sim$ 2 pc to the NW, providing a total flux of $\sim 2\ 10^{50}$ photons s^{-1} [4] (Fig. 2).

Material is concentrated in cool cores at the tips of the fingers, which have masses $\sim$ 10 to 60 $M_\odot$, which are surrounded by warmer $\sim$ 60K gas. The fingers

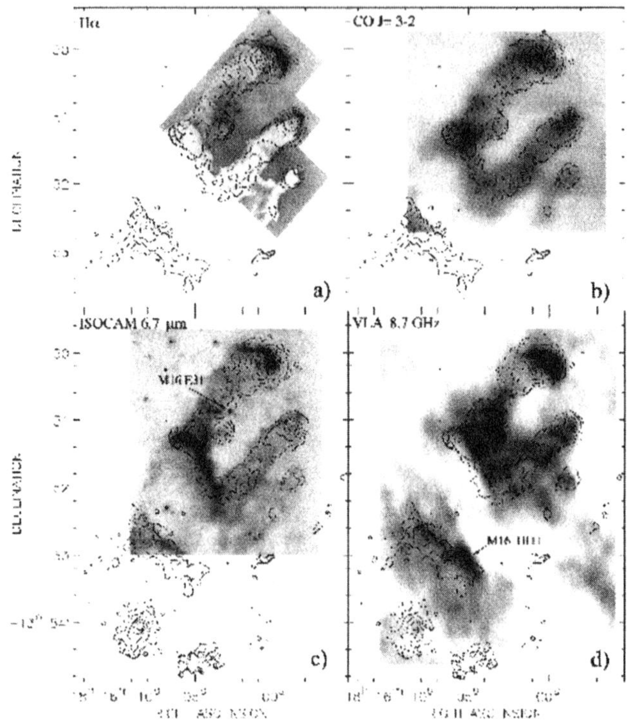

Fig. 2. The SCUBA 450 μm map shown as contours, overlaid onto a) the HST Hα image of [4], b) the CO J= 3-2 integrated emission, c) the ISOCAM 6.7 μm image, and d) the 8.7 GHz image. The contour levels on the 450 mm map are 5, 10, 20, 30, 40, 50, 60, 70 and 80 mJy per beam. M16 HH1 has an Herbig-Haro spectrum, and was the first reported submillimetre wavelength detection of an Herbig Haro object [8]

(integrated mass $\sim$ 200 $M_\odot$) contain several embedded submm continuum cores, which have $T_{dust} \sim 20$ K, compared to $T_{gas} \sim 60$ K in the fingers. The material *inside* these surface layers is compressed by the surrounding hot ionised gas.

4 Time Scales and Stability

One of the primary purposes of the work presented in this paper is to examine whether substantial star formation has occurred, or is likely to occur, in the fingers of the Eagle nebula. IR observations indicate that there are no young stellar objects embedded *inside* the dense fingertips.

The fingers have number densities typically $n(H_2) \sim 2\ 10^4$ cm^{-3}, but near the tips, the density increases to $n(H_2) \sim 2\ 10^5$ cm^{-3}. We adopt a simple model for a dense spherical clump at a fingertip, with a radius r = 0.085 pc, and a mass of M $\sim$ 31 $M_\odot$, illuminated by the radiation field described earlier. The computed density and temperature profiles are shown in Fig. 3.

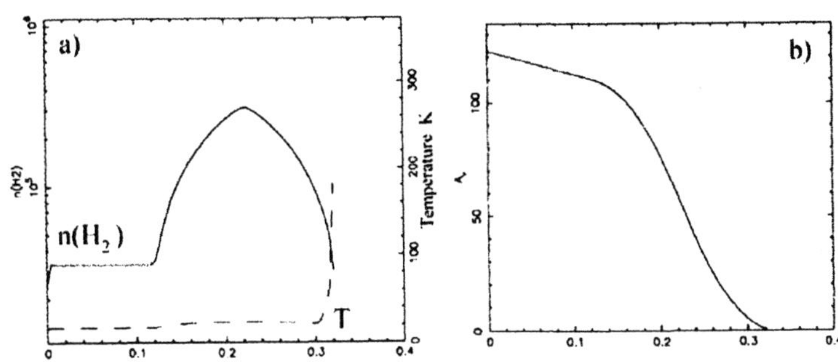

Fig. 3. a) the density and temperature, and b) the extinction along the symmetry z-axis of the modelled finger with the base of the finger on the left, and the tip at the right. The tip of the finger is located at z = 0.325 pc from the base of the model, and is shielded from the main radiation field by the core at the tip of the Finger [8]

If the pressures at the surface and just inside the dense gas are approximately equal, then the fingers will have already been compressed by an ionising-shock front (IS-front), leading to a steady *D*-type ionisation front at the surface, which will slowly eat into the cloud. If however the external pressure exceeds the internal pressure, then an IS-front will be driven into the molecular gas.

The line widths of the $C^{18}O$ spectra are $< \Delta v > = 2$ km s^{-1}. The pressure inside the cloud is composed of turbulent and thermal contributions:

$$P_{int} = \sigma^2 \rho + \frac{\Re}{\mu} \rho T \tag{1}$$

where T is the temperature, ρ is the density, μ is the mean molecular weight, and $\Re$ is the gas constant. Assuming low temperature cores, then the thermal component of the line-width is negligible and $\sigma^2 = < \Delta v > 2 \, / \, (8 \, ln \, 2)$ represents the square of the inferred turbulent velocity in the cloud. Both the observations and the thermal and chemical modelling, indicate that the temperature in the interior of the cloud is ~ 15 - 20 K and the average number density $n(H_2) \sim 2 \; 10^4$ cm^{-3} However the density in the clumps is estimated to be somewhat larger, with $n(H_2) \sim 2.2 \; 10^5$ cm^{-3}. We take the upper value of the internal density to be $n(H_2) \sim 2 \; 10^5$ cm^{-3}, giving an estimate for the total internal pressure of $P_{int} \, / \, k = 3.5 \; 10^7$ cm^{-3} K. The thermal pressure at finger tips, estimated by [4], is $P_i \, / \, k = 6 \; 10^7$ cm^{-3} K – almost double the internal pressure. We therefore conclude that equilibrium does not exist between the cloud surface and the interior, and that an IS-front is currently propagating into the column. The conservation conditions across a *D*-critical ionisation front predict that the pressure just ahead of the front is twice the thermal pressure just behind it (i.e. $P_n = 2 \, P_i$, where the subscripts are n for 'neutral' and i for 'ionised'. This arises because ionised material leaves the front at a velocity $v_i = c_i$.

If pressure equilibrium has in fact been attained between the ionisation front and the cloud interior, and the additional internal pressure required is provided by a magnetic field, then the field required is:

$$B = \sqrt{8\pi\,(P_{ext} - P_{int})} = 5.4 \quad 10^{-4}\ \mathrm{G} \tag{2}$$

assuming equipartition between the magnetic and kinetic energies, so that the turbulent line widths should be Alfvénic. If a magnetic field is responsible for providing an internal pressure able to balance the pressure at the ionisation front, then equipartition could not hold, since the Alfvén speed corresponding to a field of 5.4 10^{-4} G is $v_A = 1.9$ km s^{-1}. Alfvénic motions would then lead to an observational linewidth of:

$$\langle \Delta v \rangle = \sqrt{8 \ln 2 v_A^2} = 4.4\ \mathrm{km\ s^{-1}} \tag{3}$$

or about double that of $C^{18}O$ lines seen towards the finger tip cores.

The large scale magnetic field pressure is anisotropic, providing a stress perpendicular to the field direction. Alfvén waves are however able to provide an isotropic pressure:

$$P_\omega = \frac{\delta B^2}{8\pi} \tag{4}$$

where δB is the perturbation to the mean magnetic field associated with the Alfvén wave. The fluid velocity perturbation of the travelling wave is:

$$\delta v = \frac{\delta B}{\sqrt{4\pi\rho}} \quad \text{and hence} \quad \frac{\delta \nu}{\nu_A} = \frac{\delta B}{B} \tag{5}$$

where B is the mean magnetic field strength, and v_A is the Alfvén speed associated with the mean field. For $\delta v << v_A$, the mean field dominates over the random field component, and internal stresses provided by the magnetic field are strongly anisotropic. The observations of the fingers indicate a substantial degree of cylindrical symmetry – we conclude that a large-scale ordered magnetic field does not produce an internal pressure that can balance the pressure of the ionisation front. A magnetic field in which the disordered component was comparable to the ordered component would yield internal motions $\sim \nu_A$ since:

$$\frac{\delta B}{B} = \frac{\delta v}{\nu_A} \cong 1 \tag{6}$$

These motions would lead to linewidths $< \nu > \sim 4.4$ km s^{-1}, which are not seen in the data, thus the internal pressure in the fingers of the Eagle nebula appears to be insufficient to balance the pressure at the ionisation front – implying that an ionisation-shock front is currently being driven into the cloud.

The presence of a shock front currently propagating into the columns indicates that the dense clumps located towards their tips have probably not formed from a radiatively driven implosion, and were more likely part of a larger, dense structure that pre-existed the expansion of the HII region.

The shock propagation velocity may be derived from the usual shock discontinuity jump conditions, leading to the equation

$$V_s^2 = \frac{(P_{sn} - P_n)}{\rho_n} \left(1 - \frac{\rho_n}{\rho_{sn}}\right)^{-1} \tag{7}$$

where P_{sn} and ρ_{sn} are the pressure and density of the shocked, neutral material and P_n and ρ_n are the pressure and density of the pre-shocked neutral material. If we assume that $2 \leq \rho_{sn}/\rho_n \leq \infty$ then:

$$V_s^2 = \alpha \frac{(P_{sn} - P_n)}{\rho_n} \tag{8}$$

where $1 < \alpha < 2$, so that the maximum error that we incur in our estimate of V_s, as a result of guessing the value of ρ_{sn}, is a factor of $\sqrt{2}$. We take $P_{sn} = 2P_i$ and $\rho_n = 2\ 10^5\ m(H_2)$, where $m(H_2)$ is the mass of a hydrogen molecule, which then leads to a shock velocity of $V_s \approx 1.3$ km s^{-1}. The time-scale for this shock to propagate through the top 0.2 pc of the finger tip is then $\tau_{sh} = 0.2$ pc / $V_s \approx 1.5\ 10^5$ years.

This shock crossing time is considerably shorter that the estimated ages of the O-stars in the Eagle nebula (i.e. $\sim$ 1 Myr), indicating that the structures being observed now have only been exposed to the ionising radiation of the nearby stars for a relatively short time (an upper limit for the time taken for an R-type ionisation front to reach the elephant trunks during the initial expansion of the Strömgren sphere is given by $\tau_R = (2$ pc / $2\ c_i\ ) = 8.8\ 10^4$ years (i.e. a relatively short time after the switch-on of the O stars)).

The cores appear to be at a very early stage of *pre-protostellar* development: there are no embedded infrared sources or molecular outflows present. The pressure inside the cores is just less than that of the surrounding gas, allowing them to be compressed by the external pressure. The cores are probably just starting the final stages of collapse, which will lead to the formation of a condensed, warm object. *The cores in the tips of the Eagle Nebula's fingers have characteristics similar to those expected to occur in the earliest stages of protostellar formation.*

References

1. Bertoldi, F: ApJ 346, 735, (1989)
2. Bertoldi, F. & McKee, C.: ApJ 354, 529 (1990)
3. L. Hillenbrand et al: AJ, 1906, 106 (1993)
4. J. Hester et al: Astron. J., 111, 2349 (1996)
5. B. Lefloch, & B. Lazareff: A&A 289, 559 (1994)
6. G. L. Pilbratt et al: A&A 333, L9 (1998)
7. K. Sugitani et al: ApJS 92, 163 (1994)
8. G.J. White, R. P. Nelson et al: A&A 342, 233 (1999)
9. G.J.White, B. Lefloch et al: A&A 323, 931 (1997)

Ices in Star-Forming Regions: First Results from VLT-ISAAC

Ewine F. van Dishoeck[1,4], E. Dartois[2], W.F. Thi[1], L. d'Hendecourt[2], A.G.G.M. Tielens[3], P. Ehrenfreund[4], W.A. Schutte[4], K. Pontoppidan[1], K. Demyk[2], J. Keane[3], and A.C.A. Boogert[5]

[1] Leiden Observatory, P.O. Box 9513, NL-2300 RA Leiden, The Netherlands
[2] 'Astrochimie Expérimentale', Université Paris XI, Bat. 121, F-91405 Orsay, France
[3] Kapteyn Institute/SRON Groningen, P.O. Box 800, NL-9700 AV Groningen, The Netherlands
[4] Raymond & Beverly Sackler Laboratory for Astrophysics, Leiden Observatory, P.O. Box 9513, NL-2300 RA Leiden, The Netherlands
[5] Div. of Physics, Mathematics & Astronomy, Caltech, Pasadena, USA

Abstract. The first results from a VLT-ISAAC program on the infrared spectroscopy of deeply-embedded young stellar objects are presented. The advent of 8-m class telescopes allows high S/N spectra of low-luminosity sources to be obtained. In our first observing run, low- and medium-resolution spectra have been measured toward a dozen objects, mostly in the Vela and Chamaeleon molecular clouds. The spectra show strong absorption of H_2O and CO ice, as well as weak features at '3.47' and 4.62 μm. No significant solid CH_3OH feature at 3.54 μm is found, indicating that the CH_3OH/H_2O ice abundance is lower than toward some massive protostars. Various evolutionary diagnostics are investigated for a set of sources in Vela.

1 Introduction

Interstellar matter provides the basic building blocks from which new solar systems like our own are made. The formation of stars and planetary systems begins with the collapse of a dense interstellar cloud core, a reservoir of dust and gas from which the protostar and circumstellar disk are assembled. In this cold and dense phase, molecules freeze-out onto the grains and form an icy mantle surrounding the silicate and carbonaceous cores. The ices can contain up to 40% of the condensible material (i.e., C and O) (Whittet et al. 1998, d'Hendecourt et al. 1999). Much of this material is incorporated into the circumstellar disks and ultimately in icy solar system bodies such as comets (Ehrenfreund et al. 1997). A central quest in star formation and astrochemistry is to understand the evolution of these species from interstellar clouds to planetary bodies, and use them as diagnostic probes of the thermal history and physical processes (van Dishoeck & Blake 1998, Ehrenfreund & Charnley 2000, Langer et al. 2000).

In recent years, much progress has been made in our understanding of the chemical evolution during star formation through combined submillimeter and infrared observations. Submillimeter observations using telescopes such as the JCMT, CSO, IRAM 30m and SEST probe the gas-phase composition of the warm and dense envelopes around deeply-embedded protostars. Many different species with abundances down to 10^{-11} with respect to H_2 can be probed through their pure rotational transitions. Because of the high spectral resolution of the heterodyne technique ($R = \lambda/\Delta\lambda > 10^6$), the line profiles are resolved and provide information on the location of the molecules (e.g., quiescent gas vs. outflow). Since the lines are in emission, a map of their distribution can be made.

Infrared spectroscopy provides complementary information: at these wavelengths the vibrational modes of both gas-phase and solid-state species can be observed, but only down to abundances of $\sim 10^{-7}$ with respect to H_2 at typical resolving powers $R \approx$ a few thousand. Symmetric molecules, such as H_3^+, CH_4, C_2H_2 and CO_2 have no dipole-allowed rotational transitions, and can therefore only be probed through their strong infrared vibrational transitions. Moreover, PAH emission features appear throughout the mid-infrared range, and the dominant interstellar molecule, H_2, has its pure rotational transitions at these wavelengths. The cold gases and ices are usually observed in absorption toward an embedded YSO, where the hot dust in the immediate circumstellar environment provides the continuum. This technique samples only a pencil beam line of sight. The *Short Wavelength Spectrometer* (SWS) on the *Infrared Space Observatory* (ISO) has opened up the mid-infrared wavelength region and has obtained spectra of more than a dozen protostars without atmospheric interference (see van Dishoeck & Tielens 2001 for review). However, ISO was limited to the most luminous, massive YSOs ($L \approx 10^4 - 10^5$ $L_\odot$). A major goal of our VLT-ISAAC program is to extend this work to lower-mass objects ($L \leq 10^3$ $L_\odot$) representative of our proto-Sun.

Together, the submillimeter and infrared data have led to the following scenario for high-mass objects. In the cold pre-stellar cores and collapsing envelopes, gas-phase molecules freeze-out onto the grains and form an icy mantle. Here the abundances can be further modified by grain surface reactions and, perhaps, photoprocessing of ices. In particular, the hydrogenation and oxidation of accreted C, O, N and CO can lead to CH_4, H_2O, NH_3, H_2CO, CH_3OH and CO_2, respectively (Tielens & Charnley 1997). Most of these species have been firmly identified in interstellar ices (e.g., Whittet et al. 1996, d'Hendecourt et al. 1996, Gibb et al. 2000). Once the protostar has formed, its luminosity can heat the surrounding grains to temperatures at which the ices evaporate back into the gas phase, resulting in enhanced gas/solid ratios (e.g., van Dishoeck et al. 1996, Dartois et al. 1998, Boonman et al. 2000). The sublimation temperatures range from ~20 K for pure CO ice to ~90 K for H_2O-rich ice under typical conditions. These freshly evaporated molecules can then drive a rich and complex chemistry in the gas (called the 'hot core' phase) until the normal ion-molecule chemistry takes over again after $\sim 10^5$ yr (Charnley et al. 1992). The different solid-state and gas-phase species therefore serve not only as physical diagnostics, but also as probes of the evolution of the region.

2 Our VLT-ISAAC Program

In 1999, we proposed a large VLT-UT1 ISAAC program to probe the origin and evolution of ices in southern star-forming regions through a spectroscopic survey of 30 – 40 objects in the 2.7–5.1 μm L and M-band atmospheric windows. The aim was to obtain high-quality spectra ($S/N > 50$ on continuum) such that 3σ limits of species with abundances down to 2–4% of H_2O ice — the dominant ice component — can be obtained. The program focuses on low- and intermediate-mass southern YSOs, but covers a range of evolutionary stages from background stars to T-Tauri stars with disks. Also, several different environments (Vela, Chamaeleon, Ophiuchus, Corona Australis, ...) will be probed. The program was awarded 14 nights of VLT time. However, due to technical problems with the long-wavelength arm of ISAAC, the first observing session did not take place until January 2001.

The January 2001 observations totalled 5 nights and were carried out under mediocre conditions with high humidity. Nevertheless, good low-resolution (LR, $R = 600 - 800$) spectra were obtained for $\sim$ 15 objects and medium-resolution (MR, $R = 3000 - 5000$) spectra for $\sim$ 5 objects, covering $\sim$one-third of our project. With the new 1024×1024 Aladdin array, the low-resolution spectra can be obtained in a single spectral setting per atmospheric window, whereas the medium resolution spectra require 6/3 settings to cover the entire L/M band, respectively. Typical exposure times are $\sim$ 30 minutes per atmospheric window in LR for a L$\approx$ 7 mag object. Standard stars were observed immediately before or after the YSOs, within 0.05–0.1 airmass. The spectra were obtained using both chopping (by 15″ along the slit) and nodding of the telescope. Daily arcs and flat-fields were provided by the observatory staff. In the M-band, the atmospheric CO lines were used for wavelength calibration. The data were reduced using IDL routines developed in-house by E. Dartois and W.F. Thi.

Because the ice features are weak and broad and are superposed on a strong continuum, many tests were carried out to check the reliability of the spectra. Specifically, spectra were taken for objects also observed with ISO or UKIRT, and features were checked on different nights at the VLT and between the LR and MR modes on the VLT. In general, the reproducibility of the features is excellent (see Fig. 1).

A particularly nice capability of ISAAC is the possibility to rotate the slit on the sky to obtain spectra of more than one object simultaneously. In several cases, more than one bright object at L or M-band was discovered in the acquisition image within $\sim$ 15″. The slit was then rotated to include the additional object. An example is shown in Fig. 2 for the high-mass source GL 961, where the E and W components are separated by $\sim$5.5″.

Laboratory data such as those obtained in the Sackler Laboratory for Astrophysics in Leiden and at the Astrochimie Expérimentale Laboratory in Paris play an essential role in the analysis of the infrared spectra. First, they lead to definite identification of the molecules and to quantitative estimates of their abundances. Second, they provide an indication of the ice environment and ice components. For example, a CO molecule embedded in an H_2O-rich matrix ('polar ice') has a

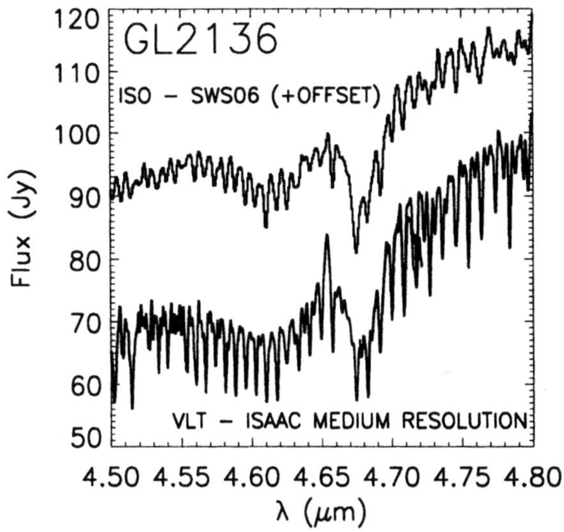

Fig. 1. VLT-ISAAC MR M-band $R \approx 2000$ spectrum of GL 2136 compared with the ISO-SWS spectrum ($R \approx 1500$). The sharp features are due to gas-phase CO, and become stronger in the higher-resolution VLT spectrum. Broad absorption features at 4.62 μm due to solid OCN^- and at 4.67 μm due to solid CO are seen as well. The H I 7–5 emission line at 4.6538 μm is stronger in the VLT spectrum due to the smaller slit

different spectral shape compared with that in a CO-rich mantle ('apolar ice'). Such analyses have shown that the ice mantles in the protostellar environment are not homogeneous, but consist of several components. These phases may reflect differential accretion of atomic H-rich versus H-poor gas and/or different degrees of outgassing of the more volatile species (Schutte 1999). Third, the ice spectra and the gas/solid ratios can provide clear evidence for heating of the ices in the more evolved objects (Ehrenfreund et al. 1998, Boogert et al. 2000, van Dishoeck et al. 1996).

3 Initial Results

The prime targets in the January 2001 run were a set of low- and intermediate mass YSOs in the Vela and Chamaeleon clouds, together with a few well-known southern high-mass protostars. In the first case, the power of the VLT+ISAAC is used to observe weaker and lower luminosity sources than possible previously, whereas for the high-mass sources the aim was to obtain higher S/N and higher spectral resolution data than provided by ISO or previous ground-based observations.

The L- and M-band windows include the following main features: 3 μm (H_2O ice), 3.47 μm (unidentified), 3.54 μm (CH_3OH ice), 4.08 μm (HDO ice), 4.62 μm

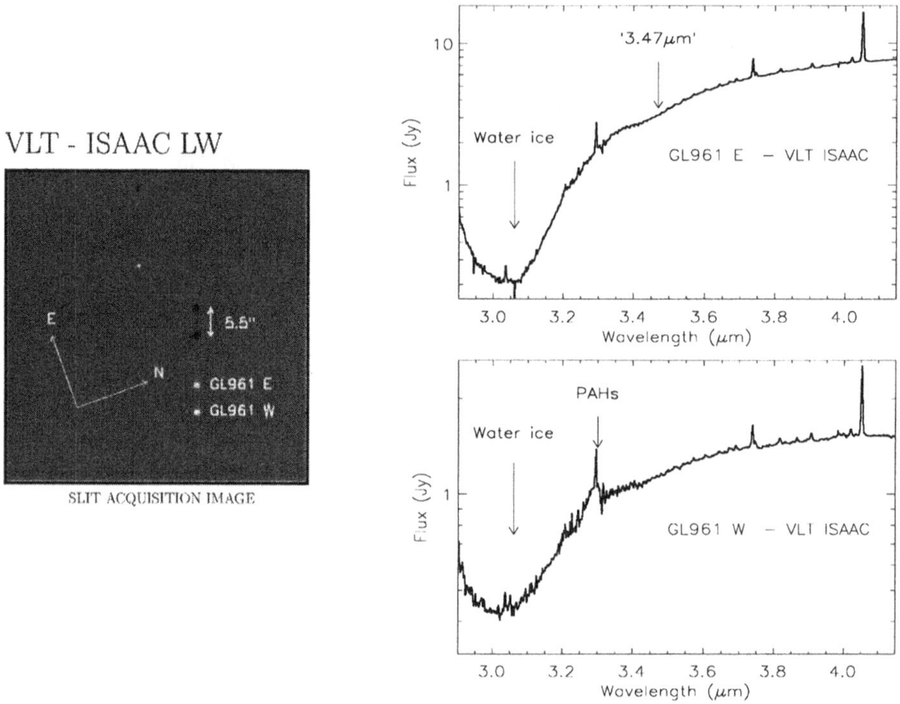

Fig. 2. VLT-ISAAC spectra obtained toward GL 961 E and W. The 3.3 μm PAH feature may be affected by poor cancellation of an atmospheric feature

(OCN^- ice) and 4.67 μm (CO gas and ice). The strong solid H_2O and CO bands are detected in most objects, although with varying amounts. Typical LR and MR spectra are shown in Figures 1–4. Because of the strong atmospheric features at the relatively low altitude of Paranal, the quality of the data around the solid OCS feature at 4.9 μm is low.

Several of these features are excellent indicators of the thermal history and energetic processing of the ices. Specifically, we can use the following diagnostics of the physical conditions and evolution of our sources: (i) the solid H_2O profile, where the peak position gives an indication of the ice temperature; (ii) the solid CO profile, where the shape indicates the 'apolar' vs. 'polar' ice fraction; (iii) the CO/H_2O abundance ratio, with the more volatile CO molecule having a lower ice abundance at high temperature; (iv) the gas/solid CO ratio and the gas-phase CO excitation temperature; and (v) the presence of the OCN^- feature, which is thought to be a tracer of energetic processing (ultraviolet irradiation or particle bombardment) (Schutte & Greenberg 1997, Demyk et al. 1998). In the following, a few specific initial results are presented.

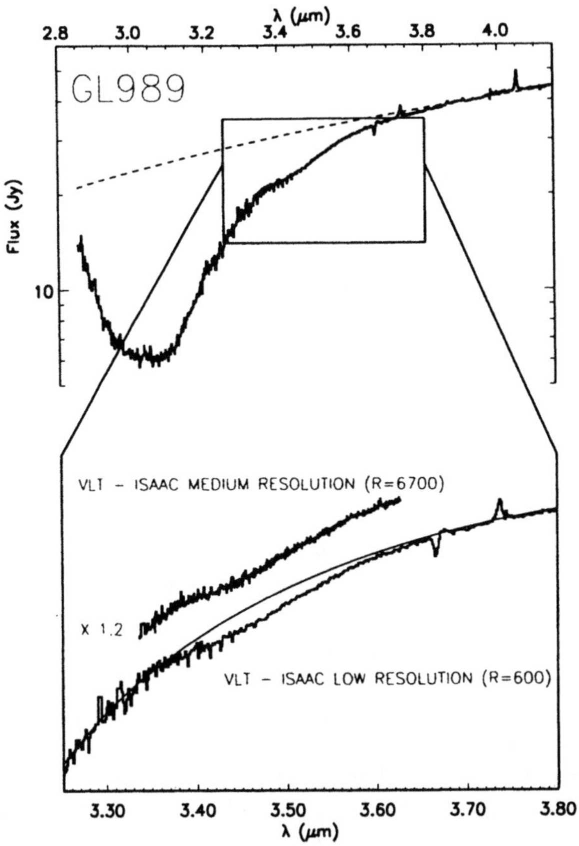

Fig. 3. VLT-ISAAC LR and MR L-band spectra obtained toward GL 989. Note the lack of substructure in the 3.47 μm feature (Dartois et al., in prep)

3.1 The 3.47μm Feature

The 3.47 μm feature, previously detected by Allamandola et al. (1992) and Brooke et al. (2000), is observed in several of our sources at high S/N. The comparison of the LR and MR spectra for the bright source GL 989 shows good agreement in the shape of the spectra (Figure 3). Also, no substructure is apparent at the higher spectral resolution. The data are currently being compared with different laboratory ice mixtures, in particular mixtures involving H_2O and NH_3 (Dartois et al., in prep.). Theoretical models predict that NH_3 is an important component of interstellar ices formed by hydrogenation of atomic N, but unfortunately the strongest NH_3 bands are blended with H_2O at 3 μm and with the silicate band at 9.6 μm. A tentative detection of the 9.6 μm feature toward one (northern) object has recently been claimed by Lacy et al. (1998), suggesting high NH_3 abundances up to 10% with respect to H_2O ice. The analysis of our VLT data indicates lower NH_3 abundances.

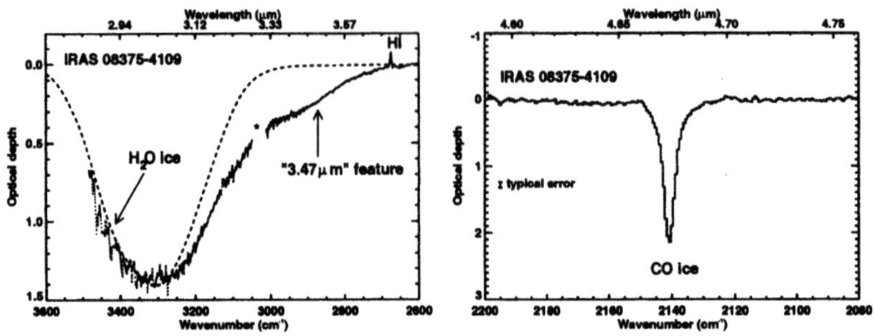

Fig. 4. VLT-ISAAC LR L- and M-band spectra obtained toward IRAS 08375 –4109 in the Vela molecular cloud (Thi et al., in prep)

3.2 Vela Sources

LR L- and M-band spectra of 5 intermediate mass YSOs ($\sim 300 - 700\ L_{\odot}$) have been obtained in the previously unexplored Vela molecular cloud (Thi et al., in prep.). The objects were selected from the list of 'class I' objects of Liseau et al. (1992), based on their spectral energy distribution. For two sources, an additional object was found in the field within a few ″, providing 'off source' information on the ices. MR spectra in the region of the solid and gas-phase CO band have been taken as well for a few cases.

H_2O ice has been detected in all objects, and CO ice in half of the objects. The H_2O profile indicates that the bulk of the ice is very cold. The solid CO band toward IRAS 08375 –4109 is one of the strongest and sharpest CO bands observed in any source: at LR, the feature is unresolved and its true depth can only be obtained from the MR spectrum (Figure 4). In spite of the low overall temperature, clear differences in the solid CO/H_2O ice abundances are observed, which correlate with the bolometric temperature of the source.

Several of the sources show the 3.47 μm feature, but the presence of the 3.54 μm band is less clear. This limits the solid CH_3OH abundance to a few % of that of H_2O ice, significantly less than the 50% found toward some massive protostars (Dartois et al. 1999). The OCN^- feature is detected in at least one high temperature source.

3.3 The Circumstellar Disk Around L1489

L1489 has been shown by Hogerheijde (2001, this volume) to be a transitional object between the class I and II phases. It is surrounded by a large 2000 AU radius rotating circumstellar disk, which must be on the verge of shrinking to the $\sim$ 100 AU size disks seen around T Tauri stars. Thus, it provides an excellent opportunity to probe the chemical composition of the gas and dust just when it is being incorporated into the disk. The VLT spectra of L1489 show strong absorption by H_2O and CO ices, but no evidence for CH_3OH and OCN^-

features, providing limits on their abundances of a few % with respect to H_2O ice.

In summary, the initial data show that the VLT-ISAAC is a powerful instrument to obtain high-quality 2.9–5 μm spectra of low-luminosity embedded YSOs, and that such data can provide an important step forward in our understanding of the physical and chemical evolution of ices in low-mass young stellar objects and their incorporation into new planetary systems.

References

1. Allamandola L.J., Sandford S.A, Tielens A.G.G.M., Herbst T.M.: ApJ **399**, 134 (1992)
2. Boogert A.C.A. et al: A&A **353**, 349 (2000)
3. Boonman, A. et al.: In ISO beyond the peaks, eds. A. Salama et al., ESA-SP 456, p. 67 (2000)
4. Brooke, T.Y., Sellgren, K., Geballe, T.R.: ApJ **517**, 883 (1999)
5. Charnley S.B., Tielens A.G.G.M., Millar T.J.: ApJ **399**, L71 (1992)
6. Dartois E., d'Hendecourt L., Boulanger F. et al.: A&A **331**, 651 (1998)
7. Dartois E., Schutte W.A., Geballe T.R. et al.: A&A **342**, L32 (1999)
8. Demyk K., Dartois E., d'Hendecourt L. et al.: A&A **339**, 553 (1998)
9. d'Hendecourt L., Jourdain de Muizon M., Dartois E. et al.: A&A **315**, 365 (1996)
10. d'Hendecourt L., Jourdain de Muizon M., Dartois E. et al.: In *The Universe as seen by ISO*, ESA-SP (1999)
11. Ehrenfreund, P., Charnley, S.: ARA&A **38**, 427 (2000)
12. Ehrenfreund P., d'Hendecourt L., Dartois E. et al.: Icarus **130**, 1 (1997)
13. Ehrenfreund P., Dartois E., Demyk K., d'Hendecourt L.: A&A **339**, L17 (1998)
14. Gibb, E. et al.: ApJ **536**, 347 (2000)
15. Hogerheijde M.R.: ApJ **553**, 618 (2001)
16. Lacy J.H., Faraji H., Sandford S.A., Allamandola L.J.: ApJ **501**, 105 (1998)
17. Langer W.D., van Dishoeck E.F., Blake G.A. et al.: In *Protostars & Planets IV*, eds. V. Mannings et al. (Univ. Arizona), p. 29 (2000)
18. Liseau R., Lorenzetti D., Nisini B., Spinoglio L., Moneti A.: A&A **265**, 577 (1992)
19. Schutte W.A. In *Solid Interstellar Matter: The ISO Revolution*, eds. L. d'Hendecourt, C. Joblin, and A. Jones (EDP, Springer), p. 183 (1999)
20. Schutte W.A., Greenberg J.M.: A&A **317**, L43 (1997)
21. Tielens A.G.G.M., Charnley S.B.: Origin of life and evol. of biosphere **27**, 23 (1997)
22. van Dishoeck E.F. et al.: A&A **315**, L349 (1996)
23. van Dishoeck E.F., Blake G.A.: ARA&A **36**, 317 (1998)
24. van Dishoeck E.F., Tielens, A.G.G.M.: to appear in *The Century of Space Sciences*, eds. J. Bleeker et al. (Kluwer, Dordrecht) in press
25. Whittet D.C.B., Schutte W.A., Tielens A.G.G.M. et al.: A&A **315**, L357 (1996)
26. Whittet D.C.B. et al.: ApJ **498**, L159 (1998)

Tyler Bourke, Achim Tieftrunk, and David Wilner

The Spectacular BHR 71 Outflow

Tyler L. Bourke

Harvard-Smithsonian Center for Astrophysics,
60 Garden Street MS 42, Cambridge MA 02138, USA.

Abstract. BHR 71 is a well isolated Bok globule located at ~200 pc, which harbours a highly collimated bipolar outflow. The outflow is driven by a very young Class 0 protostar with a luminosity of ~9 $L_{\odot}$. It is one of a very small number that show enhanced abundances of a number of molecular species, notably SiO and CH_3OH, due to shock processing of the ambient medium. In this paper the properties of the globule and outflow are discussed.

"In the darkness, there'll be hidden worlds that shine"
– Bruce Springsteen, Candy's Room 1977

1 Introduction

In 1977 Arge Sandqvist published a catalogue of southern "dark dust clouds of high visual opacity" (Sandqvist 1977 – 95 entries, numbered 101-195), an extension of an earlier paper with Lindroos (Sandqvist & Lindroos 1976) in which they presented H_2CO absorption line studies of 42 dust clouds (#1-42). Number 136 on Sandqvist's list (Sa 136) is a very opaque Bok globule located near the Coalsack, later catalogued as DC 297.7-2.8 by Hartley et al. (1986) and as entry 71 in the globule list of Bourke, Hyland & Robinson (1995a – BHR 71). Mark McCaughrean in his opening address at this conference highlighted a number of important events that occurred in 1977, in particular IAU Symposium 75 on Star Formation whose proceedings appeared that year. It is fitting that BHR 71, which is featured in a beautiful VLT optical image in the frontpiece (& poster) of these proceedings, can trace its origins in the literature to that same year.

2 Globule Properties

The globule properties have been determined by Bourke et al. (1995b, 1997). Spatially and kinematically BHR 71 is associated with the Coalsack, at an assumed distance of 200 pc though it may be as close as 150±30 pc (Corradi et al. 1997). Large scale ^{12}CO & ^{13}CO maps of the globule give a size of ~0.5 pc and mass $40M_{\odot}$, while C^{18}O observations which trace high column density gas imply a size $\sim 0.3 \times 0.15$ pc and a mass of $12M_{\odot}$. The high density (n> 10^4 cm^{-3}) core traced in ammonia is $\sim 0.2 \times 0.1$ pc in size with a mass of $3M_{\odot}$. The globule velocity is $V_{lsr} \sim -4.5$km s^{-1}.

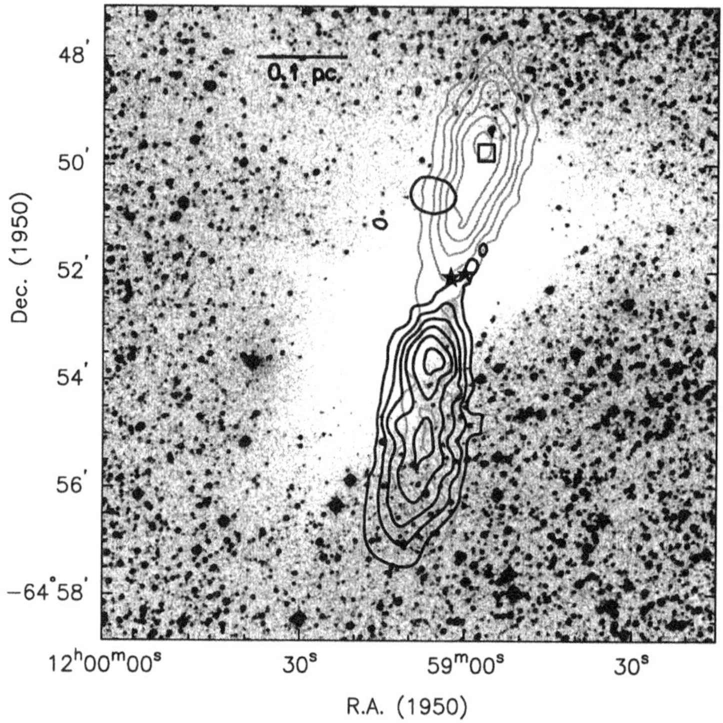

Fig. 1. Digital Sky Survey R-Band image of BHR 71, overlayed with contours of ^{12}CO $J = 1 \rightarrow 0$ emission. The black contours are blue-shifted emission, and the grey contours are red-shifted emission. The two ISO mid-infrared sources are indicated with star symbols. IRS 1, to the east, is the driving source of the large outflow. The square marks the position of the red outflow spectra shown in Fig. 4

3 CO Outflow Properties

The properties of the large scale molecular outflow have been determined by B97. As can be seen in Fig. 1, the outflow lobes are well separated on the sky, and extend ~0.3pc from their origin with an opening angle of ~15°. B97 find that the velocity structure is consistent with a steady flow with constant velocity (Cabrit et al. 1988). With this assumption the inclination of the outflow to the line-of-sight is determined to be 85°. The CO excitation temperature in the line core is greatest at the outflow peaks, indicating that the ambient gas there has been heated by interactions with the outflow.

Correcting for inclination, optical depth, and emission hidden within the line core, B97 determine the mass in the lobes to be $\sim 1.0 M_\odot$ (red lobe) and $\sim 0.3 M_\odot$ (blue lobe). Considering the different methods used to determine the outflow momentum P, kinetic energy E_k, and mechanical luminosity $L_{\rm mech}$ (upper and lower limit methods) B97 find $P = 11 M_\odot {\rm km\,s^{-1}}$, $E_k = 60 M_\odot {\rm km^2 s^{-2}}$, and $L_{\rm mech} = 0.5 L_\odot$. There is less mass in the blue lobe, which may be a result of this lobe breaking out of the globule, indicated in Fig. 1 by the conical reflection nebulosity just south of the protostars (see the beautiful colour VLT in the frontpiece of these proceedings for a more detailed view).

4 Two Protostars – Two Outflows

Near-infrared (NIR) images from the AAT are shown in Figure 2 (Bourke 2001). Most of the non-stellar emission is due to the emission in the H_2 v=1-0S(1) line, most likely due to shocks in the outflowing gas (Eislöffel 1997). The NIR emission is well aligned with the large scale CO outflow (Fig. 3).

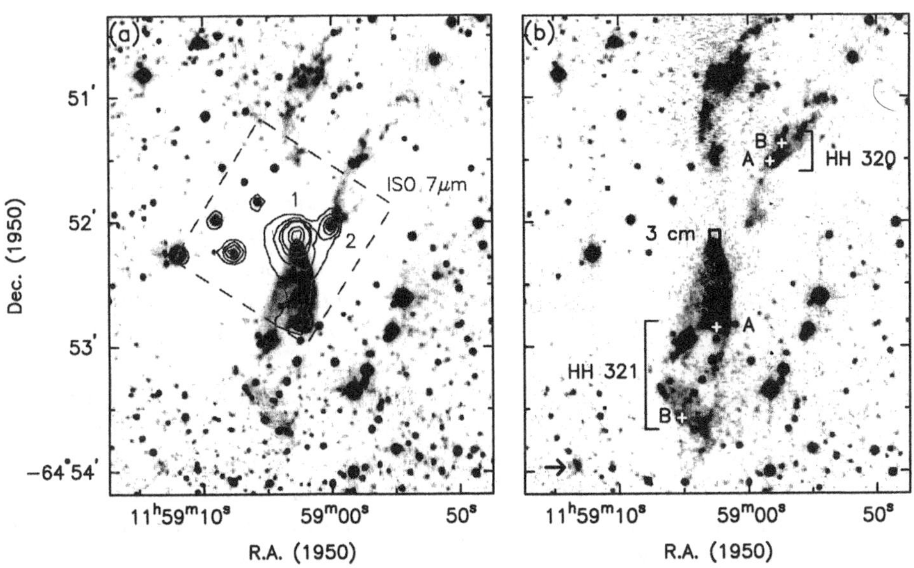

Fig. 2. (a) – K′ image of BHR 71 (greyscale) overlayed with ISO LW2 contours (5.0–8.5 μm). The embedded protostars IRS 1 ("1") and IRS 2 ("2") are labelled. (b) – Narrowband 2.12μm + continuum image (greyscale). The positions of HH 320 and HH 321 are marked with crosses, and the position of the 3 cm continuum source is marked with an unfilled box

Mid-infrared (MIR) emission in the ISO LW2 band is overlayed on Fig. 2(a). Two of the 7μm sources appear to be located at the apexes of NIR emission, strongly suggesting that they are associated with the emission. Source "1" (hereafter IRS 1) lies at the apex of the reflection nebulosity seen also in Fig. 1 and is co-incident with the position of the mm source BHR 71-mm, also known as IRAS 11590-6452 (B97). The 7μm flux from IRS 1 is an order-of-magnitude greater than from IRS 2. The NIR feature coincident with IRS 2 in Fig. 2(a) is non-stellar, by comparison of its PSF with stars in the same image.

A cm continuum source (indicated on Fig. 2(b)) is detected toward BHR 71 IRS 1, at both 3 and 6 cm (Wilner et al. 2001, in prep). The spectral index is consistent with a flat or rising spectrum due to free-free emission, a signpost of protostellar origin (Rodríguez 1994). Corporon & Reipurth (1997) discovered two Herbig-Haro associations in BHR 71 – HH 320 and HH 321, and their locations are shown on Fig. 2(b). It can be seen that HH 320 (HH 321) is coincident with the NIR emission associated with IRS 2 (IRS 1).

Bourke (2001) has shown that IRS 2 also drives a CO outflow which is more compact and much less energetic than the IRS 1 outflow. The northern part of the IRS 2 outflow is blue-shifted (and associated with HH 320) which is the opposite of the IRS 1 outflow and allows them to be separated spatially. The red lobe is confused by the IRS 1 outflow, though it is probably seen in the NIR (arrowed emission in Fig. 2(b)). Bourke (2001) suggested that IRS 1 & 2 may form a binary protostellar pair (separation ~3400AU) though the kinematic evidence for or against is lacking.

5 Outflow Chemistry

The BHR 71 IRS 1 outflow is one of a handful that show significant abundance enhancements in molecules such as SiO and CH_3OH (G98). Figure 3 shows the spatial distribution of SiO and CO in the outflow, compared to the NIR H_2 emission (the CO data is of lower spatial sampling than Fig. 1). Figure 4 shows spectra at two locations, the red lobe (as indicated by the box in Fig. 1) and at the position of IRS 1.

The spectral line profiles in the outflow and the velocity of the outflowing gas ($< 30\text{km s}^{-1}$) indicate that C-shocks dominate the flow (G98). The shocks are sufficiently strong to release molecules and atoms into the gas phase via evaporation of icy grain mantles (e.g., CH_3OH) and sputtering of grain cores or grain-grain collisions (e.g., Si, which rapidly forms SiO). Other molecules detected in the outflow include CS, H_2CO, SO, HCN, HNC, HCO^+ with SEST and H_2O with SWAS (Bourke et al. 2001, in prep). SiO is removed from the gas phase in about 10^4 years indicating that the outflow is quite young.

In the red lobe G98 determined abundance enhancements of ~350 in SiO and ~40 in CH_3OH. One particularly striking feature is the spatial distribution of SiO compared to CO in the outflow. Because SiO is the result of Si liberation it is usually only detected at the ends of outflows (where the shock interaction is greatest) or as a narrow jet along the outflow axis possibly due to interactions in a turbulent boundary layer (Garay 2000). The wide-spread distribution of SiO in the BHR 71 outflow is unique. This suggests that the SiO enhancement takes place in a shell-like structure produced by the dynamical interaction between the ambient cloud and an underlying wide-angle wind or wind driven shell (Garay 2000), or perhaps by a wandering jet. However, it has not been shown that an interaction between a wind and the ambient material can produce sufficient Si for this to be a viable explanation. If sufficient Si-bearing species are present in grain mantles then the wind model becomes attractive (Schilke et al. 1997).

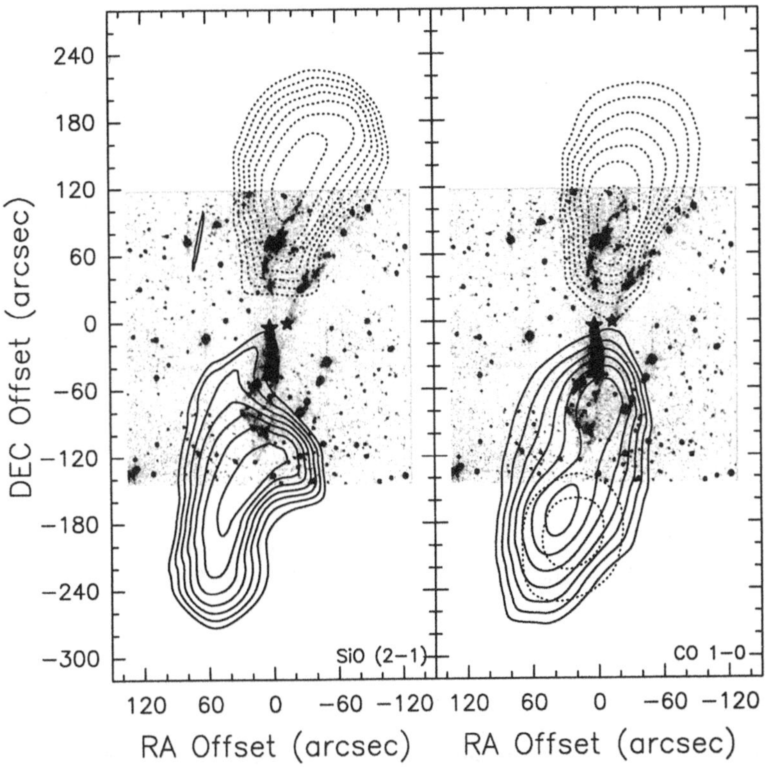

Fig. 3. SiO and CO integrated emission over the blue (solid lines) and red (dotted lines) lobes of the BHR 71 outflow, overlayed on the H_2 image from Fig. 2. The protostars IRS 1 (large star) and IRS 2 (small star) are indicated

6 The VLT Image

An optical composite image taken with the VLT is shown in the frontpiece of these proceedings. This image hints at the spectacular results we can expect from the VLT in the coming years. There is evidence in this image of both a wind component and a jet component to the IRS 1 outflow in the blue lobe. Extending from the reflection nebulosity which protrudes from the globule, one can trace out an elongated bubble, with its edges defined by enhanced extinction. This is characteristic of a wide-angle wind component. In addition, enhanced extinction is also seen along the axis of the bubble and extending beyond its southern tip. This may be an indication of the underlying jet which is probably driving this young outflow. Modelling of this one image may help answer some of the remaining questions about the spectacular BHR 71 outflow.

I thank my many collaborators on this project, in particular Guido Garay. A big hug to João for letting me present my work on this beautiful object.

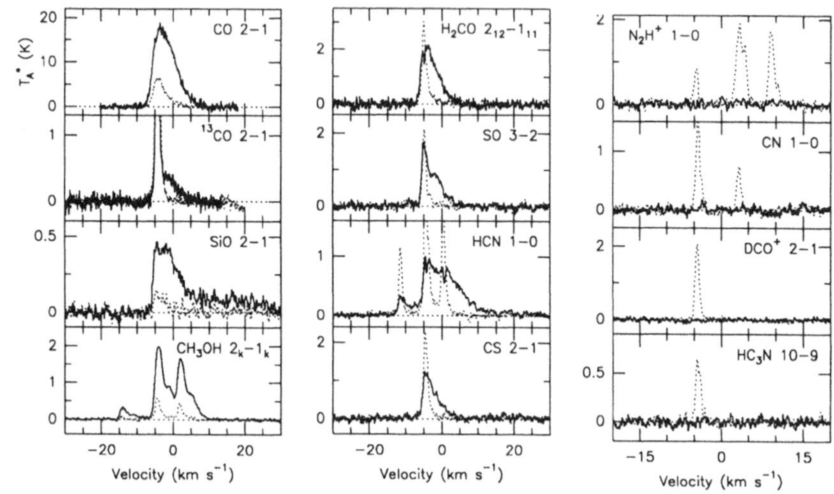

Fig. 4. Line profiles observed toward the red shifted lobe (continuous lines - indicated by the square on Fig. 1) and IRS 1 (dotted lines) of the BHR 71 outflow

References

1. Bourke, T. L., 2001, ApJ, 554, L91
2. Bourke, T. L., Garay, G., Lehtinen, K. K., Köhnenkamp, I., Launhardt, R., Nyman, L-Å, May, J., Robinson, G., & Hyland, A. R. 1997, ApJ, 476, 781 (B97)
3. Bourke, T. L., Hyland, A. R., & Robinson, G. 1995, MNRAS, 276, 1052
4. Bourke, T. L., Hyland, A. R., Robinson, G., James, S. D., & Wright, C. M. 1995b, MNRAS, 276, 1067
5. Cabrit, S., Goldsmith, P. F., & Snell, R. L. 1988, ApJ, 334, 196
6. Corradi, W. J. B., Franco, G. A. P., Knude, J., 1997, A&A, 326, 1215
7. Corporon, P., & Reipurth, B. 1997, in Poster Proceedings of IAU Symp. 82, Low Mass Star Formation - from Infall to Outflow, ed. F. Malbert & A. Castets, 85
8. Eislöffel, J., 1997, in IAU Symp. 182, Herbig-Haro Flows and the Birth of Low Mass Stars, ed. B. Reipurth & C. Bertout, 93
9. Garay, G, 2000, in IAU Symp. 197, Astrochemistry: From Molecular Clouds to Planetary Systems, ed. Y.C.Minh & E.F.van Dishoeck, 203
10. Garay, G., Köhnenkamp, I., Bourke, T. L., Rodríguez, L. F., & Lehtinen, K. K., 1998, ApJ, 509, 768 (G98)
11. Hartley, M., Tritton, S. B., Manchester, R. N., Smith, R. M., Goss, W. M., 1986, A&AS, 63, 27
12. Rodríguez, L. F., 1994, Rev. Mex. Astron. Astrofis., 29, 69
13. Sandqvist, Aa., 1977, A&A, 57, 467
14. Sandqvist, Aa., & Lindroos, K. P., 1976, A&A, 53 179
15. Schilke, P., Walmsley, C.M., Pineau des Forêts, G., & Flower, D.R. 1997, A&A, 321, 293

High Angular Resolution Analyses of Herbig-Haro Jets

Francesca Bacciotti[1,2], Thomas P. Ray[2], Reinhard Mundt[3], Jochen Eislöffel[4], and Josef Solf[4]

[1] Osservatorio Astrofisico di Arcetri, L.go E. Fermi 5, I-50125 Firenze, Italy
[2] Dublin Institute for Advanced Studies, 5 Merrion Square, Dublin 2, Ireland
[3] Max-Planck-Institut für Astronomie, Königstuhl 17, D-69117 Heidelberg, Germany
[4] Thüringer Landessternwarte Tautenburg, Sternwarte 5, D-07778 Tautenburg, Germany

Abstract. In present days, high angular resolution is making the difference in our understanding of Herbig–Haro jets associated with forming stars. To support this statement, we illustrate very recent analyses of Hubble Space Telescope 0.″1 angular resolution data. For one object (the optical flow from DG Tau) 2D velocity *channel maps* at the base of the flow show for the first time that the jet is denser, more excited and more collimated as the velocity gets higher, in an overall onion-like structure. We also have tentative evidence for *rotation* in the same region. These results appear to confirm magneto-hydrodynamic (MHD) acceleration models. The excitation properties, however, still need to find a satisfactory description. We discuss how using VLT and VLTI we can further shed light on the aspects of the picture that are still unclear.

1 Introduction – The Need for High Angular Resolution

The spectacular Herbig-Haro (HH) jets, optically emitting collimated mass flows associated with young stellar objects (see, e.g., [14]), have widely been recognized as an essential ingredient of the star formation process. For example, they are believed to contribute to the removal of excess angular momentum from the system, to disperse the infalling envelope and to contribute to the support of the parent cloud against gravitation. Despite their key role in the formation process, however, the origin of jets remains obscure, although it is now widely accepted that the acceleration process should combine rotational and magnetic effects. The most accreditated models roughly belong to two broad classes: (i) stationary magneto-centrifugal star/disk winds (e.g. [12], [15]) and (ii) unsteady magnetic field twist models (e.g. [11], [10]). The elaborated theories, however, have not yet been tested observationally, since the process occurs on very small scales. Depending on the models, the launching region in the star/disk system may go from less than 0.01 to a few AU from the star, that at the distance of, e.g., the Taurus star forming region correspond to less than 0.″03, a limit which is beyond current instrumentation. At present, the best spatial resolution is offered by the Hubble Space Telescope (HST), that allows for an angular resolution of 0.″05 – 0.″1 in the optical. Thus with HST we should be able to investigate, if not the launching region itself, at least the acceleration and collimation region of the flows, which according to theories extends to ~ 100 AU above the disk. Another severe practical problem is the very high extinction towards many young stellar objects. Observing the jet in IR lines or in radio continuum is often feasible, but one looses the great diagnostic advantages offered by optical forbidden lines. A key opportunity in this respect is to observe the less powerful flows coming from more evolved 'Class II' optically visible TTauri stars (TTSs), because of the window they offer us on the engine itself. With these ideas in mind we have observed a number of outflows in Taurus, among which a few flows from TTSs, with imaging and spectrographic instruments on-board HST. Results from this campaign have been published already ([14], [2], [3]) or are currently being analysed ([17], [4]). Here we concentrate mainly on the outflow from DG Tau (Sects. 2 and 3), the first HH jet to have been studied with the Space Telescope Imaging Spectrograph (STIS). Our results appear to confirm some of the predictions of popular MHD models, but other aspects are of difficult interpretation. In Sect. 4 we will discuss how using VLT and VLTI we can further improve our understanding of the inflow/outflow mechanisms related to star formation.

2 The Flow from DG Tau: Kinematics at the Jet Base

The active TTS DG Tau and its optical outflow have been widely studied in the past (e.g. [16]). We concentrated our study on the initial portion of the jet, within 1.″5 from the source. We took a series of seven spectra of the system with HST/STIS, keeping the slit parallel to the outflow axis, but stepping it in the transverse direction by 0.″07 each time. The spectra, that included several optical forbidden lines plus Hα, were subsequently merged together to form *channel maps* in different velocity intervals. An example is provided by Fig. 1, where we present some of the obtained synthetic images in different lines and in broad and narrow velocity intervals (for more details see [3]). We stress that to obtain such channel maps has been possible only thanks to the fact these are the first spectral observations in the optical to resolve the initial part of the jet in the transverse direction to the flow. Within 0.″5 from the star, and at the highest velocities, the initial jet is narrowly confined to the central axis, while it broadens and becomes less collimated in the lower velocity components (LVCs). The curved bowshock seen at 1″, that marks the tip of a limb-brightened 'bubble', probably traces a recent ejection event. In the jet channel, the flow appears to have an onion-like kinematic structure, with faster and more collimated flow continuously bracketed in spatially wider and slower flow. This property continues to hold in the LVC, and down to the narrowest velocity intervals (Fig. 1b). Also, the LVC emission is concentrated near the star (a few tenths of an arcsecond at most), while the highest velocity component extends far away.

Even more interestingly, we have found tentative evidence for *rotation* in the LVC of the flow, from a detailed analysis of the line profiles in four distinct regions of the jet, within the first 100 AU above the disk. For each pair of slits opposed symmetrically w.r.t. the jet axis we have determined the differences in the velocity of the peaks of the LVC emission. We have then corrected them by the offset caused by the uneven illumination of the STIS slit and other possible instrumental effects that may contaminate the data. According to our results, the SE side of the jet appears to move toward the observer faster then the corresponding NW side, and the average value found for the shift is about 10 km s^{-1}. The situation is schematically illustrated in Fig. 2 : here the map is projected

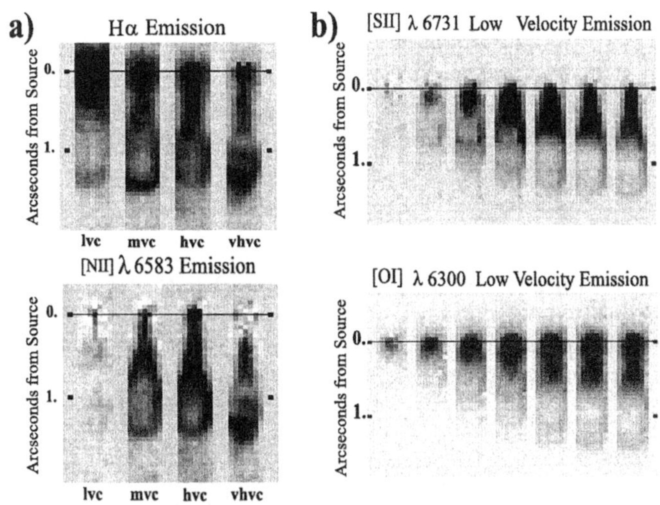

Fig. 1. Channel maps of the jet from DG Tau. (**a**) Reconstructed Hα and [NII]λ6548 images in four broad radial velocity components (low (LVC) medium (MVC) high (HVC) and very high (VHVC) velocity), from +50 to -450 km s^{-1}, each approximately 125 km s^{-1} wide. (**b**) Reconstructed [SII]λ6731 and [OI]λ6300 channel maps, in seven narrow low radial velocity bins, from + 60 to about -120 km s^{-1}, each bin being about 25 km s^{-1} wide

onto the plane of the sky, and, for reference, 0.″1 correspond to about 23 AU when deprojected onto the jet meridian plane. If we interpret these findings in terms of *rotation of the flow*, they would imply that the jet is rotating clockwise looking from the flow towards the source. Taking into account the inclination of this system with respect to the line of sight, one would derive an apparent toroidal speed of ~5 to ~15 km s^{-1} at a few tens of AU from the axis, and between 20 to 90 AU above the disk plane. We also estimate that the angular momentum transfer rate in the LVC would be about 3.2 10^{-5} $M_\odot$ yr^{-1} AU km s^{-1} [4]. Indeed, the determined velocities could be in the right range if compared with the predictions of MHD theories, although the peripheral regions of the outflow, that we are beginning to resolve, are generally less accurately modeled than the axial portions. We are now seeking if we find similar evidences in our other HST targets. Finally, we mention that following [5], rotation may also have been observed in the H_2 knots of the HH 212 jet, although on much larger scales than those discussed here.

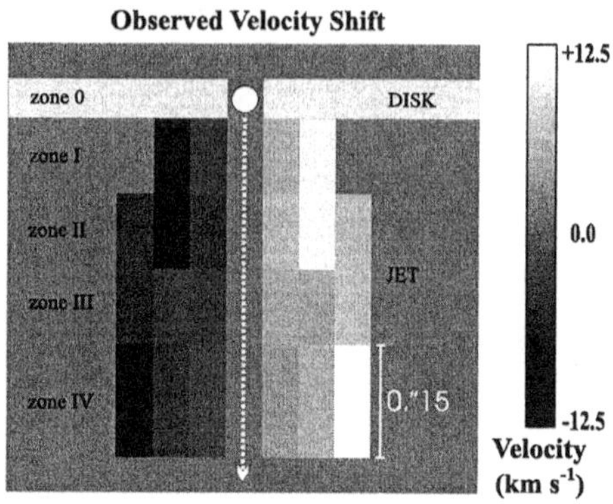

Fig. 2. Possible evidence for ***rotation*** (at low velocities) in the jet from DG Tau. The map shows the relative radial velocity shifts derived for the SE and NW sides of the flow, assumed symmetrically distributed w.r.t. the axis of symmetry. The velocity scale is linear, with the background grey corresponding to zero shift

3 Excitation Properties at the Jet Base

Recently, we have elaborated a new spectroscopic diagnostic technique that allows one to find, besides the electron density n_e, the hydrogen ionization fraction x_e (and hence the total density, n_H), and the average electron temperature T_e in the flow from the analysis of the line ratios, in a model-independent way. Details of this so-called 'BE technique' can be found in [1]. Here we present, in Fig.3, some of the results obtained for the DG Tau jet, in its first 200 AU from the source. For this case we could construct, for the first time, ***bidimensional*** maps of the excitation parameters, and in each of the four broad velocity components of Fig. 1a [4]. Figure 3a shows that in this dense flow, n_e is higher in the axial region, and in the more collimated and higher speed components. This result again is consistent with many MHD models. In addition, Fig. 3b shows the derived ionization fraction, for the MVC and HVC, which are less affected by noise. Here we see a gradual increase of the ionization with distance from the star, up to a maximum of about 0.4 – 0.5, reached at $1''$, in the interior of the rarefied bubble. Afterwards x_e decreases again. At the same time the temperature in the MVC (not shown here) rapidly decreases from 2 10^4 K to less than 10^4 K within the first $0''.5$. T_e stays constant at 10^4 in the bubble, and rises again at the bow shock, apparently reaching 3 – 4 10^4 K. Contrary to the electron density, neither x_e nor T_e appear to be higher in the axial region. Both quantities, however, increase with flow velocity and collimation. We also estimate that the ratio between mass loss and mass accretion rates is about 0.15, in rough agreement with the predictions of MHD wind models. More details can be found in [4]; here we point out that similar results, (averaged across the jet section and over velocity) were obtained by us for the HH 30 jet observed with HST/WFPC2 [2], and by [13], [6], who applied the BE technique to somewhat lower (0.″25) angular resolution data of the jets from RW Aur and DG Tau itself. Summarizing, the gas gets gradually ionized very close to the source, and then the ionization 'freezes', probably as a consequence of the rapid expansion at the base of the flow. Several groups are trying to model these recent observational findings in the framework of MHD winds (see, e.g. [8]).

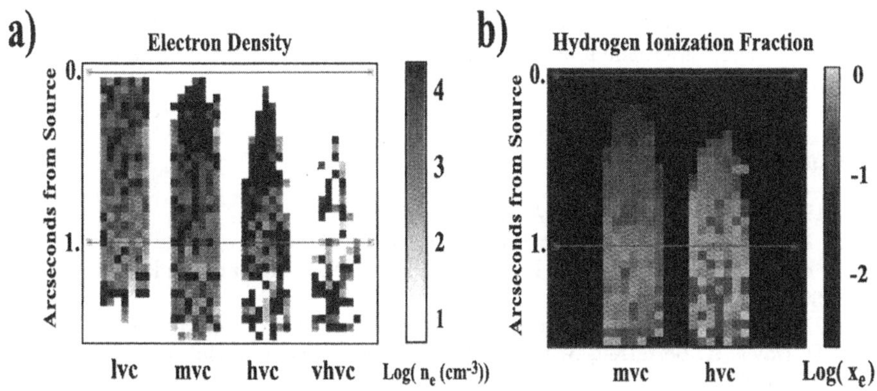

Fig. 3. Excitation conditions in the first 1.″5 of the jet from DG Tau (see text). (**a**) Electron density n_e derived from the ratio of the [SII] doublet. (**b**) Hydrogen ionization fraction x_e derived with the BE technique.
Figure available in colour on the CD-ROM

4 Observing with VLT and VLTI

As discussed above, high angular resolution finally allows us to test the acceleration models for jets of both stellar and galactic origin. Here we wish to outline how the facilities offered at the Paranal Observatory can contribute to advance along these lines of research. Many stellar jets, especially in the southern hemisphere, have not yet been analysed with the BE technique. Thus we could first carry out a study of their excitation properties averaged over velocity, using FORS2 at VLT Kueyen. To investigate the detailed kinematics in the faint beam of the flow, however, one would require higher spectral resolution: this can be achieved by UVES, that allows for a RS product of about 39000 in the red range. The *spatial* resolution of such results is of course dependent on the development of adaptive optics at the UTs. The real goal in this respect, however, is to probe the regions close to the central engine using the VLTI interferometer. With the expected angular resolution of 2 mas (corresponding to $\sim$ 0.3 AU for objects in Taurus), which is a factor 20 – 50 higher than the one achievable with HST, we can access the inner regions of the disk, where the acceleration of the most powerful components of the jets takes place. With VLTI it will be possible, for example, to test the *shape* of the flow at its base by measuring its opening angle, which can differ among the various models. An example of model visibilities of a jet having two different opening angles (10° and 90°) is given in [7]: the two cases are clearly distinguishable already with baselines of 50 – 100 m. Detailed kinematics can then be investigated by testing with the VLTI/AMBER spectrograph the emission in suitable IR lines (see [9]). The study of the excitation conditions at VLTI resolution, however, will have to wait for the second phase of AMBER, in which the spectral window of the instrument will be extended to the red.

References

1. F. Bacciotti, J. Eislöffel: A&A., **342**, 717 (1999)
2. F. Bacciotti, J. Eislöffel, T.P. Ray: A&A, **350**, 917 (1999)
3. F. Bacciotti, R. Mundt, T.P. Ray, J. Eislöffel, J. Solf, M. Camezind: ApJ, **537**, L49 (2000)
4. F. Bacciotti, T.P. Ray, R. Mundt, J. Eislöffel, J. Solf, in preparation
5. J.D. Davis, A. Berndsen, M.D. Smith, A. Chrysostomou, J. Hobson: MNRAS **314**, 241 (2000)
6. C. Dougados, S. Cabrit, C. Lavalley-Fouquet In: *Emission Lines from Jet Flows*, ed. by W. Henney, W. Steffen, L. Binette, A. Raga, (RVMXAA Series de Conferencias, 2001, in press)
7. J. Eislöffel, C. Dougados: 'Outflows from Young Stars' In: *Science with the VLT Interferometer, ESO workshop*, (Springer-Verlag, ESO Astrophysics Symposia, Berlin 1997) p.240
8. P. Garcia, J. Ferreira, S. Cabrit, L. Binette: 'The excitation properties of Disk Winds', In: this book, poster proceedings
9. P. Garcia: 'Into the Twilight Zone: Reaching the jet engine with the VLTI/AMBER.' In: this book
10. A.P. Goodson, R.M. Winglee, K.H. Böhm: ApJ, **489**, 199 (1997)
11. M.R. Hayashi, K.P. Shibata, R. Matsumoto: ApJ **468**, L37 (1996)
12. A. Konigl, R.E. Pudritz: 'Disk Winds and the Accretion-Outflow Connection'. In: *Protostars and Planets IV*, ed. by V. Mannings, A.P. Boss, S.S. Russell (University of Arizona Press, Tucson 2000) pp. 759–787
13. C. Lavalley-Fouquet, S. Cabrit, C. Dougados: A&A **356**, L41 (2000)
14. T.P. Ray, R. Mundt, J.E. Dyson, S.A.E.G. Falle, A.C. Raga: ApJ **468**, L103 (1996)
15. F.H. Shu, J.R. Najita, H. Shang, Z.-Y. Li: 'X-Winds Theory and Observations' In: *Protostars and Planets IV*, ed. by V. Mannings, A.P. Boss, S.S. Russell (University of Arizona Press, Tucson 2000) p. 789
16. J. Solf: 'Spectroscopic Signatures of Microjets', In: *Herbig-Haro Flows and the Birth of Low Mass Stars, IAU Sympos. 182*, ed. by B. Reipurth, C. Bertout (Kluwer Academic Publishers, Dordrecht 1997), pp. 63–72
17. J. Woitas, J. Eislöffel: 'The environment of FS Tau observed with HST WFPC2' In: this book, poster proceedings

Structure of Magnetocentrifugal Disk-Winds: From the Launching Surface to Large Distances

Ruben Krasnopolsky[1], Zhi-Yun Li[2], and Roger D. Blandford[3]

[1] Department of Astronomy and Astrophysics, University of Chicago, Chicago, IL 60637, USA
[2] Department of Astronomy, University of Virginia, Charlottesville, VA 22903, USA
[3] Theoretical Astrophysics, Caltech, Pasadena, CA 91125, USA

Abstract. Protostellar jets and winds are probably driven magnetocentrifugally from the surface of accretion disks close to the central stellar objects. The exact launching conditions on the disk, such as the distributions of magnetic flux and mass ejection rate, are poorly unknown. They could be constrained from observations at large distances, provided that a robust model is available to link the observable properties of the jets and winds at the large distances to the conditions at the base of the flow. We discuss the difficulties in constructing such large-scale wind models, and describe a novel technique which enables us to numerically follow the acceleration and propagation of the wind from the disk surface to arbitrarily large distances and the collimation of part of the wind into a dense, narrow "jet" around the rotation axis. Special attention is paid to the shape of the jet and its mass flux relative to that of the whole wind. The mass flux ratio is a measure of the jet formation efficiency.

1 Basic Mechanism and Previous Work

The magnetocentrifugal mechanism is the leading candidate for producing the jets and winds observed around young stellar objects. The basic principle is relatively simple, and has been understood for a long time [20]. It envisions parcels of fluid element being lifted off and accelerated centrifugally along rapidly rotating open magnetic field lines anchored firmly on an accretion disk. Beyond a certain point where the energy densities in the bulk flow motion and magnetic field are comparable, the field lines can no longer enforce rigid rotation, and the field becomes increasingly toroidal. It is the "hoop stress" associated with the toroidal field that is thought to be responsible for the wind collimation and jet production. The quantitative properties of the jet expected from this mechanism are poorly determined however, even though the MHD equations that govern the wind structure and jet formation are well known. Our understanding of the jet properties has been hampered to a large extent by the mathematical difficulties associated with obtaining wind solutions.

1.1 Time-Independent Wind Solutions

There are two basic approaches in obtaining wind solutions. The first is to solve for the steady-state wind structure directly from the time-independent MHD equations. These equations can be cast into a second order partial differential equation (the Grad-Shafranov equation). It is well known that, for a cold wind that we are interested in here, the equation changes its type from being elliptic inside the fast (magnetosonic) surface to hyperbolic outside. Computationally, the structure of the inner part of the wind inside the fast surface can be solved first by relaxation, and that beyond the fast surface later by the method of characteristics [15,17]. The fact that the position of the fast surface, where the poloidal flow speed matches the fast magnetosonic speed, is unknown a priori poses a problem. To obtain a converged solution, one needs to have a good initial guess of the fast surface position, which is generally difficult to obtain.

1.2 Time-Dependent Numerical Simulations

A more flexible approach is to numerically follow the time evolution of a wind to steady-state, if such a state exists. This approach has been taken by several groups [13,1,6]. It is also the approach that we took [2]. Our simulations are based on the ZEUS MHD code, and treat the Keplerian disk as a (lower) boundary, on which an open magnetic field of a prescribed distribution is anchored and from which cold material is injected into the wind at a prescribed rate. A novel feature of our simulation is the treatment of the region near the rotation axis, where the magnetocentrifugal mechanism is ineffective. The reason is that to launch a cold parcel of fluid element centrifugally from a Keplerian disk the field line must incline an angle of at least 30° away from the disk normal [20]. This condition is not met in the axial region since the field line along the axis must be exactly vertical by symmetry. In reality, the axial region may be filled with a normal stellar wind from the central object or bundles of open field lines from the stellar magnetosphere. We are thus motivated to inject a light fluid with little mass flux at a speed fast enough to escape from the potential well along those field lines that fail to operate magnetocentrifugally. The light axial flow provides a plausible inner boundary to the magnetocentrifugal disk-wind, the focus of our investigation.

A typical example of the steady-state disk-wind solutions obtained from time-dependent simulations is given in Fig. 1. The wind is driven off all of the (equatorial) disk surface. Flow acceleration is apparent along all field lines except those near the axis where a fast initial injection is imposed. Note that the field (and stream) lines collimate gradually, as expected. What was not expected was the great care that went into designing the shape of the simulation box, so that a steady state could be reached at all. If we were to cut the box shown in Fig. 1 in half or to elongate the box horizontally instead of vertically, and restart the simulation with the same initial and boundary conditions, the wind would become chaotic. The sensitive solution dependence on the simulation box has also been noted by others [1]. It is a major concern for the time-dependent approach to finding disk-wind solutions.

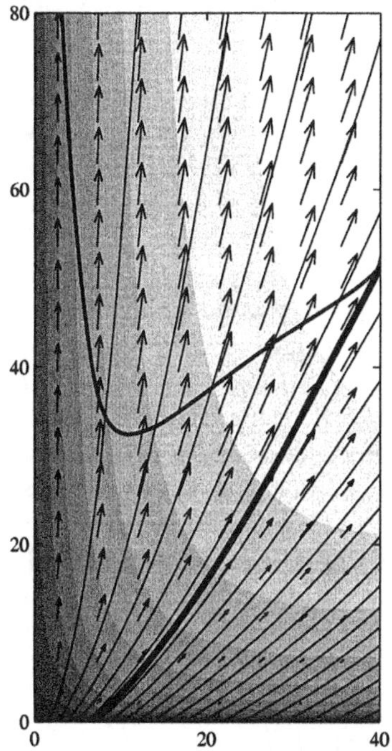

Fig. 1. A representative magnetocentrifugal wind launched from a Keplerian disk (in arbitrary units; taken from [2]). Shown are the magnetic field lines (*light solid lines*), velocity vectors (*arrows*), density contours (*shades*), and the fast magnetosonic surface (*solid line of medium thickness*). The thickest solid line divides the portion of the wind that becomes super fast-magnetosonic inside the simulation box (*above*) from the portion that does not (*below*)

2 Magnetocentrifugal Winds From Inner Accretion Disks

We believe that the simulation box dependence described above comes from the fact that a large fraction of the wind remains completely sub fast-magnetosonic in the computational domain, as shown in lower-right corner of Fig. 1. Information on the sub-fast outer boundary can propagate upstream all the way to the disk surface and interfere with the wind launching. The reason for the region to remain sub fast is simple: the wind coming off the outer part of the disk encounters the edge of the simulation box too soon; it simply does not have enough room to get accelerated to the fast speed. This situation remains as long as the wind is driven off from *all* of the (equatorial) disk plane (as assumed in Fig. 1 and other previous time-dependent disk-wind simulations) regardless of the box size. It motivates us to restrict the wind launching to only the inner region of an accretion disk, and focus on inner-disk driven winds for which the simulation box dependence disappears.

2.1 Inner-Disk Driven Winds: Simulation Setup

Physically, the wind launching may be limited to the inner region of a protostellar disk where the temperature is high enough (greater than $\sim 10^3$ K) that thermal ionization of alkali metals can provide enough charges to couple the magnetic field to the disk matter. For typical parameters, this occurs inside a radius of order 1 AU. Numerically, we set up the simulation as sketched in Fig. 2. To fill all available space above (and below) the equatorial plane, we demand the last field line anchored at the outer radius of the launching region R_0 to lie exactly on the equatorial plane. Wind plasma sliding along this last (horizontal) field line will become super fast-magnetosonic in the computational realm, provided that the size of the simulation box is sufficiently large. Once the whole fast surface is completely enclosed inside the simulation box, the size and shape of the box would have minimal effects on the structure of the wind, especially near the launching surface, since information cannot propagate upstream in a super fast region. In this way, we should be able to study the wind structure up to arbitrarily large distances from the source region, limited only by computer time.

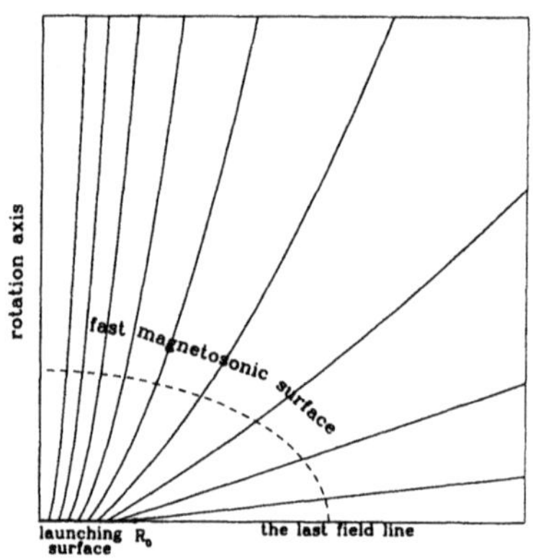

Fig. 2. Schematic view of a cold magnetocentrifugal wind launched from a limited, inner disk region

2.2 Large-Scale Wind Structure: Numerical Results

For illustration, we consider a specific example. We adopt as the launching conditions on the disk a power-law distribution of the vertical field strength with radius as $B_z \propto R^{-1.5}$ and a mass injection rate (per unit area) $j \propto R^{-0.5}$ between 0.1 and 0.8 AU. The inner radius is chosen to mimic the disk truncation radius due to the stellar magnetosphere. Inside this radius, we inject a fast light flow as described earlier. At the edge of the wind launching region, taken to be $R_0 = 1$ AU for simplicity, we impose the condition that $B_z = j = 0$ since the last field (and stream) line must be horizontal. A cubic polynomial is used to connect smoothly the values of B_z (or j) between 0.8 and 1 AU. As before, we follow the time evolution of the wind numerically to a steady state. The steady wind solution, from the launching surface (inside 1 AU) all the way to a large distance of 10^2 AU, is displayed in Fig. 3 on two scales.

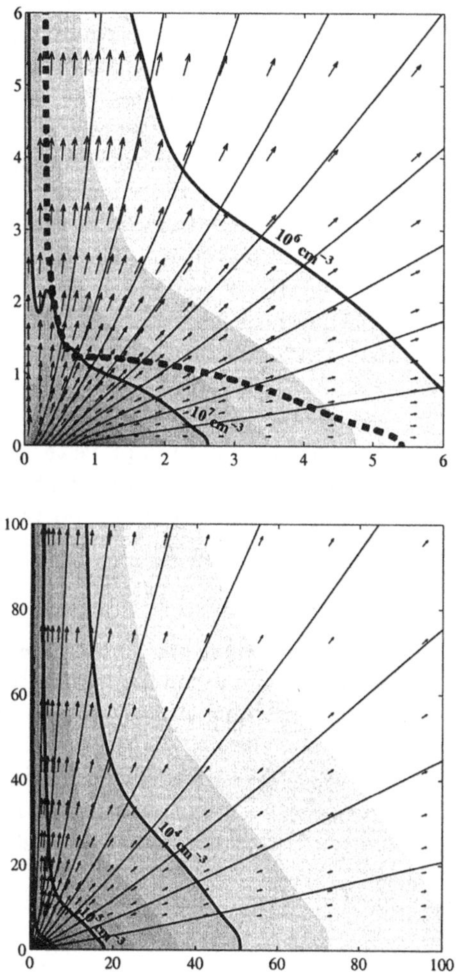

Fig. 3. Streamlines (*light solid lines*) and density contours (*heavy solid lines and shades*) of a representative steady wind driven from the inner region of an protostellar disk in a 6 AU (*top panel*) and 10^2 AU (*bottom panel*) box. The fast surface (*dashed line*) is also plotted in the smaller box. The streamlines divide the wind into 10 zones of equal mass flux, with a total mass flux of $10^{-8} M_{\odot}\mathrm{yr}^{-1}$ per side of the disk. The gray scale shows the log of the hydrogen number density, with three shades per decade. The density contours correspond to 10^4, 10^5, 10^6 and 10^7 in units of cm^{-3}

Several features are worth noting. First, the fast surface shown in the smaller box closes on the equatorial plane, as advertised. This closure enabled us to continue the wind solution on to large distances without having to worry about the effects of box size. Second, the wind speed is anisotropic, with a value roughly 3 times higher in the axial region than in the equatorial region. The anisotropy appears to be even stronger in density, which is stratified more or less cylindrically (or jet-like) near the rotation axis, as expected. In the more equatorial region, the density contours bulge outward prominently, retaining some memory of the nearly horizontal shape of the initial density contour. This non-cylindrical shape of density contours is significant because the wind emission in forbidden lines such as [SII]$\lambda\lambda$6716,6731 is sensitive to the density [3], and the shape of the jet may resemble to a zeroth order the shape of the density contour at some fiducial value. We choose to represent the outer boundary of a "jet" by a fiducial density contour of 10^4 cm^{-3} (the outermost contour in the larger box of Fig. 3). The "jet" so defined has a width of $\sim$ 30 AU at a height of 10^2 AU, comparable to that observed in HH 30 jet. The bulging out at the "jet" base is not observed, however. Furthermore, the "jet" contains only about a quarter of the total wind mass flux, making its formation rather inefficient. These undesirable "jet" features demonstrate that not all combinations of the launching conditions are capable of producing cylindrical jets that contain the majority of the wind mass flux. We find that one way to improve the jet shape *and* increase its mass flux fraction is to make the mass injection rate j on the disk decrease more steeper with radius. The details will be presented in a forthcoming paper.

3 Conclusion

To summarize, by limiting the wind launching to the inner part of an accretion disk, we are able to obtain using time-dependent simulation steady-state wind solutions that extend from the launching surface to large distances. Combined with a detailed calculation of the thermal structure and emission properties, these large-scale wind solutions can be used to yield constraints on the launching conditions from the properties of jets and winds observed at the large distances. The constraints may provide clues to the origin of the disk magnetic fields that launch the jets and winds.

References

1. R. D. Blandford, D. G. Payne: MNRAS **199**, 883 (1982)
2. T. Sakurai: AA, **152**, 121 (1985)
3. J. R. Najita, F. H. Shu: ApJ, **429**, 808 (1994)
4. R. Ouyed, R. E. Pudritz: ApJ, **482**, 712 (1997)
5. G. V. Ustyugova, A. V. Koldoba, M. M. Romanova, V. M. Chechetkin, R. V. E. Lovelace: ApJ, **516**, 221 (1999)
6. S. Bogovalov, K. Tsinganos: MNRAS, **305**, 211 (1999)
7. R. Krasnopolsky, Z.-Y. Li, R. D. Blandford: ApJ, **536**, 631 (1999)
8. H. Shang, F. H. Shu, A. E. Glassgold: ApJ, **493**, 91 (1998)

The Residencia (outside and inside)
(Photos by Massimo Tarenghi)

Paulo Garcia and Zhi-Yun Li (foreground),
Dan Harvey and Tyler Bourke (background)

Into the Twilight Zone*: Reaching the Jet Engine with AMBER/VLTI

Paulo J.V. Garcia[1,2], Renaud Foy[3], and Eric Thiébaut[3]

[1] Centro de Astrofísica da Universidade do Porto, Portugal
[2] CRAL, UMR 5574, France
[3] Observatoire de Lyon/Centre de Recherche Astronomique de Lyon, UMR 5574, France

Abstract. It is currently unknown if pre-main-sequence jets originate in the star, the star-disk interaction zone or the disk. A major constraint for pre-main-sequence jet origin models is the measurement of the jet opening angle near its source. However current angular resolutions offered by adaptive optics and by HST are not capable of reaching the jet inner zones where models differ most.

We present observational predictions for the AMBER instrument (a high spectral resolution beam combiner) at the VLT Interferometer. We show that visibilities alone are able to constrain the jet opening angle. We further compute the signal-to-noise budget for the obtained visibilities and show that they are within reach of VLTI auxiliary telescopes.

1 The Problem

One of the fundamental open problems in star formation is the understanding of the physical mechanisms by which mass is ejected from a protostellar system and collimated into jets. Locally jets will regulate the system angular momentum and therefore its evolution during the first $\sim 10^6$ years. At large scale jets inject momentum into the cloud thus affecting its star formation efficiency and evolution. Although jets are seen in a wide variety of environments, from active galactic nuclei to X-ray binaries, it is in pre-main-sequence (PMS) stars that the engine has a larger *angular size* [18] making them ideal targets for observations.

Jet activity decreases with age, however the youngest protostars are the most embedded ones with extinction limiting our ability to reach the engine. Therefore we have to select objects with relatively low extinction and still harboring jet activity. These objects exist [13] and are part of a broader class of pre-main-sequence stars: the classical T Tauri stars.

The open questions regarding the jet engine are three:

• *Where?* Is ejection taking place from the star (eg., [15]); the disk (eg., [2], and Li in these proceedings); or the star-disk interaction zone (eg., [17])?

• *When?* Is ejection a stationary phenomenon (all above references) or a time dependent one (eg., [14])?

• *How?* Is ejection magneto-centrifugal (matter simply flows out along sufficiently bent rotating magnetic field lines) or is there a disk corona heating the gas, so that matter can now cross a potential barrier like for the solar wind (eg., [4])?

It turns out that from all the available above possibilities only two models do make observational predictions; both assume that ejection is stationary and magneto-centrifugal. Their difference lies in the region where the jet originates: the disk for the disk-wind [3] or the star-disk interaction for the X-wind [16]. This difference has two consequences: 1) jets are hollow; 2) jets have different opening angles[1]. Far from the engine the above models tend to produce similar results, their largest differences being near the engine. Current angular resolutions (adaptive optics and HST $\sim 0.2''$) cannot resolve the engine. Therefore there is a clear need for higher angular resolutions such as offered by the VLTI. However angular resolution is not enough, jets emit faint emission lines and therefore high spectral resolution is needed to have enough contrast between the jet and the strong stellar emission. The AMBER instrument at the VLTI meets these constraints and is therefore ideally suited for these studies.

A first discussion of the application of the VLTI to jets, in the visible domain, was presented by [5]. In the following sections we introduce the AMBER/VLTI instrument, then in Section 3 we compute observational predictions and show how the opening angle of hollow jets can be directly measured by the interferometer using the visibilities alone, and that the signal-to-noise ratio (SNR) requirements are compatible with AMBER and known southern jet sources. Finally in Section 4 we present the conclusions.

2 Interferometric Measurements with AMBER

Readers not familiar with interferometry techniques could read the outstanding introductory course by [11].

The VLTI is described by [9]. In a first phase the VLTI will do two telescope interferometry, this means that we can only measure visibilities (V) for a given two telescope combination, and therefore no imaging capabilities are present. Available telescopes are four 8m unit telescopes (UTs) at fixed positions, and three 1.8m auxiliary telescopes (ATs) which can be placed in any of 30 positions of the VLTI sub-array (VISA). Baselines can therefore be up to 130 m for the UTs and 200m for the ATs.

AMBER is described in detail by Fabien Malbet in these proceedings and therefore we will only summarize the important properties of this instrument: it operates in the NIR (1.1μm - 2.4μm) and the fringes are dispersed up to resolutions of $R = 10^4$. This last property translates into a further advantage, allowing the measurement of differential phase ($\phi(\lambda)$) and visibility ($V(\lambda)$) shifts from emission lines to the continuum. It is important to underline that in a first phase we will only have access to visibilities and not to images of the sources, therefore the astrophysical programs must be able to draw conclusions from the visibility data alone.

[1] The opening angles could be those of the isodensity contours. Near the engine disk winds have horizontal isocontours while the X-wind has more vertical ones.

3 Observational Predictions

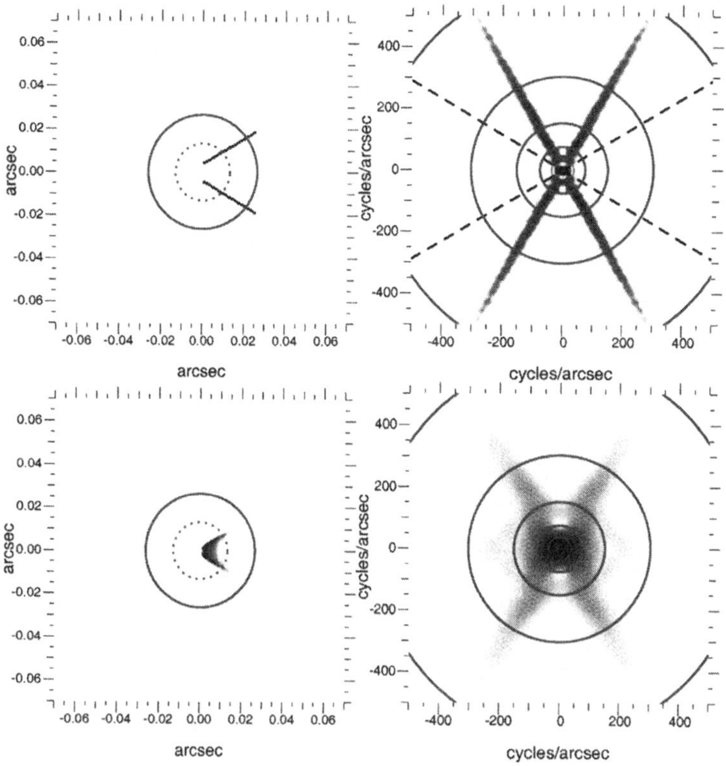

Fig. 1. Upper left: Toy model of a hollow edge-on jet with opening angle 30° in the image space. The inner (dashed) circle is the FWHM of a 8.2m diffraction limited image at 1.28 μm, the outer the first zero of the Airy function. Field of view is roughly the size of an AT diffraction limit at 1.28 μm. **Upper right:** Toy model visibility. For guidance (see text) we plot the jet opening angles as dashed lines. The circles plot the cutoff frequencies for 10m, 20m, 40m, 80m and 160m baselines. **Lower left:** Physical model of a disk jet in the image space. **Upper right:** Physical model visibility

The computation of model observational predictions proceeds as follows. First we generate an image then the image is inverse Fourier transformed to the (u, v) plane. In general we will obtain an map where each pixel intensity is a complex number z determined by an amplitude (also called visibility) V and a phase ϕ, i.e. $z = Ve^{-2\pi i\phi}$. Then we compute the projected baseline tracks in the (u, v) plane as the Earth rotates[2], which are a function of baseline, object, wavelength and hour angle. Finally we measure only the visibility V, the phase ϕ being lost because of atmospheric randomness.

[2] See Melvin Dyck's chapter in [11]

It is important to start with a toy model because it illustrates how the visibilities are enough to constrain the jet inclination angle. In Figure 1 we present images for a hollow jet toy model. The visibility has an X shape. We have a non-zero visibility to very high frequencies because the jet arms thickness is unresolved. Each arm of the visibility X shape is perpendicular to each hollow jet arm in the image space (dashed lines). This clearly shows that we can measure the jet opening with the visibility alone, provided we detect the visibility X arms in the (u, v) space.

In the bottom part of Figure 1 we present synthetic images of physical jet models from [8]. These images show that the same pattern is present although the effect is now fainter (the jet is less hollow).

3.1 Observing Strategy and Program Design

We considered in our simulations the case of RU Lupus ($m_J = 8.6$), the brightest jet source in the southern hemisphere [19] with known jet position angle. However the well studied source DG Tau ($m_J = 9.0$) could be considered as well, since it is also observable from Paranal.

In order to detect the visibility pattern the choice of baselines must be optimized such that their path in the (u, v) space crosses the visibility arms. Therefore it is critical to know in advance the jet position angle in the sky and to select baselines which are perpendicular to it. The lack of such information, obtained at lower angular resolutions (AO/HST), means that we will search the wrong region of (u, v) space and even if we detect structures in the visibility it will be very difficult to interpret them.

The choice of the emission line that best probes the jet is a critical issue. As we approach the jet base the density increases (eg., [1]) therefore we cannot use the classical forbidden lines in the NIR ([Fe] 1.64μm) because of quenching (low critical densities). We have therefore to use other lines. Contrary to their more embedded counterparts, the spectrum of CTTS harboring jets does not present H_2 emission. Hydrogen recombination lines are probably the best tracers of jet emission in the inner zone, indeed as we approach the jet base both the density and temperature increase ($T > 10^4$) and [19] finds jet Hα emission at 20 mas scales. Paβ is probably the best observationally studied [7] hydrogen recombination line in the NIR and therefore we will select it.

3.2 Error Budget

AMBER should attain a 0.01 error in the visibility measurement. Such low errors have been obtained by FLUOR, a fiber optics beam combiner, at the IOTA interferometer (45 cm apertures) [10]. Therefore in our calculations we assume such a precision in the visibility.

In order to verify our program feasibility we must be able: 1) to obtain a SNR in the visibility of $\sim$ 100 and; 2) to have visibility changes large enough (> 0.01) to be detected by AMBER.

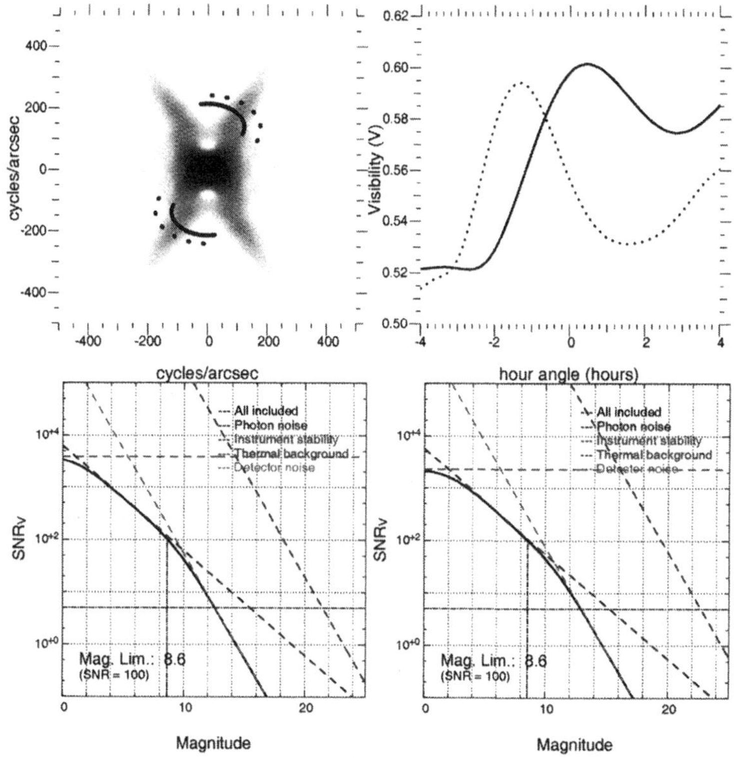

Fig. 2. Top left: we plot the telescope tracks in the (u, v) plane for UT1-UT2 (filled) and B3-G0 (dotted). The baselines were selected in order to cross the visibility arms (assumed jet PA=0°). **Top right:** we present the visibility V versus hour angle for UTs (filled) and ATs (dotted). **Bottom:** SNR predictions for ATs (**left**) and UTs (**right**).
Figure available in colour on the CD-ROM

We then disperse the fringes with a 1000 spectral resolution, fringe tracking is on (the star is bright enough) and therefore we can integrate for several minutes. We further assume a tip-tilt wavefront correction for the ATs and 64 actuator AO for the UTs and good seeing conditions 0.5″.

Veiling determinations show that both the star and, mostly, the disk contribute to the continuum emission near Paβ [6] for CTTS jet sources. However simulations by [12] for the disk of CTTS find it unresolved at 1 μm. The contribution of the unresolved star+disk in the visibility is just a constant in the visibility. We assume in the visibility calculations that at a given wavelength half the emission comes from the jet and the other half comes from the unresolved star+disk component.

In Figure 2 we present SNR calculations making use of a code (kindly provided by Fabien Malbet) that models the instrument. The figure shows that to obtain a SNR=100, we need for the ATs three 10 min. frames and one 3 min. frame for the UTs. We therefore conclude that interferometric observations of jets are feasible in the first phase of AMBER.

We expect to attain higher precisions by using differential visibilities between the line and the continuum.

4 Conclusions and Further Prospects

We have analyzed the scientific case for interferometric observations of jets and showed that in a first phase, using only the visibilities, it is possible to measure the jet opening angle. This parameter is powerful enough to disentangle an X-wind from a disk wind jet origin.

The most interesting southern object for jet interferometry is RU Lupus, but the more well known source DG Tau is also visible from Paranal. The lines to use are still an open issue but recombination lines of hydrogen and in particular Paβ are good choices. A spectroastrometric survey of southern sources at selected lines would be a good way to select target objects and lines for future interferometric observations.

The dispersion of fringes in AMBER allows further differential phase and visibility studies between the lines and the continuum.

Acknowledgments

We thank very much G. Duvert for providing us with his optical interferometry version of the (u, v) coverage software GILDAS.

References

1. F. Bacciotti and J. Eislöffel. *A&A*, 342:717–735, Feb. 1999.
2. R. D. Blandford and D. G. Payne. *MNRAS*, 199:883–903, June 1982.
3. S. Cabrit, J. Ferreira, and A. C. Raga. *A&A*, 343:L61–L64, Mar. 1999.
4. F. Casse and J. Ferreira. *A&A*, 361:1178–1190, Sept. 2000.
5. J. Eislöffel and C. Dougados. In *Science with the VLTI*, p. 240, 1997.
6. D. F. M. Folha and J. P. Emerson. *Ap&SS*, 261:147–150, 1998.
7. D. F. M. Folha and J. P. Emerson. *A&A*, 365:90–109, Jan. 2001.
8. P. J. V. Garcia. *Thése de Doctorat.* 12 1999. Université Lyon I, France.
9. A. e. a. Glindemann. *SPIE*, 4006, 2000.
10. P. Kervella, V. Coudé du Foresto, et al. *A&A*, 367:876–883, Mar. 2001.
11. P. R. Lawson, editor. *Principles of long baseline stellar interferometry.* JPL Publication 00-009, 7 2000. `http://sim.jpl.nasa.gov/michelson/iss.html`.
12. F. Malbet and C. Bertout. *A&A Supplement, v.113, p.369*, 113:369, Nov. 1995.
13. R. Mundt and J. Eislöffel. *AJ*, 116:860–867, Aug. 1998.
14. R. Ouyed and R. E. Pudritz. *ApJ*, 484:794+, July 1997.
15. C. Sauty and K. Tsinganos. *A&A*, 287:893–926, July 1994.
16. H. Shang, F. H. Shu, and A. E. Glassgold. *ApJl*, 493:L91, Feb. 1998.
17. F. Shu, et al. *ApJ*, 429:781–796, July 1994.
18. H. C. Spruit. In *NATO ASI Series C.*, volume 477, p. 249, Kluwer, 1996.
19. M. Takami, J. Bailey, et al. 323:177–187, May 2001.

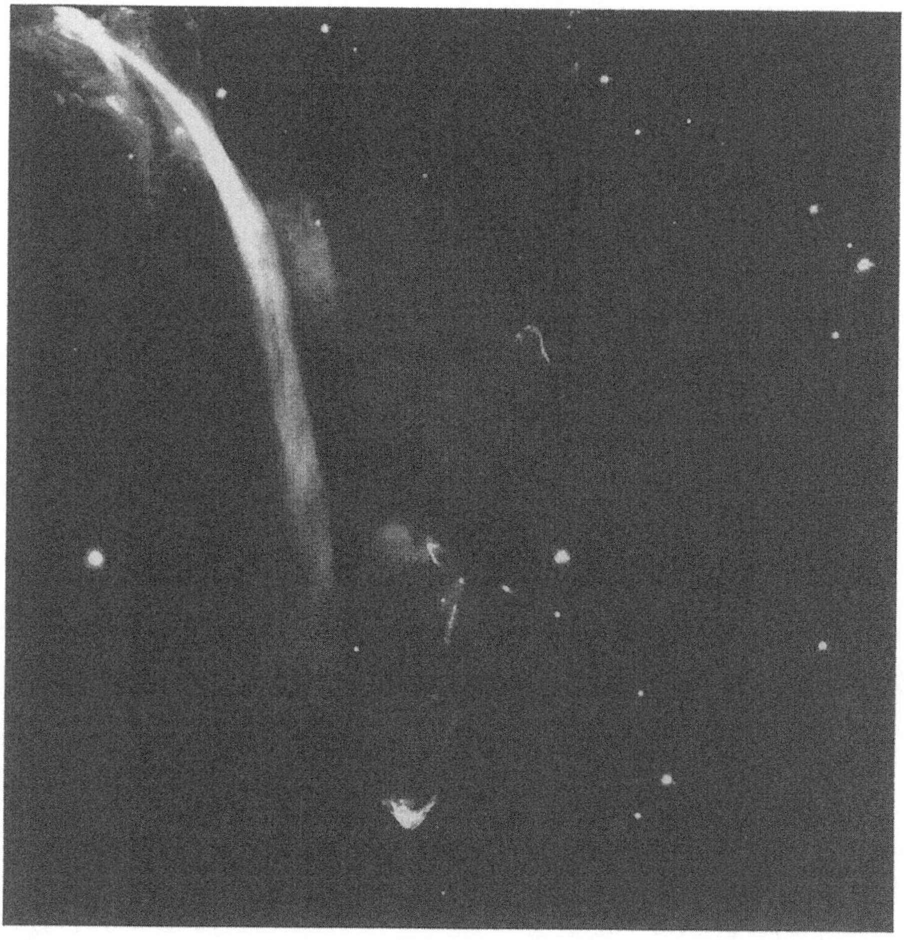

HH 34
(Kueyen/UT2 + FORS2)

Martino Romaniello

T Tauri Stars in the Large Magellanic Cloud: A Combined HST and VLT Effort

Martino Romaniello[1], Nino Panagia[2], Salvatore Scuderi[3], Roberto Gilmozzi[4], Eline Tolstoy[5], Fabio Favata[6], and Robert P. Kirshner[7]

[1] European Southern Observatory – Garching bei München (Germany)
[2] Space Telescope Science Institute – Baltimore (USA)
[3] Osservatorio Astrofisico – Catania (Italy)
[4] European Southern Observatory – Paranal Observatory (Chile)
[5] Gemini Support Group – Oxford (United Kingdom)
[6] ESTEC – Noordwijk (The Netherlands)
[7] Harvard-Smithsonian Center for Astrophysics – Boston (USA)

Abstract. The combination of the unprecedented spatial resolution attainable with WFPC2 on board HST and of the large collecting area of the VLT makes it possible to study in detail the low mass pre-Main Sequence stars in galaxies other than our own. Here we present the results of our studies of two star forming environments in our closest galactic neighbor, the Large Magellanic Cloud: the region around Supernova 1987A and the double cluster NGC 1850.

1 Stellar Populations in the Large Magellanic Cloud

When it comes to studying stars in galaxies other than our own Milky Way the Large Magellanic Cloud (LMC) is, for several reasons, an obvious starting point:

- With a distance of 52 ± 1 kpc [1] it is the closest galaxy we can look at from the outside and it is fairly easy to reach down to stars of 1 $M_\odot$, corresponding to $m_V \simeq 24$, or less.
- All of the stars are at one and the same distance.
- Our view is not severely obstructed by Galactic extinction: $E(\mathrm{B-V})_{\mathrm{Galaxy}} = 0.05$ [2].
- The stars in the LMC span a wide range of ages and physical conditions from Globular Cluster-like to star forming environments.
- Low metallicity: $Z \simeq Z_\odot/3$ corresponds to the mean metallicity of ISM at $z \simeq 1.3$ [3] at which the overall star formation rate is highest [4].

The location in the LMC of the two regions we have studied, the surroundings of SN 1987A and the double cluster NGC 1850, are shown in Fig. 1 on a Digitized Sky Survey image of the galaxy.

2 HST-WFPC2 Imaging

2.1 The Region of Supernova 1987A

The first star forming region we have considered in our search for low mass pre-Main Sequence (TTauri) stars is the one around SN 1987A. The pre-Supernova

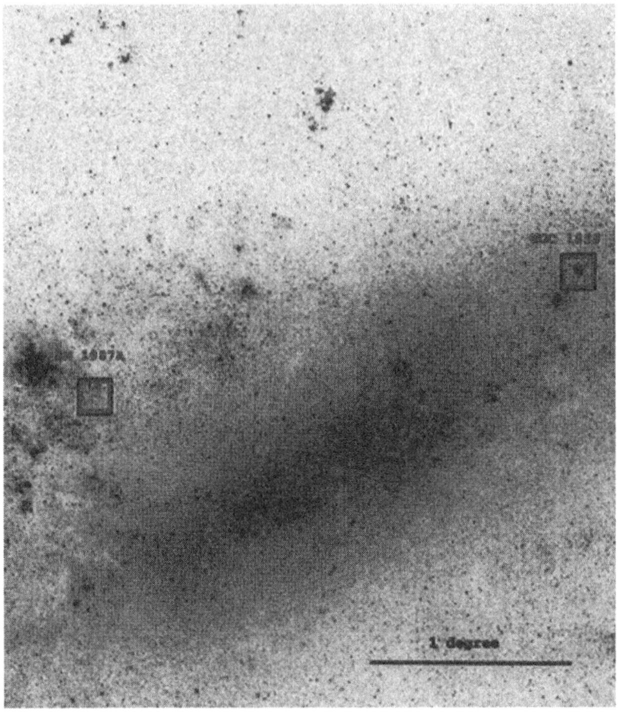

Fig. 1. The location of SN 1987A and NGC 1850 in the LMC on a DSS image of the galaxy. The scale is shown by the horizontal bar

evolution of its progenitor, Sk -69 202 [5], is estimated to have lasted 10-12 Myr [6] and one can expect to find a similarly young population, born together with it.

From 1994 the region was imaged almost every year with the WFPC2 as a part of the long term **S**upernova **IN**tensive **S**tudy led by Bob Kirshner. This resulted in the coverage in 6 wide bands, from 2500 to 8500 Å, plus OIII 5007 Å, Hα and NII 6548 Å, of a circular region with a radius of 30 pc centered on the Supernova remnant. The HR diagram for the 21,955 stars we have identified in our multiband WFPC2 frames and for which we have derived accurate luminosities and temperatures with a new technique based on photometry alone [7] is shown in Fig. 2.

As shown in Fig. 2 there are stars of very different ages, ranging from a few million to several billions years. In particular, the location of the most massive stars in the field, except for the one highlighted in the circle, is consistent with them being coeval to the progenitor of SN 1987A which, indeed, was not born in isolation, but, rather, in a loose cluster [11]. These massive ($M \simeq 12 M_{\odot}$), bright ($\log(L/L_{\odot}) \simeq 4.5$) stars are easy to identify even in a region of complex star formation such as this one. Unfortunately, though, this is not the case for the corresponding low mass population. The expected location in the HR diagram

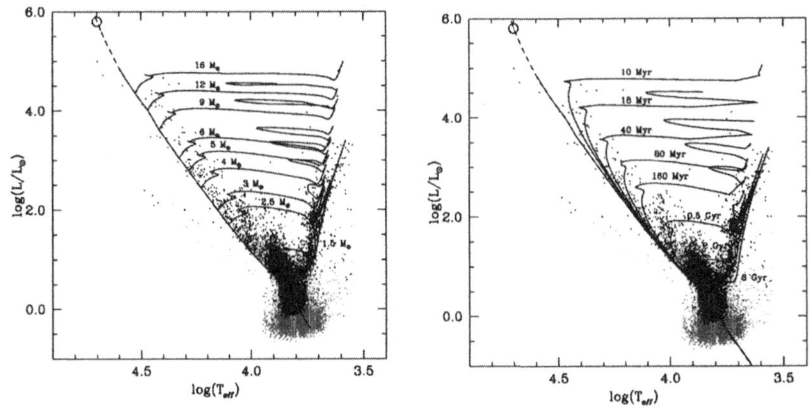

Fig. 2. HR diagram for the stars in the field of SN 1987A. Black dots are stars with $\delta \log(T_{eff}) < 0.05$, while the circle highlights the most massive star in the field. Evolutionary tracks (left panel) and isochrones (right panel) by [8] and [9] are overplotted to the data. The dashed line is the upper Main Sequence as computed by [10]

for these stars ($\log(T_{eff}) \simeq 3.8$, $\log(L/L_\odot) \simeq 0.2$), which are still contracting towards the Main Sequence, overlaps with the one of the (much more numerous) field sub-giants that, with a similar mass of a few solar masses, but an age of several billion years, have just left it.

The fundamental issue, which will present itself each and every time a star forming region is projected onto a much older population, then, is to find a way to identify the TTauri stars and disentangle them from the sub-giants. Luckily, the so-called Classical TTauri stars, which are thought to have a disk around them, have at least two clear, distinctive and correlated characteristics in the optical: a U-band excess when compared to a photosphere of an evolved star of the same spectral type (see, for example, [12]) and an Hα emission which can amount to several tens of Angstroms (see, for example, [13]).

Using these diagnostic tools we identified 850 TTauri candidates with U-band excess and 488 candidates with Hα emission. The vast majority of these latter ones also show an excess in the U. Let us state here very clearly that *both the criteria mentioned above will for sure underestimate the real number of TTauri stars.* On the one side, the detection level will in both cases depend on the depth of the exposures: a very shallow Hα image, for instance, will only allow to identify stars with a strong emission line. This effect is hard to quantify, as TTauri stars are variable and the features we use will vary significantly at different times. In addition, and more importantly, X-ray studies in the Milky Way [14] showed that Classical TTauri stars represent a minority of all low mass pre-Main Sequence objects. Unfortunately the so-called Weak TTauri stars do not have any clear photometric signature of their nature and they can be identified only either in the X-rays or with spectroscopy.

An example of the effects of the incompleteness in identifying TTauri stars is illustrated in Fig. 3. There the Initial Mass Function between 1 and 10 $M_\odot$ is plotted for the two recipes to identify pre-Main Sequence stars described above: Hα emission or U band excess. The derived slope is $\Gamma = -1.55$ in the first case (the classical Salpeter value is $\Gamma = -1.35$) and as steep as $\Gamma = -1.87$ in the latter one!

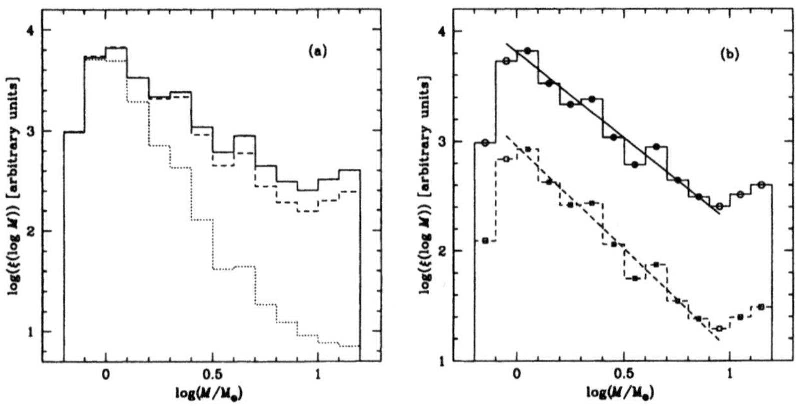

Fig. 3. Initial Mass Function in the neighborhood of SN 1987A. *Panel (a):* the IMF derived including as TTauri stars only the stars with Hα excess is shown as a full line, the one computed including also the stars with U-band excess as a dashed line. The Present Day Mass Function is also shown as a dotted line. *Panel (b):* power-law fit to the IMFs of panel (a). The bins used for the fit are marked with dots yielding a slope of $\Gamma = -1.55$ if only the stars with Hα emission are included and $\Gamma = -1.87$ if also the ones with U-band excess are considered. An arbitrary shift is applied to better show the data

A complete discussion on the young population around SN 1987A can be found in [11].

2.2 The Double Cluster NGC 1850

NGC 1850 is a double cluster in the outskirts of the LMC bar (see Fig. 1). According to our early WFPC2 investigation [15], the main component, NGC1850A, has an age of 50 ± 10 Myr and the slope of the IMF is $\alpha = -1.4 \pm 0.2$, *i.e.* considerably flatter than the Salpeter value of -2.35. NGC1850B, on the other hand, is extremely young, 4 ± 1 Myr, and is characterized by a much steeper Initial Mass Function: $\alpha = -2.6 \pm 0.1$. In addition, there are the usual LMC field stars, as clearly indicated by the presence of the Red Clump at F439W$\simeq$ 20, F439W$-$F814W$\simeq$ 2.2. The color-magnitude diagram for NGC 1850 is shown in Fig. 4.

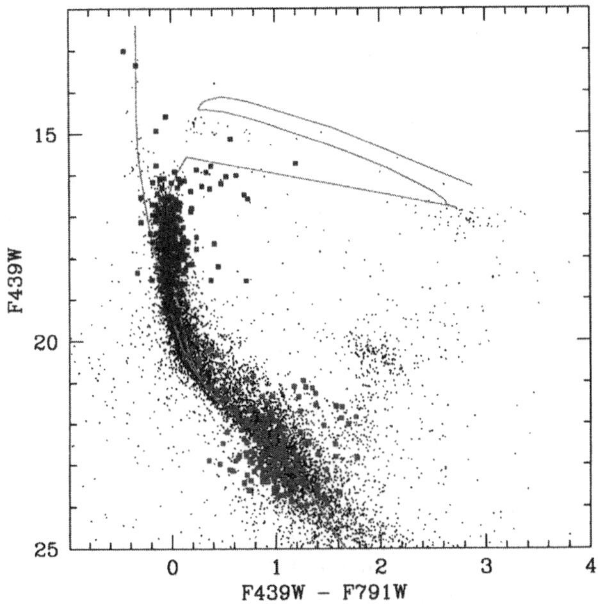

Fig. 4. F439W vs (F439W–F814W), *i.e.* roughly B vs (B-I), color-magnitude diagram of NGC 1850. The squares are Hα-emitting stars and the isochrones for 4 and 50 Myr are displayed as full lines

Once again, we know that there *must* be TTauri stars associated with the young cluster, but, as before, they are drowned in the much older sub-giant population and broad band photometry only allows for statistical arguments, but not an identification on a star to star basis. However, the addition of Hα photometry, again with WFPC2, allowed us to discover 230 Classical TTauri candidates (and 350 Be stars belonging to the older cluster). Let us stress again that the sample is by far incomplete and this number surely is a lower limit to the real content of low mass pre-Main Sequence stars. Once again, follow-up spectroscopy is needed to shed light on the low mass star population.

The full analysis of NGC 1850 will appear in [16].

3 VLT-FORS1 Spectroscopy

To recapitulate, in order to fully characterize young stellar populations projected on old field stars one has to find a way to distinguish TTauri stars from field subgiants. Both methods we have used, U-band excess and Hα emission, have allowed us to identify several hundred TTauri candidates in the two regions we have targeted. However, neither criterion yields a complete census of low mass pre-Main Sequence stars. In particular, only Classical TTauri stars can be identified, while *all* Weak TTauri stars will be missed.

Ideally, a suitably deep X-ray survey would provide a complete sample, but, unfortunately, the current generation of X-ray instruments does not have enough sensitivity to detect TTauri stars beyond the Milky Way in a reasonable integration time. In this case, then, even the Large Magellanic Cloud is too distant! Thus, in order to understand the biases introduced by the selection criteria we had to adopt on the WFPC2 imaging data, we have applied for, and were granted, two Visitor Mode nights with FORS1 on the VLT Antu (UT1) telescope in its Multi Object Spectroscopy mode. The grism we have chosen, GRIS 300V, covers a wide spectral region centered roughly at 5000 Å and including Hβ, Hα and Li I 6707 Å. The sample selected for follow-up spectroscopy consists of 20 candidate TTauri stars and as many stars that fall in the same region of the HR diagram, but without neither Hα emission nor U-band excess.

The observations were designed to fulfill two main goals. First, the spectra would provide a critical test of our selection criterion based on Hα emission and, second, they would allow to accurately determine the characteristics of our putative TTauri stars (spectral type, amount of veiling, line profile and equivalent width of the Balmer lines, etc).

Regrettably, though, we were not able to fulfill any of the proposed goals. The unfortunate combination of *El Niño* and the Bolivian winter at the beginning of the year 2000 resulted in our two nights having a seeing variable between 1.5 and 2″: way too much for spectroscopy of $V = 21.5 - 22$ objects in a crowded field, even with an 8-meter telescope!! As a partial consolation, among other things, in those nights we did obtain several narrow band images of the region of SN 1987A and of NGC 1850, which we have used to complete our understanding of them by studying, in addition to the stars, also the interstellar medium entwined with them. As for our original goal, at the time of writing we have resubmitted the proposal to the ESO OPC for Period 68. This time, though, in Service Mode ...

References

1. M. Romaniello, M. Salaris, S. Cassisi, N. Panagia: ApJ, **530**, 728 (2000)
2. P.B.W. Schwering, F.P. Israel: A&A, **246**, 231 (1991)
3. Y.C. Pei, S.M. Fall, M. Hauser: ApJ, **522**, 604 (1999)
4. P. Madau, H.C. Ferguson, M.E. Dickinson, M. Giavalisco, C.C. Steidel, A. Fruchter: MNRAS, **283**, 1388 (1996)
5. N. Sanduleak: *Contr. Cerro Tololo Interam. Obs*, **No. 89** (1969)
6. S. Van Dyke, M. Hamuy, M. Mateo: In *SN 1987A: Ten Years Later*, ed. by M.M. Phillips, N.B. Suntzeff (ASP Conference Series 1998)
7. M. Romaniello, N. Panagia, S. Scuderi, R.P. Kirshner: AJ, in press (2001)
8. E. Brocato, V. Castellani: ApJ, **410**, 99 (1993)
9. S. Cassisi, V. Castellani, O. Straniero: A&A, **282**, 753 (1994)
10. D. Schaerer, G. Meynet, A. Maeder, G. Schaller: A&ASS, **98**, 523 (1993)
11. N. Panagia, M. Romaniello, S. Scuderi, R.P. Kirshner: ApJ, **439**, 197 (2000)
12. E. Gullbring, L. Hartmann, C. Briceno, N. Calvet: ApJ, **492**, 323 (1998)
13. S. Edwards, P. Hartigan, L. Ghandour, C. Andrulis: AJ, **108**, 1056 (1994)
14. F.M. Walter: ApJ, **306**, 573 (1986)
15. R. Gilmozzi, E.K. Kinney, S.P. Ewald, N. Panagia, M. Romaniello: ApJ, **435**, L43.
16. M. Romaniello, R. Gilmozzi, N. Panagia, E. Tolstoy: in preparation

NGC 1850 in the LMC
(Antu/UT1 + FORS1)

Tom Greene

High-Resolution Near-IR Spectroscopy of Protostars with Large Telescopes

Thomas Greene

NASA / Ames Research Center, M.S. 245-6, Moffett Field, CA 94035, USA

Abstract. It is now possible to measure absorption spectra of flat-spectrum and Class I protostars using D$\geq$ 8m telescopes equipped with sensitive cryogenic IR spectrographs. For the first time ever, our latest high-resolution ($R \sim 20,000$) Keck NIRSPEC data reveal that Class I protostars are indeed low-mass stars with pre-main-sequence or dwarf-like features. However, Class I and flat-spectrum protostars differ from T Tauri stars in that protostars have much higher IR veilings ($r_k \geq 1-3+$) and they are rotating quickly, v sin $i > 20$ km s^{-1}. Interestingly, absorption spectra of low-mass protostars show stellar – not disk – absorption features. Our observations should also be able to probe whether protostellar photospheres have different physical structures than T Tauri stars, as predicted by theoretical models which account for accretion effects. There is much work to be done in this field, and the VLT can contribute immensely.

1 Introduction

Thanks to phenomenal progress in the performance of IR detectors, near-IR spectrographs have improved over the past decade to the point where their data have been useful for determining the spectral types, veilings, and by extension the masses and ages of optically invisible low-mass pre-main sequence (PMS) stars embedded in nearby star-forming clouds (e.g. [1] and [2]). Using moderate-resolution ($R \equiv \lambda/\delta\lambda \sim 1000$) single-object spectrographs on D $\sim$ 4m telescopes, 100 or more embedded PMS stars have been observed in each of several nearby dark clouds, providing unique insight to the mass functions and ages of their hidden populations (e.g. [3] and [4]). The first multi-object near-IR spectrographs [5] [6] are now being delivered, and these should allow even more thorough studies of the stellar and substellar populations hidden in nearby clouds.

However, moderate-resolution spectroscopy has not been successful in discerning much about the natures of protostars, which are younger and more deeply embedded than PMS stars. Casali and Matthews first showed that although the near-IR spectra of Class II young stellar objects (YSOs) generally show (photospheric) absorption lines, Class I YSOs do not [7]. Instead, Class I YSOs generally have featureless near-IR continua which are very reddened and veiled by the emissions of warm circumstellar dust. This was confirmed in the larger sample of Class I – III YSOs studied by Greene & Lada [8], and others have generally found this to be true also (e.g. [4]).

Very little is known about the actual *stellar* characteristics of embedded protostars since they are so difficult to observe. Are their spectral types and luminosities similar to those of T Tauri stars, or does their accretion of mass substantially change these properties [9]? What can be discerned about the disk structures and accretion properties of these young stars? Are they actually accreting at the rates predicted by their natal cloud temperatures and density structures, and how does the matter actually get to the protostar from the circumstellar envelope through the disk; is the accretion along magnetic field lines or is there some other mechanism at work? Finally, what is the angular momentum of an accreting protostar, and is it fundamentally different from that of a slowly-rotating Classical T Tauri star (CTTS)? We will show in the remainder of this paper that sensitive, *high-resolution* near-IR spectroscopy with large telescopes is a powerful observational technique which can make progress in addressing these questions.

2 High-Resolution Near-IR Stellar Spectral Diagnostics

High-resolution ($R \sim 20000$) near-IR spectra are very diagnostic of stellar temperatures, surface gravities, and rotation velocities. There are many lines in the $1 - 2.5$ μm region which are quite sensitive to these properties in late-type stars, and several of these features are located quite close to each other, easily within the small spectral range provided by single-order near-IR spectrographs with 256×256 pixel detectors ($\Delta\lambda \sim \lambda/(100 - 400)$). Figure 5 shows the spectra of a few dwarfs, giants, and a supergiant in one of these regions in the K band, $2.2055 - 2.2100$ μm.

There are also several good spectral features in the near-IR which are well-suited to diagnosing the rotation velocities and profiles of stars. The K band 2.2935 μm CO v = 0–2 bandhead region is very good for diagnosing the rotation of PMS stars because this feature is strong in late-type stars. Its features (bandhead edge and many nearby rotation-vibration lines) are very sensitive to both rotation velocity as well as rotation profile; i.e. the shape of this feature can indicate whether it arises in a stellar atmosphere or a circumstellar disk. Its sensitivity to stellar rotation is shown in Figure 2.

3 Properties of Flat-Spectrum Protostars

We have begun a program to diagnose the stellar and circumstellar properties of embedded protostars using the high-resolution spectral diagnostics in the K-band Na and CO regions shown in Figures 5 and 2. We selected K-band features because protostars are generally too faint to be observed with high resolution spectroscopy at shorter wavelengths (J or H bands) with even the world's largest telescopes. The disadvantage to working in the K band is that excess continuum emission from circumstellar dust is strong enough to veil gaseous absorption features at those wavelengths.

Our observational studies using the CSHELL spectrograph on the IRTF 3.0-m telescope were successful at revealing absorption lines of several flat-spectrum YSOs in the ρ Oph dark cloud [10] [11]. These objects are more embedded than PMS stars but are not strongly accreting like Class I protostars (see [12]). Our

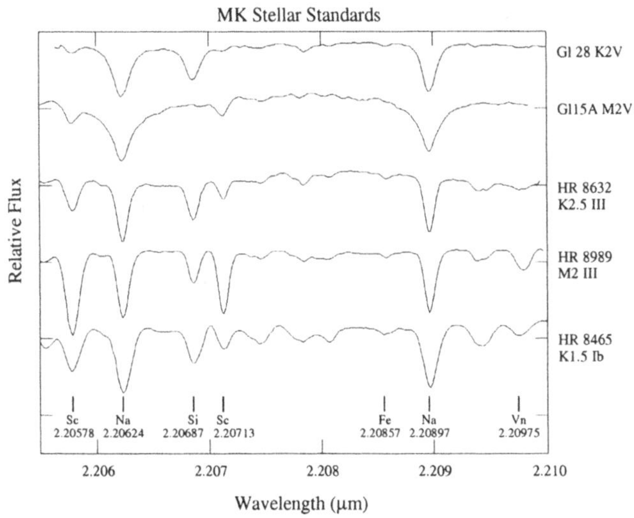

Fig. 1. Features in high-resolution near-IR spectra are very sensitive to the effective temperatures and surface gravities of late-type stars. Note that Si line strength decreases and Sc strength increases with later spectral type within each luminosity class (III and V). Sc, Vn, and molecules increase in strength as surface gravities decrease (V to III to I luminosity classes) at similar effective temperature

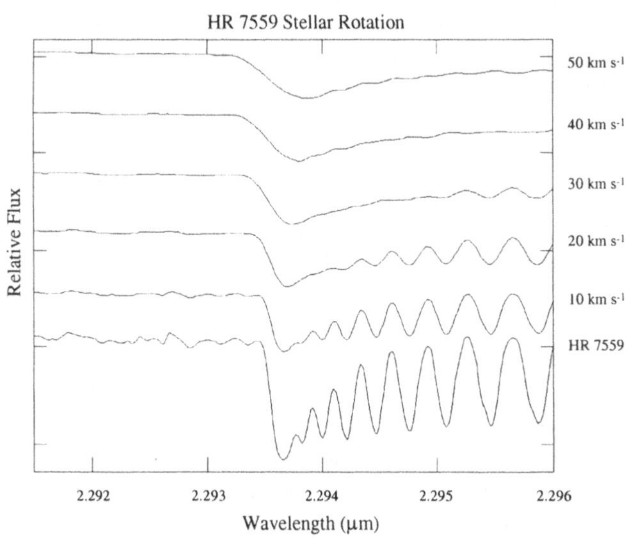

Fig. 2. The 2.2935 μm CO v = 0–2 bandhead region is very sensitive to stellar rotation. The spectrum of the slowly-rotating K5III star HR7559 is shown at the bottom of the figure, and it is convolved with limb-darkened stellar rotation profiles in the range 10 km $s^{-1} \leq v \sin i \leq 50$ km s^{-1} in the other traces

spectra show that their K-band spectral lines form in late-type stellar photospheres which are rotating relatively rapidly, $v\sin i > 20$ km s^{-1} (Figs. 3 and 4). There is no evidence that these lines form in disks: the line profiles, velocities, and the derived surface gravities all indicate stellar photospheric origins. The spectra are also highly veiled, typically $r_k \simeq 1$. This means that the flux of the excess circumstellar emission is approximately equal to the stellar fluxes of these objects at K-band wavelengths. This veiling decreases the line depths relative to the continuum, making the lines weaker and more difficult to detect.

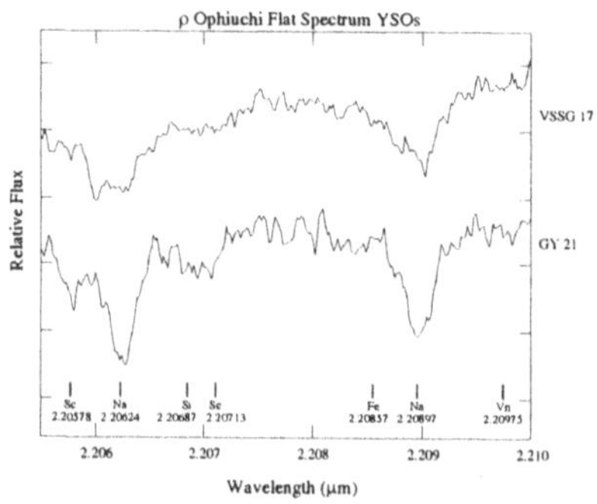

Fig. 3. 2.21 μm region plot of flat-spectrum protostars VSSG17 and GY21 in the ρ Oph cloud. These spectra are dominated by Na lines and are similar to the spectra of late- type dwarfs in Fig. 5 except they have high veilings, $r_k \sim 1$.

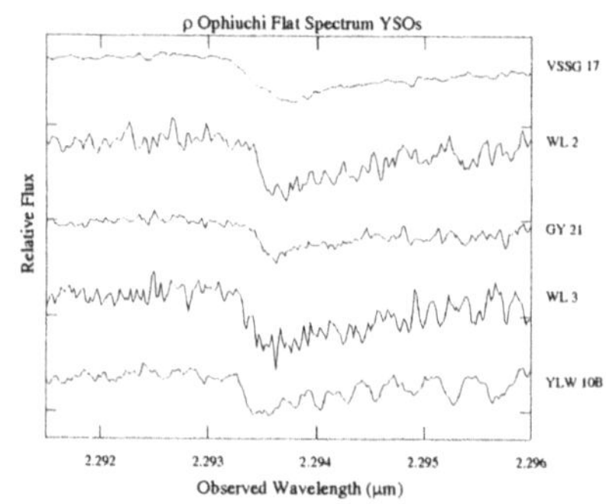

Fig. 4. v=0–2 CO bandhead region spectra of Flat-spectrum protostars in the ρ Oph cloud. The rotation velocities of these objects span the range 20 km s^{-1} $\leq v\sin i \leq$ 50 km s^{-1}, derived by fitting these spectra to the templates of Fig. 2

It is interesting that the flat-spectrum protostars have PMS or dwarf-like spectral features (consistent with luminosity class IV/V) and high rotation velocities. The models of Calvet et al. [13] predict that strong 2.3 μm CO band absorption bands should form in the circumstellar accretion disks of protostars, which should have very low surface gravities (like supergiants or lower). FU Orionis stars (Fuors) are the only young stars known to show evidence of both low-gravity line formation (e.g. in the 2.21 μ region) and rapid rotation, typically fit in the 2.3 μm CO feature. The near-IR spectra of flat-spectrum protostars clearly do not show evidence of massive Fuor-type disks. The high rotation velocities ($v\sin i > 20$ km s^{-1}) of the flat-spectrum protostars are also puzzling because many CTTSs in Tau-Aur and near ρ Oph rotate slowly, $v\sin i < 20$ km s^{-1}. The slow rotation of CTTSs in these regions has been interpreted to be caused by magnetic coupling between the stars and their circumstellar disks [14]. The faster rotation of the flat-spectrum protostars can be understood if either: 1) the stars are not fully coupled to their disks; or 2) the stars are coupled to their disks but at closer radii than for CTTSs.

4 Class I Protostellar Spectra

Our original attempts to obtain near-IR absorption spectra of Class I protostars using CSHELL on the IRTF were unsuccessful. We were able to obtain data with continuum signal-to-noise ratios of 40 – 100 on objects as faint as K=10.5 mag, but this was not adequate to detect these object's absorption lines which suffer very strong continuum veiling, $r_k \geq 3$ [11].

However, using NIRSPEC [15] on the Keck II telescope we have recently obtained higher signal-to-noise (S/N > 150) spectra of several Class I protostars which show absorption features. Like flat-spectrum YSOs, the IR spectra of these objects indicate that their absorption features form in late-type stellar photospheres that are rotating quickly. Line ratios in the 2.21 μm region also match those of luminosity class IV/V stars, and the rotation velocities of the few Class I protostars which we have observed are $v\sin i \geq 20$ km s^{-1}. One major difference between Class I and flat-spectrum protostars is that the Class I objects have much higher near-IR continuum veilings, $r_k \geq 3$. Sample Class I spectra are shown in Figure 5.

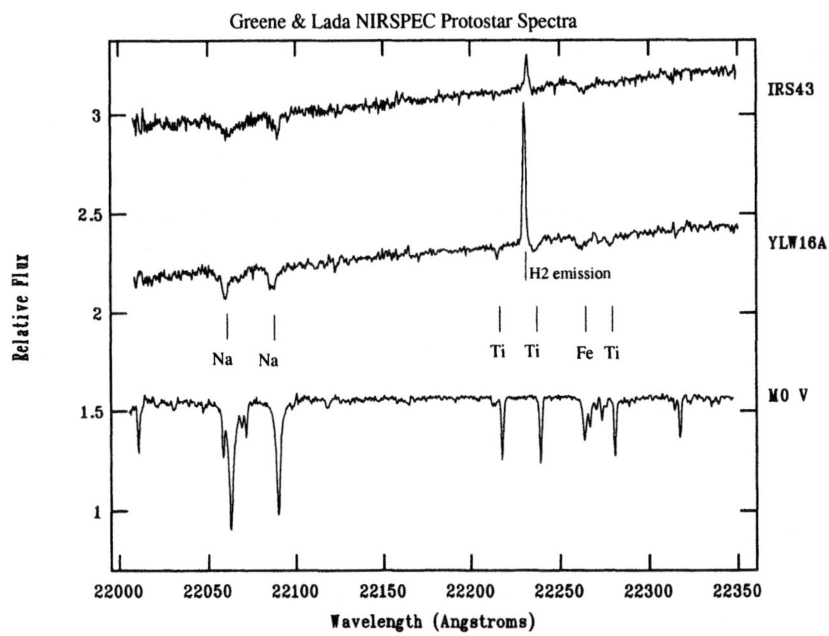

Fig. 5. A single order of Keck II NIRSPEC data is shown for the ρ Oph Class I YSOs IRS43 and YLW16A. The spectrum of a M0V stellar standard is shown for comparison. Note that this order spans 7 times the spectral range of CSHELL at this wavelength (see Fig. 5), and this is only one of the 6 orders which were obtained simultaneously

Figure 5 shows the first spectra ever which reveal the physical conditions of the photospheres of accreting, embedded low-mass protostars. It is especially interesting that even these extremely veiled spectra show no evidence of line formation in disks; instead the luminosity class IV/V line strengths indicate that protostellar radii are similar to those of PMS stars, which is also consistent with star formation theory [16]. The high veilings indicate considerable accretion from either a circumstellar disk or envelope (see [8]). The measured rotation rates are consistent with those of flat-spectrum objects, but appear to be considerably less than breakup velocity (about 1/4 of breakup for a 60° disk inclination).

5 Opportunities for the VLT

We have only recently just reached the stellar surfaces of accreting protostars with NIRSPEC on the Keck II, and we had found that this was impossible to do with smaller telescopes. There is much opportunity for the VLT to make considerable progress in this field. More high-resolution near-IR spectra should be acquired with higher signal-to-noise in order to build a large enough sample of Class I protostars in several nearby clouds. This sample is needed to measure the mass accretion rates through continuum veiling, to determine the initial stellar angular momentum (overcoming rotation axis projection effects with large numbers), and establish the zero-point of stellar evolution by placing the objects in H-R diagrams. The VLT CRIRES instrument under development will be very well-suited to the task, and it should make immediate contributions to this area. In the future, spectra could be acquired much more efficiently with new instruments which could observe multiple objects within a large field and if they could acquire the entire 2.20 – 2.30 μm region in a single spectral order. This should just be possible at R=20000 using a 2048 × 2048 pixel detector array.

References

1. K.-W. Hodapp, J. Deane: Ap. J. Supp. **88**, 119 (1993)
2. T. P. Greene, M. Meyer: Ap. J. **450**, 233 (1995)
3. K. L. Luhman, G. H. Rieke, C. J. Lada, E. A. Lada: Ap. J. **508**, 347
4. K. L. Luhman, G. H. Rieke: Ap. J. **525**, 440 (1999)
5. R. Elston: Proc. SPIE **3354**, 404 (1998)
6. `http://www.eso.org/instruments/nirmos/`
7. M. M. Casali, H. E. Matthews: MNRAS **258**, 399 (1992)
8. T. P. Greene, C. J. Lada: A. J. **112**, 2184 (1996)
9. C. A. Tout, M. Livio, I. A. Bonnell: MNRAS **310**, 360 (1999)
10. T. P. Greene, C. J. Lada: A. J. **114**, 1703 (1997)
11. T. P. Greene, C. J. Lada: A. J. **120**, 430 (2000)
12. N. Calvet, L. Hartmann, S. J. Kenyon, B. Whitney: Ap. J. **434**, 330 (1994)
13. N. Calvet, A. Patino, G. C. Magris, P. D'Alessio: Ap. J. **380**, 617 (1991)
14. S. Edwards et al.: A. J. **106** 372 (1993)
15. I. McLean et al.: Proc. SPIE **3354**, 566 (1998)
16. S. W. Stahler, F. H. Shu, R. E. Taam: Ap. J. **242**, 226 (1980)

ISAAC on Antu/UT1

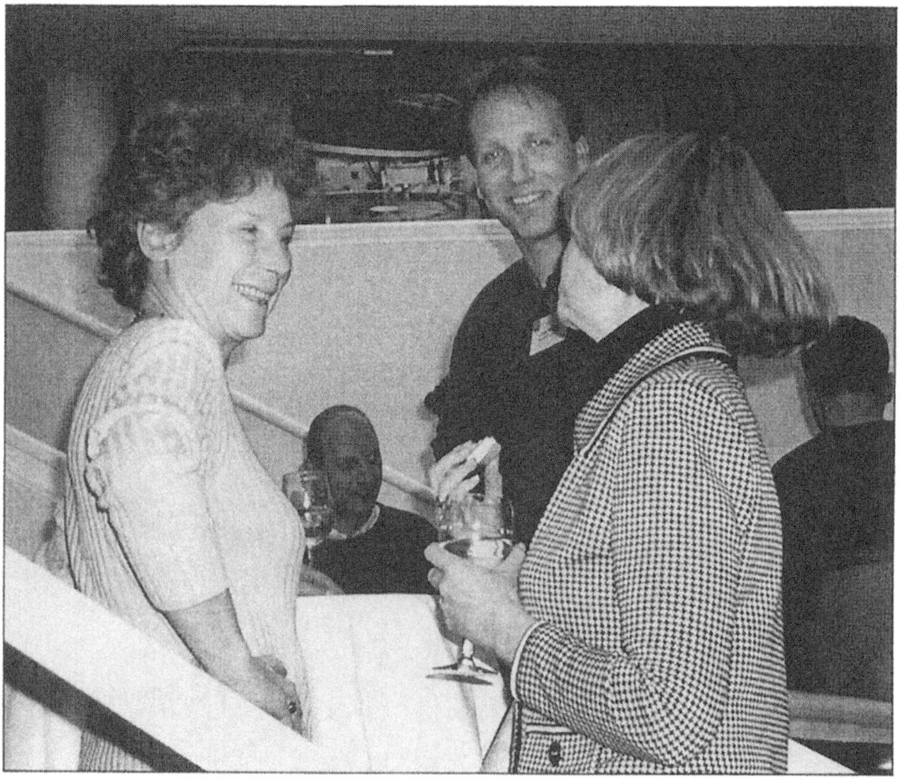

Jacqueline Bergeron, Lex Kaper, and Anneila Sargent

VLT/ISAAC Spectroscopy of Young Massive Stars Embedded in Ultra-Compact H II Regions

Lex Kaper[1], Arjan Bik[1], Margaret M. Hanson[2], and Fernando Comerón[3]

[1] Astronomical Institute, University of Amsterdam, Kruislaan 403, 1098 SJ Amsterdam, The Netherlands
[2] Dept. of Physics & Astronomy, University of Cincinnati, USA
[3] European Southern Observatory, Garching, Germany

Abstract. Using ISAAC mounted at the *Very Large Telescope*, we have obtained medium-resolution K-band spectra of newly formed massive stars, which are deeply embedded in ultra-compact H II regions (UCHIIs). Candidate young massive stars were selected on the basis of their near-infrared luminosity and colour measured from narrow-band images obtained in a survey of 45 southern UCHIIs. This strategy turned out to be very successful: follow-up spectroscopy confirmed the OB-star nature of 36 embedded stars, among them O stars of very early spectral type. The K-band spectra of over a dozen stars do not show photospheric absorption lines, but include a strong and broad Brγ emission line. These stars might represent an early phase in the evolution of massive stars, when they are still surrounded by a circumstellar disk. Our ultimate goal is to better understand the formation process of the most massive stars.

1 Introduction

The mechanism by which the most massive stars form is poorly understood. Contrary to the general picture applied to the formation of low-mass stars (Lada 1987, Shu et al. 1987), it is by no means clear whether very young, massive stars are initially surrounded by circumstellar disks, let alone how they are formed in the first place (e.g. Palla & Stahler 2000, Bonnell et al. 1998). This is in a large part due to the lack of observations covering the formation and early evolutionary stages of massive stars. The radio and infrared spectral regions are the only ones accessible for observations of the earliest stages of massive-star formation, because of the tens to hundreds of magnitudes of visual extinction. Another complication arises from the expectation that massive stars form on a short timescale, probably less than 100,000 years (cf. Hanson 1998 for a review on ZAMS O stars).

Ultra-compact H II regions (UCHIIs) represent the earliest recognizable stage in the life of massive stars. Historically, they have been identified using radio surveys, which search for their signature of a compact thermal source with a high turnover frequency, or in the mid-infrared bands, where they show unique IRAS colors (e.g. Churchwell 1991). The fact that they are tracers of ongoing massive star formation, together with their high luminosity and the transparency of the galactic disk at the wavelengths where they radiate most of their energy, make them extremely valuable in addressing a wide variety of issues related to the star-formation process in our galaxy and to the galactic structure (e.g. Comerón & Torra 1996).

Despite large observational campaigns in the radio and infrared (e.g. Wood & Churchwell 1989a, Kurtz et al. 1994, Walsh et al. 1998), fundamental measurements of UCHIIs, and, more importantly, of the embedded massive stars that ionize the H II region, remain unreliable and limited, principally due to the uncertain distances and to the indirect methods used to transform the observed radio continuum or mid-infrared fluxes into properties of the ionizing, young OB star(s). So far, only in one case the photospheric spectrum of such a deeply embedded, massive star has been detected in the near-infrared (Watson & Hanson 1997). What is attractive about working at these wavelengths (1–5 μm, accessible from the ground) is that the extinction due to the surrounding gas and dust is strongly reduced compared to that at optical wavelengths ($A_V/A_K \sim 10$). Furthermore, at just slightly longer wavelengths, the thermal emission of the dust contained in these star-forming regions becomes dominant.

The spectra of massive stars do not include many spectral features in the near-infrared wavelength domain. However, the small number of lines present in K-band spectra of OB-type stars show a clear dependence on spectral type (Hanson et al. 1996). Early O-type stars show unique emission features due to C IV and N III, as well as He II absorption lines. For late-O and early-B stars He I absorption lines provide a temperature diagnostic. Also L-band spectra of OB-type stars can be used for spectral classification (cf. Lenorzer et al. 2001).

Here we report on the first results of an observational campaign aimed at the direct detection of the photospheric spectra of recently formed massive stars, deeply embedded in ultra-compact H II regions. To achieve this, we first performed a near-infrared survey of southern UCHIIs to identify the candidate ionizing stars. Follow-up K-band spectroscopy has resulted in the detection of the photospheric spectrum of several tens of these stars, confirming their OB-star nature. The spectra are used to perform a detailed spectral classification, to study the circumstellar envelope (stellar wind, disk?), to search for evidence of binarity, and to measure their rotation rates.

2 Near-IR Survey of Southern Ultra-Compact H II Regions

We performed a near-infrared survey of 45 southern UCHII regions with SOFI mounted at ESO's *New Technology Telescope*, with the aim to first detect and select the deeply embedded, candidate ionizing stars. Due to the limited coverage of southern UCHIIs at radio wavelengths, we mainly had to rely on their specific IRAS colours (Wood & Churchwell 1989b) when selecting the sample. For the northern part of the sample we selected sources also included in the radio sample of Kurtz and coworkers (e.g. Kurtz et al. 1994); at more negative declinations we chose sources from the IRAS point-source catalogue with UCHII colours and strong 12 μm flux. All sources in our sample are comprised in the CS mm-survey by Bronfman et al. (1996).

Fig. 1. A "colour-composite" of three narrow-band images (red: nebular Brγ emission, green: stellar K continuum, blue: stellar J continuum) of the (ultra-)compact H II region IRAS 10049-5657 obtained with NTT/SOFI. North is to the left, East is to the top of the figure; the size of the image is $5' \times 5'$. The horizontal stripes are due to strong exposure. The image reveals a cluster of several tens of stars at the center of the nebular emission. The visual extinction is estimated to be more than 15 magnitudes (cf. Fig. 2). A detailed analysis of this region is subject of a poster by Bik et al. (this volume). *Figure available in colour on the CD-ROM*

We used narrow-band filters centered at strong (nebular) emission lines (Pβ 1.28 μm, H_2 2.12 μm, Brγ 2.17 μm), as well as two narrow-band continuum filters in the J and K band. The latter were used to obtain the J and K magnitudes of the target stars, thus avoiding contamination by nebular emission. Thanks to the excellent seeing ($\sim$ 0.5 arcsec) the nIR morphology of the UCHIIs can be studied at high spatial resolution. Almost all sources show nebular Brγ emission and contain one or more stars, in some cases a whole cluster (see Fig. 1 for an example). Often the UCHII is close to, or surrounded by a molecular cloud,

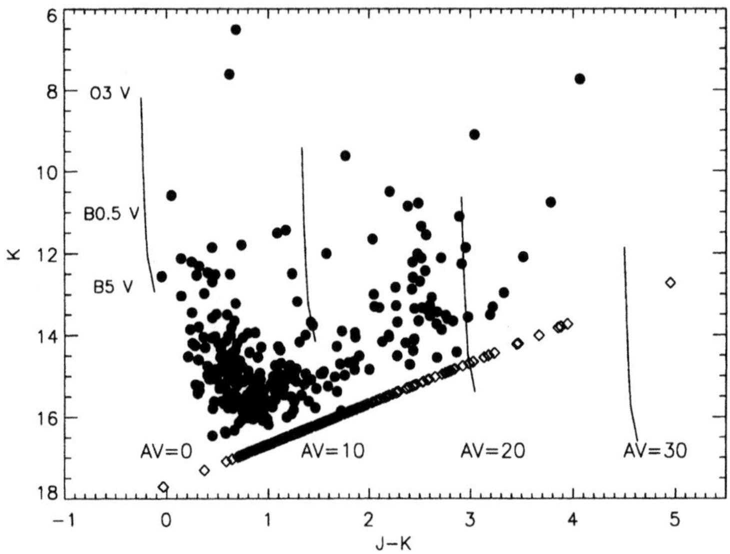

Fig. 2. The (K,J-K) diagram of the stars in the field of IRAS 10049-5657. The top of the theoretical ZAMS is drawn in assuming a distance of 7.1 kpc, for different values of the interstellar extinction (A_V). The intrinsic (J-K) colour of OB-type stars is around -0.2 (Hanson et al. 1997). A reddened main sequence, containing the (massive) stars in the embedded cluster (Fig. 1), is clearly visible. Using these diagrams we selected the candidate massive stars that provide the source of ionization of the UCHII

and partly obscured by dust lanes. The nebular (Brγ) emission of sources for which a radio counterpart was known in advance is mainly confined to a small spatial scale, as expected for ultra-compact H II regions. The more southern sources selected only on the basis of their IRAS colours often correspond to embedded H II regions of much larger spatial extent. Apparently, the limited spatial resolution of the IRAS satellite ($\sim 1'$) does not allow to discriminate between ultra-compact, compact, or even giant H II regions in several cases. The near-IR imaging survey of southern UCHIIs is presented in Kaper et al. (2001).

For the stars contained in each field we constructed a (K,J-K) diagram (Fig. 2). An estimate of the distance was obtained by comparing the radial velocity of, e.g., the observed CS emission of the UCHII (Bronfman et al. 1996) to a prediction of a flat rotation model of our galaxy (note that in some directions a distance ambiguity exists, which is removed as soon as a precise spectral classification of the embedded stars is available). With the adopted distance the zero-age main sequence (ZAMS) of massive stars, for different amounts of reddening, can be drawn in the diagram. The candidate ionizing OB stars should be located close to the top of the (reddened) ZAMS. We also selected some stars with extreme (J-K) colour and large K-band flux. These stars potentially are massive young stellar objects (YSOs) with strong infrared excess.

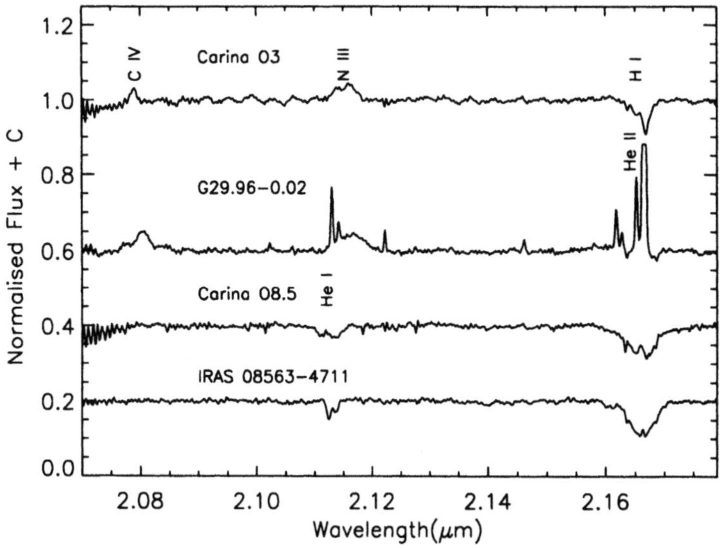

Fig. 3. Medium resolution ISAAC K-band spectra of two young O stars embedded in the UCHIIs G29.96-0.02 and IRAS 08563-4711, respectively. For comparison, spectra of "naked" O stars of similar spectral type are displayed. The massive star in G29.96-0.02 appears to be among the hottest stars in the galaxy. The narrow emission line in the blue wing of Brγ is identified as He II 2.1652 μm. The O star in IRAS 08563-4711 is of later spectral type. Note the close similarity between the young and supposedly more evolved O stars, at least in this wavelength domain

3 Follow-Up VLT/ISAAC K-Band Spectroscopy

We obtained medium-resolution ($R \sim 8,000$) K-band spectra of about 75 selected stars using VLT/ISAAC. High spectral resolution and signal-to-noise (≥ 100) are required to be able to correct for the contaminating nebular emission, as well as for telluric emission and absorption lines. Therefore, the large aperture of the VLT and the high sensitivity of ISAAC was needed.

Our strategy appeared to be very successful; only about one third of the target stars turned out to be late-type (i.e. later than OBA) foreground stars. The majority of K-band spectra show characteristic OB-star features. In Fig. 3 we show two examples. The massive star embedded in G29.96-0.02 is of very early spectral type. The spectrum is compared to that of a "naked" O3 star in Carina. Note the broad emission features due to C IV and N III, typical for early O-type main sequence stars. The K-band spectrum also includes nebular emission lines of singly and doubly-ionized helium, supporting the very early spectral classification of the embedded star. The other star displayed in Fig. 3 is a late-O star, showing photospheric He I and Brγ absorption lines.

The K-band spectra of the massive stars embedded in UCHIIs are very similar to those of supposedly older, "naked" OB main-sequence stars (cf. Hanson et al. 1996). The fraction of early O-type stars in our sample is relatively high: some of them possibly as early as spectral type O3. So far, only a few O3 stars have been found in our galaxy. Obviously, our sample should be biased towards young massive stars, which is consistent with this result. The photospheric spectra indicate that these embedded massive stars, though young, have already developed into normal OB main-sequence stars.

About 20 objects, most of them having a large (J-K) colour and a strong K-band flux, do not show any photospheric absorption lines. However, their K-band spectra include a strong and broad (FWHM up to a few 100 km s^{-1}) Brγ emission line, and in some cases broad emission features of other species (He I, Fe II, Mg II) as well. In one case the Brγ line is double-peaked, often it is asymmetric. These spectra are similar to those of proposed massive YSOs found in M17 (Hanson et al. 1997) and to precursors of Herbig Ae/Be stars (Ishii et al. 2001). For two objects in this class the CO bands at 2.3 μm are clearly in emission. The CO band-head emission, and the strong infrared excess, indicate that these objects are surrounded by dense circumstellar material, possibly a disk. For the remaining stars in this class CO observations are planned. We believe that the Brγ emission-line objects represent a very early phase in the life of massive stars (cf. Bik et al. 2001). We continue our studies with the aim to reveal their true physical nature.

References

1. Bik, A., Kaper, L., Hanson, M.M., et al. 2001, to be submitted to A&A
2. Bonnell, I., Bate, M.R., Zinnecker, H. 1998, MNRAS 298, 93
3. Bronfman, L., Nyman, L.-A., May, J. 1996, A&AS 115, 81
4. Churchwell, E. 1991, in "The physics of star formation and early stellar evolution", Eds. Lada & Kylafis, p. 221
5. Comerón, F., Torra 1996, J. 1996, A&A 314, 776
6. Hanson, M.M. 1998, in ASP Conf. Ser. 131, Ed. I.D. Howarth, p. 1
7. Hanson, M.M., Conti, P.S., Rieke, M.J. 1996, ApJS 107, 281
8. Hanson, M.M., Howarth, I.D., Conti, P.S. 1997, ApJ 489, 698
9. Ishii, M., Nagata, T., Sato, S., et al. 2001, AJ 121, 3191
10. Kaper, L., Bik, A., Comerón, F., Hanson, M.M. 2001, in preparation
11. Kurtz, S., Churchwell, E., Wood, D.O.S. 1994, ApJS 91, 659
12. Lada, C.J. 1987, in Proc. IAU Symp. 115, p. 1
13. Lenorzer, A., Vandenbussche, B., De Koter, A., et al. 2001, submitted to A&A
14. Palla, F., Stahler, S.W. 2000, in Proc. 33rd ESLAB Symposium "Star formation from the small to the large scale" Eds. Favata, Kaas, Wilson, p. 179
15. Shu, F.H., Adams, F.C., Lizano, S. 1987, ARAA 25, 23
16. Walsh, A.J., Burton, M.G., Hyland, A.R., Robinson, G. 1998, MNRAS 301, 640
17. Watson, A.M., Hanson, M.M. 1997, ApJ 490, L165
18. Wood, D.O.S, Churchwell, E. 1989a, ApJS 69, 831
19. Wood, D.O.S, Churchwell, E. 1989b, ApJ 340, 265

NGC 3603 IRS 9 : The Revealment of a Cluster of Protostars and the Potential of Mid-IR Imaging with VLT + VISIR

D. Nürnberger[1,2], L. Bronfman[3], M. Petr-Gotzens[4], and Th. Stanke[4]

[1] Institut für Theoretische Physik und Astrophysik, Universität Würzburg, Am Hubland, D-97074 Würzburg, Germany
[2] Institut de Radio-Astronomie Millimétrique, 300 Rue de la Piscine DU, F-38406 St. Martin-d'Hères, France
[3] Departamento de Astronomía, Universidad de Chile, Casilla 36-D, Santiago, Chile
[4] Max-Planck-Institut für Radioastronomie, Auf dem Hügel 69, D-53121 Bonn, Germany

Abstract. In the framework of a multi-wavelengths study of NGC 3603 we have performed deep infrared imaging with the VLT Antu + ISAAC to investigate the nature and the evolutionary status of reddened sources (IRS 9) associated with one of the molecular cloud cores adjacent to the OB cluster. Here we present first results and address the question whether these sources constitute a cluster of protostars in its own right. Our study also outlines the importance of dedicated mid-IR instrumentation (VISIR) which will allow the VLT to significantly contribute to our understanding of the earliest phases of high mass star formation.

1 Introduction

Molecular line and dust continuum observations (e.g. [5], [4]) have shown that young OB stars are usually deeply embedded in and obscured by massive environments of gas and dust. They contract to main sequence, hydrogen burning temperatures and densities on time scales which are shorter than typical accretion time scales. Thus, in contrast to low mass stars, young stellar objects of higher mass ($\mathcal{M} > 8\,\mathcal{M}_{\odot}$) are in general not detectable at optical wavelengths throughout their pre-main sequence phase.

Because they emit a large amount of hydrogen-ionizing Lyman-Continuum photons young OB stars ionize their circumstellar material resulting in the formation of ultracompact H II regions (UCHIIs; e.g. [16], [26]). They appear as compact, luminous infrared (IR) sources – similar to the Becklin-Neugebauer (BN) object in Orion – and their spectral energy distributions (SEDs) show a characteristic steep increase with IR wavelengths. For evolutionary stages prior to the UCHII phase, the massive parental cloud cores even cause near-IR radiation to be absorbed and reradiated at longer (mid-IR, far-IR and sub-mm) wavelengths.

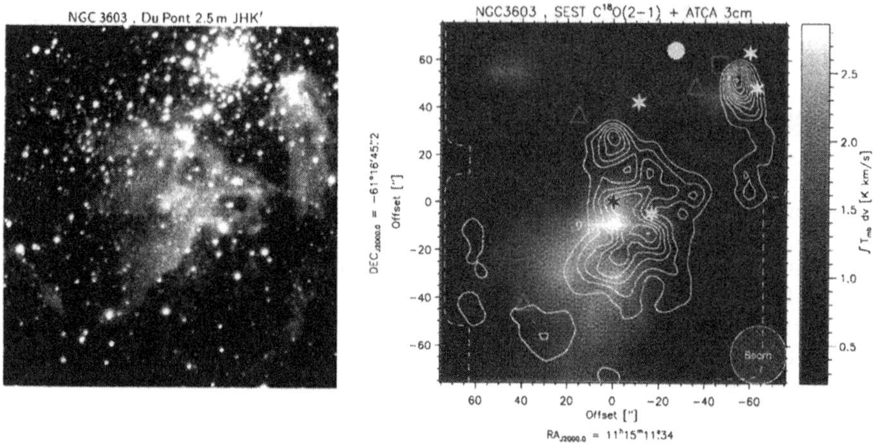

Fig. 1. 3-colour composite JHK′ image (left) as well as SEST $C^{18}O$ (2–1) and ATCA 3.4 cm data (right). In both panels the field of view is $150'' \times 150''$ centered on NGC 3603 IRS 9. North is up, East is left. In the right frame 3.4 cm emission originating from ionized material is displayed in contour lines on top of the $C^{18}O$ (2–1) emission from cold molecular gas. Additionally, the SEST beam (HPBW $\sim 24''$) at the frequency of the $C^{18}O$ (2–1) line is given in the lower right corner. The ATCA data are kindly provided by Chris De Pree [9].
Figure available in colour on the CD-ROM

In order to circumvent the problem of high extinction we here investigate a scenario where young high mass stars form in the violent neighbourhood of a cluster of early type main sequence stars. The presence of already evolved OB type stars provides a wealth of energetic photons as well as powerful stellar winds, which are capable of evaporating and dispersing the surrounding interstellar medium, setting nearby young stars free at a relatively early evolutionary stage.

2 NGC 3603 OB Cluster and its Adjacent Molecular Cloud

NGC 3603 – located in the Carina spiral arm at a distance of about 7 kpc (e.g. [18], [8], [9], [21]) – is one of the most luminous, optically visible H II regions in our galaxy. It is powered by a massive cluster of OB stars ([13]) which is a scaled down but 7 times closer galactic version of the R 136 cluster, the primary source of excitation in 30 Dor. Together with 30 Dor, NGC 3603 plays a key role in the understanding of extragalactic starburst regions.

The NGC 3603 cluster (see Fig. 1, left panel) shows the highest density of high mass stars known in the Galaxy [18], [19], [10], [15]. The combined effort of these hot stars has a severe impact on the surrounding gas and dust: on the one hand by providing a huge amount of ionizing photons (Lyman continuum flux $\sim 10^{51}\,s^{-1}$; [17], [10]) and on the other hand by (further) compressing the

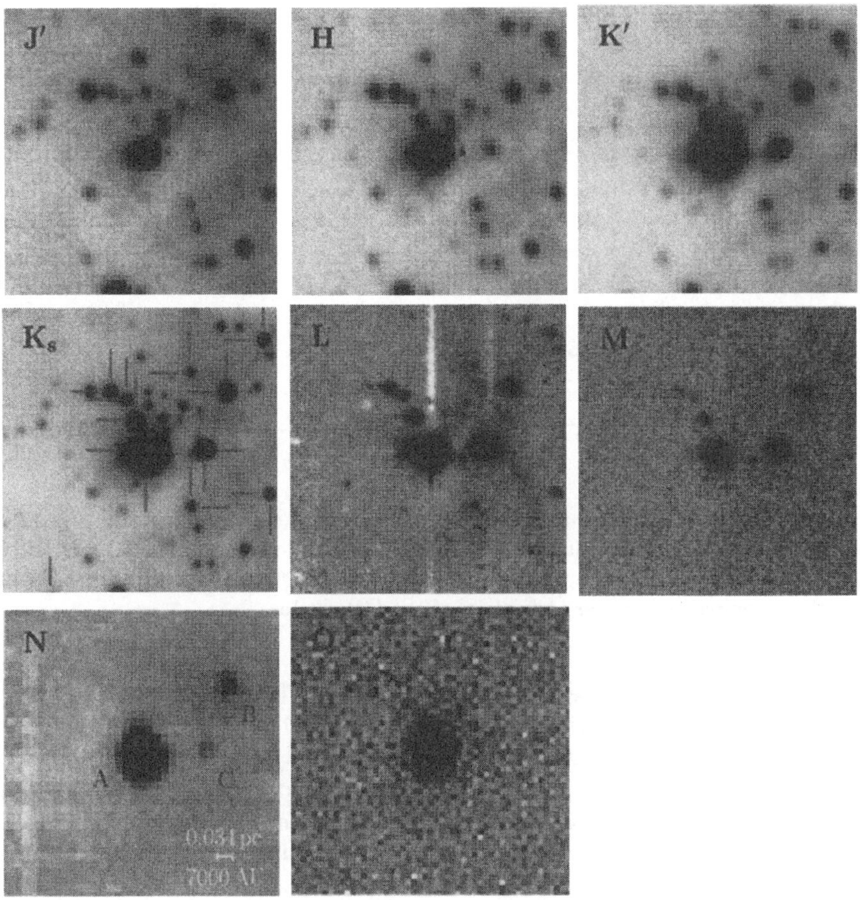

Fig. 2. Near-IR and mid-IR data obtained with the Las Campanas 2.5 m telescope (top row), the VLT Antu (middle row) and the La Silla 3.6 m telescope (bottom row). Each image is centered on IRS 9A and covers an area of $18\overset{''}{.}1 \times 18\overset{''}{.}1$, which corresponds to $0.6\,\mathrm{pc} \times 0.6\,\mathrm{pc}$ at the distance of 7 kpc. In the K_s image markers indicate sources with significant K band excess as identified by [25]

adjacent molecular cloud cores through fast stellar winds (velocities up to several $100\,\mathrm{km\,s^{-1}}$; [1]).

The synergy of both evaporates and disperses density enhancements of the interstellar medium, giving rise to proplyd- and/or pillar-like phenomena [3] and driving a gas-free "bubble" into the ambient medium [1], [6], [7]. The association of adjacent molecular clumps with several prominent IR and maser sources (i.e. typical signatures of active star forming regions) supports the idea of an ongoing star formation process within the NGC 3603 H II region.

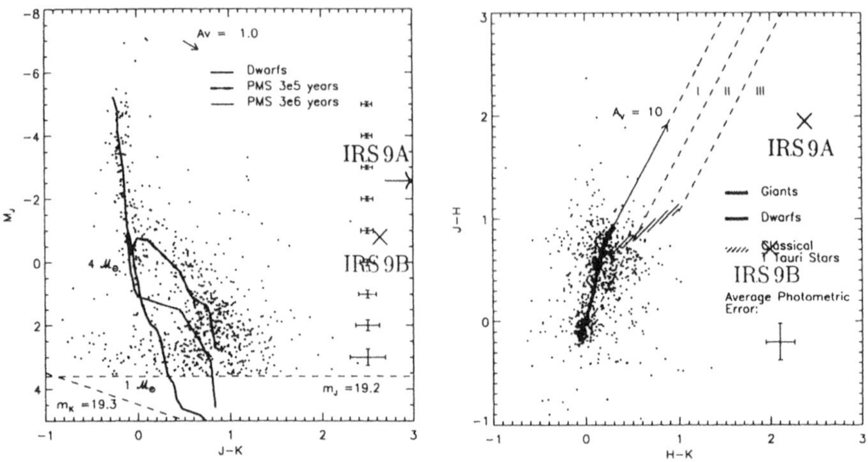

Fig. 3. Near-IR colour-magnitude and two-colour diagrams of the central 12″ × 12″ of the NGC 3603 OB cluster ([11]) with the corresponding positions of IRS 9A and IRS 9B indicated. Both diagrams impressively demonstrate the extreme youth of IRS 9A and IRS 9B in comparison to the population of the OB cluster

One particular molecular clump is located about 1.′3 towards the south of the OB cluster (see Fig. 1, right panel). From the $C^{18}O$ (2–1) data we estimate its total mass with about 400 $\mathcal{M}_\odot$ (Nürnberger, in prep.). On the OB cluster facing side of this clump – at the periphery of the gas-free cavity – a surprisingly large number of stars exhibits significant excess emission in the K band and obviously clusters around the prominent source IRS 9 [25]. It appears that these IR sources were recently revealed from their natal environment.

3 NGC 3603 IRS 9 – A Cluster of Protostars

In Fig. 2 we present near-IR and mid-IR data obtained with the Las Campanas 2.5 m telescope (J′HK′), the VLT Antu (K_sLM) and the La Silla 3.6 m telescope (NQ). Each image covers a field-of-view of 18.″1 × 18.″1, which corresponds to about 0.6 pc × 0.6 pc at the distance of 7 kpc, and is centered on the brightest source of IRS 9, which – following to the nomenclature introduced by [12] – is named IRS 9A here. In the K_s panel the excess sources are marked.

On near-IR colour-magnitude and two-colour diagrams the loci of these excess sources are clearly offset from those of the OB cluster members, as demonstrated by IRS 9A and IRS 9B in Fig. 3. According to [24] low mass sources possessing near-IR excess emission usually populate three distinct regions of the two-colour diagram: sources associated with low excess emission (WTTS) are found in zone I, while zones II and III feature sources with high excess emission caused by circumstellar disks (CTTS) and surrounding envelopes (protostars), respectively. The same or at least a similar sequence should hold for intermediate

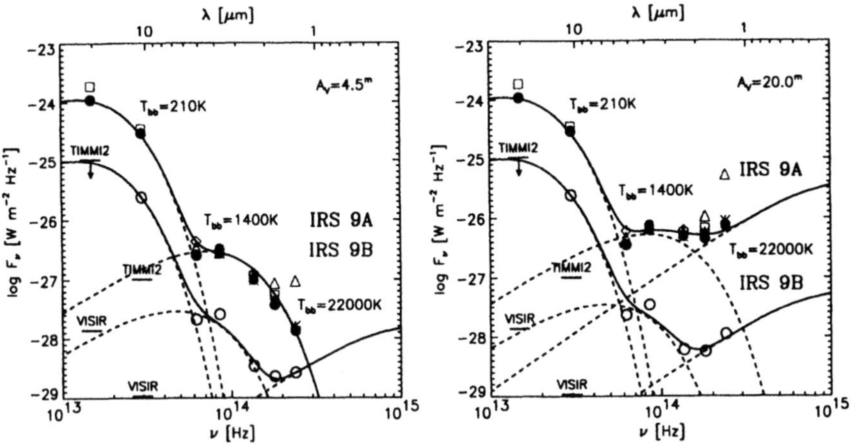

Fig. 4. Spectral energy distribution of IRS 9A after dereddening by $4^{\mathrm{m}}.5$ of foreground extinction (left) as towards the center of the OB cluster and additionally by $15^{\mathrm{m}}.5$ (IRS 9A) and $5^{\mathrm{m}}.5$ (IRS 9B) of intrinsic extinction (right). The SEDs can be reproduced by a combination of three Planck functions at temperatures 210 K, 1400 K and 22000 K, respectively. Detection limits of ESO's current (TIMMI 2) and future (VISIR) mid-IR instrumentation are indicated. The data points are taken from [12] = □, [22] = △, [23] = ◇, [25] = ∗ and this work = •, ∘.
Figure available in colour on the CD-ROM

and high mass stars. Therefore, the loci of IRS 9A and IRS 9B suggest a quite early evolutionary stage (ages $\leq 10^5$ yr), consistent with that of protostars.

This picture is confirmed by our mid-IR observations using TIMMI 2. Three sources (IRS 9A, IRS 9B and IRS 9C) stand out as strong 10 μm emitters. IRS 9A is even detected at 20 μm. This indicates the presence of massive circumstellar envelopes of gas and dust, which are gravitationally bound to and internally headed by the central source. As the mid-IR emission is optically thin, our measurements can be used to estimate the mass of the hot dust [14], [20]. Assuming typical parameters for the dust grains (size, density and emissivity as given by [2]) we obtain a gas and dust mass of about $4.2\,\mathcal{M}_\odot$ and $0.4\,\mathcal{M}_\odot$ for IRS 9A and IRS 9B, respectively.

In Fig. 4 we present the SEDs of IRS 9A and IRS 9B emerging from our IR data. In the left panel the fluxes of both sources are only dereddened for a foreground extinction of $A_V = 4^{\mathrm{m}}.5$, while in the right panel an additional intrinsic extinction of $A_V \sim 15^{\mathrm{m}}.5$ (IRS 9A) and $5^{\mathrm{m}}.5$ (IRS 9B) is taken into account. In each case the overall SED is fitted by a combination of three Planck functions at temperatures of 22000 K, 1400 K and 210 K. In full agreement with their loci in the colour magnitude diagram, a (lower) value of 22000 K for the stellar temperature qualifies the central sources as high mass stars of spectral type B2 or earlier. Moreover, positive spectral indices $\alpha_{2.2-10\mu m}$ (4.7 for IRS 9A and 8.3 for IRS 9B; using fluxes dereddened for $4^{\mathrm{m}}.5$ of foreground extinction) are indicative for protostars.

4 VISIR – VLT's Mid-IR Imaging Capabilities

As can be seen in Fig. 2 all sources with K_s excess are also detected in the L band data and nearly all of them have counterparts in the M band image. But only the three brightest sources are detected at mid-IR wavelengths (if the marginal 10 μm detection of IRS 9C is taken into account). Hence, our TIMMI 2 data are clearly flux limited. Sensitivity limits of TIMMI 2 for an integration time of 15 minutes are provided in Fig. 4.

Comparison with those values expected for the VLT's future mid-IR instrument VISIR promise a huge gain in sensitivity by several orders of magnitude. If one assumes for the weaker K band excess sources SED slopes which are similar to those of IRS 9A and IRS 9B, all of them should be detectable by VISIR. As soon as VISIR is installed and operational we will be eager to test this hypothesis. Meanwhile, our example of the IRS 9 cluster of protostars may demonstrate, that the VLT's mid-IR imaging capabilities will offer important new insight on high mass star formation processes.

References

1. B. Balick, G.O. Boeshaar, T.R. Gull: ApJ 242, 584 (1980)
2. E.E. Becklin, S. Beckwith, I. Gatley, et al. : ApJ **207**, 770 (1976)
3. W. Brandner, E.K. Grebel, Y.-H. Chu, et al. : AJ **119**, 292 (2000)
4. L. Bronfman, L.-Å. Nyman, J. May: A&A **115**, 81 (1996)
5. R.S. Chini, Th. Henning, W. Pfau: A&A **247**, 157 (1991)
6. C.A. Clayton: MNRAS **219**, 895 (1986)
7. C.A. Clayton: MNRAS **246**, 712 (1990)
8. P.A. Crowther, L. Dessart: MNRAS **296**, 622 (1998)
9. C.G. De Pree, M.C. Nysewander, W.M. Goss: AJ **117**, 2902 (1999)
10. L. Drissen, A.F.J. Moffat, N.R. Walborn, M.M. Shara: AJ **110**, 2235 (1995)
11. F. Eisenhauer, A. Quirrenbach, H. Zinnecker, R. Genzel: ApJ **498**, 278 (1998)
12. J.A. Frogel, S.E. Persson, M. Aaronson: ApJ **213**, 723 (1977)
13. W.M. Goss, V. Radhakrishnan: Ap Letters **4**, 199 (1969)
14. R.D. Hildebrand: QJRAS **24**, 267 (1983)
15. K.-H. Hofmann, W. Seggewiess, G. Weigelt: A&A **300**, 403 (1995)
16. D. Hollenbach, D. Johnston, S. Lizano, F.H. Shu: ApJ **428**, 654 (1994)
17. R.C. Kennicutt: ApJ **287**, 116 (1984)
18. J. Melnick, M. Tapia, R. Terlevich: A&A **213**, 89 (1989)
19. A.F.J. Moffat, L. Drissen, M.M. Shara: ApJ **436**, 183 (1994)
20. A. Natta, N. Panagia: A&A **50**, 191 (1976)
21. A.K. Pandey, K. Ogura, K. Sekiguchi: PASJ **52**, 847 (2000)
22. P. Persi, M. Tapia, M. Roth, M. Ferrari-Toniolo: A&A **144**, 275 (1985)
23. M. Roth, M. Tapia, M.T. Ruiz, P. Persi, M. Ferrari-Toniolo: 'New infrared observations of NGC 3603' In: *Star Forming Regions, IAU Symp. 115, 1985*, ed. by M. Peimbert, J. Jugaku (Kluwer, Dordrecht, 1987), p. 182
24. K.M. Strom, S.E. Strom, K.M. Merrill: ApJ **412**, 233 (1993)
25. M. Tapia, J. Bohigas, B. Perez, M. Roth, M.T. Ruiz: RMxAA **37**, 39 (2001)
26. H.W. Yorke, A. Welz: A&A 315, 555 (1996)

Sub-Arcsecond Millimeter Imaging of Disks and Envelopes: Probing the Density Structure

Leslie W. Looney[1], Lee G. Mundy[2], and W.J. Welch[3]

[1] Max-Planck-Institut für extraterrestrische Physik, Garching, Germany
[2] University of Maryland, College Park, MD, USA
[3] University of California, Berkeley, CA, USA

Abstract. Sub-arcsecond observations at millimeter wavelengths are providing spectacular images necessary to study directly the circumstellar disks and inner envelopes of young stellar objects. The millimeter data probe the bulk material in the circumstellar environment, complementing the beautiful optical and infrared images of young disk systems being acquired by the VLT and HST. We present an overview of the BIMA millimeter sub-arcsecond survey of deeply embedded and T Tauri systems, focusing on the morphology of the structures, the inner density in the envelope, and the youngest disks.

1 Introduction

Over the last few decades a general picture of isolated star formation as been realized. Protostars are formed deep within molecular cloud cores, which not only provide the raw materials for star formation, but also influence the evolution and eventual outcome of the system. These young stellar systems are initiated via the collapse of the core, which forms the circumstellar envelope of the protostar. Standard collapse solutions of the singular isothermal sphere [1–4] predict a density structure of $\rho \propto r^{-2}$ that evolves into a $\rho \propto r^{-3/2}$. As the collapse proceeds, a small circumstellar disk forms around the young protostar, growing larger and more massive as the envelope dissipates.

These forming stellar systems exhibit excess infrared and millimeter emission that arises primarily from the two circumstellar dust structures: envelopes and disks. We have completed a millimeter interferometric survey of the youngest, deeply embedded protostars, the so-called Class 0 sources, as well as the older, more evolved Class I/II sources, resolving the disks and the envelopes. In this paper, we will discuss the structures and density of the envelope and their implication in the star formation process.

2 The BIMA Sub-Arcsecond Survey

We have observed 24 young stellar objects [5] in 11 fields using the BIMA millimeter array [6] in the $\lambda = 2.7$ mm continuum. These multi-configuration observations utilized the high dynamic u,v range of the BIMA array: fully sampled spatial scales ranging from $0\farcs4$ to $60''$. This unique feature of BIMA allows the first consistent comparison of dust emission structures in a variety of systems [5]. The images show a diversity of structure and complexity. The youngest systems are the deeply embedded sources (L1448 IRS3, NGC1333 IRAS2, NGC1333 IRAS4, VLA 1623, and IRAS 16293-2422) which have continuum emission dominated ($\geq 85\%$) by the extended envelope, typical mass of 1 $M_\odot$(See Fig. 1). The

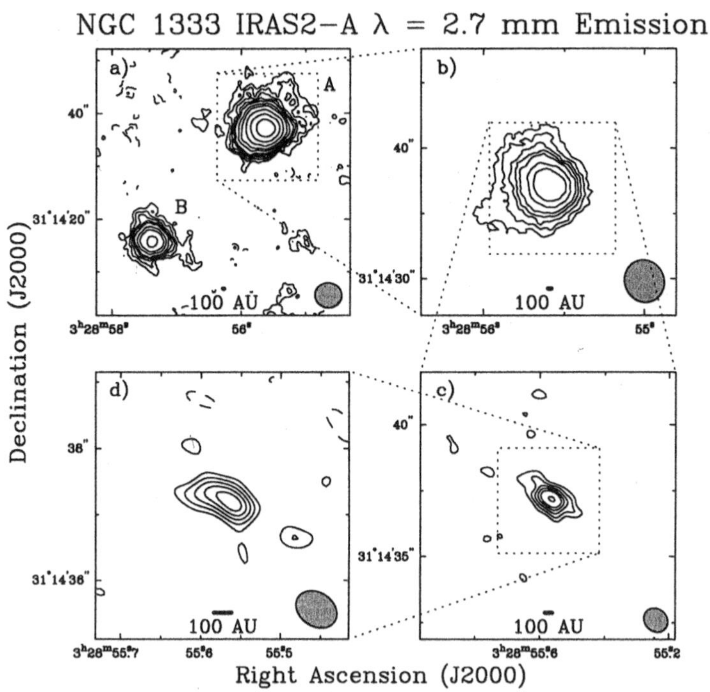

Fig. 1. NGC 1333 IRAS 2 A maps of the $\lambda = 2.7$ mm continuum emission. Panel (a) is contoured in steps of (-4 -3 -2 2 3 4 5 6 8 10 14.14 20 28.28 40 56.56) $\times$ the rms of panel (a) of 1.3 mJy/beam. Panels (b) through (d) use a rms noise of 2.7 mJy/beam. (a) $\sigma = 1.3$ mJy/beam; beam is $5''.40 \times 4''.70$ P.A. = 86 ° (b) $\sigma = 1.2$ mJy/beam; beam is $3''.36 \times 3''.16$ P.A. = 45 ° (c) $\sigma = 1.7$ mJy/beam; beam is $1''.02 \times 0''.87$ P.A. = 57 ° (d) $\sigma = 2.7$ mJy/beam; beam is $0''.69 \times 0''.52$ P.A. = 60 °. Note how the emission is resolved away in higher resolution

older, less embedded systems which are detected in the near-infrared (SVS13 and L1551 IRS5) have continuum emission that is both extended and compact. The oldest systems in our survey are the optically visible T Tauri stars (DG Tauri, HL Tauri, GG Tauri, and GM Aurigae), which have continuum emission dominated by compact ($\lesssim 1''$) circumstellar disks, typical mass of 0.02 $M_\odot$(See Fig. 2).

All of the surveyed embedded objects are in multiple systems with separations on scales of $\sim 30''$ or less. Based on the system separation, we place the

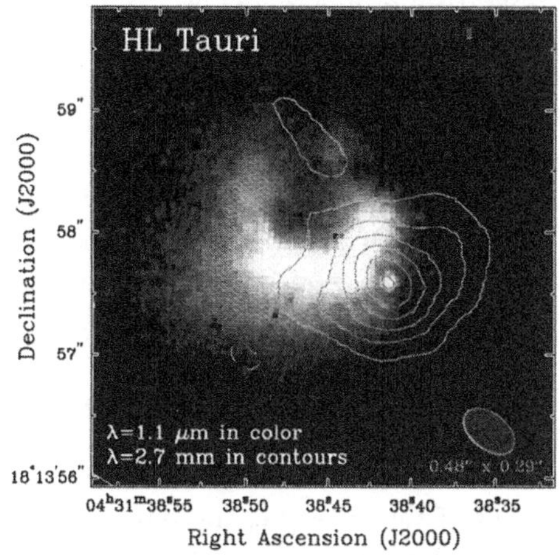

Fig. 2. BIMA sub-arcsecond imaging of the circumstellar disk around HL Tauri in contours [5]; in color is the HST near-infrared image of the reflection nebula [7]. The contours are in steps of (-4 -2 2 4 6 8 10 12) × 4 mJy/beam. With a beam size of 0.″48 × 0.″29, the disk is clearly resolved.
Figure available in colour on the CD-ROM

objects into three categories: separate envelope (separation $\geq$ 6500 AU), common envelope (separation 150-3000 AU), and common disk (separation $\leq$ 100 AU)[5]. These three groups can be linked with fragmentation events during the star formation process: separate envelopes from prompt initial fragmentation and the separate collapse of a loosely condensed cloud, common envelopes from fragmentation of a moderately centrally condensed spherical system, and common disk from fragmentation of a high angular momentum circumstellar disk.

3 Envelope Modeling

We have performed extensive modeling of the inner envelope structure in many of the deeply embedded sources in our general survey [8]. Our main goal in the modeling is to address two main questions: (1) Can we constrain the power-law of the density to either a $\rho \propto r^{-2}$ or $\rho \propto r^{-3/2}$? (2) What mass constraints can we place on the circumstellar disks in these systems?

To compare the observations directly to models of the dust structure, we have to simplify the models by making assumptions about the dust properties and source morphology. Our source model for the embedded sources is comprised of a spherical dust envelope with an embedded point source meant to represent an unresolved disk. Since the embedded systems in our survey [5] have spherical symmetry, we assume spherical envelopes. In addition, we use a power-law for the density structure, as suggested by solutions of the singular isolated sphere collapse solutions e.g. [1–4].

We adopt the temperature profile model of Wolfire and Cassinelli [9], which self-consistently solves for the temperature throughout the envelope. Past modeling of envelopes have used the simple power-law temperature distribution of optical thin dust. However, for somewhat massive envelopes or steep density power-law indices, the envelope can become optically thick at radii < 200 AU [8], a spatial scale at which our survey is sensitive. We also use a modified version of the MRN [10] and the Draine & Lee [11] optical constants to describe grain properties. Studies have suggested that dust in circumstellar disks and dense cores has a λ^{-1} dependence at submillimeter wavelengths [12–14] rather than the λ^{-2} dependence. Our hybrid model preserves the optical and infrared properties of the MRN dust grain model, while forcing the long wavelength λ^{-1} behavior and $\kappa_\nu(2.7\ \mathrm{mm}) = 0.009\ \mathrm{cm}^2\ \mathrm{g}^{-1}$.

With these assumptions, the model parameters are the inner and outer radii of the envelope, the mass of the envelope, the power-law indices of the density, and the flux of the embedded circumstellar disk. The interferometric data are binned in u,v annuli around the source location and averaged vectorially (see Fig. 3). For the six sources modeled, we fit density power-laws of p = 1.5 and p = 2.0. The other parameters are explored over grids of parameter space: inner radii of 10, 30, 50 and 100 AU, outer radius from 1000 AU to 10000 AU, and a central point source flux in steps starting at no point source, up to a maximum of the amplitude in the last u,v distance bin. The models were treated in the exact same method as the observational data. For each model, the χ^2 was minimized with respect to the envelope mass to produce the best fit to the data.

All six of the sources can be fit with a density power-law index of p = 2.0; in half of the systems, those with the highest signal to noise ratios, p = 1.5 power-law models can be rejected with a > 95% confidence level. What do these fits tell us about the nature of density structure in the systems? In order to understand the constraints that higher signal to noise systems place on descriptions of the envelope collapse, it is instructive to examine the Larson-Penston (LP) and Shu solutions in more detail.

3.1 Modeling Discussion

In the Shu model, an expansion wave travels from the inner region outwards, the "inside-out" collapse. The head of the expansion wave is roughly where the $\rho \sim r^{-2}$ region transitions to the $\rho \sim r^{-3/2}$ region, and is only dependent upon the time after the formation of the core and the local sound speed. Thus, for a sound speed, a = 0.2 km/s, the head of the expansion wave will be located at 400 and 4000 AU for ages of 10^4 and 10^5 yrs, respectively.

In the LP solution, there is no expansion wave since the initial velocity profile is already collapsing, not at rest as in the Shu solution. Nonetheless, there is an infall scale for the interface between the $\rho \sim r^{-2}$ and $\rho \sim r^{-3/2}$ regions. Since the LP infall velocity is 3.3 times that in the Shu solution, the effective spatial scale is 3.3 times as large. Thus, the radius at which one expects a p = 1.5 is $\sim$ 1300 and 13000 AU for ages of 10^4 and 10^5 yrs, respectively. The larger mass

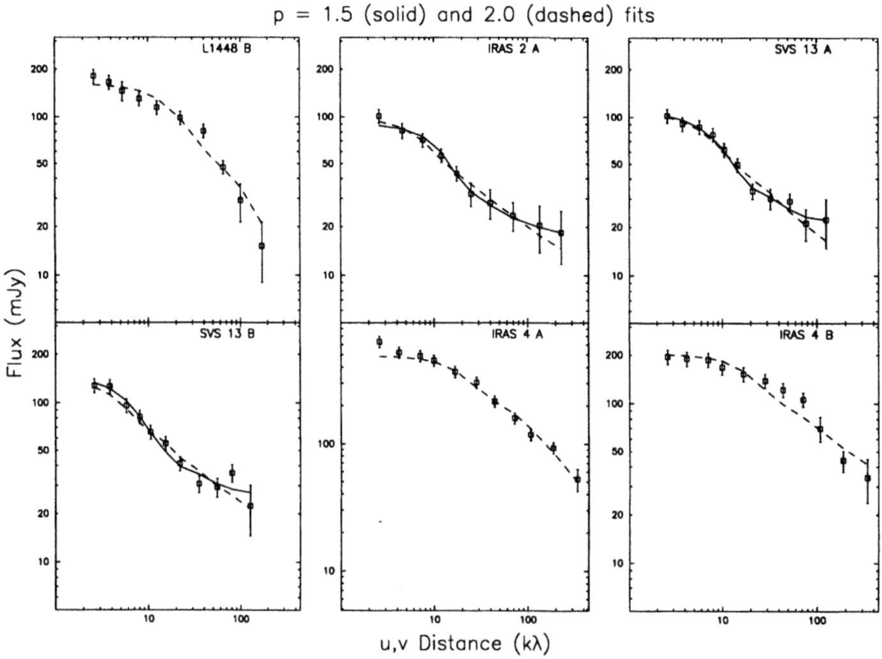

Fig. 3. Model fits for power-law density indices of p=1.5 and p=2.0 presented in u,v space, the separation of antennas. The data are effectively the Fourier transform of the sky brightness distribution. The error bars are statistical error bars based on the standard deviation of the mean of the data points in the bin with a minimum of 10%, reflecting the uncertainty in the overall calibration. For IRAS 4 A, the first data point was not used in the modeling, since it most probably traces larger scale emission that is not part of the inner envelope

infall rate creates a higher density envelope; the mass of the envelope at the onset of the core formation is 4.4 times larger than in the Shu solution.

To match the $\rho \sim r^{-2}$ power-law fits, the infall regions in these models must not dominate the radial density profiles. This means that they must be young, which is consistent with their classification as Class 0 sources. The youthfulness of the systems is also supported by the dynamical age estimates for the outflows. For example, L1448 IRS3 has the largest outflow, 14.5′ [15]. For an assumed jet speed of $\sim$ 200 km/s, the minimum age of L1448 IRS3 is $\sim$ 6,200 yrs. The corresponding times for the other systems are shorter, but it must be stressed that these are minimum ages as the outflow region should propagate out at a fraction of the basic jet speed. On the other hand, the luminosity of the system can also constrain the minimum age. In the above example of L1448 IRS3, which has a luminosity of 6.8 $L_\odot$, the Shu model for that age would require an increase of sound speed by 75%. For the LP solution, this is less of a problem since the mass accretion is nearly 46.9 times larger.

We have performed more complicated modeling of the sources using the LP and Shu solutions directly. We find that they predict similar ages, since the Shu model requires much larger sound speeds to match the density. In these cases, however, the Shu solution can still not match the luminosity. In our modeling, the LP solution provides a better match to the data, but neither solution is the ideal fit. This is not surprising since numerous computational studies of the isothermal collapse have shown that there is no strong tendency to assume a self-similar form (c.f. [4,16]). Indeed, the computational simulations of the collapse of isothermal spheres demonstrates that it is incorrect to think of the envelopes of young stars as the realization of either collapse model. The truth is likely to be between the two solutions, and our modeling substantiates this.

Finally, one of the primary strengths of this study is the ability of the interferometer to separate large scale emission from compact emission, allowing us to probe for the youngest circumstellar disks embedded within the envelope. In our modeling, all systems have acceptable models with no central point source, but typically, point source flux values are in the range of 0 to 40 mJy.

We can make a mass estimate for the embedded disk using the circumstellar disk of HL Tauri as a standard. A HL Tauri type disk (disk mass $\sim 0.05\ M_\odot$ [5]) has a flux of $\sim$ 100 mJy at the distance of Taurus (140 pc). If placed at the distance of Perseus (350 pc), the flux of HL Tauri would be 16 mJy. So the range of acceptable point source fluxes would represent circumstellar disk masses of 0 to 0.12 $M_\odot$. This is a small fraction of the circumstellar envelope mass, typically $\sim 1\ M_\odot$. In other words, since we do not see disks that are over-massive when compared to the Class I/II systems, there is not a large mass reservoir and as the system evolves from the deeply embedded stage to the more disk dominated stage, the disk must efficiency process material onto the protostar.

References

1. Larson, R.B. 1969, MNRAS, 145, 271
2. Penston, M.V. 1969, MNRAS, 144, 425
3. Shu, F.H. 1977, ApJ, 214, 488
4. Hunter, C. 1977, ApJ, 218, 834
5. Looney, L.W., Mundy, L.G., & Welch, W.J. 2000, ApJ, 529, 477
6. Welch, W.J. et al. 1999, PASP, 108, 93
7. Stapelfeldt, K.R., et al. 1995, ApJ, 449, 888
8. Looney, L.W., Mundy, L.G., & Welch, W.J. 2002, in preparation
9. Wolfire, M.G., & Cassinelli, J.P. 1986, ApJ, 310, 207
10. Mathis, J.S., Rumpl, W., & Nordsieck, K.H. 1977, ApJ, 217, 425
11. Draine, B.T., & Lee, H.M. 1994, ApJ, 285, 89
12. Beckwith, S.V.W., & Sargent, A.I. 1991, ApJ, 381, 250
13. Beckwith, S.V.W. et al. 1986, ApJ, 309, 755
14. Weintraub, D.A., Sandell, G., & Duncan, W.D. 1989, ApJ, 340, L69
15. Bally, J., Devine, D., Alten, V., Sutherland, R.S. 1997, ApJ, 478, 603
16. Foster, P.N. & Chevalier, R.A. 1993, ApJ, 416, 303

Four VLT enclosures

David Wilner and Dan Harvey

VLA Studies of Disks Around T Tauri Stars

David J. Wilner

Harvard-Smithsonian Center for Astrophysics, 60 Garden St., Cambridge, MA 02138

1 Motivation

Much activity is devoted to characterizing the properties of disks around young stars to extract information on the physics of mass accretion and of planet building. Early analysis of the spectral energy distributions of disks indicated outer radii of 10's to 100's of AU (Adams, Lada & Shu 1987) and masses sufficient to form Solar Systems like our own (Beckwith et al. 1990). Statistical studies suggest that disks dissipate on timescales of order 10 Myr (Strom et al. 1989), compatible with the standard view of giant planet formation by dust coagulation, planetesimal formation, and core accretion of nebular gas. But many issues remain in this empirical outline of disk evolution. Which disks provide fertile ground for planets? When does dust settle and how do grains grow? How is nebular gas lost? Can we see the signatures of planetary bodies forming within disks, for example tidal gaps and holes? While improved modeling of spatially unresolved observations continues to provide insight (e.g. Chiang et al. 2001), direct imaging of disks at many wavelengths is essential to address these questions.

Making resolved images of disks presents challenges. For the large sample of T Tauri stars in nearby dark clouds like Taurus and Chameleon at $\sim$ 140 pc, the 10 AU diameter of a Jupiter orbit subtends only 70 milliarcseconds. In addition, the disk material beyond a few stellar radii is cool, well below 1000 K, and a large fraction of disk emission emerges in the far-infrared, a spectral range difficult to access from the ground and lacking large apertures in space. In the optical and near-infrared, disks may be detected in scattered light, and the necessary resolution is provided by the Hubble Space Telescope and large ground-based telescopes like the ESO VLT employing adaptive optics. At these wavelengths, high contrast with stellar photospheres means that very careful point-spread-function subtraction or coronography must be used, and the innermost regions of the disks remain difficult to probe at all.

Observations at millimeter wavelengths offer some advantages (see the review by Wilner & Lay 2000) because: (1) the dust emission is (almost) entirely optically thin and probes the full disk volume, so the observed emission is proportional to mass, weighted by temperature; (2) emission from circumstellar dust dominates emission from the stellar photosphere, and contrast is not a problem; (3) the technique of interferometry allows for *imaging* observations that obtain the necessary high angular resolution. The main problem today with observations at millimeter wavelengths is that very high sensitivity and very good atmospheric conditions are required to realize imaging observations at the highest resolutions available.

2 Imaging Disks with the Very Large Array

The Very Large Array (VLA) of the National Radio Astronomy Observatory in New Mexico, USA, provides two unique capabilities for disk structure studies. First, the angular resolution for imaging thermal emission is much better than can be achieved with any other millimeter array. For the standard "A" configuration, the synthesized beam size at the shortest operating wavelength of 7 mm is about 40 milliarcseconds. Work recently completed on a real-time optical fiber connection between the VLA and the Very Long Baseline Array antenna located at Pie Town, New Mexico, effectively doubles the resolution of the VLA while maintaining excellent imaging characteristics, especially for northern sources. The resolution is similar to that expected for VLTI in the thermal infrared. Second, the long millimeter wavelengths accessible to the VLA are especially advantageous at very high angular resolution because dust opacity is low and emission emerges from within the highest surface density regions of the inner disks.

To achieve sub-arcsecond imaging with the VLA at 7 mm is not always easy. Phase correction techniques are usually required to overcome the atmospheric seeing. Unfortunately, the disks around young stars are too faint for self-calibration, whereby phase closure relations are determined in a coherence time. Instead, a mode of "fast-switching" phase referencing to nearby calibrators is used that effectively stops the phase fluctuations for baselines longer than a few hundred meters. This technique allows for diffraction limited imaging at the longest baselines, but generally only during the winter months when the atmosphere is most stable, and especially on winter nights when coherence times are longest (several minutes or more). Since the VLA varies its resolution through repositioning of the array antennas in four basic configurations over a period of approximately 16 months, the "A" configuration required for high resolution imaging is only sometimes available at the appropriate time of year.

Dust emission drops steeply toward long wavelengths, and the strength of the 7 mm signal from the disks around T Tauri stars is modest at the highest angular resolution. For a geometrically thin disk, the flux dS from a disk element filling $d\Omega$ is $dS = B_\nu(T)(1-e^{-\tau_\nu}) \cos i d\Omega$, where $B_\nu(T)$ is the Planck function, i is the inclination, and the optical depth τ_ν is given by $\tau_\nu = \Sigma\kappa_\nu/\cos i$ where Σ is the surface density and the κ_ν is the mass opacity. Following common practice, we adopt a long wavelength opacity with power law form and adopt the normalization advocated by Beckwith et al. (1990), i.e. $\kappa_\nu = 0.1(\nu/10^{12}\text{Hz})^\beta$ cm^{2}g^{-1}, $\beta = 1$, recognizing that this expression hides many uncertainties associated with grain size, composition, and dust-to-gas ratio. For example, the emissivity must be affected by the evaporation of different grain constituents, starting with water ices at $T \sim 200$ K. For fiducial values appropriate to VLA observations, where the best sensitivity in the 7 mm band is obtained at 43.4 GHz, assuming low optical depth and the Rayleigh-Jeans limit

$$dS = 0.11 \text{ mJy} \left(\frac{T}{100\ K}\right)\left(\frac{\Sigma}{100\ g/cm^2}\right)\left(\frac{\nu}{43.4\ GHz}\right)^3\left(\frac{\theta}{40\ mas}\right)^2 \quad (1)$$

where θ is the synthesized beam Gaussian fwhm size. For the inner parts of disks where the temperature exceeds 100 K and the surface density is high, the VLA can detect the disk material in a single 8 hour track. For sources at 140 pc, this detectable signal corresponds to < 1 Jupiter mass of an interstellar mixture of gas and dust filling the ~ 5 AU beam. This high sensitivity introduces a potential problem since very small amounts of ionized gas are also detectable, and the plasma can contaminate the signal from dust. However, the plasma contribution (if any) can be estimated accurately from longer wavelength VLA data.

3 Two Recent Examples

3.1 TW Hya

The TW Hya system is almost three times closer than the T Tauri stars associated with dark clouds (56 ± 7 pc), and we selected this source for VLA imaging despite an extreme southern location that makes it a difficult target. Wilner et al. (2000) provide a more complete description of the TW Hya observations than the brief summary provided here.

Figure 1 (lower left) shows the long wavelength spectrum of TW Hya, including our observations from the VLA and the BIMA array. The spectrum shows dust emission far in excess of the stellar photosphere, and, like most T Tauri stars, this excess emission is well fitted by a family of thin disk models parameterized by radial power laws in temperature and surface density. In these models, the temperature distribution is determined by the slope of the spectrum at infrared wavelengths where the disk is optically thick. Stellar irradiation together with flaring of the outer regions tend to give $T(r) \sim r^{-q}$, with $q \approx 0.5$, consistent with the TW Hya spectrum. The solid lines in Figure 1 show spectra derived from a series of face-on disk models, assuming the usual (constant) dust opacity law, with $\Sigma(r) \propto (r/1\ \text{AU})^{-p}$ and $p = 0, 0.5, 1.0$, and 1.5 (with the mass of gas+dust adjusted from 0.044 to 0.034 $M_{\odot}$ to provide the best least squares fit). The masses depend on the millimeter mass opacity and are uncertain, especially if the gas-to-dust ratio has evolved from the standard value. In any case, the spectrum is not very sensitive to the details of the surface density distribution, as is well known. Note that the models underpredict the observed emission at 3.6 cm. It's possible that the 3.6 cm emission arises from hot plasma, either from a stellar wind or pre-main-sequence magnetic activity, though a population of very large dust grains would also provide an explanation. Even at the extreme, an ionized plasma contributes a tiny fraction of the 7 mm emission.

Figure 1 (upper left) show images of TW Hya made from the VLA 7 mm data at two different resolutions. The disk is clearly visible. These resolved 7 mm images are very sensitive to the central concentration of the disk emission, parameterized by $p+q$ in the power law models. Figure 1 (right) shows images made from a series of disk models that match the TW Hya spectrum for a range of values of p (taking $q = 0.5$). To account for the spatial filtering of the interferometer observations, the models were imaged from the (u, v) tracks obtained for TW Hya for two resolutions and deconvolved in the standard way. The best fitting value of p is near unity. While inhomogeneities are likely present in the disk, and changes in disk composition and opacity will modify the energy balance and structure close to the star, the overall structure of the TW Hya disk appears amenable to this simple power law description, at least for radii outside 3 AU. The surface density distribution is consistent with that obtained from images of near-infrared scattered light (Trilling et al. 2001), for radii beyond the coronographic mask (> 50 AU) to an outer radius of 225 AU.

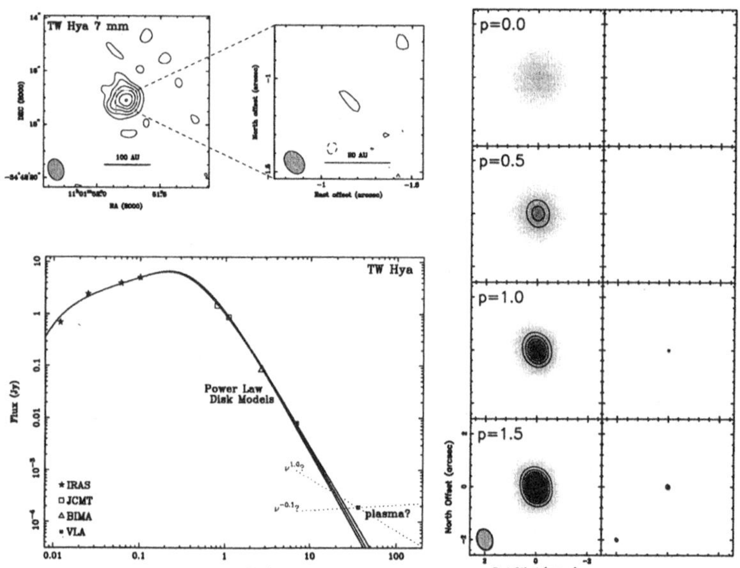

Fig. 1. A summary of results for VLA 7 mm imaging of TW Hya. *upper left:* Images made from the 7 mm data at 0.″6 resolution to emphasize the extended low brightness emission and at 0.″1 resolution where only a weak signal at size scale 10 AU remains at the center of the larger structure. *lower left:* Long wavelength spectrum showing best fit power law disk models with values of the surface density power law index of 0.5, 1 and 1.5 (the dotted lines indicate a possible plasma component). *right:* Simulated VLA images for the three model disks whose spectra are shown. A logarithmic grey scale shows low brightness emission.

3.2 DG Tau

DG Tauri is a well-studied, flat spectrum, classical T Tauri star. The 2.7 mm dust continuum emission from the disk was resolved by Dutrey et al. (1996), who measured a size of about $1''$ and an orientation nearly perpendicular to the optical jet that extends to larger size scales.

Figure 2 shows two high resolution VLA 7 mm images of DG Tau. Strong emission is detected from the inner part of the inclined disk. The most interesting aspect of the new 7 mm images is the gross departure from simple power law structure at radii < 10 AU. The highest resolution image, including the VLA-Pie Town link, shows two peaks connected by curved bridges of emission. This structure is presumably related to that seen in near-infrared lunar occultation observations that show a single star with an extended "shell" 45 ± 5 milliarcseconds fwhm ($\sim$6 AU), which accounts for about 25% of the emission (Leinert et al. 1991).

What explains the DG Tau morphology? At this high angular resolution, we can no longer precisely locate the star. Does one peak mark circumstellar dust heated from within by the star? If so, what produces the asymmetry? Does an emission peak indicate a condensing protoplanet? Or is the star located between the peaks, heating both of them? If so, then perhaps we are seeing the opening of some kind of gap, perhaps due to the dynamical effect of a protoplanet. Another explanation for the secondary peak might be an ionized blob from the jet; at these low flux levels, such a small, partly optically thick knot would remain very difficult to isolate in observations made at longer wavelengths. We hope to distinguish among these possibilities with a second epoch observation. The timescales for either orbital motion at 5 AU radius, or bipolar outflow at 10's of km s^{-1}, are both sufficiently short that a second image within a few years should show significant secular changes.

Acknowledgements

I wish to thank the NRAO staff and all of my collaborators on the VLA 7 mm disk work, especially Luis Rodriguez and Paul Ho.

References

1. Adams, F.C., Lada, C.J. & Shu, F.H. 1987, ApJ, 312, 788
2. Beckwith, S.V.W., Sargent, A.I., Chini, R. & Gusten, R. 1990, AJ, 99, 924
3. Chiang, E.I., Joung, M.K., Creech-Eakman, M.J., Qi, C., Kessler, J.E., Blake, G.A. & van Dishoeck, E.F. 2001, ApJ, 547, 1077
4. Dutrey, A., Guilloteau, S., Duvert, G. Prato, L., Simon, M., Schuster, K. & Menard, F. 1996, A&A, 309, 493
5. Leinert, C., Haas, M., Richichi, A., Zinnecker, H. & Mundt, R. 1991, A&A, 250, 407
6. Strom, K.M., Strom, S.E., Edwards, S., Cabrit, S. & Skrutskie, M.F. 1989, AJ, 97, 1451

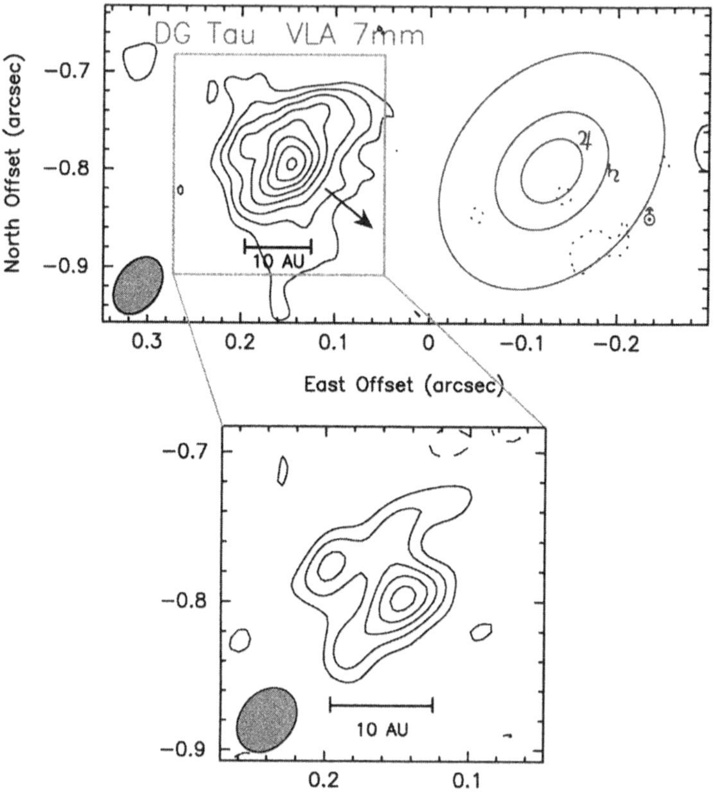

Fig. 2. High resolution VLA 7 mm images of the DG Tauri system. *upper:* The inner part of the dust disk; the arrow indicates the position angle of the optical jet observed at larger scales. The ellipses to the right show the orientation of the disk, and the spacings of the ellipses correspond to the orbits of the giant planet orbits in our Solar System. Contours levels are $(2, 3, ...) \times 0.12$ mJy. *lower* A higher resolution image made to emphasize the small scale, high brightness structure. The synthesized beam size is $\sim$ 30 milliarcseconds. The origin of the asymmetric structure is unclear, but either orbital motions or outflow should be detectable in a second epoch observation

7. Trilling, D.E., Koerner, D.W., Barnes, J.W., Ftaclas, C., & Brown, R.H. 2001, ApJ, 552, L151
8. Wilner, D.J. & Lay, O.P. 2000, in *Protostars and Planets IV*, eds. V. Mannings, A. Boss and S. Russell, p. 509
9. Wilner, D.J., Ho, P.T.P., Kastner, J.H. & Rodriguez, L.F. 2000, ApJ, 534, L101

Antu Control Room

Timo Prusti, Ralf Siebenmorgen, and Ralf Launhardt

A Close View on the Protoplanetary Disk in the Bok Globule CB 26

R. Launhardt[1], B. Stecklum[2], and A.I. Sargent[1]

[1] Astronomy Department, California Institute of Technology, Pasadena, CA 91125
[2] Thüringer Landessternwarte Tautenburg, 07778 Tautenburg, Germany

Abstract. We present new sub-arcsecond-resolution near-infrared polarimetric imaging and millimetre interferometry data on the circumstellar disk system in the Bok globule CB26. The data imply the presence of a $M \geq 0.01\,M_\odot$ edge-on-seen disk of > 400 AU in diameter, being in Keplerian rotation around a young $\sim$0.35 $M_\odot$ star. The mm dust emission from the inner 200 AU is highly optically thick, but the outer parts are optically thin and made of small dust grains. Planetesimal growth in the inner disk could neither be confirmed nor excluded. We argue that the CB 26 disk is a very young protoplanetary disk and show that it is comparable to the early solar system.

1 Introduction

Observations of protostellar systems and their prominent jets and outflows suggest that accretion disks start to form very early during the main accretion phase. These disks live much longer than the central protostar needs to build up most of its mass. When the initial protostellar core is dispersed by accretion and outflows, the central star still accretes matter at low rates from the surrounding disk. Typical disk life times around low-mass stars were shown to be at least 10^7 yrs and their masses (typically 0.01-0.1 $M_\odot$) do not seem to decrease considerably during this time [2]. Theoretical and laboratory studies show that the timescale for grain growth and planetesimal formation is shorter than the typical disk life time. Such disks should soon evolve into protoplanetary disks. Although the direct detection of planetary-mass bodies in such disks will be extremely difficult, indirect effects of larger bodies, such as a change in the dust opacity spectrum, should be observable with current or near-future techniques.

CB 26 (L 1439) is a small ($d \sim 0.15$ pc), slightly cometary-shaped Bok globule at a distance of $\sim$ 140 pc [7]. Located at the south-east rim of the globule is a small bipolar near-infrared (NIR) nebula. The central star is not visible, even at 2.2 μm. The spectral energy distribution together with the bolometric luminosity of $\geq 0.7\,L_\odot$ suggest the presence of a low-mass Class I YSO in CB 26. We observed strong submm/mm dust continuum emission from this source showing a thin extended envelope with an unresolved condensation at 10″ resolution [5], [7], [3]. No molecular outflow has been detected yet.

2 Implication of a Circumstellar Disk from NIR Data

On JHK near-infrared images obtained with the MAGIC camera at the Calar Alto 3.5-m telescope, we found a small bipolar reflection nebula in the Bok glob-

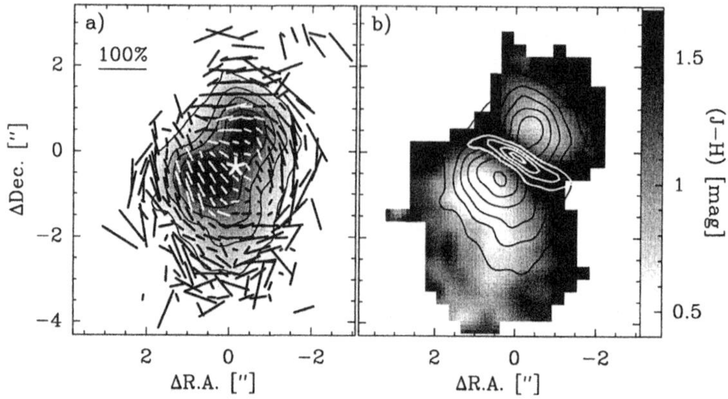

Fig. 1. **a)** K-band image of the CB 26 NIR nebula with polarization vectors superimposed. The white star marks the center of the polarization pattern, i.e., the location of the illuminating source. **b)** Grey-scale image: J − H color map of the NIR reflection nebula. Darker regions represent higher extinction. The outer boundary is due to an intensity cut-off level. Overlayed are the contours of the K-band emission from the reflection nebula (black) and of the 1.3 mm dust continuum emission from the circumstellar disk. The K-band image and color map are from [8]

ule CB 26 within the error ellipse of the IRAS source 04559+5200 [8]. Subsequent NIR imaging polarimetry confirmed the bipolar structure of this source. The two lobes are separated by an extinction lane which is most obvious in the J–H color map (Fig. 1b). Very high polarization degrees were detected in the lobes, presumably caused by single scattering at small dust grains. The orientation of the polarization vectors corresponds to a system consisting of a young star surrounded by both a circumstellar disk and a thin envelope. The polarization pattern indicates that the disk is seen almost edge-on causing the band of enhanced extinction in between the scattering lobes. At the very center, the polarization vectors are aligned linearly. This could be either due to photons scattered back from the envelope onto the disk or because of multiple scattering in the outer disk regions. The location of the illuminating source was derived to an accuracy of $0\farcs3$ by minimizing the mean square scalar product between polarization vectors in the lobes which are probably due to single scattering and their corresponding normalized radius vectors. The central source is located behind the extinction lane, i.e., at the center of the disk (Fig. 1a).

3 Direct Observations of Dust and Gas

CB 26 was observed with the Owens Valley Radio Observatory (OVRO) millimeter-wave array during 2000. We obtained wide-band continuum maps at 1.3 and 2.7 mm and a $^{13}CO(1–0)$ map with a spectral resolution of 0.17 km/s. The angular resolution of the maps is $\sim 1\farcs2 \times 0\farcs8$ at 2.7 mm and $\sim 0\farcs5 \times 0\farcs3$ at 1.3 mm, respectively. The observations are described in more detail in [7].

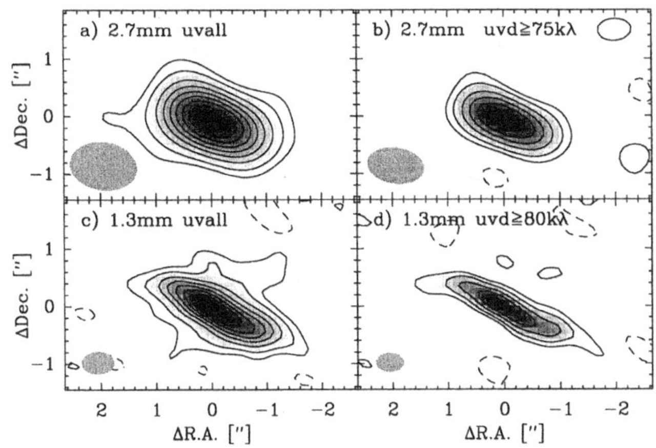

Fig. 2. Dust continuum emission from the CB 26 disk: OVRO results (from [7]). a) and c) show the 3 and 1mm maps synthesized from all uv data. b) and d) show maps derived from long *uv* spacings only. The synthesized beams (FWHM) are shown as grey ellipses at the lower left corners. Contour levels are in steps of (-2,2,4 to 22 by 3) times 1σ rms. a) σ = 0.6 mJy/ 1.″2×0.″84 beam. b) σ = 0.75 mJy/ 1.″0×0.″63 beam. c) σ = 1.3 mJy/ 0.″58×0.″39 beam. d) σ = 1.4 mJy/ 0.″50×0.″32 beam

The dust continuum maps show an elongated source with a position angle of $60 \pm 3°$ (from N to E; Fig. 2). The position and morphology of this source suggest that the emission originates from the circumstellar disk implied from the NIR data [8]. Figure 1b shows that the dust continuum emission matches well the extinction lane at the center of the NIR reflection nebula. At 3 mm we derive a projected FWHM source size of (160 ± 10) AU × < 60 AU with no signature of an extended envelope. The 1 mm images show a narrow lane and a small envelope extending perpendicular to the disk. The disk height remains unresolved at even the highest angular resolution of 0.″3, confirming that it must be seen almost edge-on. The projected FWHM size of the 1.3 mm disk is (220 ± 20) AU × ≤ 20 AU ($\frac{1}{2}$ of the HPBW in that direction). The actual scale height of the inner disk where the mm emission arises is probably much smaller. The disk has a symmetric 20° warp outside $R \sim$100 AU and is traced out to ~200 AU. The total flux densities derived from the OVRO maps are $S_{2.7\mathrm{mm}} = (22 \pm 5)$ mJy and $S_{1.3\mathrm{mm}} = (150 \pm 30)$ mJy. At 1.3 mm, the IRAM 30-m single-dish flux of the unresolved component is completely recovered (160 mJy; [5]). A simple decomposition yields 1.3 mm flux densities for the disk and envelope of $S_\nu(\mathrm{disk}) = (80 \pm 20)$ mJy and $S_\nu(\mathrm{env}) = (70 \pm 20)$ mJy, respectively. Assuming $T_\mathrm{d} = 30$ K, $\kappa_\mathrm{d}(1.3\mathrm{mm}) = 1\,\mathrm{cm}^2\,\mathrm{g}^{-1}$ of dust and $M_\mathrm{H}/M_\mathrm{d} = 110$ we derive an envelope mass of $M_\mathrm{H}(\mathrm{env}) = (0.03 \pm 0.01)\,\mathrm{M}_\odot$. The total mass of the more extended envelope seen in the single-dish maps is $(0.12 \pm 0.05)\,\mathrm{M}_\odot$ ([3], [7]). The disk appears to be considerably smaller in the 3 mm dust emission than at 1 mm. For the central $R \leq 100$ AU we derive a 1–3 mm spectral index $\alpha = 2 \pm 0.5$ (envelope subtracted). Further out α increases to reach 3.5-4 at $R \sim$ 180 AU.

At larger radii the 3 mm emission is not longer traced. This implies that most of the dust mass in the outer warped part of the disk is contained in classical, μm–mm size grains with a spectral index of the dust opacity of $\beta \sim 1.5 - 2$. The mm dust emission from the inner ~200 AU of the disk is highly optically thick. Therefore, no constraints can be made on β without better constraints on the temperature distribution and extensive modeling. Assuming that the 3 mm emission is mostly optically thin, a lower limit to the disk mass can be derived. Adopting $\kappa_\nu(1.3\mathrm{mm}) = 0.02\,\mathrm{cm}^2\,\mathrm{g}^{-1}$ of ISM, $\beta = 1$, $T_\mathrm{o} = 1000\,\mathrm{K}$ at $r_\mathrm{o} = 0.1\,\mathrm{AU}$, $T \propto r^{-0.4}$, and surface density $\Sigma \propto r^{-1.5}$ as 'typical' disk parameters, we derive $M_\mathrm{H}(\mathrm{disk}) \geq 0.01\,\mathrm{M}_\odot$, a value which is typical for disks around T Tauri stars [2].

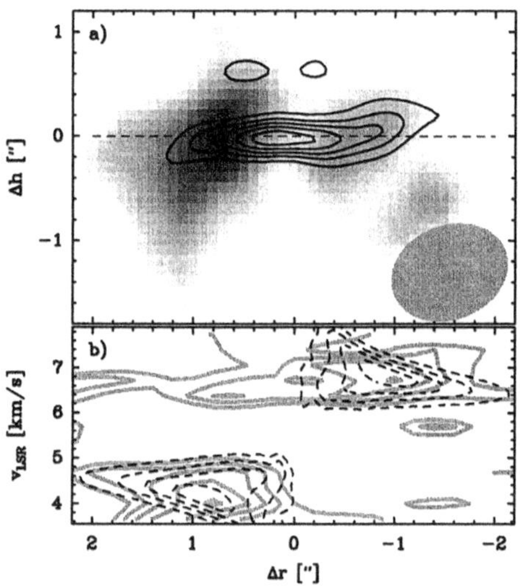

Fig. 3. ^{13}CO(1–0) emission from the CB 26 disk: OVRO results. a) grey-scale: integrated intensity of the ^{13}CO(1–0) emission. The synthesized beam (FWHM) is shown as grey ellipse in the lower right corner. Overlayed are the contours of the 1.3 mm dust continuum emission from the disk. The image is rotated by 30°. b) Position-velocity diagram of ^{13}CO along the plane of the disk (dashed line in a). Thick grey contours show the observed velocity field (27, 45, 63, 81, 99% of max). Dashed contours show the modeled emission from a Keplerian disk around a 0.35 $\mathrm{M}_\odot$ star

Strong ^{13}CO(1–0) emission ($T_\mathrm{R}^* \sim 20\,\mathrm{K}$) was detected from the outer parts of the CB 26 disk, but there seems to be a lack of emission from the central part (Fig. 3a). The line is double-peaked with the blue part coming from the N-E side and the red part from the S-W side of the disk. Emission at the systemic velocity of the envelope $v_\mathrm{LSR} = 5.5\,\mathrm{km\,s}^{-1}$ ($\Delta v \sim 0.6\,\mathrm{km\,s}^{-1}$) is self-absorbed and/or resolved out. Although the current spectral data cover only a small velocity range ($3 \leq v_\mathrm{LSR} \leq 8\,\mathrm{km/s}$), they could be well-modeled by a Keplerian disk rotating around a $(0.35\pm0.1)\,\mathrm{M}_\odot$ star (Fig. 3b). The high-velocity line wings from the inner parts of the disk expected for a Keplerian disk are not recovered in the data because of the small band width. There is some 'forbidden' red-shifted emission from the 'blue side' of the disk which may be due to infall or outflow. A more detailed study of the kinematics of this disk-envelope system is subject to follow-up observations.

4 Conclusions

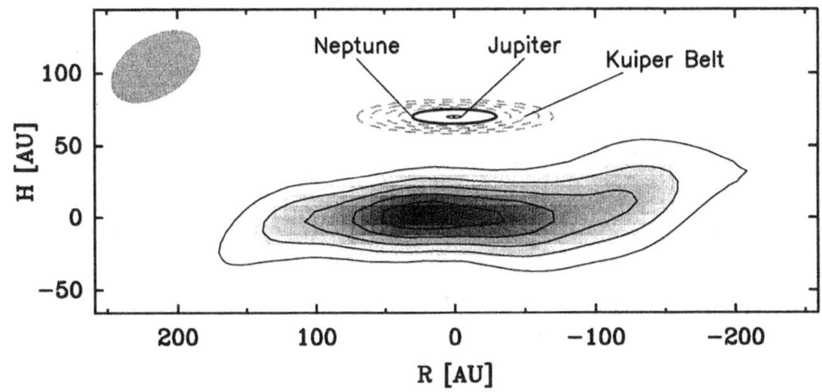

Fig. 4. The CB 26 disk compared to the solar system. The grey scale image and contours show the 1.3 mm dust continuum emission from the CB 26 disk as observed with OVRO. For comparison, the solar system is represented by the orbits of Jupiter and Neptune and the Kuiper Belt (40–70 AU)

From OVRO millimeter interferometric observations we discovered and resolved the thermal dust continuum and ^{13}CO line emission from the circumstellar disk in CB 26 which was recently implied by high-resolution NIR polarimetric observations. The disk is seen edge-on as predicted by the NIR observations and matches well the extinction lane at the center of the bipolar NIR reflection nebula. It has a FWHM diameter of (220±20) AU and a maximum (traced) diameter of ~400 AU. The (projected) scale height remains unresolved ($h \leq 20$ AU). For the first time, we directly detect a warp in a young circumstellar disk. The warp affects the disk outside $R > 100$ AU. Like in many extra-galactic disks, such a warp could be caused by a small external disturbance like, e.g., an encounter with a nearby star. However, the actual cause of the warp is not known.

The 1.3 mm dust emission from the inner $R \sim 100$ AU of the disk is highly optically thick, and only a lower limit to the total disk mass $M_{\rm H}(\text{disk}) \geq 0.01\,M_{\odot}$ can be derived from the 3 mm emission. Due to the high optical depths, no constraints on the dust opacity spectral index β and possible planetesimal growth in the inner undisturbed part of the disk can be drawn. The dust emission from the outer parts of the disk is optically thin and arises from small 'classical' grains. However, since the outer disk is obviously disturbed as indicated by the warp and has much lower particle densities, this does not exclude that particle growth in the inner disk has already taken place. The small disk envelope, which extends above and below the plane of the disk and may be related to an yet undetected outflow or a disk wind, contains $\sim (0.03 \pm 0.01)\,M_\odot$. Further $(0.12 \pm 0.05)\,M_\odot$ are contained in a more extended envelope ($\sim$ 3000 AU in diameter) which extends mainly to the north-east and may represent the remnant, nearly dispersed protostellar core from which this system has formed [3].

Strong ^{13}CO emission shows that the disk must be still gas rich, that it is rotating, and has a kinetic temperature of order 30 K outside $R \sim 100$ AU. The rotation curve is consistent with Keplerian rotation around a $(0.35 \pm 0.1)\,M_\odot$ star. With a total luminosity of $\geq 0.7\,L_\odot$, its large disk and its close connection with the parental cloud core, the CB 26 system resembles an intermediate M type T Tauri star of age $< 10^5$ yrs. The lack of a dense, centrally peaked cloud core with spectroscopic infall signatures as well as of a prominent molecular outflow indicates that the system has already passed its main accretion phase. Therefore, we conclude that this is a young protoplanetary disk.

If compared to the solar system today, the CB 26 disk is $\sim$ 2.5 times larger than the outer Kuiper Belt (Fig. 4). However, the optically thick and undisturbed part inside the warp has approximately the size of the Kuiper Belt. The total mass in the CB 26 disk (if all the gas is still there) is at least 10 times higher than the planetary mass in the solar system. However, the dust mass in the CB 26 disk could be comparable or lower than the rocky mass in the solar system. These numbers depend critically on the dust properties which are not well-known. Altogether, the CB 26 system appears like a 2–3 times lower-mass equivalent to the early solar nebula at the verge of forming planetesimals and probably planets.

References

1. S.V.W. Beckwith, Th. Henning, Y. Nakagawa: 'Dust properties and assembly of large particles in protoplanetary disks'. In: *Protostars and Planets IV*, ed. by V. Mannings, A.P. Boss, S.S. Russell (Univ. of Arizona Press 2000) pp. 533–558
2. S.V.W. Beckwith, A.I. Sargent, R. Chini, R. Güsten: AJ **99**, 924 (1990)
3. Th. Henning, S. Wolf, R. Launhardt, R. Waters: 'Measurements of the magnetic field geometry and strengths in Bok globules', ApJ (submitted)
4. R. Launhardt: Star formation in Bok globules, PhD thesis, University of Jena (1996)
5. R. Launhardt, Th. Henning: A&A **326**, 329 (1997)
6. R. Launhardt, N.J. Evans II, Y. Wang, D.P. Clemens, Th. Henning, J.L. Yun: ApJ **119**, 59 (1998)
7. R. Launhardt, A.I. Sargent, H. Zinnecker: 'A protoplanetary disk in the Bok globule CB26?', ApJ (submitted)
8. B. Stecklum, O. Fischer, R. Launhardt, Ch. Leinert: 'Discovery of a circumstellar disk in the Bok globule CB26', ApJ (submitted)

Detecting Gaps in Protoplanetary Disks with MIDI at the VLTI

Sebastian Wolf[1], Thomas Henning[2], and Gennaro D'Angelo[2]

[1] Thüringer Landessternwarte Tautenburg, Sternwarte 5, D–07778 Tautenburg, Germany
[2] Astrophysikalisches Institut und Universitäts-Sternwarte Jena, Schillergässchen 2-3, D–07745 Jena, Germany

Abstract. Numerical simulations show that a Jupiter mass planet is able to dig the gas material along its path. Resulting from this, a density gap is produced within the disk which may represent a large-scale density feature.

Based on these hydrodynamical simulations we investigate the possibility to trace planets in circumstellar disks by means of infrared interferometry. On the basis of three-dimensional self-consistent radiative transfer calculations, images of this system were created which provide the basis for the simulation of the VLTI equipped with MIDI. We find that MIDI can be used to distinguish between different disk models – disks with or without an embedded planet – on the basis of the observed visibilities.

1 Introduction

Based on studies of the evolution of protoplanets in protostellar disks, it has been established that - depending on the hydrodynamic properties of the planet and the disk - giant protoplanets may open a gap and cause spiral density waves in the disk (see, e.g., Kley 1999, Kley et al. 2001). The gap may extend up to several AU in width. Thus, the question arises if one can find such a gap as an indicator for the presence of a protoplanet with present-day or near-future techniques. In order to study this possibility, we used hydrodynamical simulations of a protostellar disk with an embedded planet (§ 2) to compute the expected brightness distributions, using a 3D radiative transfer (RT) code. Based on subsequent calculations of the visibility of this source we show that interferometers such as the VLTI of the European Southern Observatory equipped with MIDI (Midinfrared Interferometric Instrument) provides the necessary basis to detect gaps in circumstellar disks (§ 3).

2 Hydrodynamical Simulations

In the hydrodynamical simulations we assume the disk to be flat and non-self-gravitating. The mutual gravitational interaction between the planet and the central star, and the gravitational torques of the disk acting on the planet and the central star are included. The disk is treated as a viscous medium by using a value of the α-parameter equal to 4×10^{-3}. The three-dimensional density structure is Gaussian in the vertical direction where for the scale height $H(r)$ we assume a constant ratio of $H/r = 0.05$. The undisturbed (no gap/planet) surface density distribution is given by $\Sigma \propto r^{-1/2}$. The mass of the star is assumed to be 1 $M_\odot$ whereas the mass of the planet is 1 Jupiter mass. The circular orbit of the planet has a radius of 5.2 AU. More detailed information about the physical and the numerical model adopted for this kind of computations can be found in Kley (1999) and Kley et al. (2001).

The presence of a massive body within a circumstellar disk affects its structure very rapidly. Already after 10 planet's revolutions the gap has developed along with the spiral patterns, as displayed in the left panel of Fig. 1. The other two panels of this Figure show the evolution of the gap and the spirals and how they appear at later times. This simulation has been performed using a smaller radial extent ([2, 13] AU) to achieve a sufficiently high resolution of both the gap and the spiral density waves. However, the density waves propagate beyond 13 AU. Figure 1 also indicates that along the orbit of the planet the gap is not dug everywhere at the same rate. After 10 orbits – while the density has already dropped to 60% in the vicinity of the planet – two bumps still exist around the Lagrangian points L_4 and L_5, where the density has reduced to only 30%. However, also this material is removed later during the evolution of the system (see Fig. 2).

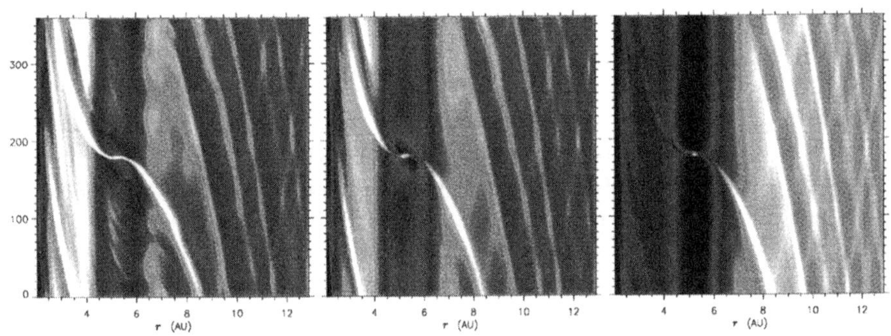

Fig. 1. Three snapshots reproducing the surface density distribution after 10 (left), 50 (centre) and 300 (right) orbital periods. At the location of the planet r_p, the resolution of the grid is such that $\Delta r = r_p \, \Delta\phi = 0.078$ AU.
Figure available in colour on the CD-ROM

The azimuthally averaged depth of the gap settles to a constant value after approximately 100 orbital periods. Anyway, the disk inside the orbit of the planet keeps on depleting because material is allowed to flow out of the computational domain through the inner radial border. This choice is intended to simulate the drifting motion of the gas towards the central star.

The formation of the gap basically depends on the balance of two competing processes. On one hand there are the gravitational torques, exerted by the planet, which reduce the angular momentum of the inside-orbit material (at $r \lesssim r_p$) and

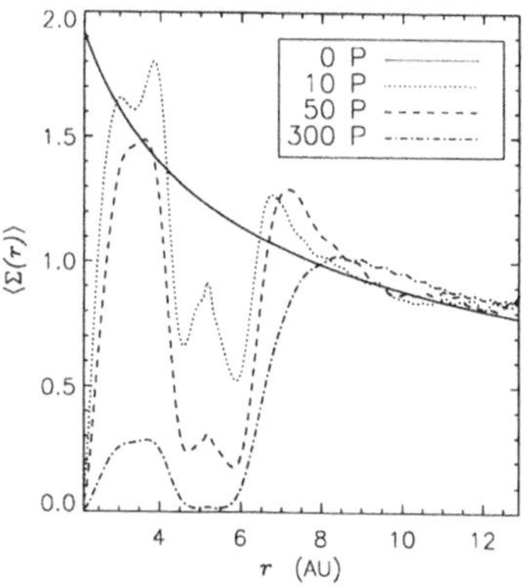

Fig. 2. Evolution of the azimuthally averaged surface density distribution. The peaks within the gap are due to the material lingering around the two triangular Lagrangian points located on the orbit of the planet, 60° away from it. This happens because these points represent positions of stable equilibrium for the restricted three-body problem. At 50 orbital periods they nearly vanish

increase that of the outside-orbit one (at $r \gtrsim r_p$). As a result, the gas is cleared out along the planet's path. On the other hand the viscous torques smooth the surface density and thus they oppose to the former process.

As the mass of the perturber reduces, the efficiency of the gravitational torques reduces too, hence the gap becomes shallower. This can be clearly seen in Fig. 3, where both the surface density and its azimuthal mean are compared for models with planets of different masses.

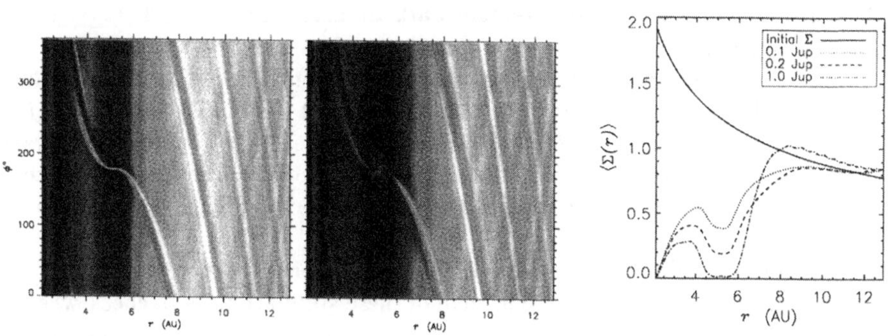

Fig. 3. Gap from models with planets of different masses: 0.1 (left) and 0.2 (centre) Jupiter masses. The averaged surface density $\langle \Sigma(r) \rangle$ reported in the right panel, clearly shows how viscous torques fill the gap as the planet's mass reduces.
Figure available in colour on the CD-ROM

3 Simulations of Observations with MIDI

For the present purposes, we use models with a larger radial extent of the simulated region such that the diameter of the disk measures 104 AU. This is comparable to the size of the disks seen in silhouette against the Orion Nebula (McCaughrean & O'Dell 1996). Furthermore, because of the inner extension of the disk, in these simulations matter is prevented from drifting out of the computational domain. Additionally, we assume that the mass of the disk is 0.05 $M_\odot$ and for the dust-to-gas mass ratio we take the classical value of 1:100.

Based on the resulting density distribution, we simulated the dust continuum emission of the dust. The RT has been performed with a 3D continuum code which is based on the Monte Carlo method. The RT is simulated self-consistently taking into account both the initial temperature of the dust due to viscous heating and the additional energy input of the central star. The dust density and temperature structure are defined on a grid which is chosen to be identical with that of the hydrodynamical simulations. Thus, any additional discretization error resulting from the "transfer" of the density/temperature structure to the RT code is avoided. For further information concerning the RT code, we refer to Wolf et al. (1999) and Wolf & Henning (2000).

In addition to the results from the hydrodynamical simulations, we introduce the following RT parameters: spherical dust grains consisting of "astronomical" silicates (optical data from Draine & Lee 1984, radius 0.12 μm); star: effective temperature $T_{\rm eff} = 5500\,$K, luminosity $L = 1\,L_\odot$; wavelength range for the simulation of the radiative transfer: 0.03...2000 μm.

As the first step towards visibility calculations we simulated images of the disk in the near-infrared... submillimetre wavelength range. Due to the huge intensity peak caused by warm/hot dust located very close to the star (1...2 AU), the outer regions of the disk remain dark at near/mid-infrared wavelengths. The gap located at 5 AU can be hardly seen. In contrast to this, in the sub-millimetre/millimetre wavelength range the brightness distribution is much smoother. This can be explained by the behaviour of the Planck function $B_\lambda(T)$. The ratio $B_\lambda(T_1)/B_\lambda(T_2)$ ($T_1 > T_2$) strongly decreases with increasing wavelength. Thus, the flux ratio and the brightness contrast between regions characterized by different dust temperatures (T_1 and T_2) decreases with increasing wavelength λ.

In Figure 4 visibility curves at λ=10 μm for different inclinations of the disk are shown (with/without gap). The goal of these simulations is to check whether MIDI at the VLTI can be used to detect gaps. The visibility V has been calculated based on the spatial intensity distribution $I(x, y)$ derived beforehand:

$$\Gamma(u,v) = Ve^{i\Phi_{\rm V}} = \int_{-\infty}^{+\infty}\int_{-\infty}^{+\infty} I(x,y)\exp\left[-2\pi i(ux+vy)\right]dx\,dy\,. \qquad (1)$$

Here, $\Phi_{\rm V}$ is the phase function and $\Gamma(u, v)$ is the complex visibility of the radiative source (see Pearson & Readhead 1984 and Shao & Colavita 1992 for details).

In the case of a disk with a gap, the visibility shows a steeper decrease than without a gap. This decrease strongly depends on the inclination of the disk i and the relative orientation between the direction of the baseline between the two telescopes and the major axis of the ellipse resulting from the projection

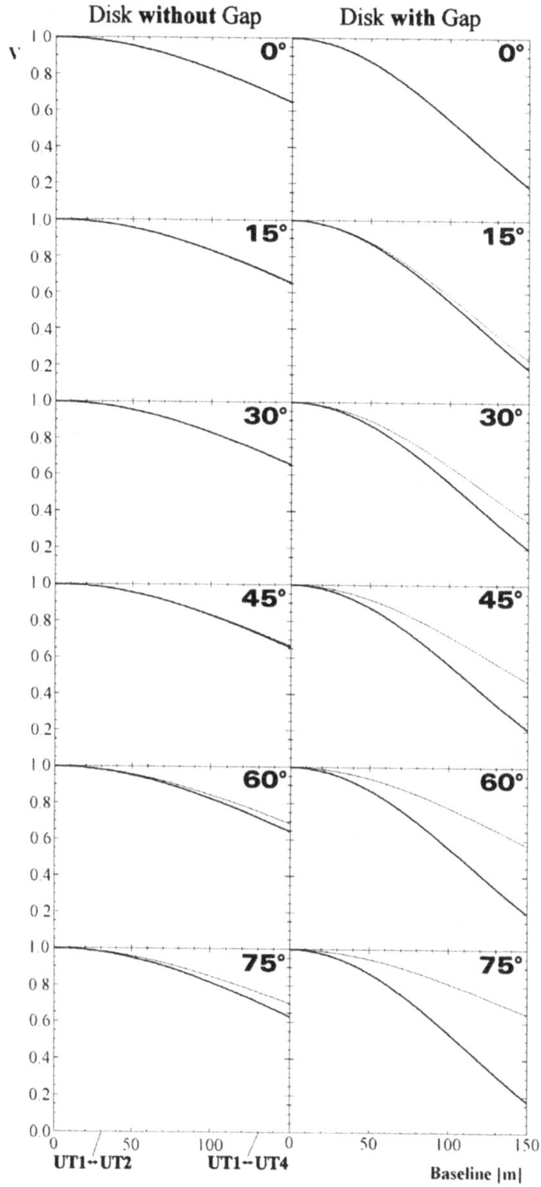

Fig. 4. Normalized visibilities at $\lambda=10\,\mu$m for disks with an inclination of i = 0° (face-on), 15°, 30°, 45°, 60°, and 75° (right column; assumed distance: 140 pc). For comparison the same profiles in the u-v plane but for a disk without a gap (but with the same mass and inclinations) are also given (left column). The visibilities marked by the thick and the thin line are oriented perpendicular to each other in the u-v plane (parallel to the major/minor axis of the ellipse resulting from the projection of the disk onto the plane of the sky). We would like to mention that the exact shape of the visibility profile depends on the details of the hydrodynamical simulations

of the disk onto the plane of the sky. The differences in the visibility profiles at a fixed inclination i result from the different density distributions in the inner, warm region of the disk (within a radius of 3 AU around the star) caused by the planet's motion and not from the presence of the gap. Furthermore, in Figure 4 the distances between the telescope UT1 and UT2 resp. UT1 and UT4 of the VLTI are indicated. Taking into account uncertainties of "real" measurements using MIDI at the VLTI, a distinction between different disk models (with/without gap) based on significant differences between the visibility profiles is possible only by a beam combination of the most distant telescopes of the VLTI.

4 Conclusions

We investigated the influence of a massive body – such as a protoplanet – on the density structure in a protoplanetary disk. We showed that the planet – depending on its mass – clears out the material on its path and thus opens a gap in the disk. Although the spatial extent of the gap is much larger than the planet itself, the gap is too small to be mapped using present-day observational techniques (assuming the disk to be located in nearby star-forming systems).

However, based on the hydrodynamical simulations and subsequent 3D RT simulations we derived the visibilities at 10 μm as a function of the baseline assuming a two-beam interferometer. It turned out that MIDI at the VLTI can be used to distinguish between different disk models (with/without a gap) on the basis of the observed visibilities. However, the longest possible baseline is required.

Acknowledgments

We wish to thank W. Kley for providing his results from hydrodynamical simulations discussed in § 3.

References

1. Draine, B.T., Lee, H.M., 1984, ApJ, 285, 89
2. Kley, W., 1999, MNRAS, 303, 696
3. Kley, W., D'Angelo, G., Henning, Th., 2001, ApJ, 547, 457
4. McCaughrean, M.J., O'Dell, C.R., 1996, ApJ, 111, 1977
5. Pearson, T.J., Readhead A.C.S, 1984, ARA&A 22, 97
6. Shao, M., Colavita M.M., 1992, ARA&A 30, 457
7. Wolf, S., Henning, Th., 2000, Comp. Phys. Comm. 132, 166
8. Wolf, S., Henning, Th., Stecklum, B., 1999, A&A, 349, 839

The Evolution of Circumstellar Disks: Lessons from the VLT and ISO

Wolfgang Brandner[1], Dan Potter[2], Scott S. Sheppard[2], Andrea Moneti[3], and Hans Zinnecker[4]

[1] European Southern Observatory, Karl-Schwarzschild-Str. 2, D-85748 Garching, Germany
[2] Institute for Astronomy, University of Hawaii, 2680 Woodlawn Dr., Honolulu, HI 96922, USA
[3] Institut d'Astrophysique Paris, 98bis Blvd Arago, F-75014 Paris, France
[4] Astrophysikalisches Institut Potsdam, An der Sternwarte 16, D-14482 Potsdam, Germany

Abstract. There is strong evidence that the planets in the solar system evolved from a disk-shaped solar nebula ≈4.56 Gyr ago. By studying young stars in various evolutionary stages, one aims at tracing back the early history of the solar system, in particular the timescales for disk dissipation and for the formation of planetary systems. We used the VLT & ISAAC, and ESA's Infrared Space Observatory & ISOCAM to study the circumstellar environment of young low-mass stars.

1 Disk-Origin of the Solar System

The coplanarity of planetary orbits and their moons, and the preferentially prograde rotation direction led [14] and [17] to the conclusion that the solar system evolved out of a flattened disk. Meteoritic evidence from carbonaceous chondrites suggests that the Solar System formed 4.56 Gyr ago within a time span of $\lesssim$10 Myr. Studies of circumstellar disks around young stars aim at establishing the initial conditions and the exact sequence of events leading to the formation of giant and terrestrial planets in the solar system.

2 How to Find Circumstellar Disks?

Until the late 20th century, the existence of circumstellar disks around young stars could only be deduced indirectly. IR and UV excess from young stars were interpreted as disk and accretion signatures [19,1]. Submm and mm measurements provided the first estimates on disk masses and disk lifetimes [2,13]. Measurements of the frequency of IR excess, in particular L-band excess of stars in young clusters reveal that up to 100% of all young ($\lesssim$1 Myr) stars show evidence for circumstellar disks [12].

Physical Parameters of star-disk systems were determined from their spectral energy distribution (SED) and the comparison to models [1,3]. Unfortunately, solutions derived from this SED fitting turned out to be not necessarily unique [15]. This highlights the need for spatially resolved observations. The size of the solar system is ≈100 AU, which corresponds to $\leq 1''$ at the distance of the nearest star forming regions. Thus high-spatial resolution observations are required. We used VLT/ISAAC [21], complemented by high-sensitivity ISO/ISOCAM [10] observations for our studies.

3 High-Resolution Antu/ISAAC Survey for Edge-On Disks

The main observational difficulty in identifying circumstellar disks in the visual and near-infrared is that they are typically 10^6 times fainter than the central star. Circumstellar disks have been detected and spatially resolved in the Orion HII region as dark silhouettes seen against the bright background [20]. Another possibility to detect disks is if they are oriented close to edge-on (within 5° to 10°) and hence act as natural occulting bars, which block out star light [28,8,25].

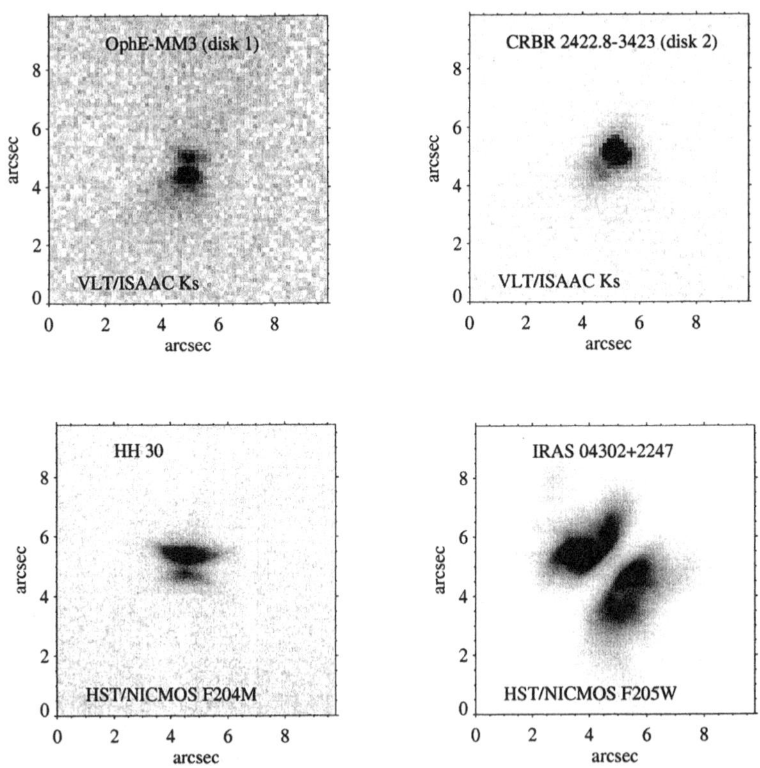

Fig. 1. Top: Edge-on circumstellar disk sources in Ophiuchus detected with VLT/ISAAC. In comparison to isolated disks in Taurus (bottom, shown as observed with HST/NICMOS), the Ophiuchus disks and their reflection nebulosities are more compact. For the disk sources in the Ophiuchus region, North is up and East is to the left

With the aim to establish a sample of edge-on circumstellar disk sources, which is suitable for detailed follow-up studies with VLT, VLTI and ALMA, we employed VLT & ISAAC to observe southern Class I IRAS sources associated with faint nebulosities in the optical or NIR.

16 sources were observed with Antu & ISAAC in JHKs in April 1999 under 0.35″ seeing conditions, which corresponds to a spatial resolution of 50 AU at a distance of 140 pc, i.e., comparable to the radius of the Kuiper Belt. The central dark lane in the Chamaeleon IR Nebula is for the first time nicely resolved [29]. Furthermore, two disks seen close to edge-on could be identified in the ρ Oph star forming region (Figure 5, [6]). Disk 1 (OphE-MM3) was previously classified as a starless core or isothermal protostar [22]. It is located 50″ west and north 10″ of Elias 2-29 – one of the most prominent and most luminous IR sources in Oph. Disk 2 (CRBR 2422.8–3423), which was originally identified as an IR source by [11], is located 30″ west and 10″ south of WLY 2-43.

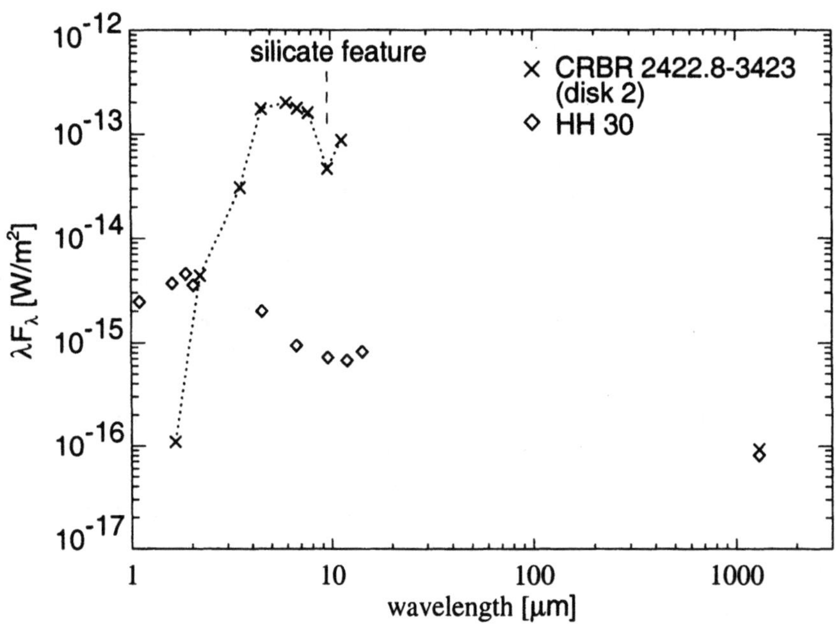

Fig. 2. Spectral energy distribution of CRBR 2422.8–3423 (disk 2) and HH 30. Both sources exhibit about the same flux at 2.2 μm and 1.3 mm. The slightly larger inclination of CRBR 2422.8–3423's disk allows the warm, inner disk to become detectable at NIR to MIR wavelengths. The dip at 9.6 μm can be explained by absorption due to the silicate feature. The spectral energy distribution of HH 30, whose disk is seen closer to edge-on than in the case of CRBR 2422.8–3423, is dominated by scattered light out to wavelengths of 10 μm

The mm measurements by [22] yield disk masses of $\approx$0.01 $M_\odot$, which is also the mass of the "minimum solar nebula" [9]. The disks in the ρ Oph region appear to be more compact (100 AU × 40 AU) than the majority of edge-on disks in Taurus. Due to the prevalence of forward scattering, the brightness ratio of the two parts of the bipolar reflection nebulosity in each disk directly yields information on the disk inclination. Disk 1, which in the NIR is also the fainter of the two disks, is seen closer to edge-on than disk 2 (inclination angle 85° vs. 75°).

ISOCAM observations between 3 μm and 12 μm of disk 2 (inclination $\approx$75°), and of the edge-on disk source HH 30 (inclination $\geq$80°) confirm theoretical predictions that a slight change in the viewing angle of a disk leads to a dramatic difference in the spectral energy distribution of YSOs [4,26]. HH 30 and disk 2 have about the same integrated brightness in K and exhibit the same 1.3 mm continuum flux. The SED of disk 2 is steeply rising between 2.2 μm and 6.0 μm, and a silicate absorption feature is visible at 9.6 μm. The SED of HH 30 is rather flat, and appears to be dominated by scattered light out to a wavelength of 10 μm [27,6].

4 How Long Do Disks Survive?

Disk survival times and the sequence of events leading to the formation of planetary systems are closely linked. Giant planets in the solar system possess a core of higher density material surrounded by a shell of metallic hydrogen and an outer atmosphere. According to one model, a higher density (rocky) core with a mass of $\approx$10 $M_\oplus$ has to form first, before noticeable amounts of nebular gas can be accreted by the proto-giant planet. Simulations indicate that at least 10^6 yr are necessary to form a 10 $M_\oplus$ rocky core [18], and that another 10^7 yr are required for the 10 $M_\oplus$ core to accrete 300 $M_\oplus$ of nebular gas. It is still unknown if massive circumstellar disks can indeed survive for such an extended period. A second model, recently reinstated by [5], suggests that gravitational instability of a protoplanetary disk leads directly to the formation of a giant gaseous protoplanet on time scales as short as 10^3 yr. The rocky core then forms due to the settling of dust grains initially acquired, and by further accretion of solid bodies in the course of the subsequential 0^5 yr. The difference in timescales for the formation of giant planets in the two models provides observational means to decide for or against either model by studying the circumstellar environment of stars with ages $\lesssim$ 15 Myr.

A sub-mm study of Lindroos binaries with ages between 3 and 150 Myr indicates that dust depletion in circumstellar disks occurs within the first 10 Myr [13]. Similarly, the percentage of sources with near-infrared and L-band excess in young clusters drops sharply within the first $\sim$ 6 Myr (e.g., [12]).

Is this the same period, which defines the "birth" of the Solar System 4.56 Gyr ago? Do dust and gaseous disk components vanish within 5 to 10 Myr? This would imply a potential timescale problem for the formation of giant planets.

ISO studies of nearby star forming regions with ages of the order of 1 Myr to 10 Myr also suggest a gradual decrease of the amount of circumstellar material

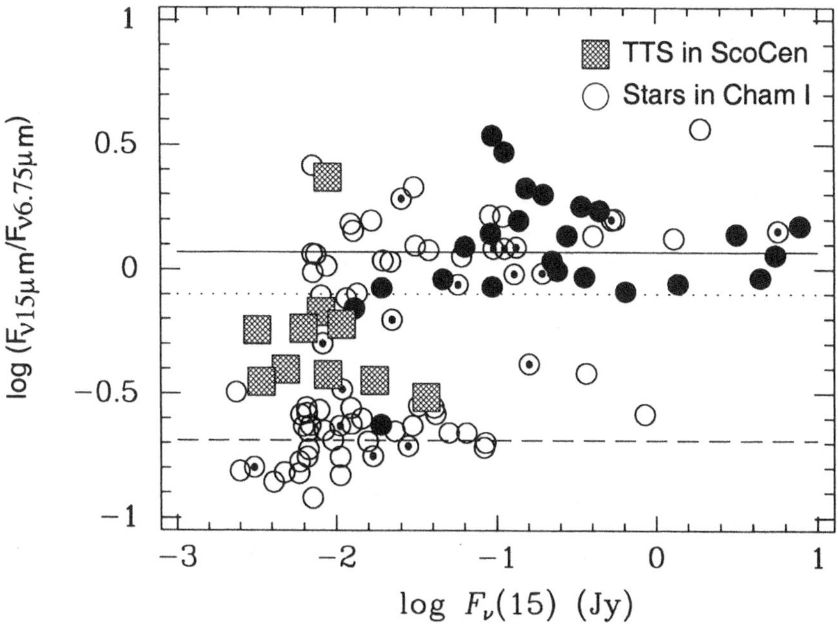

Fig. 3. Color-magnitude diagram based on ISOCAM observations of young stars in Chamaeleon (circles) and Scorpius-Centaurus (squares). Black filled circles indicate previously known Young Stellar Objects and classical T Tauri stars (CTTS), circles with a central dot previously known weak-line T Tauri stars (WTTS), and open circles new sources detected with ISO by [23]. The dashed line indicates the location of pure stellar photospheres, the dotted and solid line the location of flat and flared circumstellar disks, respectively, as predicted by the models from [16]. Post-T Tauri stars in Scorpius-Centaurus show a spectral index intermediate between main-sequence stars and CTTS and WTTS in Chamaeleon [7]

(Figure 3), in particular a depletion of smaller size dust grains, as stars evolve towards the main-sequence [23,24,7].

The diminished infrared excess can be explained by disk dissipation or by changes in the global dust opacities due to, e.g., grain growth on timescales of 5 to 15 Myr.

5 Outlook

Detailed images of the distribution of scattered light and polarization maps can now be obtained with adaptive optics systems at 8m class telescopes (e.g., Hokupa'a & QUIRC at Gemini, and NAOS & CONICA at VLT). The diffraction limit of 50mas in the H-band corresponds to <8 AU at a distance of 150 pc, thus the inner regions of circumstellar disks (and potential "protoplanetary systems") become resolvable. In particular dual imaging methods using a Wollaston prism (in order to eliminate speckle noise from unpolarized light) appear to be predestined for these kind of studies. Detailed polarization maps combined with refined theoretical models will enable us to determine physical properties of young disks, such as disk geometry, density structure, or dust properties.

High-spectral resolution studies in the infrared (e.g., with CRIRES at VLT), CRIRES, could probe and (weigh in) the gaseous disks around the post-T Tauri stars by searching for H_2 features seen in absorption against the stellar photosphere. This should provide additional information on the dispersal time scales for (gaseous) disks and the formation time scales of giant planets.

References

1. F.C. Adams, C.J. Lada, F.H. Shu: ApJ **312**, 351 (1987)
2. S.V.W. Beckwith, A.I. Sargent et al.: AJ **99**, 924 (1990)
3. C. Bertout, G. Basri, J. Bouvier: ApJ **330**, 350 (1988)
4. A.P. Boss, H.W. Yorke: ApJ **469**, 366 (1996)
5. A.P. Boss: ApJ **503**, 923 (1998)
6. W. Brandner, S.S. Sheppard et al.: A&A **364**, L13 (2000)
7. W. Brandner, H. Zinnecker et al.: AJ **119**, 292 (2000)
8. C.J. Burrows, K.R. Stapelfeldt et al.: ApJ **473**, 437 (1996)
9. A.G.W. Cameron: ICARUS **1**, 13 (1962)
10. C.J. Cesarsky, A. Abergel et al.: A&A **315**, L32 (1996)
11. F. Comerón, G.H. Rieke et al.: ApJ **416**, 185 (1993)
12. K.E. Haisch, E.A. Lada, C.J. Lada: AJ **121**, 2065 (2001)
13. D.C. Jewitt: AJ **108**, 661 (1994)
14. I. Kant: *Allgemeine Naturgeschichte und Theorie des Himmels*, (Leipzig 1755)
15. S.J. Kenyon, L. Hartmann: ApJ **323**, 714 (1987)
16. S.J. Kenyon, L. Hartmann: ApJS **101**, 117 (1995)
17. P.-S. Laplace: *L'Exposition du système du monde*, (1796)
18. J.J. Lissauer: ICARUS **69**, 249 (1987)
19. D. Lynden-Bell, J.E. Pringle: MNRAS **168**, 603 (1974)
20. M.J. McCaughrean, C.R. O'Dell: AJ **111**, 1977 (1996)
21. A. Moorwood, J.-G. Cuby et al.: The Messenger **94**, 7 (1998)
22. F. Motte, P. André, R. Neri: A&A **336**, 150 (1998)
23. L. North, G. Olofsson et al.: A&A **315**, L185 (1996)
24. G. Olofsson, M. Huldtgren et al.: A&A **350**, 883 (1999)
25. D.L. Padgett, W. Brandner et al.: AJ **117**, 1490 (1999)
26. C. Sonnhalter, T. Preibisch, H.W. Yorke.: A&A **299**, 545 (1995)
27. K.R. Stapelfeldt, A. Moneti: In: *The Universe as Seen by ISO*, ed. by P. Cox, M.F. Kessler, ESA-SP 427, p. 521 (1999)
28. B.A. Whitney, L. Hartmann: ApJ **395**, 529 (1992)
29. H. Zinnecker, A. Krabbe et al.: A&A **352**, L73 (1999)

NAOS-CONICA at VLT Yepun

Richard Edgar and Cathie Clarke

The Destruction of Circumstellar Discs

C.J. Clarke

Institute of Astronomy, Madingley Road, Cambridge, CB3 OHA, U.K.

Abstract. We present the results of Nbody simulations of the Orion Nebula Cluster which keep track of the destruction of discs by star-disc collisions and photoevaporation. We find the former process to be negligible, but that photoionisation is highly significant, although planet forming discs should still be able to survive in the outer reaches of the cluster. We find that orbital dynamics *cannot* solve the proplyd frequency problem.

We also present the results of a new 'ultraviolet switch' model for the dispersal of discs around T Tauri stars, which combines photoevaporation by the central star with viscous evolution. This model successfully reproduces the observed 'two timescale' nature of disc dispersal and promotes the rapid ($\sim 10^5$ year) demise of the disc within $5-10$ AU. Observational tests and some implications for planet formation are discussed.

1 Introduction

It is currently unclear what processes disperse the discs around young stars. In principle, these processes may be extrinsic - such as star-disc interactions (Clarke and Pringle 1991; Armitage and Clarke 1997) or photoionisation by massive stars (Johnstone et al 1998; Storzer and Hollenbach 1999) - or else may involve a range of processes intrinsic to the star-disc system, such as viscous evolution (Hartmann et al 1998), the effect of a stellar wind (Elmegreen 1979) or photoionisation provided by the central star (Hollenbach et al 1994, Yorke and Welz 1996; Richling and Yorke 1997). Another possibility that is often invoked (e.g. Brandner et al 2000) is that disc dispersal results from dust coagulation during planet formation. Although this is often cited as a motive for studying disc dispersal, it may be of more interest if other processes are in fact responsible, since then the opportunity for planet formation is limited by the efficiency of competing processes that disperse the disc.

Here we focus on two aspects of disc dispersal. Firstly, in Section 2, we review recent dynamical studies of disc destruction in Orion. Then in Section 3 we set out how the 'two timescale' nature of disc dispersal sets some severe constraints on viable mechanisms, and outline a new 'ultraviolet switch' model which satisfies these constraints.

2 Dynamical Simulations of the ONC

The Orion Nebula Cluster (henceforth ONC) provides an ideal laboratory for studying extrinsic disc destruction processes. Its high core density (comparable with that of globular clusters) indicates a possibly important role for star-disc collisions (Larson 1989), whilst the presence of several OB stars also suggests that disc photoionisation by the ultraviolet radiation of these stars might be significant (Armitage 2000). Moreover, its relative proximity, high galactic latitude and relatively low extinction have permitted the most detailed studies of any young cluster, so that its stellar - and associated disc - population is quite well characterised (McCaughrean and Stauffer 1994, Hillenbrand and Hartmann 1998, Hillenbrand and Carpenter 2000).

Recently, Scally and Clarke (2001a) performed Nbody simulations of the ONC in which the history of disc destruction via these processes is logged along each star's orbit. The advantages of this approach over estimates based on local cluster properties (e.g. Clarke and Pringle 1991, Bonnell et al 2001) are two-fold: it takes into account the fact that stars' orbits take them into regions of the cluster with different mean properties and can also model the effect of any global cluster evolution. The study of stellar encounters in the gravitationally focused regime (corresponding to within $\sim$ 160 AU in the case of the ONC) has however had to await the advent of codes (such as Aarseth's Nbody6) that can integrate large numbers of particles with no softening of the gravitational field, since softening artificially suppresses close encounters.

2.1 Star-Disc Collisions

Scally and Clarke (2001a) analysed the distribution of closest encounter distances for stars as a function of their radius in the cluster and found that after a couple of Myr, around 1% of stars in the ONC had suffered encounters within a few tens of AU. Simulations of star-disc encounters (e.g. Clarke and Pringle 1993, Hall, Clarke and Pringle 1995, Watkins et al 1998) show that encounters strip discs down to a radius of about half the pericentre distance. Thus an immediate implication of these results is that ***very few stars have suffered encounters that would perturb protoplanetary discs.*** It was also found that the distribution of closest encounter distances was very broad at any cluster radius, so that one would not expect any gradients in disc or stellar properties to result from collision processes. We also note in passing that all stars in the ONC have had an encounter with another star within $\sim 10^4$ AU within the first couple of Myr of the cluster's life: clearly, a primordial Oort cloud would not survive in this environment. However, the existence of a popular alternative model for the origin of the Oort cloud (cometary scattering from the planet forming zone after $\sim 10^8$ years; Duncan et al 1987), means that this does not place any firm constraints on possible birthplaces for the Sun.

With the benefit of hindsight, it was found that the first advantage of the Nbody approach mentioned above in fact has little effect on the closest encounter distribution. When the results of the Nbody simulation are compared with a Monte Carlo encounter model (in which each star is forced to reside at fixed radius, selecting its encounters from a pool of companions with fixed density and velocity dispersion) it is found that the Nbody simulations produce relatively few outliers (stars with an anomalously large or small encounter distance given their position in the cluster). This shows that radial mixing plays a rather minor role. However, the global evolution of the cluster *does* turn out to have a significant effect, even in a simulation like this one which is globally bound initially: since the initial conditions are not an exact solution of the collisionless Boltzmann equation, the high density core relaxes somewhat over a few crossing times. In consequence, the encounter distribution after 10^7 years is not very different from that after a couple of Myr. ***We thus conclude that most of the close encounters in the ONC have already happened.***

An important caveat in interpreting these results is that this simulation of the ONC is assumed to have originated as a smooth distribution with similar density profile to that observed. However, a variety of tightly sub-clustered initial conditions can dissolve within a few Myr so as to produce a density profile similar to the ONC (Scally and Clarke 2001b). If the ONC indeed originated as an ensemble of such mini-clusters, then star-disc encounters could have been much more significant in the past.

2.2 Photoionisation

The presence of OB stars in the core of the ONC inflicts damage on discs in the central region owing to the effects of photoevaporative mass loss. Scally and Clarke integrated this mass loss rate (as parameterised by Hollenbach et al 2000) along stellar orbits in order to construct a distribution of disc mass lost per star over the $\sim$ 2 Myr lifetime of the ONC. For stars within 0.3 pc of the most luminous cluster OB star, $\theta^1 C$ Ori, the photoevaporative mass loss rate is $\sim 10^{-7} M_\odot$/yr for a 100 AU disc, with a much lower mass loss rate beyond this sphere of influence. The integrated mass losses are considerable, extending to nearly a solar mass for systems that have remained in the central region throughout their lives. Thus photoionisation is likely to have a major effect on planet formation (note that although photoevaporative flows cannot operate directly on the tightly bound material in the planet forming region of the disc, their effect on material at larger radii can prevent re-supply of the inner disc). These findings are in line with the suggestion of Armitage (2000) that photoionisation of discs may explain the lack of 'hot Jupiters' in the globular cluster 47 Tuc (Gilliland et al 2000). However the simulations show that the peak of the mass loss distribution is considerably less ($0.04 M_\odot$), with there being a tail of stars for which the mass loss is very low (comparable with the minimum mass solar nebula). Thus we would expect some planetary formation to be viable in the outer regions of the ONC and suggest likewise that the outer regions of open clusters may harbour planetary systems.

Our simulations also shed some light on the 'proplyd frequency problem.' Simple estimates of mass loss from photoionisation suggest that discs that spend several Myr in the core of the ONC should lose several tenths of a solar mass of material. As a result one would not still expect to see discs in the core of the ONC. However, proplyd activity (presumed evidence for photoevaporating discs) is seen in virtually all stars in this region (Bally et al 2000), spawning a number of suggestions as to how such discs could survive for so long. Among these, Storzer and Hollenbach 1999 argued for a population of stars on radial orbits that 'light up' as proplyds as they pass through the core region, but which have spent most of their lives in the less destructive environs of the outer cluster. Our simulations found no evidence for such an effect, either in the case of initially virialised or initially cold clusters. We thus concur with previous authors (Bally et al 1998; Henney and O'Dell 1999), that the most likely explanation is that the massive stars in the ONC switched on relatively recently.

3 Intrinsic Dispersal Mechanisms

Despite the obvious interest in studying disc survival in the extreme environs of the ONC, it is also necessary to explain why disc diagnostics disappear on a timescale of $\sim 10^7$ years even in environments like Taurus, where neither star-disc collisions nor photoionisation can play a significant role. An important feature of disc destruction is that there appear to be *two* timescales associated with the process: the overall lifetime of discs ($\sim 10^7$ years) and the timescale on which the disc makes the transition between disc possessing and discless status, which is one to orders of magnitude less than this. The *relative* values of these timescales is set purely by the relative scarcity of 'transition objects' and is thus insensitive to any uncertainties in establishing stellar ages.

It turns out, however, to be remarkably difficult to reproduce this 'two timescale' behaviour in models of disc dispersal. Viscous evolution, for example is inadequate, since the roughly power law decline of all quantities in this model (e.g. Hartmann et al 1998) imparts the system with no second timescale. This problem is vividly demonstrated by the failure of such models to reproduce the distribution of T Tauri stars in the infrared (K-L, K-N) plane: sources are bimodally distributed in this plane (Kenyon and Hartmann 1995), being either red (disc possessing) or blue (discless) systems (these classes corresponding to the spectroscopic categories of Classical T Tauri stars - CTTs - or Weak Line T Tauri stars - WTTs). Between these two groups lies a virtually empty 'forbidden region'. Viscous models, however, linger a long time in this forbidden zone, as the disc becomes progressively optically thin as a result of viscous clearing. Magnetospheric clearing models hardly help the situation (Armitage, Clarke and Tout 1999). Though they promote the rapid demise of emission associated with the magnetospheric cavity (i.e. around 5μm), they have relatively little effect on longer wavelength emission which originates at larger radius. The fading of this emission is still controlled by viscous communication between the cavity and the disc at larger radius, so the models now linger in the forbidden region with excessively high K-N values .

A recent model that circumvents these difficulties has been proposed by Clarke, Gendrin and Sotomayor (2001). This involves the combination of photoionisation by the central star, as first elucidated by Hollenbach et al 1994, with viscous evolution. The photoevaporative mass loss is effective only at radii beyond the gravitational radius (r_g) where photoionised gas is unbound, which lies at $5 - 10$ AU. Given the modest photoionising fluxes of T Tauri stars, the associated mass loss rate is only a few $\times 10^{-10} M_{\odot}/\mathrm{yr}$, which is negligible compared with the initial accretion rate through the disc and scarcely perturbs the system at this stage. However, as the disc drains, the accretion rate falls, eventually approaching the low photoionisation mass loss value. At this point, essentially all the material viscously flowing in towards r_g is blown away and the inner disc becomes detached from its reservoir of re-supplying material. Consequently, the inner disc is starved and all diagnostics associated with this region (infrared excess out to $\sim 20\mu\mathrm{m}$, accretion indicators on the central star) fade on the *short* viscous timescale of this inner disc. Subsequently, the outer disc is progressively eroded by continued photoevaporative mass loss.

This 'ultraviolet switch' model is potentially attractive inasmuch as it is the first model that successfully reproduces the observed 'two timescale' behaviour (the two timescales in the model being set by the viscous timescale of the disc at large radius, which controls the overall fading timescale, and the viscous timescale within r_g which fixes the timescale of the eventual rapid turn-off). There are, however, several aspects of the model that can be probed by future observations. Firstly, it predicts that although the inner disc turns off very quickly, the disc beyond r_g is eroded on a much longer timescale. This brings the possibility of conflict with observations of submillimetre dust emission around Weak Line T Tauri stars: the very low upper limits (2.5mJy) recently obtained by Duvert et al (2000) are potentially problematical, although given the small size of Duvert's sample it is unclear whether this represents a serious challenge to the model. Further high sensitivity submillimetre observations are required to clarify the issue. Secondly, the viability of the model hinges on the requirement that the photoionising flux in T Tauri stars does *not* derive from accretion energy, since otherwise the source of the feedback would fade as the disc declined. Despite recent re-analyses of archival IUE data (Costa et al 2000, Johns-Krull et al 2000), the situation is still rather unclear, due largely to the small number of Weak Line T Tauri stars observed by IUE. The fact that Weak Line stars emit if anything a higher level of Xrays than Classical stars (Damiani et al 1995; Stelzer 2001) obviously points to an important energy source in these systems that does not derive from accretion.

Finally, it is worthwhile speculating about the implications of such a model for planet formation scenarios. Shu et al (1993) first considered this issue in the case of non-viscous photoevaporation models; since all the direct mass loss occurs from outside r_g, they focused on the implications for the formation of planets at larger radius (Saturn and beyond). However, when viscosity is included, our models clearly show that the dramatic depletion of surface density occurs *inside* r_g. The impact on planet formation models of such a sudden end to the inner disc is yet to be explored.

References

1. Armitage, P.J, A&A **262**,968 66 (2000)
2. Armitage, P.J., Clarke, C.J., MNRAS **285**,540 (1997)

3. Armitage, P.J., Clarke, C.J., Tout, C.A., MNRAS **304**,425 (1999)
4. Bally, J., Testi, L., Sargent, A., Carlstrom, J., AJ **116**, 854 (1998)
5. Bally, J., O'Dell, C.R., McCaughrean, M.J., AJ **119**,2919 (2000)
6. Bonnell, I.A., Smith, K.W., Davies, M.B., Horne, K., MNRAS **322**,859 (2001)
7. Brandner, W. et al, AJ **120**,950 (2000)
8. Clarke, C.J., Gendrin, A., Sotomayer, M., MNRAS submitted (2001)
9. Clarke, C.J., Pringle, J.E., MNRAS **249**, 584 (1991)
10. Clarke, C.J., Pringle, J.E., MNRAS **261**, 190 (1993)
11. Costa, V.M., Lago, M.T.V.T., Norci, L., Meurs, E.J.A., A& A **354**,621 (2000)
12. Damiani, F., Micela, G., Sciortino, S., Harnden, F.R., ApJ **446** 331(1995)
13. Duncan, M., Quinn, T., Tremaine, S., AJ **94**,1330 (1987)
14. Duvert, G. et al, A& A **355**,265 (2000)
15. Elmegreen, B.G., A&A **80**,77
16. Gilliland, R. et al, ApJ **545**,47 (2000)
17. Hall, S.M., Clarke, C.J., Pringle, J.E., MNRAS **278**,303 (1993)
18. Hartigan, P. et al, ApJ **354**, 25 (1990)
19. Hartmann, L., Calvet, N., Gullbring, E., d'Alessio, P., ApJ **495**,385 (1998)
20. Henney, W.J., O'Dell, C.R., AJ **118**, 2350 (1999)
21. Hillenbrand, L.A., Carpenter, J.M., ApJ **540**,236
22. Hillenbrand, L.A., Hartmann, L.W., ApJ **492**,540 (1998)
23. Hollenbach, D., Johnstone, D., Lizano, S., Shu, F., ApJ **428**, 654 (1994)
24. Hollenbach, D., Yorke, H.W., Johnstone, D., in *Protostars and Planets IV*. ed. by V. Mannings, A.P. Boss, S.S. Russell, (Tucson:University of Arizona Press), pp 401-428
25. Johns-Krull, C., Valenti, J., Linsky, J., ApJ **539**,815 (2000)
26. Johnstone, D., Hollenbach, D., Bally, J., ApJ **499**,758 (1998)
27. Kenyon, S.J., Hartmann, L., ApJS **101**, 117 (1995)
28. Larson, R.B., in *Physical processes in fragmentation and star formation*, eds R. Cappuzzo-Dolcetta, C. Chiosi, A. di Fazio, (Kluwer:Dordrecht), pp 389-399 (1989)
29. McCaughrean, M.J., Stauffer, J.R., AJ **104**, 1382 (1994)
30. Scally, A., Clarke, C.J., MNRAS in press (2001a)
31. Scally, A., Clarke, C.J., MNRAS submitted (2001b)
32. Shu, F.H., Johnstone, D., Hollenbach, D., Icarus **106**,92 (1993)
33. Stelzer, B., to appear in *Xray Astronomy 2000*, eds. R. Giacconi, L. Stella, S. Serio, ASP Conf Series, in press (2001)
34. Storzer, H., Hollenbach, D., ApJ **515**, 669 (1999)
35. Watkins, S.J. et al, MNRAS **300**,1189 (1998)
36. Yorke, H.W., Welz, A. A&A **315**,555 (1996)

VISIR and the Formation of Stars and Planets

Pierre-Olivier Lagage

DSM/DAPNIA/Service d'Astrophysique, CEA/Saclay, Bat 709,
F-91191 Gif-sur-Yvette Cedex, France

Abstract. The mid-infrared wavelength range is particularly well suited to study warm circumstellar environments, in relation with star and planet formation theories. VISIR is the VLT instrument dedicated to this wavelength range. It will be installed in 2002 at the Cassegrain focus of Melipal, the VLT telescope unit number 3. This cryogenic instrument, optimized for the two mid-infrared atmospheric windows (N and Q bands), combines imaging capabilities at the diffraction limit of the telescope (0.3 arcsec at 10 μm) over a field up to 51 arcsec, and long-slit (32 arcsec) grating spectroscopy capabilities with various spectral resolutions up to 25000 at 10 μm and 12500 at 20 μm. In this paper, the status of the instrument and some examples of observing programs are discussed, with emphasis on the high spectral resolution observing mode, which will give access, in a unique way, to observations of circumstellar warm molecular hydrogen gas from the South hemisphere.

1 Introduction

Since the beginning of the VLT project, ESO has set up a comprehensive instrumentation plan that provides a wide range of imaging and spectroscopic facilities from near-ultraviolet to mid-infrared (see R. Siebenmorgen, these proceedings, for the VLT imaging capabilities, and A. Moorwood, these proceedings, for the VLT spectroscopic capabilities). VISIR is the VLT mid-IR instrument to be installed at the Cassegrain focus of Melipal. It is built by a French-Dutch consortium of institutes, led by the Service d'Astrophysique of Commissariat à l'Energie Atomique (CEA, Saclay, France). The Dutch partner is the Netherlands Foundation for Research in Astronomy (NFRA, Dwingeloo, The Netherlands; Co-PI: J.W. Pel). For technical information about VISIR, the reader is referred to paper [1].

Here we just list the VISIR observing modes which are:

• imaging with a choice between 3 magnifications and 24 narrow and broad-band filters. The three Pixel Fields Of View (PFOV) are 0.075 arcsec, 0.127 arcsec, 0.2 arcsec. The corresponding fields of view are 19.2 $\times$ 19.2 arcsec^2, 32.5 $\times$ 32.5 arcsec^2 and 51.2 $\times$ 51.2 arcsec^2.

• long-slit grating spectroscopy with various spectral resolutions (R=$\lambda/\delta\lambda$), given for an entrance slit width set to 2λ/D, (where, as usual, λ is the wavelength and D the telescope diameter). Three spectral resolutions are available in the N band: low (R = 350 at 10 μm), medium (R = 3200 at 10 μm) and high (R = 25000 at 10 μm). Two spectral resolutions are available in the Q band: medium (R = 1600 at 20 μm) and high (R = 12500 at 20 μm). The PFOV along the slit is 0.127 arcsec and the slit length on the sky is 32 arcsec.

Fig. 1. VISIR cryostat mounted on its integration support in the Saclay test hall

2 VISIR and Circumstellar Disks

Studying disk around stars at various stages of evolution is crucial for our understanding of star and planet formation.

For example, disks are key ingredients to planet formation, not only as material

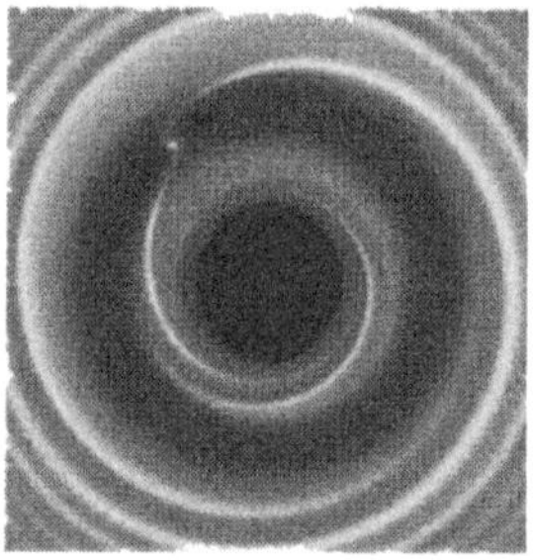

Fig. 2. Image obtained from a numerical simulation calculating the migration of a newly born planet interacting with the disk in which the planet is still embedded; the planet can open gaps in the disk. (By courtesy of F. Masset)

reservoir, (whose evolution determines the time scale for planet formation), but also to determine the orbit of the planet (migration inward (e.g. [2]), but also outward (e.g. [3])). Physical information on "protoplanetary" disk (surface density, thickness...) are needed to determine the evolution of a newly born planet. Protoplanetary disks disappear with time. However, collisions between planetesimals or comet sublimation can generate a second generation dust disk; these disks are usually quoted as "debris" disks. The presence of a planet in such a disk can deeply modify the morphology of the disk through gravitational perturbations of the dust orbit (e.g. [4]). Thus, studying the morphology of debris disks is a way to search for "footprints" of planets.

2.1 Dust Imaging

Dust disk imaging in the mid-infrared has started in 1993 ([5], see also figures 3 and 5) with the first image of the β-Pictoris dust disk, obtained with the ESO TIMMI camera. Up to now, only a few objects have been imaged [e.g. 6 and references therein], but we can expect than an instrument like VISIR will allow to enlarge significantly the number of resolved disks. VISIR low spectral resolution mode in the N band and medium resolution in the Q band will be used to determine the nature of the dust (by observing silicate features ...).

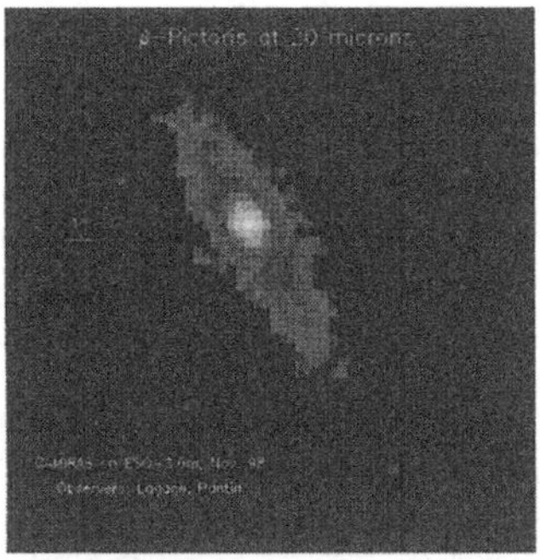

Fig. 3. β-Pictoris dust disk as observed at 20 μm (raw image). North is up; East is left. The PFOV is 0.3 arcsec, which means 5.7 AU at the distance of β-Pictoris (19.3 pc). At this wavelength, the disk emission dominates over the star emission.
Figure available in colour on the CD-ROM

2.2 Gas Imaging

The warm H_2 gas in disks has not yet been imaged. But the prospects for VISIR being able to do so are very good. H_2 is the most abundant molecule in the Universe, but it is difficult to detect. It is a symmetrical molecule, so that all rotational, vibrational transitions within the electronic ground state are quadripolar with low spontaneous emission coefficient. In addition, in normal conditions (i.e. M_{dust}/M_{gas} =1/100, $T_{dust} = T_{gas}$, dust size = 0.1 μm), H_2 lines are deeply embedded within dust continuum emission, so that very high spectral resolution are needed to disentangle the lines. However, in special conditions (dust depletion, dust and gas at a different temperature), the lines can be detected even with moderate spectral resolution. The first detections of H_2 in protoplanetary and debris disks have been achieved with the ISOSWS instrument [7]. But the angular resolution of ISOSWS was too poor (14×27 arcsec^2) to image the H_2 emission.

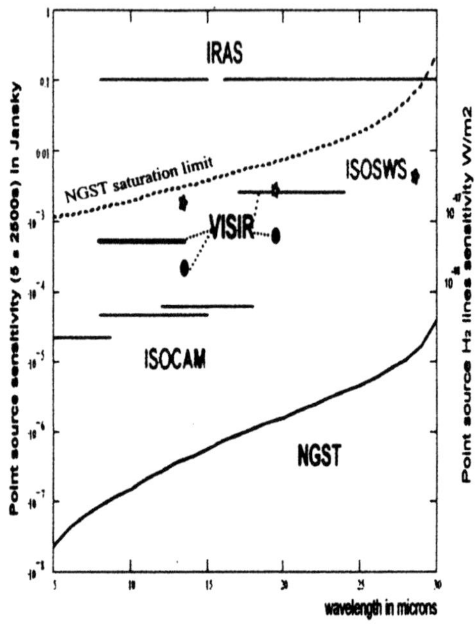

Fig. 4. Point source VISIR sensibility to continuum emission and H_2 line emission at 12.3 and 17 μm, as compared to IRAS, ISO and NGST. In imaging, VISIR will be less sensitive than ISOCAM, but with a better angular resolution, which is essential to undergo programs such as disk around stars (see Fig. 5). For observations of H_2 lines, VISIR should be more performant than ISOSWS and with a much better angular resolution (from $14'' \times 27''$ to $1'' \times 0.5''$), opening new perspectives in the area of imaging warm gas in disks. Within a time frame of 7 years, NGST will overpass VISIR

Among the H_2 lines present in the mid-IR, two lines are accessible from the ground: the pure rotational lines at 12.28 μm and 17.03 μm. To have access to these lines, observations at high (> 10000) spectral resolution are needed in order not to be contaminated by atmospheric lines (from H_2O or CO_2). H_2 lines from Orion [8] and from the supernova remnant IC 443 [9] have already been detected from the ground, thanks to the pioneering IRSHELL instrument [10]. VISIR will be almost 2 orders of magnitude more sensitive than IRSHELL and will have a sensitivity good enough to detected H_2 lines from disks, located up to 100 pc away and containing down to 1 Jupiter mass of H_2 at 100K. Among the first targets for VISIR, we can quote β-Pictoris, HD135344, GG Tau.

3 VISIR and the Direct Detection of Giant Exo-Planets

The direct detection of giant exo-planets with an instrument like VISIR appears possible for a limited, but plausible, range of parameters [11]. When considering

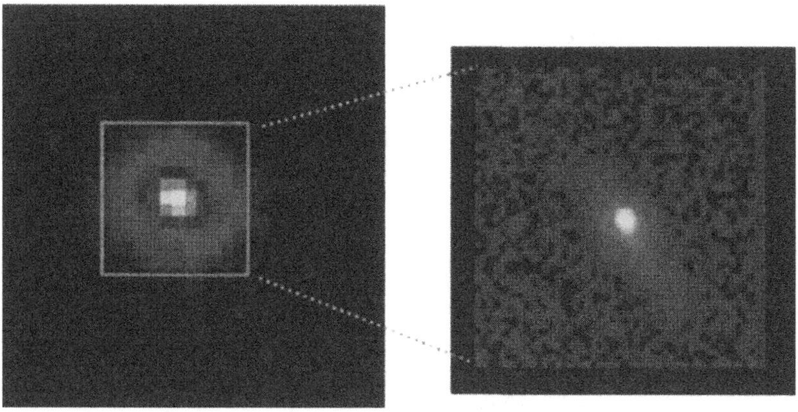

Fig. 5. Two images of the same object, β-Pictoris, to illustrate the gain in angular resolution when going from a 60-cm telescope to a 3.6-m telescope; left: image of β-Pictoris obtained with ISOCAM (PFOV: 1.5 arcsec); the disk is not resolved; it is embedded in the diffraction pattern from the star. Right: raw image of β-Pictoris, in the same wavelength range (11-13 μm), but obtained in 1993 with the TIMMI ESO camera at the 3.6 m La Silla telescope, (PFOV: 0.3 arcsec). The noise is higher, but the angular resolution is much better and the disk is resolved [5]

the direct detection of an exo-planet, two main issues have to be discussed, the sensitivity and the contrast in brightness between the hosting star and the planet. Given the sensitivity limits shown on Fig. 4, the detection of exo-planets with VISIR will be limited to giant planets (7-12 M_{jup}), at a relatively high temperature (450-350 K), and orbiting a star located at a distance of less than 10 pc from the Earth. Given the angular resolution, (telescope diffraction limited), the contrast should not be an issue when searching those giant planets orbiting at a distance greater than 5-10 AU from a M-dwarf (see Fig. 6); from the radial-velocity planet search technique, we know that planets can indeed be present around M-dwarfs (see the case of Gliese 876 [12 and references therein]). When the hosting star is of solar-type, then the exo-planet is embedded into the point spread function of the star and more sophisticated instruments (for example featuring a nulling coronograph) may be needed.

4 Conclusion

VISIR will be a great instrument to study star and planet formation. The detection of warm molecular hydrogen will be of particular interest, and VISIR will be a unique facility in the South hemisphere to provide the spectral resolution required to perform such detections. The only other instrument with a high spectral resolution mode in the mid-infrared is the Michelle instrument [13] to be mounted on UKIRT this year (2001) and on GEMINI next year. The shipment

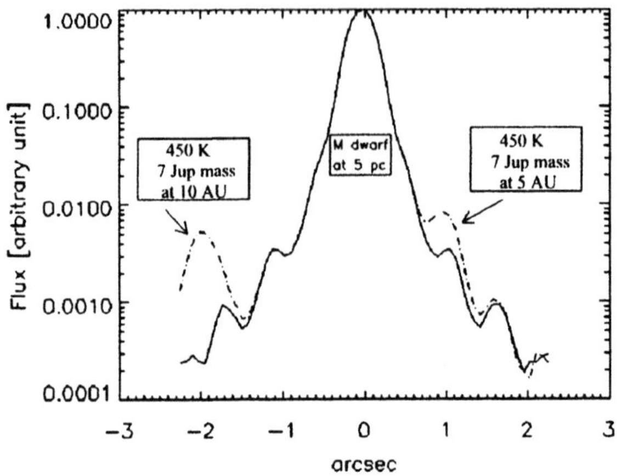

Fig. 6. Signal expected in VISIR from a system made of two 7 M_{jup} exo-planets at 450 K, orbiting respectively at 5 AU and 10 AU from a M-dwarf located at 5 pc from us. (The plot shows the pixel signal along a line containing the star and the two planets peaks). The diffraction by the telescope and a seeing of 0.4″ in the visible have been taken into account, as well as the background noise generated by the atmosphere and the telescope. Solid line: pattern from the star alone; dashed-dotted line: star + planets

of VISIR to Paranal is scheduled during summer 2002; the first observations on the sky are scheduled for the end of 2002. First target? β-Pictoris!

References

1. P.O. Lagage: 'The final design of VISIR'. In: *Optical and IR Telescope Instrumentation and Detectors, I. Masanori and A. Moorwood Eds, SPIE Vol.* **4008**, pp. 1120 (2001)
2. D. Lin, J. Papaloizou: ApJ **309**, 846 (1986)
3. F. Masset, M. Snellgrove: MNRAS **320L**, 55 (2001)
4. F. Paresce: Adv Space Res **12**, 208 (1992)
5. P.O. Lagage, E. Pantin: Nature **369**, 628 (1994)
6. W.F. Thi, et al.: Nature **409**, 60 (2001)
7. C.M. Telesco, et al.: ApJ **530**, 329 (2000)
8. P.S. Parmar, et al.: ApJ **430**, 786 (1994)
9. M.J. Richter, et al.: ApJ **449**, L83 (1995)
10. J.H. Lacy, et al.: PASP **101**, 1166 (1989)
11. P.O. Lagage: 'VISIR and the detection of exo-planets'. In: *From extrasolar planets to cosmology: the VLT opening symposium, J. Bergeron and A. Renzini Eds, Springer publisher, 2000*, pp. 528
12. G.W. Marcy et al.: ApJ submitted
13. A. Glasse, et al.: 'Michelle Midinfrared Spectrometer and Imager', In: *Optical Telescopes of Today and Tomorrow, Arne L. Ardeberg Eds, SPIE Vol.* **2871**, pp. 1197-1203 (1997)

Mid Infrared Variability of Herbig Ae/Be Stars

Timo Prusti[1] and Antonella Natta[2]

[1] Space Science Department of ESA, ESTEC, Postbus 299, 2200 AG Noordwijk, The Netherlands
[2] Osservatorio Astrofisico de Arcetri, Largo E. Fermi 5, 50125 Firenze, Italy

Abstract. We present the first results of an observational ISO programme to study mid infrared variability in young intermediate mass Herbig Ae/Be stars. With the current accuracy of automatic data processing we can identify variability in one of our programme stars. SV Cep was measured 13 times between 3.3 and 100 μm over a period of slightly more than two years. Over this period, the star showed a mid infrared variation of about ±20 % above and below its average level. This mid infrared variability was in anti correlation with the optical variability over the same period. This study shows that SV Cep, in addition to IRAS variables WW Vul and AB Aur, experiences non-axisymmetric changes in its circumstellar environment causing the mid infrared flux to vary in time scales of months.

1 Introduction

Variability is one of the characteristics of stars which can be used to probe the physics of the object and its environment. For young stars variability is used to recognise their early evolutionary status. Naturally, the variability at optical wavelengths has been well studied and correspondingly the understanding of the underlying physics is most developed. An example can be the quasi periodic variability of T Tauri stars (TTS) which is believed to be due to transient spots, which, however, have long enough life times that rotation periods can be deduced by examining the periodicity.

When moving from the optical to the infrared wavelengths the number of variability studies diminishes rapidly. Nevertheless, near infrared variability is a known property of young stars and in regions of high extinction it can be used, in analogy to optical studies, to recognise the optically obscured young stellar population [6]. In a study of the near infrared variability characteristics of a large sample of TTS it was possible to explain 99 % of the cases in terms of stellar processes, well known from optical studies [1]. This is not surprising, since the near infrared emission in TTS is dominated by the direct radiation of the star itself.

In order to probe other time variable processes that originate in the circumstellar environment one has to look at wavelengths where the emission is not dominated by direct stellar light, i.e., in the mid infrared. Mid infrared variability studies are scarce due to the difficulty to obtain photometrically reliable data over a long time span. IRAS data can be used to examine the mid infrared variability characteristics of young stars. Unfortunately, the spatial resolution of IRAS was not really sufficient for variability examination of all young stars because confusion may cause variation which cannot be distinguished from real physical variability of the sources. On the other hand isolated sources can be studied extremely well over different time scales. The advantage of the IRAS data is in the well established reliability and flux consistency estimates. When variability is seen at levels above the measurement accuracy and in phase in two independent wavelengths, then a high variability probability can be deduced. This way variability of the order of 0.2 mag could be revealed at 12 and 25 μm for Herbig Ae/Be stars AB Aur and WW Vul [12].

Although the detection of infrared variability among young stars by IRAS is a very reliable result, neither the temporal nor the wavelength coverage is sufficient to obtain deeper insight into the phenomenon. Therefore we conducted a systematic programme to monitor with ISO a sample of Herbig Ae/Be stars from 3.3 to 100 μm. The sample selection was based on the visibility constraints set by the ISO orbit. We required a possibility to measure the objects over the time span of the whole mission. Furthermore, we concentrated on isolated stars in order to minimise possible problems of confusion. We have examined the measurements using the final OLP10 version of automatic processing. The clearest case of variability has been seen in SV Cep. This study presents our preliminary results on SV Cep obtained using the OLP10 pipeline products without any further manual reduction steps.

The spectral type of SV Cep has been recently determined to be AOIIIe [7] which is in agreement with earlier work. The distance is estimated to be 700 pc, but there is a high uncertainty in this value [7]. The optical light curve shows periodic wave like pattern with periodicities of 4000 and 670 d [13]. While the wave like pattern can also be seen in the long term light curve, the periodicities do not remain constant over periods of 100 years according to Discrete Fourier Analysis [4]. In addition to long term brightness decline and two long lasting minima at the beginning of the 20th century, SV Cep shows irregular deep Algol-type minima of few days duration, which classify it among UX Ori variables (UXOR). These minima, of the order of 1 mag in the case of SV Cep, are accompanied by an increase of the polarisation degree and are believed to be caused by circumstellar dust clouds occasionally passing across the line of sight to the star.

During the two year period SV Cep was monitored with ISO, the optical light curve showed clear variability. At the beginning of the observing period, the star was close to its maximum optical brightness; it showed then a slow decline in luminosity for about one year, followed by a brightening towards the end of the ISO monitoring period. At the end of it, however, the star was still dimmer that at the beginning. On top of this long term trend there was small amplitude irregular variability. However, there was only one deeper minimum, of rather small amplitude and without a clear accompanying polarisation increase [13]. Therefore we cannot relate the mid infrared variability of SV Cep to any of its UXOR deep minima, but rather examine the infrared behaviour over one of the optical 670 d cycles.

2 Observations

We have examined the ISO data reduced with the automatic OLP10 pipeline for those SV Cep observations which covered the spectral energy distribution at 3.3, 3.6, 4.8, 7.3, 10.0, 12.0, 12.8, 15.0, 20, 25, 60, and 100 μm. These observations were made with PHT03 mode and they were always accompanied with an offset measurement for a background estimate. The observations were done on 10-Feb-1996, 28-May-1996, 28-Jul-1996, 8-Oct-1996, 31-Dec-1996, 24-Jan-1997, 8-Feb-1997, 31-May-1997, 21-Jun-1997, 26-Sep-1997, 5-Dec-1997, 21-Dec-1997, and 29-Mar-1998. The respective observation numbers are 085 02136, 193 02326, 255 00128, 326 01432, 411 00734, 435 01230, 449 01744, 562 00648, 583 02204, 681 00206, 751 00108, 767 00952, and 865 00854 with the corresponding background measurement having a number which is one less than that of the on-source observation. In the observation numbers the three first digits represent the ISO revolution. As the ISO revolution was close to 24 h, the revolution number is used later to indicate time spans in days between observations. The input parameters were kept constant in order to avoid introducing any systematic effect into the series. However, over the ISO life time there were small enhancements in the internal instrument operation sequences over the mission and therefore the measurements are not executed always exactly the same way.

3 Results

The average spectral energy distribution of SV Cep based on the 13 measurements reduced with the OLP10 pipeline is displayed in Fig. 1. We see on top of the continuum maxima at about 10 and 20 μm due to silicates. In addition to amorphous silicates it seems likely that crystalline silicates (and/or FeO) are present in SV Cep to make the features broader at the long wavelength side of them. This is common among Herbig Ae/Be stars [8]. Figure 2 shows as a function of time the values of the observed fluxes normalized to the average values at three wavelengths (7.3, 25, and 100 μm) over the ISO mission. The flux variations are about 20 % at most. This is also the quoted accuracy of a PHT measurement. This means that the flux variation between any two arbitrarily chosen epochs cannot be associated with variability intrinsic to the source. However, an indication of true variability comes from the fact that the measurements show a trend over time. By examining the 7.3 μm flux in Fig. 2 one can see that it is less than average at the beginning of the observational period, it is above the average between ISO revolutions 400 and 600, and it gets fainter again afterwards. At 25 μm the same trend, although less pronounced, can be seen. To see the same trend at these two wavelengths is particularly important because the calibration for measurements between 3.3 and 15 μm is independent of that for the 20 and 25 μm measurements. The 100 μm time series has the lowest photometric accuracy due to lower fluxes (see Fig. 1) and therefore it is difficult to make strong conclusions. Nevertheless, if there is any 100 μm variability at all, then it is most likely in anti-correlation with the variability in the mid infrared.

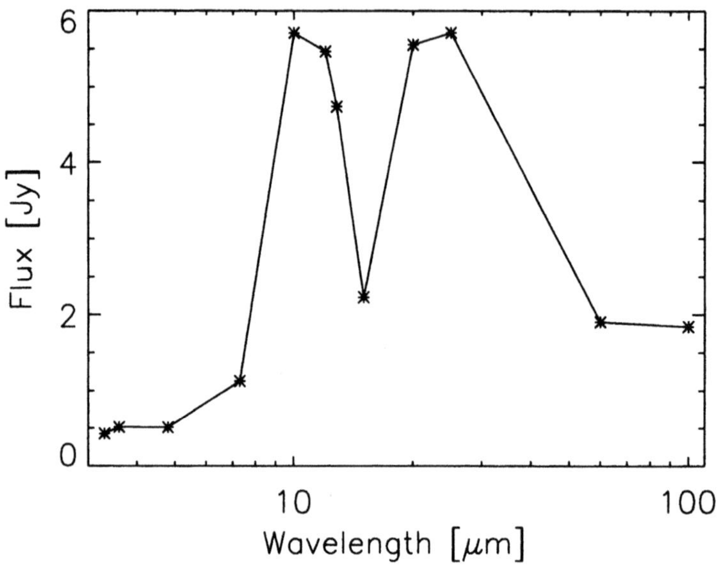

Fig. 1. Average spectral energy distribution of SV Cep based on the 13 ISO measurements reduced with the OLP10 pipeline

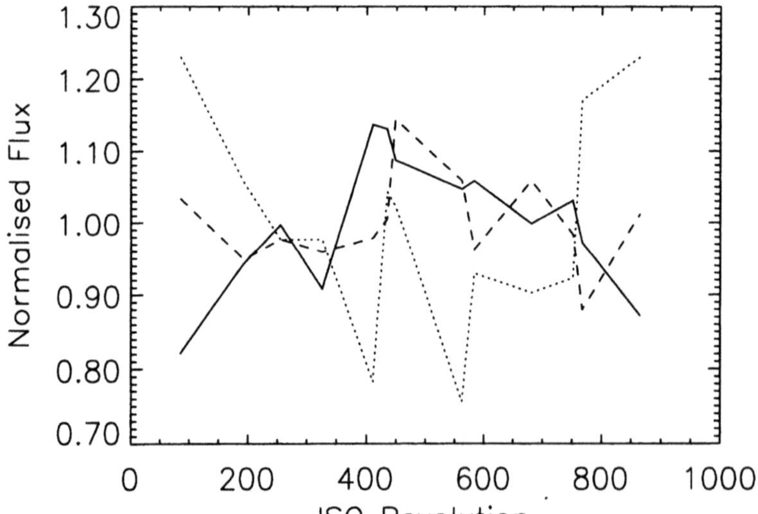

Fig. 2. Normalized OLP10 pipeline fluxes of SV Cep over the ISO mission at 7.3, 25 and 100 μm (solid, dashed and dotted lines respectively)

4 Discussion

For the following discussion, it is useful to remind what is currently the best model of the the spectral energy distribution of pre-main sequence stars of intermediate mass [11] (see Fig. 3). This model is based on a hydrostatic equilibrium disk which has naturally an optically thin disk atmosphere above the optically thick mid plane [2]. The important addition necessary to explain the more massive Herbig Ae/Be stars was the disk rim [11]. This rim is located at the distance of dust destruction. Due to the dust sublimation at distance closer to the star, the rim is fully exposed to the stellar radiation. This exposure causes the rim to heat up which in turn puffs the disk up creating a disk rim. Therefore the spectral energy distribution of a Herbig Ae/Be star can be described with the help of four components: the star, whose emission dominate at visual wavelengths, the disk rim, which dominates in the near-infrared, the disk atmosphere, where the silicate feature and mid-infrared emission originates, and the disk mid plane, which is only observable at long wavelengths.

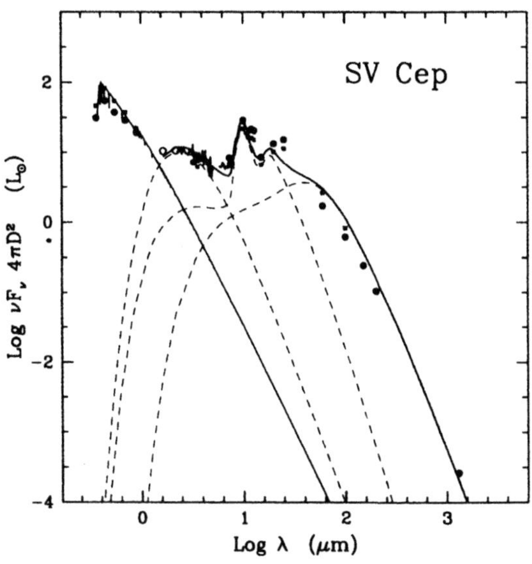

Fig. 3. Model of the spectral energy distribution. The lines represent (from short to long wavelengths) the stellar, disk rim, disk atmosphere, and disk plane contributions. The observations are in the optical wavelengths from [13], in the near infrared from [5] (open circles), between 2.5 and 200 μm ISO data from this paper, and 1.3-mm measurement from [9]. For the the variability parts the squares and circles represent optically bright and faint epochs respectively (i.e. mid infrared faint and bright epochs respectively). The PHT40 spectrum from 2.5 to 11.5 μm and PHT22 150 and 200 μm ISO measurements are shown for reference here and will be described elsewhere

Within the context of such models, the mid-IR variability should be due to a change in the disk atmosphere emission, caused by a change in its temperature and/or mass. Let us assume, for the sake of speculation, that the variability of UXORs is triggered by changes in the structure of the disk rim. Then, if, for example, the rim becomes more puffed up, we should observe an increase of the near infrared flux, a decrease of the emission of the disk atmosphere, which will be more in the shade of the rim, and, possibly, a decrease of the direct stellar radiation, if the rim puffs up enough to intercept the star [3]. This does not seem to be the case, since the visual and mid infrared radiation vary in opposite directions. We should, however, point out the many caveats of this interpretation. First of all, it is not clear that the long term variability we observe in the optical is due, as the UXORs deep minima, to a change in the line-of-sight extinction. Second, even in the case of bona-fide UXORs minima, there is no evidence that one could associate them to a variation of the vertical extension of the disk rim, since the time scale of the minima duration seems to require a location of the obscuring dust further from the star than the rim itself [10].

An interesting clue to the infrared variability mechanisms and to the disk rim/atmosphere interplay can come from the correlation of the 3.3 and 25 μm emission, where the disk rim and atmosphere dominate. There is in the ISO data a hint that they vary together. If confirmed, this will not be consistent with the expectations of the simple models described above.

While the IRAS data showed the existence of mid infrared variability the ISO data has so far only confirmed this. In order to address the physics causing the mid infrared variability, it is necessary to reduce the data set presented here carefully by hand with the hope to push down the error bars significantly.

References

1. J.M. Carpenter, L.A. Hillenbrand, M.F. Skrutskie: AJ **121**, 3160 (2001)
2. E.I. Chiang, P. Goldreich: ApJ **490**, 368 (1997)
3. C.P. Dullemond, C. Dominik, A. Natta: ApJ, in press.
4. C. Friedemann, H.-G. Reimann, J. Gürtler: A&A **255**, 246 (1992)
5. I.S. Glass, M.V. Penston: MNRAS **167**, 237 (1970)
6. A.A. Kaas: AJ **118**, 558 (1999)
7. M. Kun, J. Vinkó, L. Szabados: MNRAS **319**, 777 (2000)
8. G. Meeus, L.B.F.M. Waters, J. Bouwman, M.E. van den Ancker, C. Waelkens, K. Malfait: A&A **365**, 476 (2001)
9. A. Natta, V.P. Grinin, V. Mannings, H. Ungerechts: ApJ **491**, 885 (1997)
10. A. Natta, T. Prusti, R. Neri, W.F. Thi, V.P. Grinin, V. Mannings: A&A **350**, 541 (1999)
11. A. Natta, T. Prusti, R. Neri, D. Wooden, V.P. Grinin, V. Mannings: A&A **371**, 186 (2001)
12. T. Prusti, A.S. Mitskevich: 'Far Infrared Variability of Herbig Ae/Be Stars'. In: *The Nature and Evolutionary Status of Herbig Ae/Be Stars*, ed. by P.S. The, M.R. Perez, E.P.J. van den Heuvel (ASPC 62, 1994), pp. 257–260
13. A.N. Rostopchina, V.P. Grinin, D.N. Shakhovskoĭ, P.S. Thé, N.Kh. Minikulov: Ast. Rep. **44**, 365 (2000)

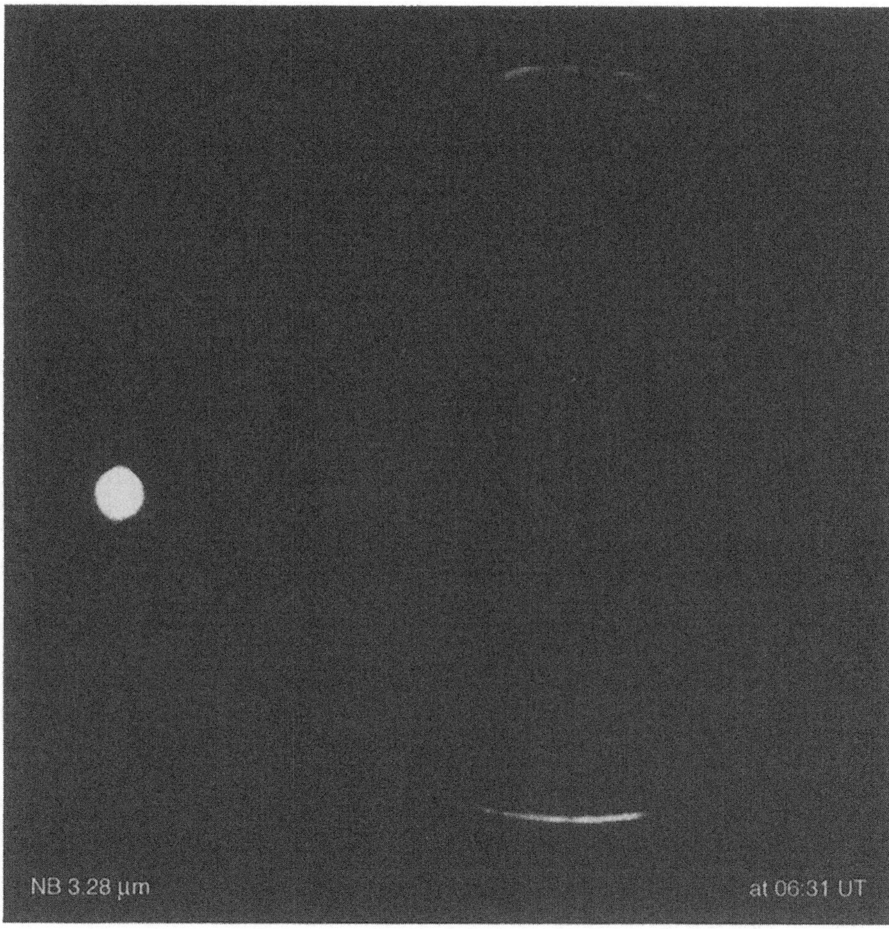

Jupiter and Io. Thermal-IR image (3.28 μm)
(Antu/UT1 + ISAAC)

Michel Mayor, Ray Jayawardhana, Mansur Ibrahimov, Matthew Bate, and Nuno Santos

Prospects for Star Formation Studies with Mid-Infrared Instruments on Large Telescopes

Ray Jayawardhana

Department of Astronomy, University of California, Berkeley, CA 94720, U.S.A.

Abstract. Imaging and spectroscopic observations in the mid-infrared wavelength range (5μm–30μm) offer valuable insight into the origins of stars and planets. Sensitive new array detectors on 8-meter class telescopes make it possible to study a wide range of phenomena, from protoplanetary disks to starburst galaxies, in unprecedented detail. I review the capabilities of ground-based mid-infrared instruments (e.g., high spatial resolution) and their limitations (e.g., poor sensitivity, small field of view) using several examples in the field of star formation, and discuss prospects for the near future.

1 Introduction

The earliest stages of star and planet formation are deeply shrouded in cocoons of gas and dust, usually impenetrable at shorter wavelengths. Mid-infrared radiation suffers relatively little extinction, and therefore provides a valuable probe for investigating these dusty beginnings. The 5μm–30μm range is ideal for observations of the thermal continuum emitted by warm dust with $T \approx$ 100-300 K, often found in the circumstellar environment, and of the spectral features due to a variety of atomic (e.g., [Ne II]), molecular (e.g., H_2, PAHs) and solid-state (e.g., silicates) species.

2 Promise

The great promise of ground-based mid-infrared instruments is their high spatial resolution. Since imaging in this wavelength regime is (almost) diffraction limited, large ground-based telescopes have a significant advantage over satellite observatories such as IRAS and ISO with their small primary mirrors.

3 Limitations

However, ground-based mid-infrared observations are challenging, to say the least, for two primary reasons:

- The atmosphere is only partially clear in the 5μm–30μm wavelength range. Since the atmospheric opacity is mainly due to absorption by water vapor and carbon dioxide, transmission significantly improves with dryness of the site and altitude.

- The background emission is dominant over the astronomical sources and variable on short timescales, making it necessary to continuously measure and subtract the sky. (A 300 K blackbody peaks at 10μm.) Integration times are very short (tens of milliseconds), and the use of chopping and nodding is essential.

As a result, observations are usually photon-noise limited and the sensitivity is relatively poor. Even on an 8-meter telescope, it is difficult to detect a point source fainter than 1 mJy (or mag$\approx$11.5) in the broad N-band centered at 10μm in a reasonable time. Cooled space-borne observatories –such as SIRTF– will have much better sensitivity.

While mid-infrared detectors have vastly improved over the past two decades, most arrays in astronomical use to date have been limited to 128 × 128 pixels, providing a small field-of-view ($\sim$ 10″ × 10″) on 8-meter class telescopes. The new mid-infrared instrument on the ESO 3.6-meter telescope, TIMMI2, marks a step forward with its 320×240 Si:As array developed by Raytheon.

The other handicap of the current generation of mid-infrared instruments is their low spectral resolution, typically $R \approx$100 at 10μm. However, COMICS on Subaru achieves $R \sim$ 2000, and VISIR on the VLT is designed to do long-slit spectroscopy up to $R \sim$ 30,000.

4 Science: Results and Prospects

Despite these limitations, many important scientific results have been obtained in recent years through the use of mid-infrared instruments on the ground. The advent of 8-meter class telescopes enhance the science prospects for the near future.

4.1 Compact HII Regions

Continuum mid-infrared imaging is useful for tracing the distribution of warm dust emission within compact HII regions while narrow-band (e.g., [NeII]) imaging can help identify ionizing fronts. For example, Smith et al. (2000) have studied the mid-infrared morphologies of the W49A complex, located on the far side of the Galaxy and thus behind some $\approx$300 mag of extinction along the line of sight. Their images have sufficient resolution to investigate the nature of the dust emission from individual sources and make comparisons with the radio observations. As another recent example, Chini et al. (2001) produced a 20μm/10μm "temperature map" of the dust in the Orion BN/KL complex, and identified the main sources of heating.

4.2 Embedded Protostars

One of the difficulties of probing the formation of protostars is a deficit of high spatial resolution observations. Large extinction ($A_v \gg$ 10 mag) usually prevents

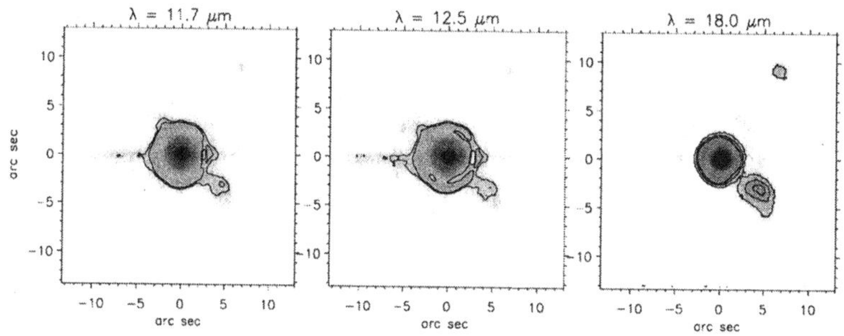

Fig. 1. IRTF/MIRAC3 images of AFGL 2591 at (a) 11.7μm, (b) 12.5μm, and (c) 18.0μm. Contours are plotted for 10 and 15 σ levels in panels (a) and (b), and for 5, 10 and 15 σ in panel (c). From Marengo et al. (2000)

optical imaging while millimeter observations typically have larger beam sizes. Mid-infrared imaging and spectroscopy can help.

For example, Marengo et al. (2000) have recently presented sub-arcsecond-resolution images of the high-luminosity young stellar object AFGL 2591 and its circumstellar environment. Their images, at 11.7, 12.5 and 18.0 μm, reveal a knot of emission $\approx 6''$ SW of the star, which may be evidence for a recent ejection event or an embedded companion star (Fig. 1). This knot is roughly coincident with a previously seen near-infrared reflection nebula and a radio source, and lies within the known large-scale CO outflow. Marengo et al. also find a new faint NW source which may be another embedded lower-luminosity star. The *IRAS* mid-infrared spectrum of AFGL 2591 shows a large silicate absorption feature at 10μm, implying that the primary source is surrounded by an optically thick dusty envelope.

4.3 Herbig Ae/Be Stars

First identified by Herbig (1960) as higher-mass counterparts to young T Tauri stars, these objects show many of the same signs of activity such as emission lines and large infrared excesses. Over the years, several authors have attempted to model their spectral energy distributions as dusty envelopes and/or disks (e.g., Miroshnichenko et al. 1999 and references therein). Recent mid-infrared imaging by Polomski (2001; Fig. 2) and others reveal companions and complex circumstellar environments around many of these sources.

4.4 Circumstellar Disks

The recent discovery of a spatially-resolved disk around the nearby 10-Myr-old star HR 4796A is a spectacular demonstration of the power of high-resolution mid-infrared imaging (Jayawardhana et al. 1998; Koerner et al. 1998). The surface brightness distribution of the disk is consistent with the presence of an inner

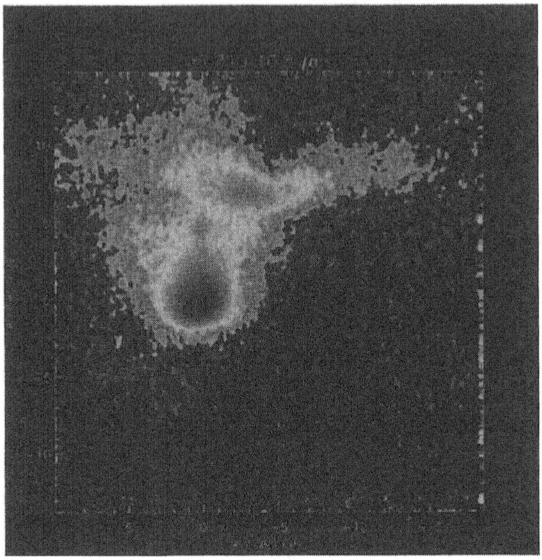

Fig. 2. IRTF/OSCIR image of AS 310 at 10.8μm, showing complex circumstellar structure. From Polomski (2001)

disk hole of ~50 AU radius, as was first suggested by Jura et al. (1993) based on the infrared spectrum, and provides an important constraint on inner disk evolution timescales. Follow-up 10μm and 18μm observations on Keck also revealed tentative evidence for a brightness asymmetry in the disk (Telesco et al. 2000; Fig. 3) which may be the result of a forced eccentricity on dust particle orbits by a companion.

Mid-infrared spectroscopy, even at a relatively low resolution of $R \approx 100$, is a useful probe of disk mineralogy. In particular, the 10μm silicate feature of some disks appears to be rather similar to that of comets in the solar system (e.g., Knacke et al. 1993). There is also some evidence, albeit limited to small samples, for evolution of the silicate feature over timescales of several million years (e.g., Sitko et al. 2000).

One of the exciting prospects for the near future is nulling interferometry at 10μm on the Keck Interferometer and VLTI, with the possibility of detecting (somewhat higher-surface-brightness) counterparts of the zodiacal cloud around nearby stars. Hinz et al. (1998) have already demonstrated nulling on an astronomical target with two segments of the MMT.

4.5 Starburst Galaxies

Observations in the thermal infrared have the potential to identify the luminosity source in so-called "ultra-luminous infrared galaxies" determining the size of the emitting region. In many cases, the sources are unresolved, rather than extended, implying that they are either very compact starbursts or dust-enshrouded

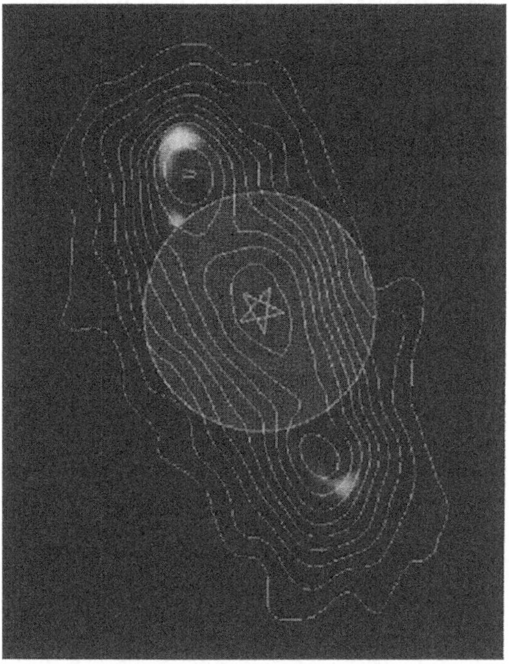

Fig. 3. Overlay of Keck/OSCIR 18.2μm contours on the 1.1μm HST/NICMOS coronagraphic image of HR 4796A disk. From Telesco et al. (2000)

active galactic nuclei (e.g., Soifer et al. 2000). In some starburst galaxies, such as NGC 7469, there is evidence for a starbursting ring surrounding an AGN (Jayawardhana et al. 1997) and in the case of NGC 253, a "super star cluster" (Keto et al. 1999).

References

1. R. Chini, et al., in preparation (2001)
2. G.H. Herbig: ApJS, **4**, 337
3. P.M. Hinz, et al.: Nature, **395**, 251 (1998)
4. R. Jayawardhana, et al.: 'Infrared Imaging of the Starburst Galaxy NGC 7469'. In: *Star Formation Near and Far*, ed. by S.S. Holt, L.G. Mundy (AIP Press, New York 1997) pp. 303-306
5. R. Jayawardhana, S. Fisher, L. Hartmann, C. Telesco, R. Piña, G. Fazio: ApJ, **503**, L79 (1998)
6. M. Jura, B. Zuckerman, E.E. Becklin, R.C. Smith: ApJ, **418**, L37 (1993)
7. E. Keto, et al.: ApJ, **518**, 183 (1999)
8. R.F. Knacke, S.B. Fajardo-Acosta, C.M. Telesco, J.A. Hackwell, D.K. Lynch, R.W. Russell: ApJ, **418**, 440 (1993)

9. D.W. Koerner, M.E. Ressler, M.W. Werner, D.E. Backman: ApJ, **503**, L83 (1998)
10. M. Marengo, R. Jayawardhana, G.G. Fazio, W.F. Hoffmann, J.L. Hora, A. Dayal, L.K. Deutsch: ApJ, **541**, L63 (2000)
11. A. Miroshnichenko, Z. Ivezić, D. Vinković, M. Elitzur: ApJ, **520**, L115
12. E. Polomski: PhD Thesis, University of Florida, Gainesville (2001)
13. M.L. Sitko, D.K. Lynch, R.W. Russell: AJ, **120**, 2609 (2000)
14. N. Smith, et al.: ApJ, **540**, 316 (2000)
15. B.T. Soifer, et al.: AJ, **119**, 509 (2000)
16. C.M. Telesco, et al.: ApJ, **530**, 329 (2000)

Adaptive Optics Search for Faint Companions Around Young Nearby Stars

Michael F. Sterzik

ESO, Alonso de Cordova 3107, Vitacura, Casilla 19001, Santiago 19, Chile

Abstract. Results of an ongoing coronographic NIR imaging survey using ADONIS at the 3.6m telescope are presented. The sample consists of young (ages $\leq$ 100Myr), and simultaneously nearby (distances $\leq 100pc$) field stars. Several new, low-mass companions are found. Their physical association has been proven by repeated observations at different epochs. Although the sensitivity allows to probe companion masses well below the hydrogen burning limit, companions with large mass ratios appear to be rare. This finding is in general agreement with several previous and ongoing brown-dwarf companion searches. I suggest a possible explanation for the scarcity of close brown-dwarf companions in the framework of stellar dynamical few-body interactions in early phases of their formation and evolution.

1 Introduction

Star and binary formation are intimately linked. An understanding of the formation process requires determinations of the stellar and brown dwarf (BD) companion frequencies as a function of separation, primary and secondary mass. Binary fractions for field dwarfs are known to vary with the mass of the primary: solar-type stars have a multiplicity fraction of about 60% [5], and the fraction decreases with decreasing mass of the primary [8,21,24]. Young stars (T Tauri stars) in some star forming regions have enhanced multiplicity frequencies [9,13].

Although the sub-stellar mass function is still very uncertain, isolated BDs appear to be quite common in the field [23]. But precision radial velocity planet search programs find with high statistical significance that BDs – although easily detectable – are rare with separations <3 AU around nearby stars (defining the so-called "brown dwarf desert" [12,16]). Contrary, at wide separations >1000 AU the sub-stellar companion frequency appears to be as high as the stellar companion frequency [10].

However the peak of the separation distribution for field dwarf binaries is located around 30 AU, and – assuming that this distribution also approximates the separations for BDs – high resolution imaging techniques with enough sensitivity should therefore be promising to detect many of these companions around nearby stars ($\leq 100pc$).

Does the "BD desert" extend to the separation regime where most of the stellar companions are found? Is the *secondary mass function* of very low mass companions in that separation range different from isolated field objects? Answering these questions provide an effective mean to constrain star formation models.

In this contribution, we summarize preliminary results of our own observing program to search for very-low mass companions around a sample of simultaneously young and nearby objects with ADONIS. Our results corroborate the trend that BD companions are rare in this separation range. Finally we sketch how stellar dynamical interactions in an early formation phase can help to understand the observed BD companion frequencies.

2 ADONIS Search for Very-Low Mass Companions

A detailed description of our approach, the sensitivities, and the results will be presented elsewhere [30]. Here only a short summary is given. Our strategy is aimed to enhance the probability to detect BD companions. Our original sample consists of 28 G-M type field stars with ages between 10 and 100 Myr estimated from chromospheric or coronal youth indicators, Lithium abundance, or age dated according to their (pre-main-sequence) position in the HRD diagram. Youth is crucial, as sub-stellar objects, not being able to sustain hydrogen burning, cool with age. The younger they are for a given mass, the brighter they are: $L \propto t^{-1.3} M^{2.24}$ [3]. Almost all of our targets have known parallaxes, with a mean distance around $30pc$. The peak of the separation distribution for binary stars is therefore well sampled when using near-infrared adaptive optics imaging like ADONIS@3.6m telescope, achieving an H-band diffraction limited resolution of $0''.2$ in a field of view of $\approx 10''$. ADONIS offers a coronographic mode that effectively enhances the contrast up to 2 magnitudes close to the mask. In terms of sensitivity the instrument set-up used is capable to detect Gl229B like companions ($\Delta H \approx 9$) as close as $\approx 2''$ from the primary. Candidate companions were confirmed in second epoch observations one year later. The known space motions of the primaries were sufficient to distinguish the relative motion of a background object from a common proper motion movement.

No new BD candidate companions could be confirmed. Several faint objects in the field of the first epoch observations, consistent with sub-stellar or even planetary masses if located at the same distance as the primary, turned out to be background stars using the second epoch information (e.g. in the case of Gl182).

On the other hand, several new *stellar* companions were found. As example in Fig. 1 two H-band images of Gl900 are shown. Gl900 is an $\approx$ 100Myr old M0-dwarf located at a distance of $19pc$. The total exposure time is 30sec each, and standard processing techniques (flat-field and dark correction, bad pixel removal, and shift and add) has been applied. The primary has a proper motion of $p.m._\alpha = .343''/yr$ and $p.m._\delta = .027''/yr$. No differential motion of the stars $> .010''/yr$ is seen, and therefore a common proper motion triple system is inferred. From the infrared colors, a spectral types of M3 of the secondary at a distance of $0''.4$ and of M4 for the tertiary can be derived.

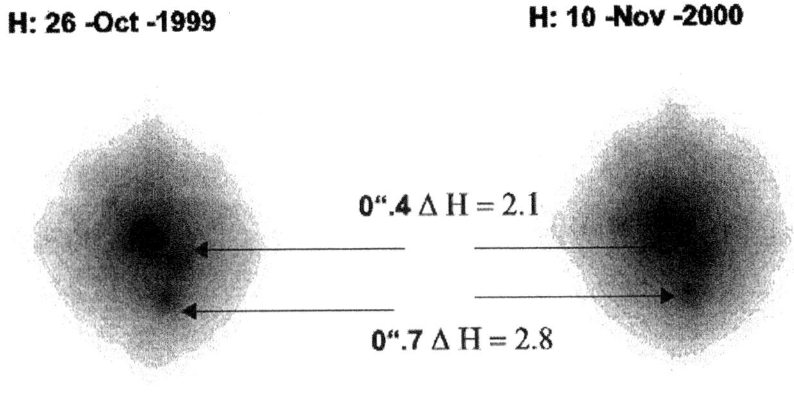

Fig. 1. ADONIS H-band images of Gl900 at two different epochs

3 Context: Scarcity of Brown Dwarf Companions

The discovery of Gl229B as first T-type brown dwarf companion to a nearby field star from the ground boosted the efforts to search for faint, very low mass companions by direct optical and infrared imaging techniques with enough sensitivity and sufficient spatial resolution to detect Gl229B-type objects.

But only a handful have been found since them. The complete, volume limited sample of 163 nearby field stars with distances <8 pc has been surveyed [21,19], but the only other BD found so far is the wide (≈ 2000 AU) companion Gl570D [2]. Recently, a sample of 23 field stars closer then 13 pc were investigated using HST-WFPC with negative results [26]. Also young open clusters have been searched for faint companions. Among 53 Hyades surveyed with the HST-PC [22], and among 66 late-type dwarfs in IC348 observed with adaptive optics [4], no BD companions were detected. Also in the Pleiades no visual BD companions could be found [17].

Although more attempts have been made, and not all *negative* results have been published, two important conclusions can be drawn from these observations in the mentioned separation and sensitivity regime. (1) The *overall BD companion frequency* of field stars and late-type dwarfs in young open clusters is probably not higher than a few percent, and (2) the *secondary mass function* is probably inconsistent with being randomly drawn from a field mass function in the sub-stellar regime.

In the following table we summarize properties of stars and BD companions found so far in the separation range of 10 to 1000 AU. The sub-stellar nature of the companion is confirmed by spectroscopy and/or by common proper motion and photometry. Most BD companions are found around late-K to early-M type primaries, with the exception of HR 7329.

Table 1. Stars and BD companions from imaging surveys with separations from 10 to 1000 AU, ordered according to year of discovery.

object	spectral types	separation	reference
GD 165	wd + L	3″.7 (120 AU)	Becklin & Zuckerman 1988
Gl 229	M1 + T	7″.8 (45 AU)	Nakajima et al. 1995
G 196-3	M2.5 + L	16″.2 (300 AU)	Rebolo et al. 1998
LHS 102	M4 + L	20″ (400 AU)	Goldman et al. 1999
TWA 5	M1.5 + M8.5	2″ (100 AU)	Lowrance et al. 1999
GG Tau	K7 + M0.5 + M5 + M7	1″.5 (200 AU)	White et al. 1999
HR 7329	A0 + M7-8	4″ (200 AU)	Lowrance et al. 2000
Gl 86	K1 + L/T + planet	1″.7 (20 AU)	Els et al. 2001

4 Predictions of Dynamical Decay Models

If stars originate in transient bound clusters of moderate size ($N < 10$), then these clusters will decay by dynamical interactions. Hard binaries form and most other, preferentially low-mass, stars are ejected. This scenario has been studied by detailed numerical simulations aiming to statistically describe the distribution function of fragment properties such as velocity dispersions, binary fractions and mass-ratios [27,28]. Depending on initial conditions, especially on the assumed mass function, the predicted distributions tend to agree with observed quantities. E.g. the peak of the semi-major axis distributions of binaries formed after chaotic interactions in decaying "mini-clusters" is between 10 and 1000 AU [29]. As discussed in [6], a particular good match of the observed trends in the binary frequencies and mass ratios can be achieved by separating the determination of the stellar mass in two distinct steps – first the choice of a cluster mass, and then the formation of individual stars within the cluster.

In general, the "dynamical decay" scenario also implies *strong biases in the secondary mass function*, i.e. deviations from a "random pairing" statistics in the selection of companion masses for a specific primary mass. For the following quantitative comparison, reference is made to a particular initial cluster model (varNbd) in [6]. Masses have been transformed to spectral type bins to allow better comparisons with observed quantities. Table 2 summarizes the results. In the second column, the total number of stars (which defines the underlying mass function) and the total number of primaries is given for the spectral type bin referred to. Column 3-8 now expand the relative binary fractions for each spectral type of the *secondary*. For comparison, the expected values for *random pairing* from the underlying mass function are calculated and given in brackets.

Note the strong deviations in the predictions of the "random pairing" and "dynamical decay" models when comparing the relative binary fractions of sec-

Table 2. Binary fractions per spectral type: Comparison of the "dynamical decay" models with "random pairing" from the mass function.

sp.type	total/primaries	L_s	late M_s	early M_s	K_s	G_s	$F+_s$
L	88680/28	1.0(1.0)	-	-	-	-	-
late M	218319/8341	.03(.30)	.97(.70)	-	-	-	-
early M	98844/25996	.05(.24)	.66(.56)	.29(.20)	-	-	-
K	68922/26835	.06(.21)	.36(.51)	.38(.18)	.20(.10)	-	-
G	22595/10930	.02(.21)	.19(.49)	.27(.17)	.40(.10)	.12(.03)	-
F+	31795/22753	.01(.20)	.08(.48)	.15(.17)	.30(.10)	.22(.03)	.25(.02)

ondaries: whereas random pairing implies that companions of L spectral type should contribute between 20 and 30% of all secondaries (for M to F types primaries), the decay models predict a strong deficit of L type secondaries by dynamical biasing effects. The overall probability for any L type object being found as companion is only about 4%. It also suggests higher chances in finding L-type companions around K and early type M primaries (see Table 1).

Dynamical decay model predict relatively low BD companion frequencies, and a vastly different secondary and primary mass function. If this is true, the majority of stars were indeed subject to strong dynamical interactions early in stellar "mini-clusters" (e.g. [25]).

5 Conclusions

Results of an ongoing near IR search for very low-mass and BD companions around young, nearby field stars using the ADONIS adaptive optics facility are presented. Although the sensitivity is high enough to detect Gl228B-type objects, no sub-stellar companions are found within the sample of 28 stars. However, several new low-mass, binary and multiple systems could be discovered, augmenting the multiplicities in the solar neighborhood.

This finding is in general agreement with other BD companion searches in the separation range of 10 to 1000 AU. Although the brown dwarf desert is well established for close (spectroscopic) BD companions, there seems to be mounting evidence that also this separation range appears to be only sparsely populated with BD. 100's of nearby stars have meanwhile been searched for sub-stellar companions with high sensitivity, but only a few have been found, implying a BD companion frequency of a few percent. Despite the uncertainties in the sub-stellar field mass function, this low companion frequency appears highly incompatible with a random pairing statistics. Instead, interactions during the disintegration of few-body clusters lead to strong dynamical biasing effects and suppress the formation of high-mass ratio binaries in general. Statistical models incorporating the results of detailed numerical simulations of decaying few-body systems allow to make quantitative predictions of binary frequencies and their dependencies on primary and secondary masses.

These models help to understand why BD companion searches turn out to be so hard.

Acknowledgements

Thanks to F. Marchis, S. Els, J.L. Beuzit, J.P. Veran and K.F. Schuster for the support in the ADONIS observations and data analysis. I am indebted to Dick Durisen for essential contributions on the theoretical side.

References

1. E.E. Becklin, B. Zuckerman: Nature **336**, 658 (1988)
2. A.J. Burgasser, J.D. Kirckpatrick, et al.: AJ **120**, 473 (2000)
3. A. Burrows et al.: ApJ **491**, 856 (1997)
4. G. Duchene, J. Bouvier, T. Simon: A&A **343** 831 (1999)
5. A. Duquennoy, M. Mayor: A&A **248**, 485 (1991)
6. R.H. Durisen, M.F. Sterzik, B.K. Pickett: A&A **371**, 952 (2001)
7. S. Els, M.F. Sterzik et al.: A&A **370**, L1 (2001)
8. D.A. Fischer, G.W. Marcy: ApJ **396**, 178 (1992)
9. A. Ghez, G. Neugebauer, K. Matthews K.: AJ **106**, 2005 (1993)
10. J.E. Giziz, J.D. Kirckpatrick, et al.: ApJ **551**, L163 (2001)
11. B. Goldman, X. Delfosse, et al.: A&A **351**, L5 (1999)
12. J.L. Halbwachs, F. Arenou et al.: A&A **355**, 581 (2000)
13. C. Leinert, H. Zinnecker, et al.: A&A **278**, 129 (1993)
14. P.J. Lowrance, C. McCarthy et al.: ApJ **512**, L69 (1999)
15. P.J. Lowrance, G. Schneider: ApJ **541**, 390 (2000)
16. G.W. Marcy, R.P. Butler: ARA&A **36**, 52 (1998)
17. E.L. Martin, W. Brandner et al.: ApJ **543**, 299 (2000)
18. T. Nakajima, S.T. Durrance et al.: Nature **378**, 463 (1995)
19. B.R. Oppenheimer, D.A. Golimowski et al.: AJ **121**, 2189 (2001)
20. R. Rebolo, M.R. Zapatero Osorio et al.: Science **282**, 5392 (1998)
21. I.N. Reid, J.E. Giziz: AJ **113**, 2246 (1997)
22. I.N. Reid, J.E. Giziz: AJ **114**, 1992 (1997)
23. I.N. Reid, J.D. Kirckpatrick, et al.: ApJ **521**, 613 (1999)
24. I.N. Reid, J.E. Giziz, J.D. Kirckpatrick, D.W. Koerner: AJ **121**, 489 (2001)
25. B. Reipurth: AJ **120**, 3177 (2000)
26. D.J. Schroeder, D.A. Golimowski et al.: AJ **119**, 906 (2000)
27. M.F. Sterzik, R.H. Durisen: A&A **304**, L9 (1995)
28. M.F. Sterzik, R.H. Durisen: A&A **304**, L9 (1995)
29. M.F. Sterzik, R.H. Durisen: 'Binary Properties in Dynamically Decaying Few-Body Clusters'. In: *Star Formation, Nagoya, Japan, June 21 – 25, 1999* . ed. by T. Nakamoto (Nobeyama Radio Observatory, Nagano, 1999) pp. 387-388
30. M.F. Sterzik, F. Marchis et al.: *in prep.*
31. R.J. White, A.M. Ghez, I.N. Reid, G. Schultz: ApJ **520**, 811 (1999)

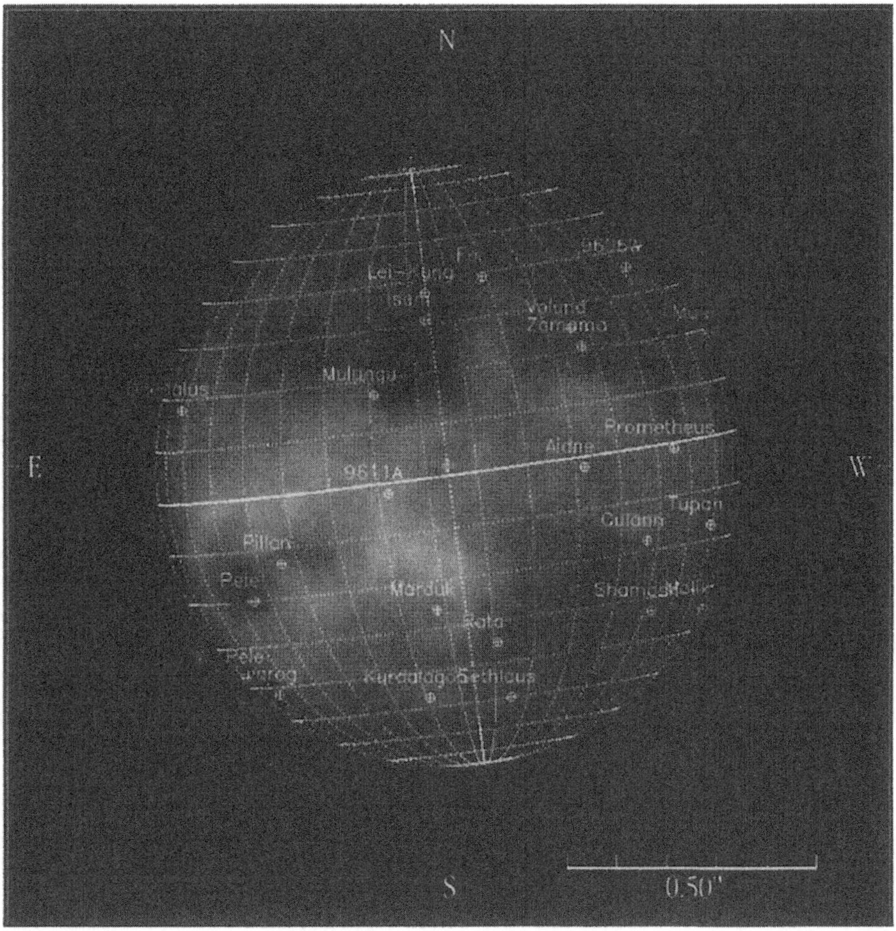

Io, Jovian Moon (IR image)
(Yepun/UT4 + NAOS-CONICA)

Michel Mayor (centre)

Extra-Solar Planetary Systems in the VLT Era

Michel Mayor and Nuno C. Santos

Geneva Observatory, 51 Ch. des Maillettes, CH-1290 Sauverny, Switzerland

Abstract. Radial Velocity surveys have revealed 63 exoplanets ($M \sin i < 10\, M_{Jup}$) and 6 multi-planetary systems (status on the 4th of April 2001). Distributions of orbital elements already have been given some constraints on the formation of planetary systems. Derived from recent high resolution spectroscopy studies, the impressive role of the stellar metallicity on the giant planet formation has been revealed. The chemical composition of the molecular cloud is probably the key parameter to form giant planets. However some evidences exist showing the possibility of accretion of matter in the stellar outer convective zone.

1 Introduction

Following the discovery in 1995 of the planet orbiting 51 Peg [20], we have witnessed a complete revolution in the field of extra-solar planets. More than 60 other exo-planets were unveiled since then, and are giving us the opportunity to reconsider the theories dealing with planetary formation and evolution.

In particular, the information gathered from the radial-velocity surveys can be interpreted as fossil traces of the planet formation and evolution processes. By looking at the planetary orbital characteristics, like the distribution of eccentricities and periods, or to the planetary masses, we are facing a lot of questions on planetary formation. Furthermore, other evidences are coming from the planet host stars themselves, namely by the fact that stars with planets are very metal-rich. The new results are showing that planet formation is not as simple as we thought. In the next Sects. we will summarize the latest results on this field.

2 The Period Distribution

One of the most interesting problems that appeared after the first planets were discovered as to do with their proximity from their host stars. Several mechanisms have been proposed to explain this fact. Current results show that *in situ* formation is very unlikely [2], and we have to invoke inward migration, either due to gravitational interaction with the disk [7], [13] or with other companions [25] to explain the observed orbital periods.

Although still strongly biased for the long period systems, the period distribution of the extra-solar planetary companions can already tell us something about the planetary formation and evolution processes. This is particularly true for the short period systems, for which the biases are not so important. In Fig. 1

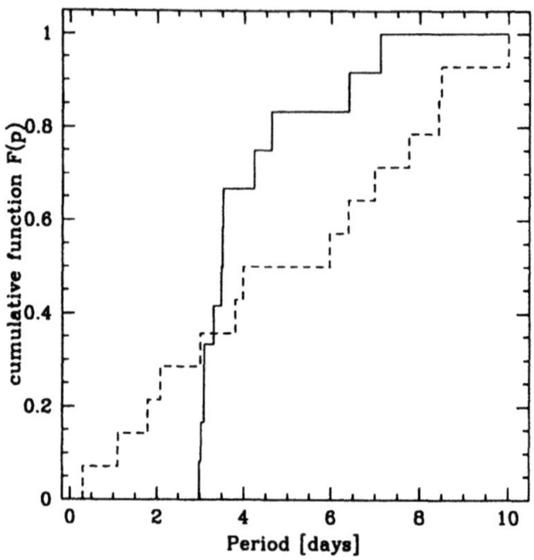

Fig. 1. Cumulative distributions of periods smaller than 10 days for planetary (solid line) and stellar companions (dashed line) to solar type dwarfs

we present the orbital period cumulative function for stars with planetary companions ($M<10\,M_{Jup}$) with $P<10$ days, as well as for solar type binary stars in the same period regime. The most impressive feature in the diagram is the clear cutoff of the planet distribution for periods shorter than ~3 days. To explain this distribution, several ideas have been presented, invoking e.g. a magnetospheric central cavity of the accretion disk, tidal interaction with the host star, Roche lobe overflow by the young inflated giant planet, or evaporation (see [27] and references therein).

3 The Mass Distribution

Another important clue concerning the nature of the now discovered planetary systems comes from their mass distribution. Clearly, given that the radial-velocity technique is more sensitive to massive companions, we could expect to find more "massive" planets when compared to their less massive counterparts. However, a look at the mass distribution tells us exactly the contrary (Fig. 2).

Several conclusions may be taken from the plots. First, the gap in the distribution, separating low mass stellar companions from the lower mass planets (often called the "brown dwarf desert") represents a strong evidence that these two populations are the result of different formation processes. Second, we can see that the planetary mass distribution has a sharp cutoff for masses lower than $\sim 10\,M_{Jup}$ [27]. This limit is clearly not related to the D-burning mass limit of $13\,M_{Jup}$. As was recently shown [12], and contrary to some recent suggestions [10] (easily refuted by statistical arguments [9], [24]), this conclusion is not an arti-

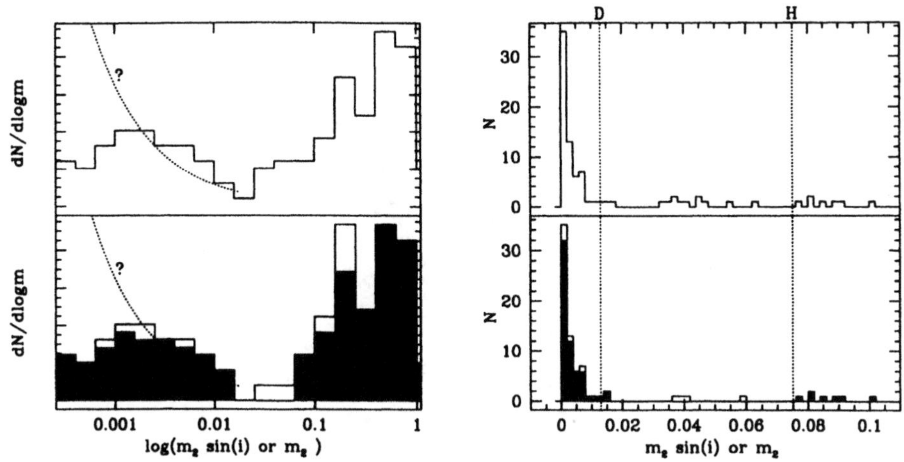

Fig. 2. Mass function of companions to solar-type stars in *log* (left) and linear (right) scales. The dotted vertical lines indicate the H- and D-burning limits. *Top:* $m_2 \sin i$. *Bottom:* composite histograms of m_2 (open part) and $m_2 \sin i$ (if $\sin i$ not known)

fact of the fact that for most of the targets we only have minimum masses, but a real upper limit for the mass of the planetary companions discovered so far.

4 The Distribution of Eccentricities

One of the most enigmatic results to date is illustrated in Fig. 3. A look at the figure shows that there are no clear differences between the eccentricity distributions of planetary and stellar binary systems. How then can this be fit into the "traditional" picture of a planet forming in a disk? For masses lower than $\sim 20\,M_{\rm Jup}$, it has been shown that the interaction with a disk has the effect of damping the eccentricity [22]. This suggests that other processes, like the interaction between planets in a multiple system, or the influence of a distant stellar companion, may play an important role in defining the "final" orbital configuration. In this respect, one particularly interesting case of very high eccentricity amongst the planetary companions is the planet around HD 80606 [21] (Fig. 4).

Although still not clear, however, a close inspection of the Fig. 3, permits to find a few differences between stellar and planetary companions eccentricities. For example, for periods in the range of 10 to 30 days (clearly outside the circularization period by tidal interaction with the star), there are a few stars with planets having very low eccentricity, while no stellar binaries are present in this region. On the other hand, for the very short period systems, we can see some planetary companions with eccentricities higher than those found for "stars". This facts may be telling us that different formation and evolution processes took place: for example, the former group may be seen as a sign for formation in a disk, and the latter one as an evidence of the influence of a longer period companion on the eccentricity (cf. case of HD 217107 [6] and HD 83443 [18]).

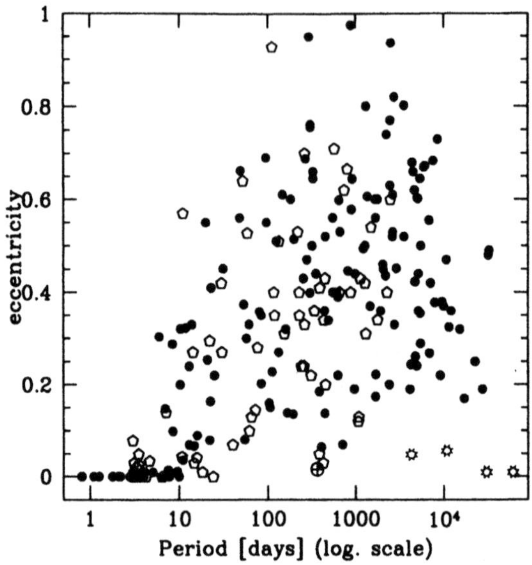

Fig. 3. The $e - \log P$ diagram for planetary (open pentagons) and stellar companions (filled circles) to solar type field dwarfs. Starred symbols represent the giant planets of our Solar System, while the "earth" symbol represents our planet

5 Multi-Planetary Systems

Another possible source of information about the formation of giant planets may came from the multi-planetary systems. To date, 6 such systems are known. Their main characteristics are summarized in Table 1. A few cases, like the resonant planets around HD 82943 [17] – see Fig 4 – and Gl 876 [14], or the planet-brown dwarf pair around HD 168443 [27], are of particular interest; their orbital configurations may provide new constraints on the planetary migration and eccentricity pumping mechanisms (cf. [1]).

Table 1. Characteristics of the known multi-planetary Systems.

Star	P_1 [days]	P_2 [days]	P_3 [days]	m_1 [M_J/sin i]	m_2 [M_J/sin i]	m_3 [M_J/sin i]	Ref. for First Planet	Ref. for System
υ And	4.62	241	1308	0.71	2.11	4.61	[4]	[3]
HD 83443	2.98	29.8	–	1.14	0.53	–	[18]	[18]
HD 168443	58.1	1667	–	7.2	15.1	–	[15]	[28]
Gl 876	61.02	30.1	–	1.89	0.56	–	[5], [16]	[14]
HD 82943	444	221	–	1.63	0.88	–	[19]	[17]
HD 74156	51.61	2300	–	1.56	>7.5	–	[17]	[17]

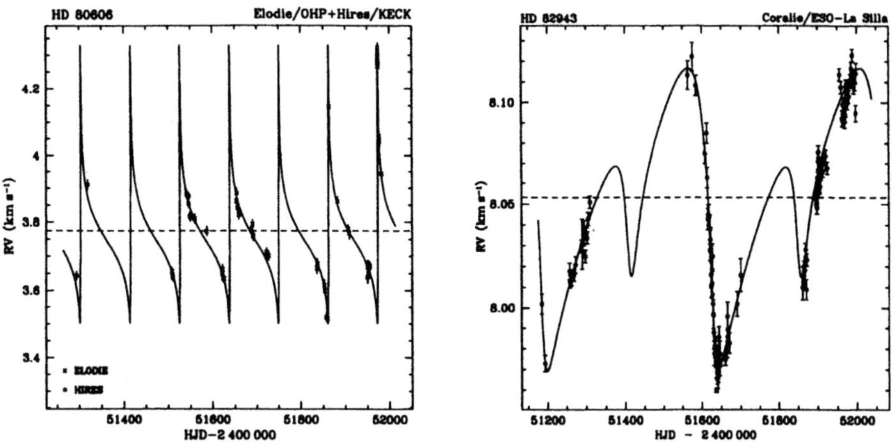

Fig. 4. *Left*: Radial-velocity measurements and best Keplerian solution for HD 80606. The discovered planet orbits the star in a highly eccentric orbit (e~0.93) with a period of 111.78-days [21]; *Right*: the same for HD 82943, a system of two resonant planets [17]. *Figure available in colour on the CD-ROM*

6 The Metallicity Correlation

One of the most promising results that became evident after the discovery of the first exo-planets is that their host stars have shown to be very metal-rich when compared with dwarf stars in the solar neighborhood [8], [26]. In Fig.5 (left) we can see a comparison between the [Fe/H] distribution of a volume limited sample of field dwarfs without planetary mass companions, and the same distribution for the stars with planets [26]. There is a remarkable difference between both distribution, as can be seen from their cumulative functions Fig. 5 (right).

There are basically two ways of interpreting this result. The first is saying that the [Fe/H] excess is the result of the accretion of planets and/or planetary material into the star. The second, is to consider that the planetary formation mechanism is dependent on the metallicity of the proto-planetary disk: according to the "traditional" view, a gas giant planet is formed by runaway accretion of gas by a ~10 earth masses planetesimal. The higher the metallicity (and thus the number of dust particles) the faster a planetesimal can growth, and the higher the probability of forming a giant planet before the gas in the disk dissipates.

Recent results seem to support the latter scenario [26]. The argument is mostly based on the fact that material falling into a star's surface would induce a different increase in [Fe/H] depending on the stellar mass, i.e. on the depth of its convective envelope (where mixing can occur). However, the data shows no such trend. Furthermore, the results also show that the [Fe/H] distribution of stars with planets steeply rises for higher metallicities, which might be interpreted as an evidence that the planet formation mechanisms are highly dependent on the metallicity of the disk. Our own Sun is in the "metal-poor" tail of the planet hosts [Fe/H] distribution!

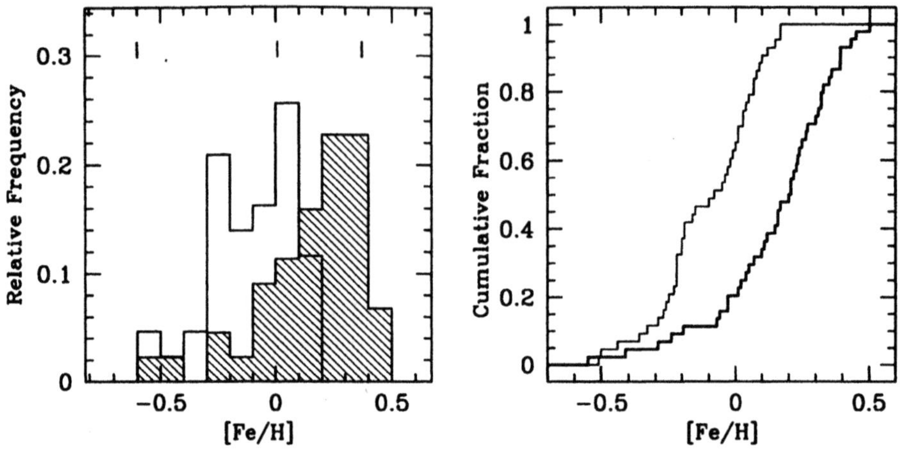

Fig. 5. *Left*: [Fe/H] distribution of stars with planets (shaded histogram) compared with the same distribution of field dwarfs in the solar neighborhood (open histogram) [26]. The vertical lines represent stars with brown dwarf candidate companions having minimum masses between 10 and 20 M_{Jup}. *Right*: The cumulative functions of both samples. A Kolmogorov-Smirnov test shows the probability of the stars' being part of the same sample is around 10^{-7}

These results do not exclude, however, that pollution may play a role (eventually important in some cases), but rather that it is not the key process leading to the observed high-metallicity of the planet host stars. In fact, the recent detection of ^{6}Li in the atmosphere of the star HD 82943 [11] (known to harbor a system of two planets [17]) is most likely an indication that this star has engulfed a planet sometime during its lifetime.

7 Conclusions and Future Perspectives

The results presented above are giving astronomers a completely different view on the formation and evolution of the planetary systems. We no longer have the Sun as the only example, and today we have to deal with the peculiar characteristics of the "new" extra-solar planets: a huge variety of periods, eccentricities, masses. After only 5 years, we can say that at least 5% of the solar type dwarfs have giant planetary companions. Furthermore, one interesting conclusion can be taken: our Solar System, with giant planets orbiting away from the star in quasi-circular orbits, is definitely not typical when compared to the presently discovered extra-solar giant planets. The question now is: is it really untypical or are these systems the exception? To help answer this question several projects are currently in the pipeline.

From *radial-velocity* searches, the current surveys, including between 2000 and 3000 stars, will continue to increase the number of known exo-planets. Several dozens are expected to be announced in the next few years. On the other

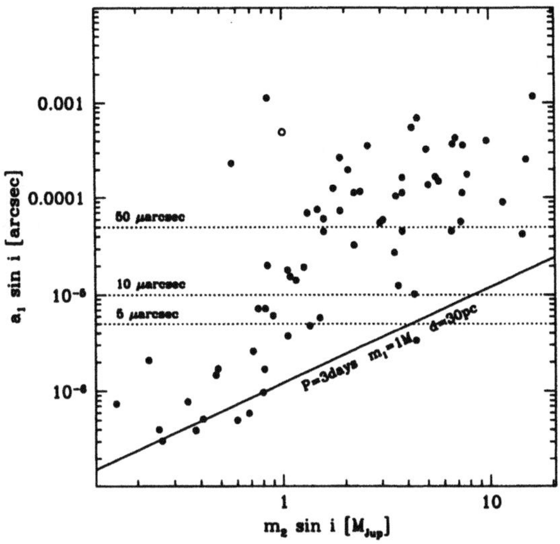

Fig. 6. Mass vs. astrometric motion diagram for stars with very low mass companions discovered by radial-velocity surveys. The dotted lines represent the limits for astrometric precisions of 50, 10 and 5 μarcsec. The open circle illustrates the position of the Sun as seen from a distance of 10pc. The solid line indicates an approximate limit imposed by the fact that no planets were found with periods shorter than ~3 days

hand, the ever increasing precision will permit to discover lighter planets, as well as to increase the number of known multi-planetary systems. In this context, newly or soon available ESO instruments (e.g. UVES/VLT or HARPS/3.6-m [23]) will certainly play an important role.

From the *astrometric* point of view, the expectations are not lower. Instruments like the VLTI or KeckI will give us the possibility to estimate real masses for many of the known planetary systems (see Fig. 6). Furthermore, space missions like GAIA or the interferometric mission SIM will completely change the current landscape by adding tens of thousands of new planets. Given that astrometry is more sensitive to longer period systems (contrary to the radial-velocity method), these projects will also permit to better cover the period distribution of the exo-planets. It will further permit to find planets around targets not accessible with radial-velocity surveys, like A or B stars, or T Tauri stars.

Further hopes will come from *photometric transit* searches, mostly based upon space missions like COROT, Eddington or Kepler. Out of the Earth's atmosphere, these satellites will achieve a photometric precision better than 0.01%, permitting the detection of transiting earths.

All these steps will permit to better understand the mechanisms leading to the formation of planetary systems like our own, and will thus somehow represent an important step towards the search for life in the universe. Two similar projects are currently directed towards this specific goal (Darwin/ESO and TPF/NASA). Using nulling interferometry techniques (to remove the light from the brighter star and leave the one coming from the planet), they will try to find traces of life in the spectrum of exo-earths. In a very close future humanity has to prepare itself to find out that the whole universe may be teeming with life.

We would like to thank the members of the Geneva extra-solar planet search group, D. Naef, F. Pepe, D. Queloz, S. Udry, our French colleagues J.-P. Sivan, C. Perrier and J.-L. Beuzit, as well as G. Israelian and R. Rebolo (from the IAC), who have largely contributed to the results presented here. We wish to thank the Swiss National Science Foundation (FNSRS) for the continuous support to this project. Support from Fundação para a Ciência e Tecnologia, Portugal, to N.C.S. in the form of a scholarship is gratefully acknowledged.

References

1. Armitage P.J., Livio M., Lubow S.H., Pringle J.E., 2001, submitted to ApJL
2. Bodenheimer P., Hubickyj O., Lissauer J.J., 2000, Icarus 143, 2
3. Butler R.P., Marcy G.W., Fischer D., et al., 1999, ApJ 526, 916
4. Butler R.P., Marcy G.W., Williams E., et al., 1997, ApJ 474, L115
5. Delfosse X., Forveille T., Mayor M., et al., 1998, A&A 338, L67
6. Fischer D.A., Marcy G.W., Butler R.P., et al., 2001, ApJ 551, 1107
7. Goldreich P., Tremaine S., 1980, ApJ 241, 425
8. Gonzalez G., 1998, A&A 334, 221
9. Halbwachs J.-L., Arenou F., Mayor M., Udry S., Queloz D., 2000, A&A 355, 573
10. Han I., Black D., Gatewood G., 2001, ApJ 548, L57
11. Israelian G, Santos N.C., Mayor M., Rebolo R., 2001, Nature 411, 163
12. Jorissen, Mayor M., Udry S., 2001, A&A, submitted
13. Lin D., Bodenheimer P., Richardson D.C., 1996, Nat. 380, 606
14. Marcy G.W., Butler R.P., Fischer D., et al., 2001 ApJ, in press
15. Marcy G.W., Butler R.P., Vogt S., et al., 1999, ApJ 520, 239
16. Marcy G.W., Butler R.P., Vogt S., 1998, ApJ 505, L147
17. Mayor M., et al., 2001, ESO press-release 07/01
18. Mayor M., Naef F., Pepe F., et al., 2001. In: Penny A., Artimowicz P., Lagrange A.-M., Russel S., "Planetary Systems in the Universe: Observations, Formation and Evolution", ASP Conf. Ser., in press
19. Mayor et al., 2000, ESO press-release 13/00
20. Mayor M., Queloz D., 1995, Nature 378, 355
21. Naef D., et al., 2001, A&A Letters, submitted
22. Papaloizou J.C., Nelson R.P., Masset F., 2001, A&A 366, 263
23. Pepe F., Mayor M., Delabre B., et al., 2000, Proc. SPIE Vol. 4008, 582
24. Pourbaix D., Arenou F., 2001, A&A, in press
25. Rasio F.A., Ford E.B., 1996, Science 274, 954
26. Santos N.C., Israelian G., Mayor M., 2001, A&A, in press
27. Udry S., Mayor M., Halbwachs J.-L., Arenou F., 2000, in "Microlensing 2000", Eds. Menzies J.W., Sackett P.D., ASP Conf. Series, in press
28. Udry S., Mayor M., Queloz D., 2001. In: Penny A., Artimowicz P., Lagrange A.-M., Russel S., "Planetary Systems in the Universe: Observations, Formation and Evolution", ASP Conf. Ser., in press

Antu enclosure rotating (afterglow after sunset)

Ralph Neuhäuser

Direct Imaging and Spectroscopy of Substellar Companions Next to Young Nearby Stars in TWA

Ralph Neuhäuser[1,2], Eike Guenther[3], Wolfgang Brandner[4], Nuria Húelamo[1], Thomas Ott[1], João Alves[4], Fernando Cómeron[4], Jean-Gabriel Cuby[4], and Andreas Eckart[5]

[1] MPI extraterrestrische Physik, D-85740 Garching, Germany
[2] University of Hawaii, Institute for Astronomy, 2680 Woodlawn Dr., Honolulu, HI 96822, USA
[3] Thüringer Landessternwarte Tautenburg, Sternwarte 5, D-07778 Tautenburg, Germany
[4] European Southern Observatory, Karl-Schwarzschild-Straße 2, D-85748 Garching, Germany
[5] Universität Köln, Köln, Germany

Abstract. Direct imaging of substellar companions is possible since several years around nearby stars (e.g. Gl 229) and young stars (e.g. TWA-5). We are searching for brown dwarfs and giant planets as companions to stars which are both very young (up to 100 Myrs) and relatively nearby (up to 100 pc), using ground-based facilities on La Silla, Cerro Paranal, Mauna Kea, and Calar Alto with infrared imaging including speckle and adaptive optics. The young nearby association of co-moving T Tauri stars around TW Hya, called the TW Hya Association (TWA), is a prime target of our observations. We will present imaging detections of substellar companion candidates around three TWA stars and their H-band spectra, showing that they are background stars. Given all the available ground-based and space-based (HST NICMOS) data obtained so far for the TWA stars, we will discuss the frequency of young brown dwarf companions, to be compared to the frequency of old brown dwarf companions, both wide pairs from imaging surveys as well as close pairs from radial velocity surveys. We find no indications for young brown dwarf secondaries to be overabundant, so that there is no evidence that many of them get ejected during young ages.

1 Introduction: Sub-Stellar Companions in TWA

The TW Hya Association (TWA) of T Tauri stars offers prime targets for the direct imaging search for sub-stellar companions, both brown dwarfs and giant planets, because these stars are very young (few Myrs) and quite nearby ($\sim$ 55 pc, as determined by Hipparcos for a few members). In total, there are now roughly 30 members published including several visual and spectroscopic companions (Webb et al. 1999, Sterzik et al. 1999, Zuckerman et al. 2001).

Using infrared images obtained with NICMOS on the Hubble Space Telescope, Lowrance et al. (1999) found a brown dwarf companion candidate around TWA-5. With ground-based optical and infrared imaging and spectroscopic follow-up observations, this companion candidate was confirmed as M9-type brown dwarf with the same proper motion as TWA-5 A (Neuhäuser et al. 2000b). Slightly later, Schneider et al. (2001) presented an HST STIS spectrum with smaller wavelength coverage confirming the spectral type published by us.

For most of the 14 TWA member systems published until last year, we have obtained deep infrared images to search for more substellar companions. Near TWA-7, we did find a very faint companion candidate (Neuhäuser et al. 2000a), which could have been a few Jupiter mass companion (according to Burrows et al. 1997 tracks and isochrones) if at the same distance and age as TWA-7, but our follow-up H-band spectrum showed that it is a K-type background star (Neuhäuser et al. 2001).

Around three of the other TWA stars, we did find new companion candidates, for which we present our imaging and follow-up spectroscopic data below (sect. 2). We did not find any substellar companion candidates down to H$\simeq$ 17 mag between about 2 and 15 arcsec ($\geq$ 100 AU at the distance of TWA) around the TWA members HR 4796, TWA-7, Hen 3-600, CoD$-29°$8887, CoD$-36°$7429, TWA-10, and RXJ1109.7-3907 (=PPM 288568) in our SOFI NTT imaging data obtained in May 2000.

More recently, we also observed several TWA members with the 36-element curvature-sensing AO instrument *Hōkūpa'a* of the University of Hawai'i at the 8.3m Gemini-North telescope on Mauna Kea, where we could obtain a FWHM of 64 mas and 25 % Strehl. Our image of the brown dwarf companion TWA-5 B around the spectroscopic binary T Tauri star TWA-5 A (shown elongated,

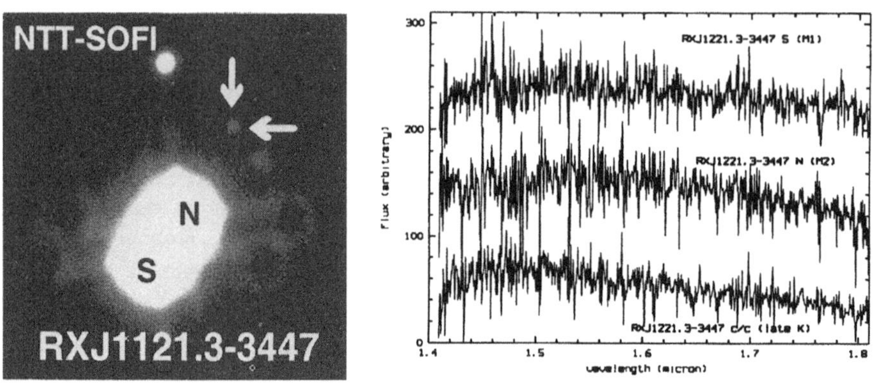

Fig. 1. RXJ1121.3-3447. left: The binary T Tauri star with its slightly brighter M1 primary to the south, the M2 secondary in the north, and the faint companion candidate (c/c 1) to the NW; another (brighter) companion candidate (c/c 2 in the table) is seen further to the north, which may be a stellar or sub-stellar component of this system; H-band image obtained with SOFI at the NTT. right: H-band spectra of the M1 primary (top), the M2 secondary (middle) and the faint companion candidate (bottom) obtained with ISAAC at Antu (VLT-UT1) showing that the companion candidate is slightly bluer than the two early M-type T Tauri stars, i.e. it is a late K-type (or possibly K7 to M0) background star not related to the T Tauri star

i.e. possibly resolved to be a 50 to 100 mas binary) is presented in Neuhäuser, Potter, and Brandner (2001).

2 Imaging and Spectroscopic Observations

We have taken images of several TWA members in 2000 using the infrared imager Son of ISAAC (SOFI) at the ESO 3.5m New Technology Telescope (NTT) on La Silla (see also Neuhäuser et al. 2001). The observations were made on 2000 May 17 and 19 with 1.0″ to 1.1″ seeing. We used the H-band filter and the small SofI field (0.147″/pix) with a total exposure time of 10 min per target (460 × 1.3 sec in auto-jitter mode). New companion candidates were found around TWA-8 (a TWA T Tauri star found by Webb et al. 1999), RXJ1121.1-3845 (=GSC 7739 2190), and RXJ1121.3-3447 (=GSC 7210 1352), originally found as new T Tauri stars by Hoff et al. (1997) and published later also by Sterzik et al. (1999). Data are listed in table 1, images are shown in fig. 1, 2, and 3 (left panels).

Table 1. Three companion candidates (= c/c) in TWA

Object designation	spectral type	H [mag]	$\Delta\alpha$ [arcsec]	$\Delta\delta$ [arcsec]	sep [arcsec]	remarks
RXJ1121.3-3447 S	M1	7.7				primary (a)
RXJ1121.3-3447 N	M2	7.8	2.8″ E	4.3″ N	5.1″	secondary
RXJ1121.3-3447 c/c 1	late K	16.8	5.4″ W	9.1″ N	10.6″	backgr. at ~2 kpc
RXJ1121.3-3447 c/c 2	?	14.0	1.8″ E	15.5″ N	15.6″	?
RXJ1121.1-3845	M2	8.4				(b)
RXJ1121.1-3845 c/c	mid K	16.5	6.0″ W	2.6″ S	6.5″	backgr. at ~2.5 kpc
TWA-8 A	M2	7.8				primary
TWA-8 B	M5	9.4	1.2″ W	13.2″ S	13.3″	secondary
TWA-8 c/c	early M	15.5	5.6″ E	8.3″ N	10.0″	backgr. at ~1 kpc

Remarks: All mag ±0.1 mag, sep ±0.1″ with respect to primaries.

(a) The southern component is the primary, slightly brighter in both V and H than the northern component. In Neuhäuser et al. (2001), we said that we would give the separation of c/c with respect to the *primary (northern) star*, which is misleading: Indeed, we gave the separation with respect to the primary, which is the southern component. (b) The companion candidate to RXJ1121.1-3845 was found also by D. Trilling & R. Jayawardhana using the cold coronograph CoCo at the NASA IRTF (priv. com.).

If the companion candidates would be bound, i.e. at ~ 55 pc and ~ 10 Myrs as TWA, their magnitudes would correspond to objects with only a few Jupiter masses, according to different sub-stellar tracks and isochrones, i.e. they would be sub-stellar. This hypothesis has to be checked by spectra of the companion candidates, which would need to show very cool atmospheres (L or T), and – if positive – also by a 2nd epoch image to be taken one or several years later, to make sure that the proper motions are identical.

We took H-band spectra using ISAAC at ESO 8.2m Antu (VLT-UT1 on Cerro Paranal, Chile) in the nights 2000 June 9 (for RXJ1121.1 and RXJ1121.3) and 21 (for TWA-8), all in service mode during sub-arcsec seeing. Each of the three spectra consists of 20 times 1 min integration. In the data reduction, we first subtracted the darks, then divided by a normalized flat field, and then did the combination and sky subtraction, all with eclipse. We then identified and extracted the correct spectra with IRAF. The companion candidates were always easily found as the brightest and most narrow spectra, because they were well separated from the T Tauri stars. Previously, in the case of the H-band spectrum of the companion candidate to TWA-7, the spectrum extraction was much more difficult because of larger dynamical range (Neuhäuser et al. 2001).

The spectra of these three possibly sub-stellar companion candidates are shown in fig. 1, 2, and 3 (right panels) together with some of the T Tauri stars also observed in the same slit at the same time.

We determine the spectral types of the companion candidates in the low S/N spectra by comparing the general shape and slope of their continua with those of the T Tauri star spectra (observed simultaneously in the same slit), by comparison with spectroscopic standards (from Meyer et al. 1999 and Greene & Lada 1996) and our own previous ISAAC H-band spectra, as well as by the presence and strength of some absorption lines line Mg, Al, and Si.

All three companion candidates are found to be mid-K to early-M type objects, i.e. they are unrelated background stars.

Because we classify the companion candidate near TWA-8 as early M-type object, which is the same spectral type as for the primary T Tauri star, there could be doubts as to whether our spectrum extraction went ok. As in all such

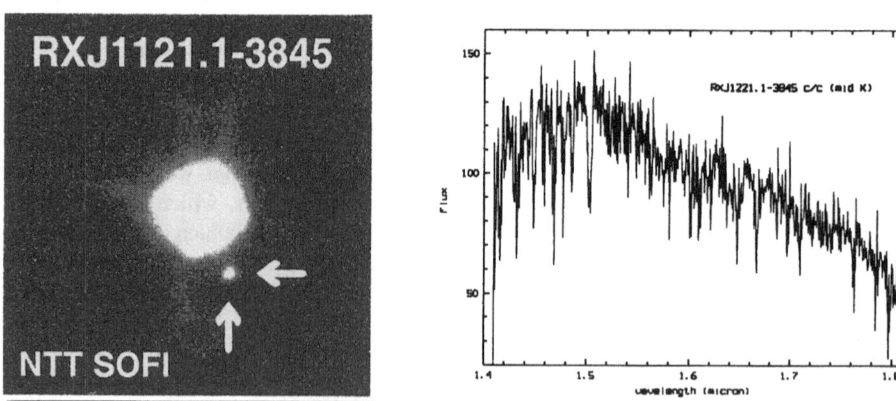

Fig. 2. RXJ1121.1-3845. left: The single T Tauri star in the center with the faint companion candidate to the SW, H-band image obtained with SOFI at the NTT. right: H-band spectrum of the faint companion candidate obtained with ISAAC at Antu (VLT-UT1) showing that this object has spectral type mid K, with the strong Mg line seen at $\sim 1.5\ \mu m$, i.e. is a background star not related to the T Tauri star

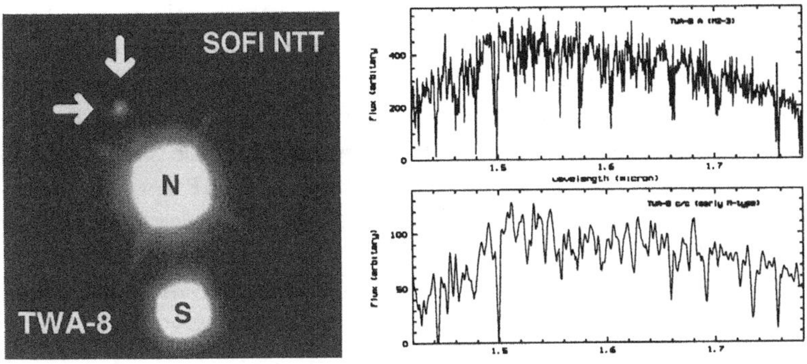

Fig. 3. TWA-8. left: The binary T Tauri star with its brighter M2 primary to the north, the M5 secondary in the south, and the faint companion candidate to the NE; H-band image obtained with SOFI at the NTT. right: H-band spectra of the M2 primary (top) and the faint companion candidate (bottom) obtained with ISAAC at Antu (VLT-UT1) showing that the companion candidate and the early M-type T Tauri primary have quite similar spectra regarding both the presence and strength of lines and the general shape and slope of the continuum; note in particular the strong Mg line at $\sim 1.5\ \mu m$, which is not present for spectral types later than $\sim$ M3.5, while the shape of the continuum for both objects is not as blue as in K-type stars. Hence, we classify both as early M. Because the spectra are still slightly different, we are quite confident that the extraction went ok. We conclude that the companion candidate is an early M-type background star not related to the T Tauri star

cases of faint objects very close to bright stars, we fit the wing of the trace of TWA-8 A for each wavelength bin. This fit is subtracted at each wavelength from the spectrum of the companion candidate. While the TWA-8 companion candidate with magnitude difference of almost 8 mag in H is separated by 10 arcsec from the primary, the separation between TWA-7 and its companion candidate was only 2.5 arcsec (at magnitude difference of 9.5 in H). Even in that case, we were able to extract the spectrum of the companion candidate, found to be K-type, i.e. background (Neuhäuser et al. 2001).

3 Conclusions: Brown Dwarf Companions Frequency

In TWA, there are 20 T Tauri systems published so far (Webb et al. 1999, Sterzik et al. 1999, Zuckerman et al. 2001). The 14 systems published until last year are made up of 22 (visually resolved) objects (including the brown dwarf TWA-5 B), not counting additional spectroscopic companions. Eight of the 14 systems are known to be multiples, which is quite typical. Among the 21 stars (excluding the brown dwarf TWA-5 B), there is only one with a known brown dwarf companion, namely TWA-5. For all others, deep infrared imaging with HST and/or ground-based facilities can exclude brown dwarf companions (with masses between the H and D burning limits) at separations outside $\sim$ 2 arcsec

(100 AU). Hence, very roughly, the fraction of TWA stars with brown dwarf companions is ~ 5 %. This is for a spectral type range of the primary stars from A0 (HR 4796 A) to M5 (TWA-8 B) and for the companions from the sub-stellar limit (about M7 at a few Myrs) down to approximatelly the D burning limit.

Recently, Gizis et al. (2001) estimated the frequency of wide, visual, old L- and T-type brown dwarf companions to normal stars to be 18 ± 14 %. Marcy & Butler (2000) found from radial velocity surveys that the percentage of close brown dwarf companions to normal stars is rather small, namely ≤ 5 %.

Apparently, all these three numbers are not inconsistent with each other. However, any such comparison is very preliminary given the large uncertainties involved in such low-number statistics.

Acknowledgements. We would like to thank the NTT team on La Silla and the ISAAC service mode observers on Cerro Paranal as well as the ESO User Support Group for the help and assistance in our observations.

References

1. Burrows A., Marley M., Hubbard W. et al.: ApJ 491, 856 (1997)
2. Gizis J.E., Kirkpatrick J.D., Burgasser A., et al.: ApJ 551, L163 (2001)
3. Greene T.P. & Lada C.J.: AJ 112, 2184 (1996)
4. Hoff W., Alcalá J.M., Sterzik M.F.: 'X-ray based survey for T Tauri stars near TW Hydrae'. In: Cool stars in Clusters and Associations: Magnetic Activity and Age Indicators, ed. by Micela G., Pallavicini R., Sciortino S., abstract booklet (1997)
5. Lowrance P.J., McCarthy C., Becklin E.E. et al.: ApJ 512, L69 (1999)
6. Marcy G.W. & Butler R.P.: PASP 112, 137 (2000)
7. Meyer M.R., Edwards S., Hinkle K.H., Strom S.E.: ApJ 508, 397 (1999)
8. Neuhäuser R., Potter D., Brandner W.: 'Observing the planet formation time-scale by ground-based direct imaging of planetary companions to young nearby stars: Gemini/*Hōkūpa'a* image of TWA-5'. In: Astrophysical ages and time-scales, ed. by von Hippel T., Manset N., Simpson C. (ASP Conf. Series 2001), in press, astro-ph/0106304
9. Neuhäuser R., Brandner W., Eckart A., Guenther E.W., Alves J., Ott T., Huélamo N., Fernández M.: A&A 354, L9 (2000a)
10. Neuhäuser R., Guenther E.W., Petr M.G., Brandner W., Huélamo N., Alves J.: A&A 360, L39 (2000b)
11. Neuhäuser R., Guenther E.W., Brandner W., Huélamo N., Ott T., Alves J., Comerón F., Eckart A., Cuby J.-G.: 'Direct imaging search for planetary companions next to young nearby stars'. In: From Darkness to Light, ed. by Montmerle T. & Andre P. (ASP Conf. Series 2001) in press, astro-ph/0007305
12. Neuhäuser R., Huélamo N., Guenther E.W., Brandner W., Alves J., Comerón F., Petr M.G., Cuby J.-G.: 'Ground-based exoplanet near-infrared search by imaging and spectroscopy: 3 new companion candidates in TWA'. In: IAU Symposium 202, ed. by Penny A.J., Artymowicz P., Lagrange A.-M., Russell S.S., (2001) in press
13. Schneider G., Becklin E.E., Lowrance P.J., Smith B.A.: In: Disks, Planets, and Planetesimals (ASP Conf. Ser. 2001), in press, astro-ph/0007330
14. Sterzik M.F., Alcalá J.M., Covino E., Petr M.G.: A&A 346, L41 (1999)
15. Webb R.A., Zuckerman B., Platais I. et al.: ApJ 512, L63 (1999)
16. Zuckerman B., Webb R.A., Schwartz M., Becklin E.E.: ApJ 549, L233 (2001)

The VLT platform at Paranal

Monika Petr-Gotzens

VLT Observations of the Young Stellar Object TMR-1

M.G. Petr-Gotzens[1], J-G. Cuby[2], M.F. Sterzik[2], P. Schilke[1], and A. Walsh[1]

[1] Max-Planck-Institut für Radioastronomie, Auf dem Hügel 69, 53121 Bonn, Germany
[2] European Southern Observatory, Casilla 19001, Santiago 19, Chile

Abstract. We present results from an observational multi-wavelength study of the circumstellar environment of the young binary TMR-1, and the still disputed TMR-1C. TMR-1C is a faint object at $\sim 10''$ distance from the protobinary TMR-1 and was suggested to be a young runaway planet. K-band spectroscopy of TMR-1C shows that the object is consistent with a background field star. Using narrow-band FeII imaging, we further discover a jet emanating from TMR-1, while the filament structure that arises from TMR-1 and points towards TMR-1C may be interpreted as the edge of an outflow cavity. Strong molecular hydrogen emission, which is shock excited, arises from the filament. From millimeter observations we find strong 1.3mm dust continuum emission around the protobinary TMR-1. Moreover, we detect first kinematical evidence for a rotating circumbinary disk using velocity resolved ^{13}CO (2−1) line observations.

1 Introduction

TMR-1 (IRAS 04361+2547) is an embedded protostellar object of luminosity $\sim 2.8 L_{\odot}$ located in the nearby Taurus star-forming region (e.g. Terebey et al. 1990). According to its IRAS color, log(f25/f60)≈−0.4, and its spectral energy distribution from 1.25-100μm, TMR-1 has been classified as a class I object (Kenyon et al. 1993). More recently, special interest in TMR-1 was raised through high spatial resolution *Hubble Space Telescope* imaging at near-infrared wavelengths (Terebey et al. 1998), which uncovered that: a) TMR-1 is a binary with a measured components' separation of $0.31''$ (hereafter, TMR-1AB) b) a long filament structure of length $9\text{-}10''$ arises from TMR-1AB c) a faint point source (TMR-1C) of K≈17.6 mag is located exactly at the tip of the filament. With the assumption of a physical association of TMR-1C with TMR-1AB, and from the analysis of their photometric measurements, Terebey et al. (1998) reported that TMR-1C may be a protoplanet of several Jupiter masses, which was ejected from the protobinary system TMR-1. Follow-up spectroscopy using the KECK telescope could, however, not confirm an intrinsic very low luminosity for the source (Terebey et al. 2000). The KECK-spectrum showed TMR-1C to be consistent with a background star of spectral type earlier than M4.5 V, but due to low signal/noise and low spectral resolution no clear spectral line signatures could be detected, preventing further classification.

In this article we discuss new VLT near-infrared imaging and spectroscopy, as well as mm-observations, in order to analyse the circumstellar environment of TMR-1, and the nature of the still disputed TMR-1C.

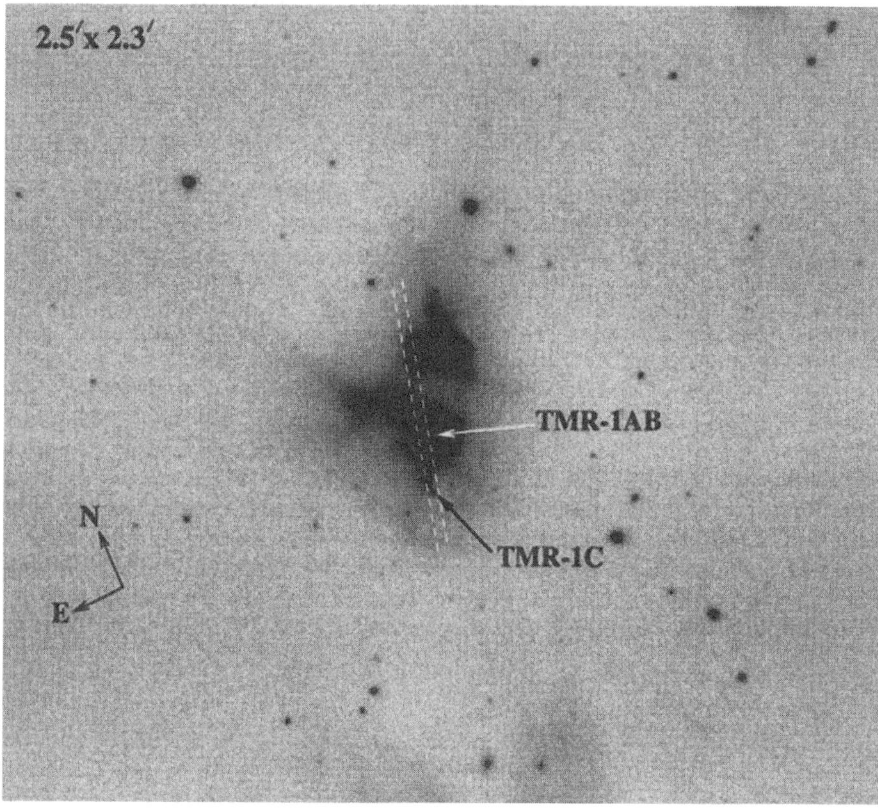

Fig. 1. ISAAC Ks (2.16μm) image of TMR-1 and its environment. The position of the slit (120″ total length) on TMR-1C and the filament is indicated by the white dashed lines

2 VLT/ISAAC Observations

Long-slit spectroscopy at low-resolution mode (R=450) was performed during an ISAAC instrument commissioning period in November 1998. The slit, of width 1″, was positioned in order to include TMR-1C and parts of the bent filament; an illustration of the slit positioning is shown in Figure 1. The resulting 2-dimensional image of the spectrum, which has a total effective exposure time of ~2 hrs, is presented in Cuby et al. (2000).

ISAAC in imaging mode was used to obtain broad-band images at K_s (2.16μm) and H (1.65μm), and images through narrow-band filters centered on the emission lines of molecular hydrogen at 2.12μm [v=1-0 S(1)], and of FeII at 1.64μm. All images were obtained under subarcsecond seeing conditions.

Figure 1 shows the reduced K_s-band image with unprecedented sensitivity and covering a large field of view. Several faint objects are detected all over the field, indicating that we are beginning to penetrate through the large amount of

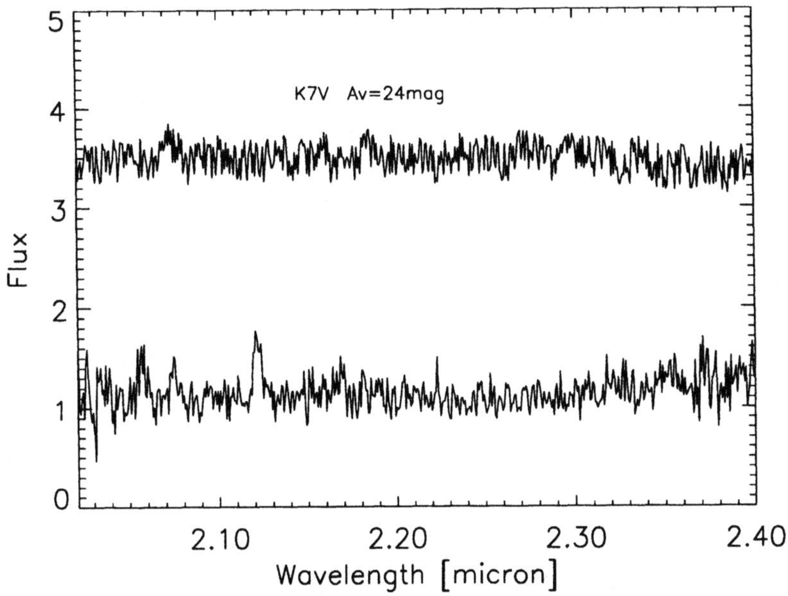

Fig. 2. *Lower spectrum:* TMR-1C. *Upper spectrum:* K7 V star spectrum from the stellar spectral library of Pickles (1998), smoothed and extincted by A_v=24 mag, and noise added. Both spectra are in arbitrary flux units, with the K7 V spectrum offset for comparison

dust of the molecular cloud core. In this respect the faintness and appearance of TMR-1C seems not to be any special, except for its location at the end of the filament. On the other hand, still striking is our detection of a "TMR-1C counterpart" object at the end of the broad filamentary structure associated with the north-western nebulosity. Although the morphology may suggest that the gap between the two nebulosities is caused by extinction, maps in cold dust emission suggest that the gap is *not* a high column density structure (Motte & André, 2001, Sec. 3 of this article, and also Whitney et al. 1997).

2.1 Spectroscopy and Imaging of TMR-1C

The spectrum of TMR-1C is shown in Figure 2. It lacks any spectral features and shows a flat spectral slope (the emission line at 2.12μm is a residual from a close-by emission knot of molecular hydrogen). The spectrum cannot be explained by a very cold object, like a T-dwarf, since those have distinct H_2O and CH_4 absorption bands that are easily detectable even in low signal/noise spectra (Cuby et al. 1999). It is neither consistent with a bullet nor dense knot of shocked molecular gas, since the spectrum is obviously pure continuum emission without any shock excited emission lines.

The VLT spectrum of TMR-1C is very similar to the Keck spectrum, and thus we come to the same conclusion as Terebey et al. (2000), namely that TMR-1C is probably a background field dwarf. Photometry of TMR-1C at H and K_s provide us with valuable constraints on required/possible values for the interstellar extinction. The range in H-K colours implied by our photometry measured on the ISAAC images and from measurements of other authors is H-K=1.55-1.75 mag, thus constraining A_v to 20-27 mag. We then compared our observed TMR-1C spectrum with extincted dwarf spectra from the spectral library of Pickles (Pickles 1998). However, as normal stars earlier than M-type don't show a significant difference in their spectral slope over the wavelength range analysed here, the spectral type of TMR-1C (if it is a background dwarf) can only be classified as $\lesssim$ M5. Figure 2, therefore, gives just an example of a conceivable fit.

Comparing the position of TMR-1C in a K versus H-K colour-magnitude diagram (presented elsewhere) with the position of many other faint sources in our field, which we expect to be mostly background stars, does also show TMR-1C to be very with a background field star.

2.2 Spectrum of the Filament and Discovery of a Jet

The spectra extracted along the filament show a wealth of emission lines, which are predominantly ro-vibrational lines of molecular hydrogen (Figure 3). We use the intensity ratio of the transitions H_2 v=1-0 S(1) at 2.12μm and H_2 v=2-1 S(1) at 2.24μm to investigate the origin of the H_2 emission. Thermal excitation via shocks should be the responsible mechanism if this ratio is $\gtrsim 5$, and gas densities are not too high ($\lesssim 10^5 \mathrm{cm}^{-3}$) (Shull and Beckwith 1982). Since we measure a 2.12μm/2.24μm ratio of ~ 8 and we do not observe any lines from transitions of high vibrational levels, we believe that the H_2 emission in the filament is due to shock excitation.

Our conclusion is nicely supported by the discovery of a high velocity jet emanating from TMR-1. VLT narrow-band images taken at 1.64μm (Fe II) show a bright emission line knot at only $\sim 3.5''$ distance from TMR-1AB, as well as an extended flow structure close to the base of TMR-1AB and in line with the bright Fe II knot (Figure 3). Furthermore, the direction of a molecular outflow observed at ^{12}CO (3–2) (Hogerheijde et al. 1998) is in very good agreement with the position angle of the Fe II jet. Typically, Fe II traces higher excitation, and higher velocities, and is in general found close to a jet axis. Hence, the detection of Fe II argues for a high velocity jet ($\gtrsim$ 40 km/s), which usually occurs in J-type shocks, and where molecular hydrogen is dissociated.

Considering this new information, i.e. the detection of a jet, interpretation of the filament may be as follows: The filament marks the edge of a cavity cleared by a lower velocity outflow, as for example observed in HH 211 (Gueth & Guilloteau 1999). The molecular hydrogen emission knots along the filament would then arise, because this lower velocity flow hits into the cavity rim, thereby creating a C-type shock. Or, the filament is intrinsic, pre-existing dense material being shocked by the outflow as it happens to be in its way. It is unlikely that

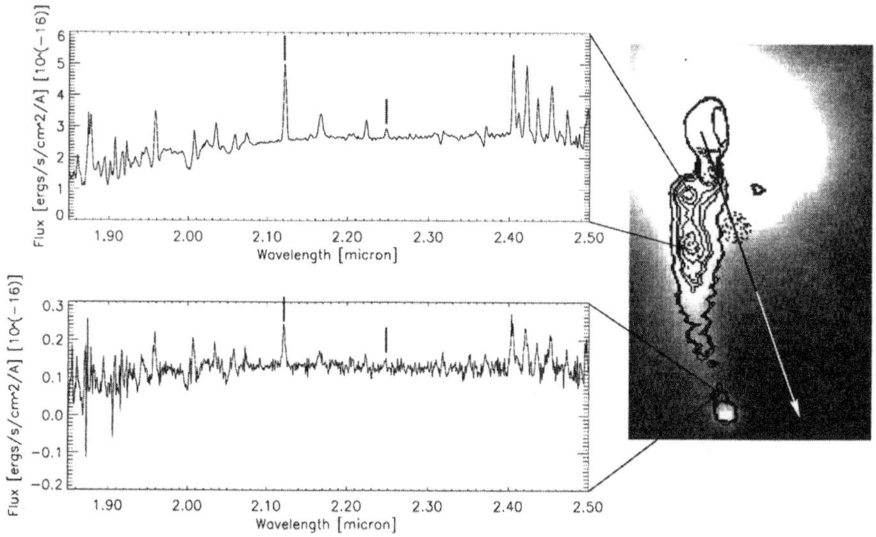

Fig. 3. Left: Two spectra extracted along the filament at positions where the filament displays knots of H_2 emission. The two H_2 lines marked are the transitions 1-0 S(1) at 2.12μm and 2-1 S(1) at 2.24μm. Right: VLT/ISAAC K_s image shown in greyscale, with continuum subtracted H_2 (*solid contours*) and Fe II (*dotted contours*) overplotted. The long arrow indicates the position angle (P.A.=170°) given for the ^{12}CO-outflow observed by Hogerheijde et al. (1998)

the filament is a jet by itself. While bent jets have been observed (Davis et al. 1994, Bally & Reipurth 2001), the significant amounts of continuum emission coming from the filament, together with a high degree of polarization (Whitney et al. 1997), and detection of high column densities associated with the filament (Hogerheijde et al. 1998, Motte & André 2001), exclude a jet.

3 Millimeter Interferometry

A molecular outflow emanating from TMR-1 has been confirmed by several studies from single-dish as well as from interferometer observations (e.g. Terebey et al. 1990, Hogerheijde et al. 1998). However, the structural details of the outflow, envelope, and circumstellar environment are far from being simple, and have not yet been fully understood.

We have used the IRAM/Plateau de Bure Interferometer in order to observe 1.3mm dust continuum and molecular line emission in ^{13}CO (2–1) (Walsh et al., in prep.). The synthesized beam size was $\sim 1.5''$, and maps were obtained with a primary beam of 25″ HPBW at this frequency. In order to directly compare the mm-observations with our K_s image, an astrometric solution was calculated for the VLT data. The final alignment uncertainty between the near-infrared and

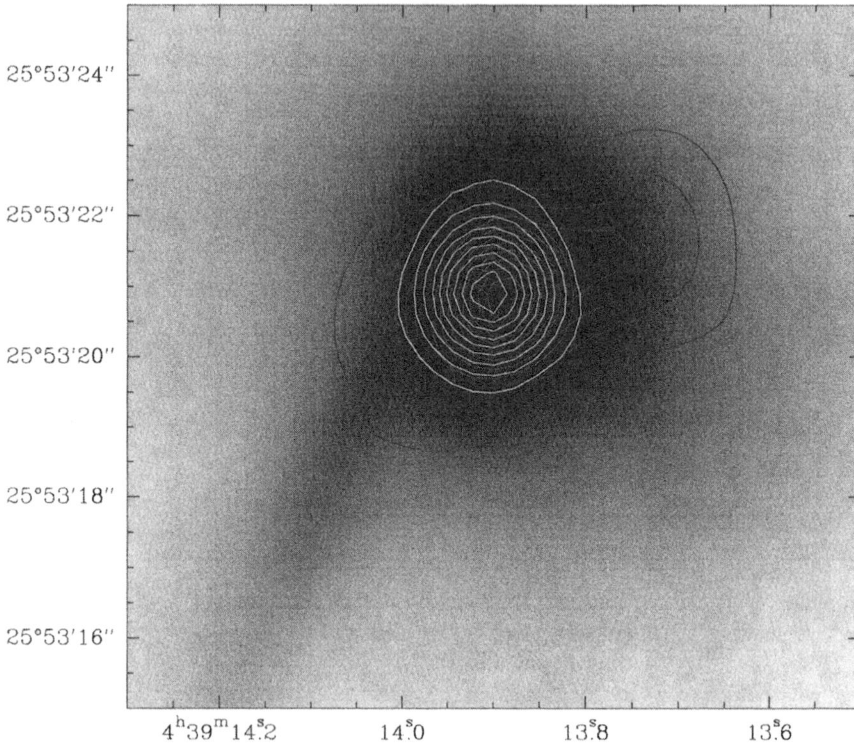

Fig. 4. Millimeter interferometry observations (contour lines) compared to the 2.2μm ISAAC image (*greyscale*). Coordinate system is J2000. White contours: 1.3 mm dust continuum, *Red contours:* ^{13}CO (2-1) integrated over [7.5,8.5] km/s, *Blue contours:* ^{13}CO (2-1) integrated over [2.0,5.0] km/s, System velocity of TMR-1 is $\sim$ 6.2 km/s. *Figure available in colour on the CD-ROM*

mm-data is $\sim 0.5''$. The comparison shows that a dust continuum source (unresolved) with 50$\pm$1 mJy/beam which traces cold dust from the circumbinary and/or circumstellar disks, is associated with the position of TMR-1AB (Figure 4). Assuming an average dust temperature of 20 K the inferred mass is $\sim 0.1 M_\odot$. No additional mm-sources are present, especially no compact source was detected in the gap between the two prominent 2.2μm nebulosities.

From velocity resolved observations at ^{13}CO (2–1) we derive very interesting kinematic information on the close circumstellar environment. The blue- and red-shifted emission in Figure 4 may be tracing rotation in a circumbinary disk, but higher spatial resolution data and more detailed data analysis is needed in order to confirm this interpretation.

Acknowledgments

We gratefully acknowledge helpful suggestions from F. Bertoldi, F. Comeron, T. Stanke, and M. Smith. Part of the data was obtained in service mode at Paranal Observatory, and thus we wish to thank the Paranal Science Operations Group for their effort and support.

References

1. Bally, J., and Reipurth, B.: ApJ **546**, 299 (2001)
2. Cuby, J. G., Saracco, P., Moorwood, A. F. M., D'Odorico, S., et al.: A&A **349**, L41 (1999)
3. Cuby, J. G., Barucci, A., de Bergh, C. et al.: 'Scientific results with ISAAC at the VLT'. In: *Discoveries and Research Prospects from 8- to 10-Meter-Class Telescopes*, ed. by J. Bergeron (Proc. SPIE Vol. 4005), 2000, pp. 212–223
4. Davis, C. J., Dent, W. R. F, Matthews, H. E., Aspin, C., Lightfoot, J. F.: MNRAS **266**, 933 (1994)
5. Gueth, F., Guilloteau, S.: A&A **343**, 571 (1999)
6. Hogerheijde, M. R., van Dishoeck, E. F., Blake, G. A., van Langevelde, H. J.: ApJ **502**, 315 (1998)
7. Kenyon, S. J., Calvet, N., Hartmann, L.: ApJ **414**, 676 (1993)
8. Motte, F., André, P.: A&A **365**, 440 (2001)
9. Pickles, A. J.: PASP **110**, 863, (1998)
10. Shull, J. M., Beckwith, S. V. W.: ARA&A **20**, 163 (1982)
11. Terebey S., Beichman, C. A., Gautier, T. N., and Hester, J. J.: ApJ **362**, L63 (1990)
12. Terebey S., van Buren D., Padgett D. L., Hancock T., Brundage M.: ApJ **507**, L71 (1998)
13. Terebey S., van Buren D., Matthews K., Padgett D. L.: AJ **119**, 2341 (2000)
14. Whitney, B. A., Kenyon, S. J., Gómez, M.: ApJ **485**, 703 (1997)

Beate Stelzer and Nuria Huélamo (foreground),
Brigitte Koenig and Artie Hatzes (background)

Searching for Planets in Stellar Clusters: Preliminary Results from the Hyades

Artie P. Hatzes[1], William D. Cochran[2], and Diane B. Paulson[2]

[1] Thüringer Landessternwarte Tautenburg, Sternwarte 5, D-07778, Germany
[2] McDonald Observatory, University of Texas at Austin, Austin, Texas, 78712, USA

Abstract. Radial velocity surveys have had spectacular success at finding giant planets around solar-type stars, but all of these discoveries have been among field stars which have a wide range of of metal abundances and ages that are poorly determined. Stellar clusters, on the other hand, represent a uniform sample of stars with the same abundance, the same age (which is well determined), and a known birth environment. These enable us to probe the process of planet formation with only one independent variable: the stellar mass. We present preliminary results from our survey of extra solar planets around Hyades dwarfs using the Keck 1 telescope.

1 Introduction

Radial Velocity surveys have had stunning success at finding giant planets in orbit around stars thus providing the driving force in this exciting new field (see review by Marcy et al. 2000). These surveys, however, have been biased toward solar-type stars (spectral type late F–K) in the field which can have diverse properties. Field stars possess a wide range of abundances of heavy elements, have ages that are poorly determined, and more importantly, come from a completely unknown birth environment. Stellar encounters may influence planet formation (Bonnell et al. 2001), but for a field star we do not know if it was born in a stellar rich, or stellar poor cluster. Disentangling the various various factors that can influence planet formation (stellar mass, abundance, dynamical effects) may be difficult using a stellar sample consisting only of field stars.

By contrast, stars in clusters represent a homogeneous sample that is coeval with a well determined age and uniform chemical abundance. The birth environment is also reasonably known. Searching for planets in cluster stars will thus enable us to probe the influence of stellar mass on planet formation with all other factors being more or less the same.

There are a number of other reasons for searching for extra-solar planets in stellar clusters. Theoretical work suggests that an investigation of planets in clusters may provide important clues to planet formation. de la Fuente Marcos & de la Fuente Marcos (2000) have argued that dynamical interactions between stars in clusters could produce planets in eccentric orbits. If true then the orbital characteristics in rich clusters should be substantially different from those in poor clusters. Photoevaporation by winds from massive stars can inhibit planet formation Armitage (2000). Thus stellar clusters with a higher number of massive stars should have fewer planets.

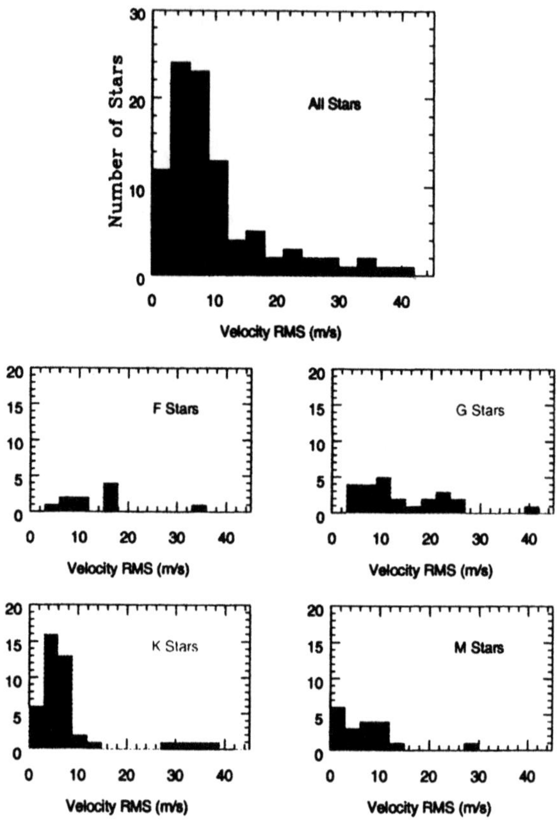

Fig. 1. Histogram of the standard deviation of Hyades RV measurements for all stars and F, G, K, and M dwarfs in the sample

The dynamical evolution of planetary systems may also be important. Lin & Ida (1996) have shown that the interaction between multiple giant planets in a system may lead to eccentric orbits and the merger of several objects into one massive planet. Tidal effects between the planet and star can cause the orbits of 51 Peg type systems to change (Trilling et al. 1998). Recently Pätzold & Rauer (2001) have argued that for F-type stars tidal forces will cause 51 Peg-type planets to spiral into the star in less than 1 Gyr. Thus older, F-type stars should be not have planets in short-period orbits. This scenario is consistent with the study of Suchkov & Schultz (2001) that argued that all known F-stars with extra-solar planets are young objects. These hypotheses can be tested by comparing the characteristics of planets around stars of various ages and this is best done in clusters.

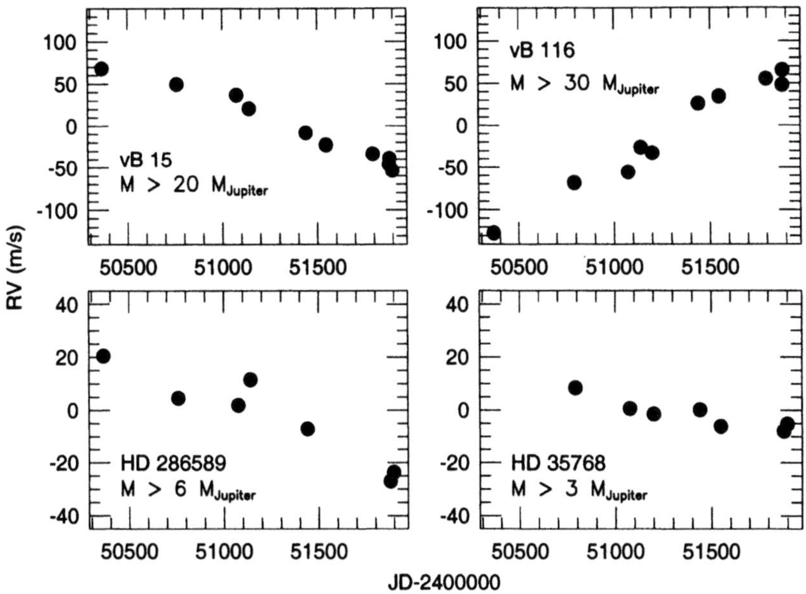

Fig. 2. Stars in the Hyades program showing long-term RV trends indicative of low-mass companions

2 The Hyades Survey

In order to probe the characteristics of planetary systems in clusters and to define further the exact role that stellar mass plays in star formation we began, in 1996, a survey for planets in the Hyades cluster using the Keck 1 10-m telescope. The Hyades are the nearest star cluster to the Sun with an age of 625 ± 50 Myr and an abundance of heavy elements that is about 50% higher than the solar value (Cayrel de Strobel 1990). Observations were made using the HIgh RESolution spectrograph of Keck I and an iodine absorption cell as the velocity metric. The program sample consists of 13 F dwarfs, 24 G dwarfs, 40 K dwarfs, and 21 M dwarfs. We typically received about 3 observing runs per season. Here we present preliminary results from this program.

There was some concern that due to the relatively young age of the Hyades that stellar activity would compromise our radial velocity precision. Active phenomenon such as spots and plage are known to increase the radial velocity noise of the star making it more difficult to detect extra-solar planets (Saar & Donahue 1999; Hatzes 1999). This does not seem to be the case for the Hyades. Figure 1 shows the standard deviation (σ) of radial velocities (after removal of any long term trends, see below). At least 70% of the stars have $\sigma < 15\ \mathrm{m\,s^{-1}}$ and 5% of the sample no variability to $\sigma < 6\ \mathrm{m\,s^{-1}}$. 30% of the stars show significant variability with most of these occuring among spectral types G and F.

Approximately 15% of the stars in our sample show significant linear trends due to the presence of low mass companions. Figure 2 shows 4 of these stars.

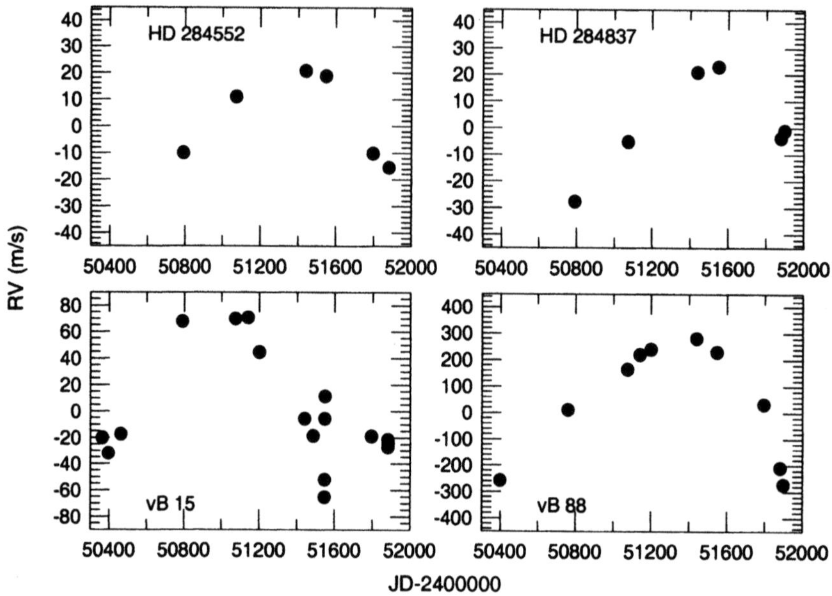

Fig. 3. Four stars in the sample showing RV trends with significant curvature

(These linear trends were removed before computing the rms RV scatter used in Figure 1). These RV trends are consistent with companion masses greater than 3–40 $M_{Jupiter}$ which would put these in the brown dwarf regime. Since our data only covers a short span of the orbit the companion objects could eventually turn out to low mass stars.

A significant fraction (6%) of our sample show significant curvature in the RV trends consistent with giant planet or brown dwarf companions. Examples of these stars are shown in Figure 3. Estimated M_{sini} for these objects range from 2–20 $M_{Jupiter}$. Definitive orbits for these objects can be found within the next 3–5 years.

Approximately 6% of our sample show weak RV periodicities when performing a periodogram analysis, but with low false alarm probabilities. These low false alarm probabilities result partly from the higher RV noise in these stars, but also due to the small number of sampling points (a poorly sampled sine wave also will have a low false alarm probability in spite of the signal definitely being present). These periods range from 6–45 days and may ultimately prove to be due to rotational modulation by active features (spots, plage). Our instrumental setup includes the Ca II H & K lines which can be used to determine timescales for rotation and activity cycles. This can confirm whether the variations are due to companions or activity.

3 Discussion

At the present time our Hyades survey shows that 30% of our sample of stars show significant RV variability. Continued monitoring of these stars will establish whether these variations are due to planetary companions, low mass stars, or stellar activity. Since the current frequency of planets among the field is only about 3–5% there is a chance that the frequency of planets in the Hyades may be significantly higher. The potentially higher frequency of extra-solar planets among Hyades stars may be related to their higher abundance (see Gonzalez et al. 2001). Planet searches among stars in other clusters could prove fruitful.

Since this is workshop is involved with "the VLT view" of planet and star formation, it is appropriate to ask: "What role can the VLT play in searching for planets among cluster stars?". Table 1 lists the visual magnitude of a K0 star in several well-known clusters (not all accessible from the VLT!). Clearly, the solar-type stars in clusters are faint and to get reasonable RV precision requires using an 8-m class telescope. Also listed is the expected standard deviation of RV measurements made on a cluster K0 star using UVES, based on its current current best performance of $\sigma < 2\ \mathrm{m\,s^{-1}}$ on brighter stars (Kürster, private communication) when using an iodine absorption cell. UVES should be able to find planets in clusters with a main sequence magnitude as faint as M67, and massive planets around main sequence stars in brighter globular clusters. According to the values in Table 1, the VLT + UVES could have been used to do follow-up RV work on F or G dwarfs in 47 Tuc, had photometric studies found hot Jupiter companions (Gilliland et al. 2000).

RV searches for extra-solar planets requires the allocation of considerable telescope time. Since cluster stars are spatially distributed over a small field of the sky, such RV searches may be best suited for multi-object spectrographs (MOS). Although multi-object spectrographs can increase the number of objects that are observed simultaneously, this feature comes only at the expense of wavelength coverage which degrades the RV precision. Given the tradeoff between wavelength coverage and number of objects, MOS searches may or may not be an efficient means of searching for planets in cluster stars.

We investigated whether FLAMES is suited for RV surveys in clusters by comparing it to the performance of UVES in single object mode. We estimate that UVES should be able to achieve an RV precision of 10 m s^{-1} in an 8 min exposure on a $V = 14.5$ mag star. Allowing for an overhead of 4 min/object an RV survey should be able to acquire data on about 5 stars per hour.

FLAMES will observe 8 objects simultaneously at a resolving power of $R =$ 45,000 and a bandpass of 200 Å. This compares to a nominal resolving power of R = 60,000 for UVES and a useful bandpass (for iodine absorption) of 1000 Å. The radial velocity precision scales as $R^{-3/2}$ and (bandpass)$^{-1}$ (Hatzes 1996). Thus in the same 8 min exposure FLAMES will achieve an RV precision of about 36 m s^{-1}. Since the RV precision also scales as (signal-to-noise ratio)$^{-1}$, FLAMES will have to observe 25 times as long or 3.3 hrs to achieve the same RV precision (this calculation does not include possible losses due to the optical fibers). This corresponds to a rate of $8/3.3 = 2.4$ stars/hour. This estimate does not include possible losses of the fiber). It seems that FLAMES will be less efficient than UVES single object mode for searching for planets in clusters. For a MOS to be competitive with single object, large wavelength coverage spectrographs it must be able to simultaneously acquire data on at least 20 objects.

Table 1. RV precision for Cluster stars

Cluster	K0 V-mag	σ (m/s)
Pleiades	11.4	3
Praesepe	12.4	5
M 67	15.5	9
47 Tuc	19.4	35

We acknowledge NSF grant AST9808980 and NASA grant NAG5-9227.

References

1. P.J. Armitage: A&A, **362**, 968, (2000)
2. I.A. Bonnell, K.W. Smith, M.B. Davies, K. Horne: MNRAS, **322**, 859 (2000).
3. G. Cayrel de Strobel: Societa Astronomica Italiana, Memorie, **61**, 613 (1990)
4. R. de la Fuente Marcos, R., C. de la Fuente Marcos: In: *Stellar Clusters and Associations: Convection, Rotation, and Dynamos*, Proceedings from ASP Conference, Vol. 198, ed. by R. Pallavicini, G. Micela, and S. Sciortino, 183 (2000)
5. R. Gilliland et al. (23 other authors): ApJ, **522**, 699 (2001)
6. G. Gonzalez, C. Laws, T. Sudhi, B.E. Reddy, AJ: **121**, 1559 (2001)
7. A.P. Hatzes: In *Precise Stellar Radial Velocities IAU Colloquium 170*, eds. J.B. Hearnshaw and C.D Scarfe, 259 (1999)
8. A.P. Hatzes: In *Proceedings of ESO Workshop on High Resolution Spectroscopy with the VLT*, ed. M.-H. Ulrich, 275 (1992).
9. D.N.C. Lin, S. Ida: ApJ, **477**, 781 (1997)
10. G.W. Marcy, W.D. Cochran, M. Mayor: In *Protostars and Planets IV*, eds. V. Mannings, A.P. Boss and S. S. Russell, Univ of Arizona Press, 1285 (2000)
11. M. Pätzold, H. Rauer: preprint (2001)
12. S.H. Saar, R.A. Donahue: ApJ, **485**, 319 (1997)
13. A.A., Suchov, A. Schultz: ApJ, **549**, L237 (2001)
14. D.E. Trilling, W. Benz, T. Guillot, J.I. Lunine, W.B. Hubbard, A. Burrows: ApJ, **500**, 428 (1998)

The Kuiper Belt As An Evolved Circumstellar Disk

David C. Jewitt

Institute for Astronomy, University of Hawaii,
2680 Woodlawn Drive, Honolulu, HI 96822, USA

Abstract. In the past decade, more than 400 objects have been discovered in orbit about the sun beyond Neptune. They constitute a vast ring of bodies the larger of which are the surviving products of accretion in the outer circumsolar disk. This so-called Kuiper Belt is a repository of the solar system's most primitive (volatile-rich) matter, a supplier of comets to the inner solar system and a source of collisionally produced dust with similarities to extra-solar dust disks.

1 Introduction

Since 1992, more than 400 trans-Neptunian bodies (objects with orbital semi-major axes greater than Neptune's 30 AU) have been discovered. Widely known as Kuiper Belt (or sometimes Edgeworth-Kuiper) Objects (KBOs), they are of scientific significance on several levels. First, dynamical chaos and (possibly) gravitational scattering in the Kuiper Belt lead some objects into planet-crossing trajectories that are dynamically short-lived. Planet-crossing KBOs that are not scattered out of the planetary system may be passed inwards to the vicinity of the earth where sublimation of trapped volatiles leads to the production of measurable coma. In this way, the Kuiper Belt feeds the short-period comet population. Long-period comets, in contrast, originate in the larger and more distended Oort Cloud.

The Kuiper Belt is also important as a repository of primitive planetary matter. Blackbody equilibrium temperatures at Neptune's orbit are near 50 K. At such low temperatures, many volatiles are either frozen solid (e.g. H_2O) or physically trapped in the interstices of other ices (e.g. CO, CO_2). Thus, KBOs are thought to be carriers of volatile molecules. They may supply a significant fraction of the volatiles now found on the terrestrial planets. Indeed, a cometary source of Earth's ocean water is often discussed (Lauffer et al. 1999).

The main importance of the Kuiper Belt is that it provides a very local example of a highly evolved circumstellar disk. Processes that are probably general to the circumstellar disks of solar-mass stars can be discerned already in the size and orbital element distributions of the Kuiper Belt. For example, resonant structure in the Kuiper Belt is probably indicative of early planetary migration. Collisions between KBOs and erosion of KBOs by interstellar dust lead to the production of dust, some of which has already been detected by spacecraft. Viewed externally, our solar system would sport a diaphanous ring, morphologically very similar to that around $HR4796A$, but fainter.

In this very brief overview, we draw attention to the main features of the Kuiper Belt and attempt to give an idea of where the research frontier currently lies. Recent, more detailed reviews may be found in Jewitt (1999), Farinella, Davis and Stern (2000), Jewitt and Luu (2000), Malhotra, Duncan and Levison (2000), and Luu and Jewitt (2002). Active and informative web sites about the Kuiper Belt include:
`http://cfa-www.harvard.edu/cfa/ps/lists/TNOs.html`
`http://www.boulder.swri.edu/ekonews/`
`http://www.ifa.hawaii.edu/faculty/jewitt/kb.html`

2 Physical Properties

The flux density scattered by a planetary body varies in proportion to $pr^2/(R\Delta)^2$, where p is the albedo, r is the body radius and R and Δ are the heliocentric and geocentric distances, respectively. At opposition, $\Delta = R-1$. When $R >> 1$, $\Delta \approx R$ and the flux density varies roughly as pr^2/R^4; the strong distance dependence renders distant objects very faint. This simple fact delayed the discovery and recognition of the trans-Neptunian solar system until the very end of the 20th century. It also explains why physical observations of KBOs remain challenging.

2.1 Sizes and Albedos

Optical observations provide a measure of pr^2, but cannot directly yield either quantity separately. Most estimates of the sizes of KBOs are based on optical measurements and an *assumed* geometric albedo $p = 0.04$. This is the albedo measured for objects believed to have escaped the Kuiper Belt, notably the Centaurs and the nuclei of certain short-period comets. Diameters derived in this way are generally larger than 100 km, as a result of bias against detecting smaller, fainter objects. The largest KBO discovered recently is (20000) Varuna, with a diameter near 900 km (Jewitt, Aussel and Evans 2001). Thermal radiation from this object has been detected at submillimeter wavelengths, allowing a determination of the albedo: (20000) Varuna has $p \approx 0.07$. By comparison, Kuiper Belt Object Pluto is 2300 km in diameter and has albedo $p = 0.6$ as a consequence of bright surface frosts deposited from a tenuous atmosphere. While Varuna appears mantled in dark, possibly carbon-rich material, the extent of frost coverage on other KBOs remains uncertain. Keck near-infrared spectroscopy shows evidence for water ice absorption at 2 μm on some but not all objects.

The geometric albedo is also a function of wavelength. KBOs range from neutral to considerably redder than the sun (e.g. $1.4 < (B-I) < 2.7$; the color of the sun is $(B-I) = 1.4$). This color dispersion indicates compositional diversity among the KBOs, the origin of which is unexplained (Luu and Jewitt 1996, Jewitt and Luu 2001).

The heat conduction time for a body of radius r is $\tau_C \approx r^2/\kappa$, where κ $[m^2s^{-1}]$ is the thermal diffusivity of the constituent material. Slightly porous dielectric solids of the type expected to make up the KBOs have $\kappa \approx 10^{-7}$ $[m^2 s^{-1}]$. Objects with $r > 100$ km have $\tau_C >$ the age of the solar system, and therefore cannot cool by conduction. Objects with $r \approx 100$ km experience only modest (few K) internal heating by trapped long-lived radioactive nuclei (notably Th, K, U). Those with $r \approx 1000$ km experience heating sufficient to mobilize internal super-volatiles (notably CO and N_2). We expect that these volatiles should migrate out of the cores towards the cooler, outer layers. We should not be surprised to find evidence of super-volatile outgassing from at least the larger KBOs. The potent radioactive nucleus of ^{26}Al, with half-life $\approx 10^6$ yr, is unlikely to have been incorporated in $r > 1$ km KBOs because of the long growth times in the outer disk.

2.2 Size Distribution

The size distribution is inferred from the brightness distribution of the KBOs by assuming a constant albedo for all objects. The distribution is well represented by a differential power law of the form

$$n(r)dr = \Gamma r^{-q} dr$$

with $q = 4.0^{+0.6}_{-0.5}$ (Trujillo, Jewitt and Luu 2001). This distribution is such that the collisional cross-section of the KBOs is dominated by the smallest objects, while the mass is spread uniformly among equal (logarithmic) mass intervals. Models of agglomeration that include the effects of collisional shattering of growing KBOs are able to match the measured sized distribution (Kenyon and Luu 1999).

The distribution at sizes << 100 km is not well constrained. Models suggest that the smaller objects could be produced by collisional shattering of KBOs larger than 100 km. If so, we expect that the size distribution should flatten, somewhat, approaching the $q \approx 3.5$ value expected for a set of colliding bodies in collisional equilibrium cascade (Dohnanyi 1969). In this picture, the kilometer-scale nuclei of comets can be identified with fragments of larger KBOs. Collisional shattering of unseen parent objects is one possible source of the dust in β Pic type systems (Dent et al. 2000).

2.3 Orbital Element Distributions

One of many surprising observational discoveries about the Kuiper Belt is that the orbital element distribution is highly structured, with evidence for division into several distinct groups (Figure 1).

- Roughly 10% of the known KBOs occupy mean motion resonances with Neptune. The 3 : 2 mean motion resonance at 39.4 AU is particularly densely occupied, and includes Pluto as a member. Pluto and some other "Plutinos" are stable even in Neptune-crossing orbits because of protection provided by the resonance. The leading theory for the population of resonances invokes planetary migration in response to torques exerted by the scattering of nearby planetesimals. In this scenario, Jupiter is the ultimate source of the angular momentum lost to planetesimals scattered out of the solar system. It drifts inwards as Neptune and the other giant planets migrate outwards. Capture into resonance is only efficient when the velocity dispersion is small and the planet drift velocity is small and steady (Malhotra, Duncan and Levison 2000). Capture must have occurred early, when the disk was still very thin. The maximum eccentricities of Plutinos (e ≈ 0.3) are well fit by outward migration of Neptune by $\Delta a \approx 7$ AU on timescales $\approx 10^7$ yr (Hahn and Malhotra 1999).

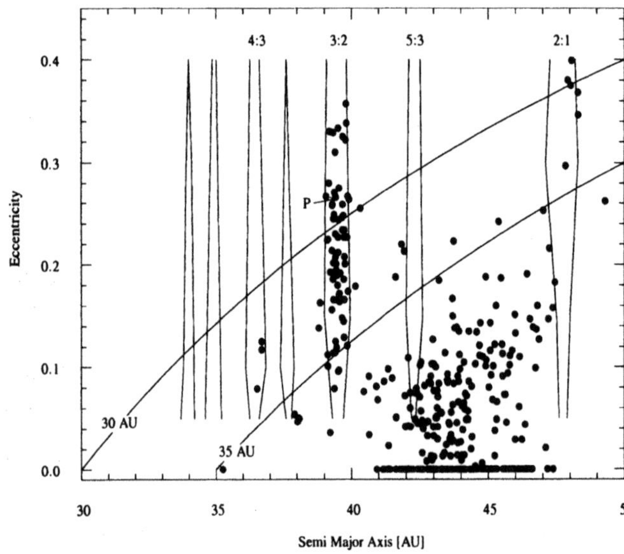

Fig. 1. Distribution of semi-major axis vs. orbital eccentricity for the Classical and Resonant Kuiper Belt Objects. The major mean-motion resonances are marked as vertical bands. Diagonal arcs show the loci of orbits having perihelia at 30 AU and 35 AU. Objects above the $q = 30$ AU line have Neptune-crossing orbits. Neptune is located at the lower left intersection of the axes ($a = 30$ AU, $e = 0$). Pluto is marked with a "P"

- A majority of the known KBOs occupy the "Classical Kuiper Belt". They have semi-major axes $42 < a < 47$ AU and modest inclinations and eccentricities. The number of $D > 100$ km Classical objects alone is $3.8^{+2.0}_{-1.5} \times 10^4$. These objects are dynamically stable on billion year timescales without resonance protection because they maintain a safe distance from Neptune. The CKBOs show a distinctly excited velocity distribution, not at all consistent with the small relative velocities that must have prevailed when these bodies accreted. The source of the excitation is unknown, but several possibilities have been considered. Large planetesimals, with masses comparable to or greater than that of Mars, could have excited the velocity dispersion if these bodies were scattered through the Kuiper Belt by Neptune (Petit, Morbidelli and Valsecchi 1998). Such objects would also have disturbed and depleted the resonant objects. Perhaps the resonant KBOs are survivors from once much larger populations.

The Classical Belt has an outer edge near $R \approx 47$ AU (Figure 2: JLT98, Allen et al. 2001, TJL2001). The smallest of the proplyd disks have radii of comparable scale (McCaughrean and O'Dell 1996) and tidal truncation has been proposed as an explanation. Similarly, tidal truncation has been invoked for the Kuiper Belt by Ida et al. 2000. In the sun's current low density environment the probability of a sufficiently close (150 - 200 AU) stellar encounter is negligible. However, the sun probably formed in a cluster and close encounters might then have been correspondingly frequent. Evidence for the sun's early cluster environment is provided by isotopic contamination patterns which suggest the early, nearby detonation of one or more supernovae. Plausible guesses of the stellar mass function suggest that the sun formed in a cluster of ≈ 2000 stars (Adams and Laughlin 2001).

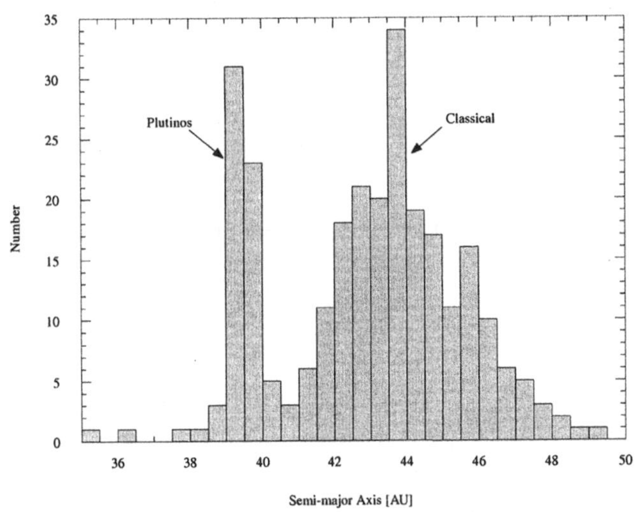

Fig. 2. Histogram of semi-major axes. Detailed consideration of the observational biasses inherent in ground-based surveys reveals that the deficiency of objects beyond 50 AU indicates a real edge to the Classical Belt

- About two dozen KBOs belong to the Scattered Object (SKBO) group (Luu et al. 1997, Figure 3). These have large orbits with perihelia near 35 AU and eccentricities and inclinations larger than typical for the other dynamical groups in the Kuiper Belt. Such orbits are unstable on billion year timescales, as a result of continuing weak interactions with Neptune near perihelion. Recent identification of SKBOs with perihelia substantially beyond Neptune's influence, however, renders the perturbing role of Neptune uncertain. Other perturbers may be responsible.

SKBOs are subject to particularly large observational selection effects. The number of SKBOs interacting with Neptune is at least $3.1^{+1.9}_{-1.3} \times 10^4$. If the SKBO perihelion distribution extends far beyond 40 AU, the total population of these objects could be much larger, possibly dwarfing the Classical and Resonant populations.

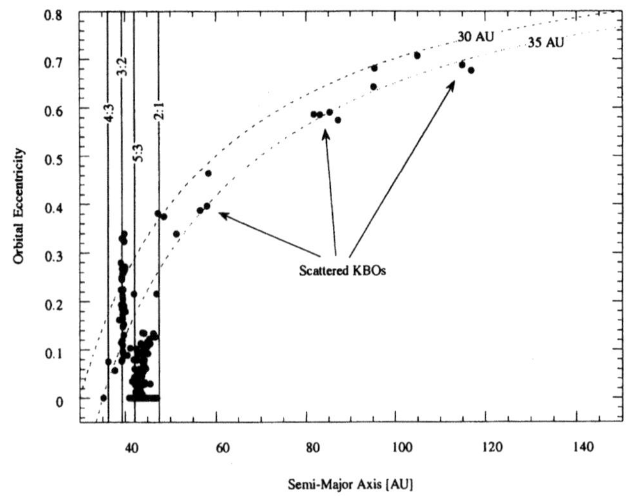

Fig. 3. Same as Figure 1 except that a wider parameter space is plotted to reveal the Scattered Kuiper Belt Objects. Note how these appear to cluster near a line of constant perihelion distance

3 Numbers and Mass

The population of KBOs is estimated by extrapolation from surveys of limited areas of the ecliptic sky. The number of KBOs of diameter $D > 100$ km in the $30 < R < 50$AU region is $N(D > 100km) \approx 10^5$. Extrapolated down to $D = 1$ km with a $q = 4$ power law, the number of KBOs is $N(D > 100km) \approx 10^{11}$.

By mass, however, the present day Kuiper Belt is insubstantial. Published estimates from optical survey data give $M \approx 0.1\ M_{\oplus}$ (Jewitt, Luu and Trujillo 1998, Trujillo, Jewitt and Luu 2001). An independent upper limit, $M < 1M_{\oplus}$, is set by dynamics (specifically by the absence of perturbations on the orbit of comet P/Halley). We note below that the original mass in the Kuiper Belt may have been much larger, perhaps near $30M_{\oplus}$.

Short-period comets enter the inner solar system at a rate near $10^{-2}\ yr^{-1}$. The Kuiper Belt appears easily able to supply this rate for the full age of the solar system. However, it is not clear from precisely where in the Kuiper Belt the short-period comets originate. Regions of chaotic instability near resonances are likely sources. Decay of SKBOs could also provide a source of planet-crossing bodies.

4 Collisions and Dust

The $\Delta V = 1.3$ km s^{-1} velocity dispersion among KBOs is larger than the gravitational escape velocity from all but the largest objects. Therefore, collisions between KBOs tend to be destructive, leading to the production of dust. Collisions with interstellar dust grains also erode KBOs and produce dust. Dust detected by the Voyager spacecraft when leaving the planetary region after their encounters with the gas giants (Gurnett et al. 1997) has been interpreted as Kuiper Belt dust (Jewitt 1999). The mean optical depth normal to the plane of the ecliptic is about 10^{-7}, three to four orders of magnitude less than the corresponding optical depth of the β Pic disk. From the distance of another star, the Kuiper Belt would be invisible with current coronagraphic technology. However, it may not always have been so. If the mass of KBOs was originally much larger than now, enhanced collision rates might have maintained much larger mean optical depths. Dust of 1 to 10 μm size drifts inwards, on 10^6 - 10^7 yr timescales, under the action of Poynting-Robertson drag (Figure 4). Many of these particles are destroyed by collisions with interstellar dust before they reach the vicinity of Earth. It is at least possible, however, that we will find Kuiper Belt grains in the stratospheric dust collections. They would be distinguished by extreme cosmic ray track densities consistent with their long transport timescales.

5 Formation

A disk of planetesimals each of radius, r [m], will develop, through mutual gravitational scattering, an internal velocity dispersion comparable to the escape velocities of the constituent planetesimals. The escape velocity from a spherical ice body of unit density is $V_e \approx 0.001r$ $[ms^{-1}]$. The circular Keplerian velocity is just $V_K = (GM_\odot/a)^{1/2}$, where G $= 6.67 \times 10^{-11} Nkg^{-2}m^2$ is the gravitational constant, $M_\odot = 2 \times 10^{30}$ kg is the solar mass and a is the semi-major axis expressed in meters. The orbital inclination resulting from the velocity dispersion is simply $\tan(i) = \Delta V/V_K$. Substituting $\Delta V = V_e$ and assuming $\Delta V << V_K$, we see

$$i[rad] \approx 0.001r(a/GM_\odot)^{1/2}.$$

At $a = 40$ AU, a disk of $r = 1$ km bodies would inflate to an average inclination near 2×10^{-4} [rad] = 0.01 [deg]. With $r = 100$ km the equilibrium mean inclination is still only 1 [deg]. For comparison, the measured, de-biassed width of the KBO inclination distribution is nearer 20 [deg] (Trujillo, Jewitt and Luu 2001).

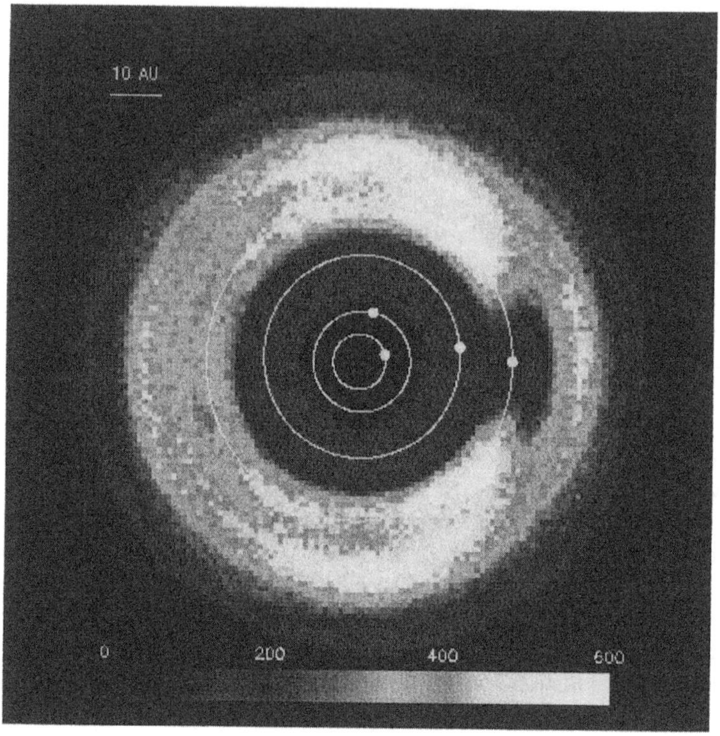

Fig. 4. Distribution of dust particles released from the Kuiper Belt and moving under the action of solar and planetary gravity, and radiation forces. This figure is intensity coded for the column density of 23 μm sized grains and includes neither the dust size distribution nor the radial variation of the impressed radiation field. Orbits and locations of the gas giant planets are marked. Figure from Liou and Zook 1998

How was the inclination distribution excited to its observed, high value? The emergence of nearby Neptune presumably had a profound influence on the structure of the Kuiper Belt. The dynamically cool accretion disk would be heated by Neptune- perturbed objects, effectively truncating the growth of the KBOs. The timescale for the accretion of Neptune is uncertain. Its high molecular weight suggests that growth was not complete until after depletion of the sun's gas nebula at 10^7 (Bryden, Lin and Ida 2000) but much longer times, 10^8 yrs, are possible (Lissauer et al. 1995). For this reason, it is widely believed that the KBOs, up to and including 1000 km scale Pluto, grew in the first 10^7 to 10^8 yrs of the solar system and that subsequent accumulation of KBOs in the inner regions was terminated by the runaway growth of Neptune. The planet would introduce less disturbance in the outer Belt, but low disk surface densities at large radii would seriously retard the growth of large objects.

The outstanding problem with this scenario is that the mass of material in the belt is too small for growth to have produced the observed KBOs even in

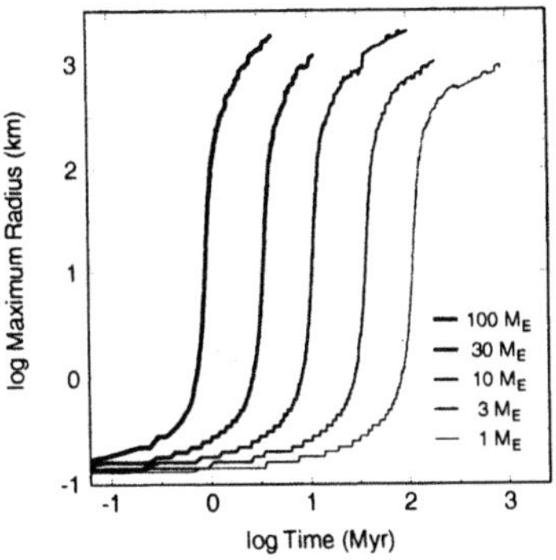

Fig. 5. Maximum radius of growth vs. elapsed time for Kuiper Belt disk masses 1, 10 and 100 $M_\oplus$, as marked. The models assume an initial eccentricity 0.001, corresponding to velocity dispersion $\Delta V \approx 10$ m s^{-1}. Figure from Kenyon and Luu 1999

10^8 yrs (Stern and Colwell 1997, Kenyon and Luu 1999). The latter authors find growth times for 1000 km scale objects $\tau[Myr] \approx 200(M/10M_\oplus)^{-1}$. To obtain the observed objects by agglomeration on 10^8 (10^7) yr timescales requires masses $M \approx 20M_\oplus$ ($200M_\oplus$), roughly 2×10^2 (2×10^3) times higher than now present (Figure 5). The inference is that the present day Kuiper Belt is a remnant of a once much more massive structure (Stern and Colwell 1997, Kenyon and Luu 1999).

How was the Kuiper Belt cleared? Dynamical clearing certainly played a role, particularly in the inner ($R < 40$ AU) regions. The disk beyond 50 AU might have been ejected by a passing star (Ida *etal.* 2000). In between, collisional grinding of KBOs into dust has been proposed as a mechanism of reducing the total mass in the Belt (Farinella, Davis and Stern 2000). Collisional grinding presumably pulverizes KBOs until the fragments are so small that radiation drag or other weak forces are able to clear the dust from the system. This scenario, still not worked out in convincingly quantitative detail, is attractive not least because it encourages us to think of the Kuiper Belt in a younger phase as a very dusty and prominent ring, perhaps similar to that around HR4796A.

Planetesimals closer to the sun were either accumulated into the growing giant planets or gravitationally scattered away. A majority of those ejected left the planetary region at greater than the local solar escape velocity. These objects, perhaps 10^{13} in number from our sun alone, perpetually roam interstellar space. If all stars ejected a similar number of planetesimals, we should expect to find interstellar comets re-entering the inner solar system once every 10 - 100 yrs or so. None has yet been seen but future survey telescopes (e.g. VISTA, VST and LSST) might find them. Many planetesimals ejected close to the local escape velocity were subsequently perturbed, by galactic tides and by gravitational impulses from passing stars into weakly bound, highly eccentric orbits (the capture efficiency would be higher if the sun were in a dense cluster; see Fernandez 1997). The 10^{12} objects now in the Oort Cloud are thought to have been captured in this way. Thus we see that, ironically, the remote, cold Oort Cloud (long period) comets in fact formed closer to the sun and therefore at higher temperatures than the comets in the Kuiper Belt.

References

1. F. Adams and G. Laughlin. Icarus, in press (astro-ph 0011326). (2001)
2. G. Bryden, D. Lin and S. Ida. Ap. J., 544, 481-495 (2000).
3. W. Dent, H. Walker, W. Holland and J. Greaves. MNRAS, 314, 702-712 (2000).
4. J. Dohnanyi. J. Geophys. Res., 74, 2531-2554 (1969)
5. P. Farinella, D. Davis and S. Stern. Formation and Evolution of the Edgeworth-Kuiper Belt. Protostars and Planets IV, edited by V. Mannings, A. Boss and S. Russell, University of Arizona Press, Tucson. pp:1255-1282.
6. J. Fernandez. Icarus, 129, 106-119 (1997).
7. D. Gurnett, J. Ansher, W. Kurth, and L. Granroth. Geophys. Res. Lett., 24, 3125 (1997)
8. J. Hahn and R. Malhotra. Astron. J. 117, 3041-3053 (1999).
9. S. Ida, J. Larwood and J. Burkert. Ap. J. 528, 351 (2000)
10. D. Jewitt, J. Luu, and C. Trujillo. Astron. J., 115, 2125-2135 (1998)
11. D. Jewitt. The Kuiper Belt. Annual Reviews of Earth and Planetary Sciences, 27, 287-312. (1999).
12. D. Jewitt and J. Luu. Physical Nature of the Kuiper Belt. Protostars and Planets IV, edited by V. Mannings, A. Boss and S. Russell, University of Arizona Press, Tucson. pp:1201-1229 (2000)
13. D. Jewitt, H. Aussel and A. Evans. Nature, 411, 446-447 (2001).
14. D. Jewitt and J. Luu. Astron. J., in press. (2001)
15. S. Kenyon, and J. Luu. Astron. J., 118, 1101-1119 (1999)
16. D. Laufer, G. Notesco, A. Bar-Nun, T. Owen. Icarus, 140, Issue 2, pp. 446-450 (1999)
17. J.-C. Liou, and A. Zook. Astron. J., 118, 580-590 (1998)
18. J. Lissauer, J. Pollack, G. Wetherill and D. Stevenson. Formation of the Neptune System. In Neptune and Triton. Ed. D. Cruikshank. Univ. Az. Press, Tucson, pp. 37-108 (1995)
19. J. Luu, and D. Jewitt. Astron. J., 112, 2310-2318 (1996)
20. J. Luu, B. Marsden, D. Jewitt, C. Trujillo, C. Hergenrother, J. Chen and W. Offutt. Nature, 387, 573 (1997)
21. J. Luu, and Jewitt. Ann. Rev. Astron. Ap., in press (2002)
22. R. Malhotra, M. Duncan, and H. Levison. Dynamics of the Kuiper Belt. Protostars and Planets IV, edited by V. Mannings, A. Boss and S. Russell, University of Arizona Press, Tucson. pp: 1231-1254. (2000)
23. M. McCaughrean and C. O'Dell. Astron. J., 111, 1977 (1996).
24. J.-M. Petit, A. Morbidelli and G. Valsecchi. Icarus, 141, 367-387 (1998)
25. S. Stern and J. Colwell. Astron. J., 114, 841 (1997).
26. C. Trujillo, D. Jewitt, and J. Luu. Astron. J., 122, 457-473 (2001)

Anneila Sargent and Charlie Lada

The VLTI and Its Instrumentation

Andrea Richichi, Andreas Glindemann, and Markus Schöller

European Southern Observatory, Karl-Schwarzschild-Str. 2,
85748 Garching b.M., Germany

Abstract. High angular resolution is one of the key factors needed to detect and investigate young stars, their circumstellar environment and the possible presence of exoplanets. In this respect, the VLTI represents one of the most powerful tools available in the short term. We describe the phased implementation of the VLTI, from the basic configuration which has recently achieved first fringes, to the full array with all subsystems and complete instrumentation that will be operational in the next years. We provide examples of techniques for the study of young stars and exoplanets.

1 Concept and Current Status of the VLTI

The ESO Very Large Telescope Interferometer (VLTI) is located on Cerro Paranal, in northern Chile. This facility consists of the four fixed 8.2 m Unit Telescopes (UT), and of a number of 1.8 m Auxiliary Telescopes (AT) which can be moved over an array of 30 stations (Glindemann et al. 2000b). A scheme of the VLTI is shown in Fig. 1.

All the AT stations, as well as the UTs, are connected by a network of underground light ducts. Central to the facility is the delay line tunnel, where optical path differences are continually adjusted to correct for both long-range effects (such as those due to sidereal motion) and fast, short-range variations such as those due to differential atmospheric piston. A number of relay mirrors feed the light from the telescopes into the tunnel, and from there to a central underground laboratory, where the beams from two or more chosen telescopes are brought together and coherently combined.

At present, only test telescopes (40cm siderostats) are being used, mainly for commissioning purposes. Recently, the major milestones of first fringes has been achieved, using the siderostat telescopes placed on the same stations on which the ATs will be used (ESO PR 06/01 2001). Details on further phases of the project are given in Sect. 3.

In the following, we will concentrate on the instrumentation which is available or is currently under construction, and we provide an overview of the developments that will take place in the course of the next few years.

2 VLTI Instrumentation

A number of first-generation instruments have been planned for the VLTI. The first one is VINCI, a K-band, 2-beam combiner which has been developed by

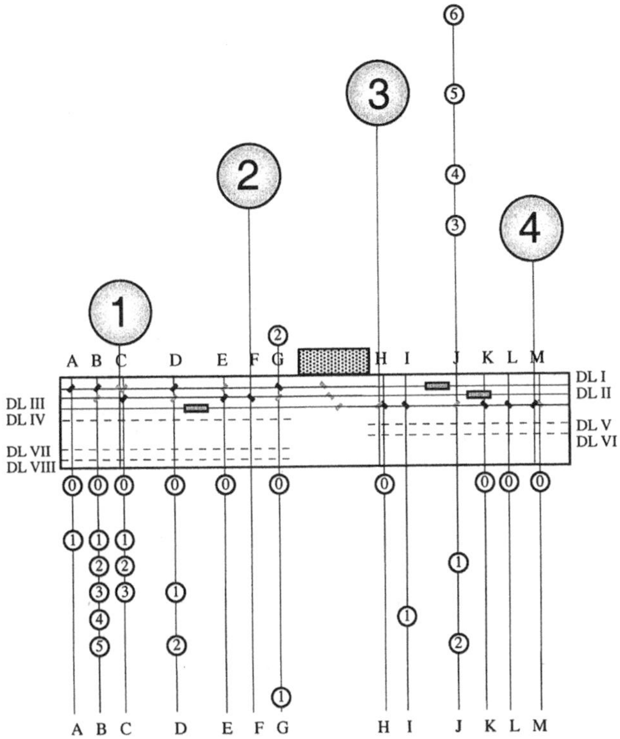

Fig. 1. General lay-out of the VLT Interferometer. Marked are the positions of the 4 UTs, the 30 AT stations, the systems of light ducts (vertical lines in the figure), the delay line tunnel, and the interferometric laboratory.
Figure available in colour on the CD-ROM

the Observatoire de Paris (France) in collaboration with ESO. Strictly speaking VINCI was designed to be used as a test instrument only. It achieved first fringe detection in March 2001 (see ESO PR 06/01, 2001). VINCI is currently being used to characterize the AT stations, as well as the optical, mechanical and software subsystems of the interferometer. It will also be used to commission new subsystems as they come on-line, such as for instance the fringe tracker and the adaptive optics system. Perhaps the main characteristic of VINCI is the fact that it uses optical fibers to combine the two beams, using a scheme successfully implemented at the IOTA interferometer (Coudé du Foresto et al. 1998). The fibers act also as spatial filters, selecting only the light within the Airy disk of each telescope, where the coherence information is higher, and they output a gaussian-shaped wavefront with only tip-tilt as a major distortion. This has the cost of sacrificing some sensitivity, but results in excellent stability and precision of the instrumental visibility, with accuracies which are expected to be at the 10^{-4} level. It should be stressed that for such performance, the main limitation will actually be imposed by the quality of the calibrators. A program to establish a database of VLTI calibrators has been started and is expected to last for some years.

MIDI is the VLTI instrument designed to operate in the mid-IR. It is being developed jointly by a consortium of German, Dutch and French partners led by the Max-Planck-Institut für Astronomie in Heidelberg (Leinert et al. 2000). It will work at 10 and 20μm, with some spectroscopic capabilities. Most of the MIDI subsystems have already been manufactured, and integration is to begin by summer 2001. It is expected that the instrument will be completed and tested by spring 2002, and transported to Paranal shortly thereafter. MIDI will have an initial test period with the siderostats. However, due to the intrinsic characteristics of the thermal IR background and detectors, it will need large apertures to perform observations of scientific significance, and it is expected to be operated mostly with the UTs (already from summer 2002), and with the ATs for a limited number of programs which target expecially bright sources. One of the challenges imposed by MIDI is the very high data rate, given that individual exposures will have to be limited to few tens of milliseconds by the background brightness. Continuous rates of about 2Mb/s are expected, and will require special care to be dealt with by the pipeline and archive facilities of ESO. It will also be one of the first interferometric beam-combiners to operate in the thermal IR in a classical way, as opposed to instruments based on heterodyne detection (the only approach used until now), such as the ISI interferometer (Monnier et al. 2000).

In the near-IR, the facility instrument which will be offered by the VLTI is AMBER, which is being built by a consortium of French, German and Italian institutes based mainly, but not only, in Nice, Grenoble, Bonn and Florence (Petrov et al. 2000). AMBER will cover the 1 to 2.4μm range, with a range of medium and relatively high spectroscopic dispersions, up to $R = 10^4$. It is noteworthy that this instrument is designed to combine 2 or 3 telescopes. In this latter case, closure phases can be measured, which are fundamental if one wishes to attempt actual imaging (as opposed to the mere measurement of visibilities which is the classical product of standard interferometers).

For what concerns sensitivities, the performance of AMBER and MIDI will be strongly dependent on the quality of atmospheric seeing and coherence time, as well as on the availability and operation of subsystems such as the fringe tracker, the AO and PRIMA (see Sect. 3). Limiting magnitudes in long (cumulative) integrations for MIDI are close to N=8 (with fringe tracker on the UTs), and close to K=13 for AMBER (AO on the UTs). These limits are for self-tracking on the scientific source. The dual feed facility PRIMA would allow to push these limits significantly.

The instruments will be installed in an underground laboratory, expecially designed to be clean (class 30,000) and with a constant temperature (to within 0.1°C). For this, all instruments are designed to be operated remotely from the control room, and will not need human intervention during the night. The same applies of course to the delay line tunnel. A layout of the interferometric laboratory is given in Fig. 2. Also a coherent reference light source (LEONARDO)

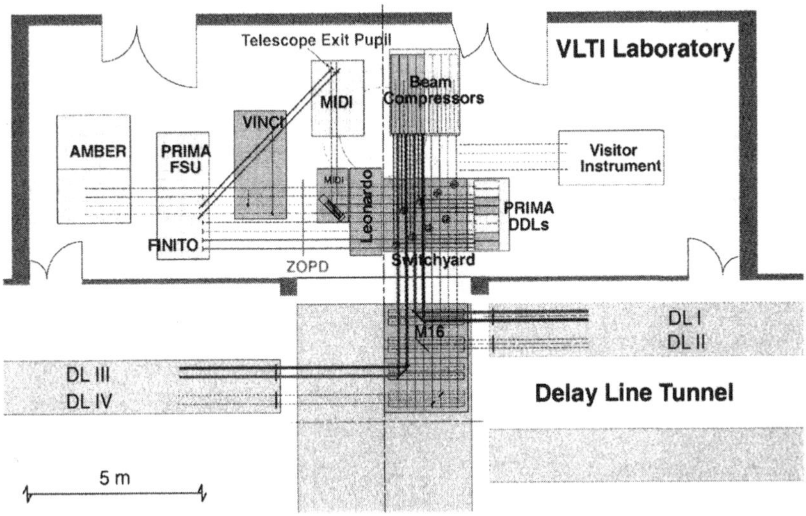

Fig. 2. Lay-out of the underground interferometric laboratory. The beams from the Delay Line tunnel (bottom) are propagated to the switchyard (center), and from there they are sent to the various optical tables by motorized mirrors.
Figure available in colour on the CD-ROM

is present. It should be noted that space has been reserved (to the right of the laboratory in Fig. 2) for the possibility of visitor instruments. One first such instrument could be the GENIE nulling instrument, which is currently being studied by the European Space Agency and that should serve as a technology demonstration for the Darwin space interferometer (Fridlund 2000).

3 Future Developments

The VLTI is following an implementation plan that is characterized by the progressive introduction of several subsystems. The main milestones in this plan are summarized in Fig. 3, and are briefly discussed in the following.

As this text is being written, the beam compressors are being installed, which will permit to make full use of the aperture of the siderostats (from first fringes until now, only about 10 cm could be used). By the end of 2001, first fringes will be attempted on the 8.2 m telescopes. It must be noted however that by this time the UTs will not have a Coudé AO system yet; therefore this milestone will have only demonstration purposes, and almost no scientific use is foreseen.

The FINITO fringe sensing unit will permit to stabilize the fringes, using light at a different wavelength than that used for the scientific observation. This will permit to use longer integration times on all instruments, thereby extending their sensitivity. It is foreseen to be completed and installed by mid-2002. By that time, also MIDI should be delivered on Paranal. By end of 2002, it will be

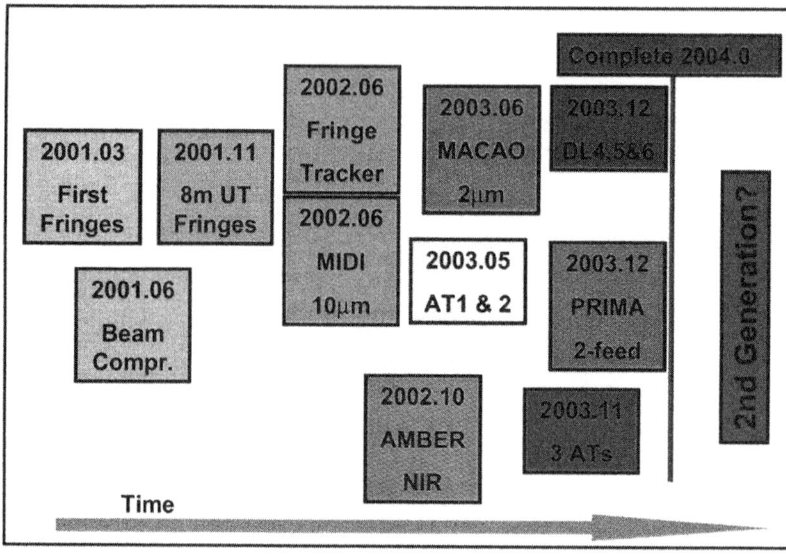

Fig. 3. Timeline of the progressive implementation of the VLTI subsystems. Concepts for second generation instrumentation are currently being examined.
Figure available in colour on the CD-ROM

joined by AMBER, which however will not be able to access the UTs until an AO system will be available. On the contrary, since it is expected that the good seeing of Paranal and fringe tracking will deliver a good image quality in the thermal IR, MIDI could use the UTs almost from the beginning.

The next step will be the delivery of the first two 1.8 m AT telescopes, which will enable AMBER to increase significantly the range of its scientific observations. A major milestone will then be represented by the introduction of the MACAO-VLTI AO system (Donaldson et al. 2000), which will provide image quality sufficient for the interferometric use of the UTs (initially UT2&4, subsequently UT1&3) also in the near-IR.

Subsequently, it is foreseen that the VLTI will be completed with the delivery of the third AT (enabling AMBER to combine three beams), the installation of additional delay lines (which will permit to extend the range of AT stations that can be combined), and finally with the completion of PRIMA. This is the VLTI dual feed facility, which will permit to select a nearby (typically $\lesssim$30 arcsecs) bright source for fringe tracking in one beam, and use a second beam to feed a fainter source to the scientific instruments. In this way, the limiting sensitivity will be pushed by several magnitudes. It is expected to reach K$\approx$20 under favorable conditions.

By the end of 2003, the VLTI should be completed in all its subsystems and operate to the full extent permitted by AMBER and MIDI. Beyond that, concepts for future extensions of the instrumentation and/or of the interferometer itself are being considered.

4 Studying Young Stars and Exoplanets with the VLTI

With its wide range of baselines, its set of instrumentation and their predicted performance, the VLTI will be a key tool for studies of young stars and exoplanets. Reviews have been given by Glindemann et al. (2000a) and Richichi (2001).

Here we only mention that one of the main areas of research in which interferometers are going to be used, is the detection and study of orbital motions in star-star and star-planet systems. Such motions are of course a proof of the existence of the companion (in the case of exoplanets, this constitutes a fundamental counterpart to the detections by radial velocities), but they also represent the only way to determine the masses of the system. Masses of young stars are until now very poorly known.

In order to detect orbital motions with some accuracy also over relatively short periods of time, a quick computation shows that it is necessary to achieve angular resolutions much smaller than the pure diffraction limit of hectometric baselines provided today by the largest interferometers.

The VLTI can achieve this, in at least two different ways. On one hand, it is possible to detect small changes in the visibility of a binary system, when high accuracy and stability is achieved. On the other hand, given a bright reference star nearby, the VLTI equipped with PRIMA will perform micro-arcsecond astrometry and detect the reflex motion of the primary in a binary system. Both methods have been outlined by Richichi (2001).

References

1. R. Donaldson et al. 2000, in SPIE Proc. Vol. 4007, Adaptive Optical Systems Technology, ed. P. L. Wizinowich, p. 82
2. V. Coudé du Foresto et al. 1998, in SPIE Proc. Vol. 3350, p. 856
3. ESO PR 06/01 2001, `http://www.eso.org/outreach/press-rel/pr-2001/pr-06-01.html`
4. C.V.M. Fridlund 2000, in SPIE Proc. Vol. 4006, Interferometry in Optical Astronomy, ed. P. Lèna & A. Quirrenbach, p. 762
5. Glindemann A., Delplancke F., Kervella P., Paresce F., Richichi A., Schöller M., 2000a, Proc. of XXIV IAU Assembly, Penny A.J. et al. (eds), in press
6. A. Glindemann et al. 2000b, in SPIE Proc. Vol. 4006, Interferometry in Optical Astronomy, ed. P. Lèna & A. Quirrenbach, p. 2
7. C. Leinert et al. 2000, in SPIE Proc. Vol. 4006, Interferometry in Optical Astronomy, ed. P. Lèna & A. Quirrenbach, p. 43
8. J.D. Monnier et al. 2000, in SPIE Proc. Vol. 4006, Interferometry in Optical Astronomy, ed. P. Lèna & A. Quirrenbach, p. 574
9. Petrov, R., et al. 2000 in SPIE Proc. Vol. 4006, Interferometry in Optical Astronomy, ed. P. Lèna & A. Quirrenbach, p. 68
10. Richichi A. 2001, in Proc. of Workshop on Young Stars Near Earth, T. Greene and R. Jayawardhana (eds.), in press

Fabien Malbet, Nicolas Lodieu, Jérôme Bouvier,
Jean-Louis Lemaire, and Mónica Rubio

Observing Young Stellar Objects with the VLT Interferometer

Fabien Malbet

Laboratoire d'Astrophysique, Observatoire de Grenoble,
BP 53, 38041 Grenoble cedex 9, France

Abstract. Young stellar objects is one of the most exciting topic to be handled with the VLTI. The magnitudes of these stars are too high for current optical interferometers limiting the studies to a few of them but are well within the capability of the VLTI. The milli-arcsecond spatial resolution of the VLTI allows to probe the very close environment of young stars down to a tenth of an astronomical units. I review the different aspects of star formation that can be tackled by the VLTI: circumstellar disks, multiplicity, jets, etc. I also present recent observations performed with the two infrared interferometers IOTA and PTI and show how the VLTI is able to extend these studies.

The general infrastructure of the VLTI is described by Richichi et al. in this volume. The VLTI is composed of 4×8-m fixed telescopes (UTs) with a maximum baseline of 130 m and 3×1.8-m movable telescopes (ATs) with a 200-m maximum baseline. For the first phase of development, the ATs are equipped with tip-tilt correction and UTs with 60-actuator curvature-sensed adaptive optics

The performances of the instruments of the first generation, AMBER the near infrared one, and, MIDI the thermal infrared one, are reported in Table 1 reported together with the characteristics of these instruments relevant to the field of star formation.

1 Areas of Investigation

Using the characteristics of the VLTI and its instruments, it is possible to draw the limits of the parameter space that we are able to study in the main star formation regions:

- Spatial scale: 0.1 – 10 AU from the central star;
- Temperature range: 300 – 4000 K;
- Velocity field: down to 50 km/s.

With these numbers in mind, a large palette of physical phenomena can be investigated.

Multiplicity. Between 50% and 80% of young stars are expected to be multiple. The VLTI fills the gap between spectroscopic binaries and sub-arcsecond binaries. Many spectroscopic binaries are resolvable by the VLTI leading to the determination of their mass.

Table 1. Characteristics of the science instruments AMBER and MIDI

	AMBER	MIDI
Spectral coverage	J, H, K' $(1 - 2.4\ \mu\text{m})$	N $(8 - 12\ \mu\text{m})$
Spectral resolution	35, 1000, 10000	100
Beams combined	3	2
Lim. mag. with UTs (ATs)	$K = 13$ (9.8)	$N = 5$ (1.8)
Field of view	$0.06 - 0.24''$	$0.26 - 1.14''$
Min. fringe spacing (λ/B)	1-2 mas	10 mas

Circumstellar disks. In the region accessible by the VLTI, the thermal emission dominates the global radiation since the scattered light is spread out over a larger area. See Sect. 3 for further details.

Stars. Young stars are in average located further than 100 pc from the Sun and therefore are barely resolved. Even if there is a slight visibility decrease at large baselines, it is difficult to disentangle it from the effect of the disks and other phenomena. However in weak-line T Tauri stars, for which the environment is believed to be clean, there is a chance to determine directly their angular diameters and therefore to constrain the photosphere effective temperature.

Ejection. Jets and outflows are often collimated at large distances from their energy source. The VLTI offers a chance to investigate the 1 AU zone around the stars and to actually see the origin of the ejection phenomena: star wind, disk wind, X wind, etc.

Accretion. The keplerian motion of the disks is already detected in the spectrum of FU Orionis objects [1]. The VLTI add to this velocity information an important spatial localization using differential interferometry. Using accretion tracers like *veiling* of near-infrared photospheric line, it is possible to identify if this accretion process occurs on the star or at the disk inner boundary.

Formation of planets. The disk around young stellar objects is believed to be the progenitor of planets. A possible process to form planets is through gravitation waves developing in the disk. The VLTI is able to detect brightness structures in disks with a dynamic range larger than 100. The newly formed planets will also create gaps in the disks that can be detected by a lack of emission.

Stellar formation scenarii. The gain in sensitivity of about 6 magnitudes compared to current infrared interferometers allows to extend the observations to a large number of objects: classical T Tauri stars which are believed to have an accretion disk, weak-line T Tauri stars for which the disk would have disappeared, FU Orionis objects that undergoes luminosity outbursts and young stars of intermediate mass like Herbig Ae/Be stars.

In conclusion, the VLTI will investigate a completely new region around young stellar objects where many physical processes occur. We will certainly obtain new constraints on the nature of the viscosity in accretion disks, clarify the role of the magnetic field, understand the growth of dust grains, and connect the accretion and ejection processes.

Many of these subjects have been addressed during the meeting and are included in this volume: for example Garcia et al. showed how VLTI with AMBER can probe the launching part of stellar jets, Guenther et al. presented a strategy to observe pre-main sequence binaries, and, Wolf et al. a prospective study for detecting gaps in protoplanetary disks with MIDI. A number of posters have also been presented at this meeting by Dullemond et al., Lachaume et al., Pascucci et al. and Weigelt et al.

2 Expected Progress with the VLTI

In terms of spatial resolution, the VLTI has performances comparable to the ones of current infrared interferometers. In terms of coverage of the (u, v) plane, the VLTI is better than others with four 8-m telescopes in a non redundant array and with its 30 stations for the auxiliary telescopes.

However, the big leap comes from the sensitivity and the use of adaptive optics in a good site. At most we expect to have a gain of 6 magnitudes. As explained by Malbet & Bertout [2], the sensitivity of an interferometer is connected to the total flux of the unresolved part of the source. In young stellar systems in the near infrared, most of the flux comes from the central part of the system which hosts the star. Its magnitude can be directly compared to the limiting magnitudes of the instruments.

Up to now, fringes have been detected only on few YSOs: FU Ori by [4]; AB Aur by [5]; T Tau, MWC 147, SU Aur and AB Aur by [6]; and 15 Herbig Ae/Be stars by [7]. All these observations have been carried out with PTI and IOTA and a limiting magnitude $H, K \leq 6$. Figure 1 shows the histogram of V and K magnitudes of stars from the Herbig & Bell Catalog [3]. The VLTI sensitivity is increased both with the ATs and the UTs: with a limiting magnitude of $K = 10$ for fringe detection without off-axis fringe tracking on the ATs, the VLTI can access to 80% of the sample of known YSOs and therefore opens an era where statistical studies can be performed. We are therefore able to start surveys and statistical studies and therefore correlate the observations with indicators of accretion, of binary frequency, veiling, etc.

A large range of issues can be addressed by the VLTI and its instruments. Almost all YSOs are observable and with a large number of different observational configurations. However the wavelength coverage does not cover the visible domain; the field of view is limited to compact objects; the data do not lead to an image like with radio interferometers. Star formation is therefore a major scientific driver for second generation VLTI instruments.

3 Protoplanetary Disks

I focus now my presentation on the particular topic of protoplanetary disks. These disks which are intermediate states between protostellar disks when the star is not yet formed but the collapsing cloud has the shape of a disk and the disk of dust that is found around main sequence stars. A typical disk is made of

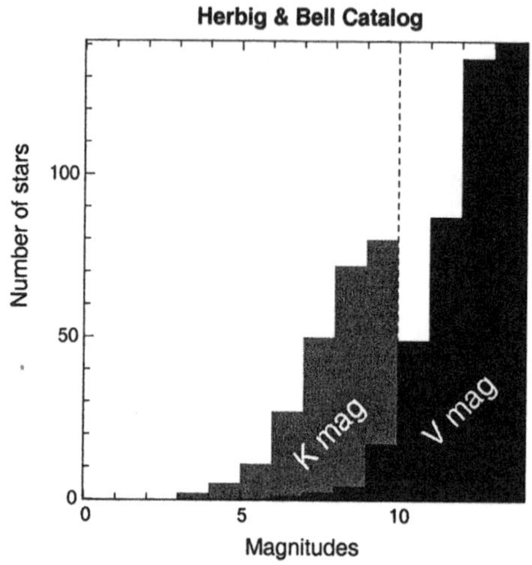

Fig. 1. Histogram of the K- (in gray) and V- (in black) magnitudes for the young stellar objects of the Herbig and Bell Catalog [3])

a mixture of gas and dust, geometrically thin but optically thick in the visible and infrared wavelengths yet optically thin in the millimeter domain. Accretion is one of the main heating process, even if reprocessing of stellar light in the disk is not negligible. Some stellar light is being scattered by the upper layer of the disk photosphere.

With visible and infrared interferometers, one probes the inner region of the disk between 0.1 and 10 AU. In this region for the infrared domain, the thermal emission of the disk dominates compared to the scattered light. We therefore observe the brightness distribution of the photospheric layer of the disk. This intensity distribution depends on the physical conditions found inside the disk and governed by the density and temperature spatial distributions.

The observables that can be extracted from interferometric are mainly visibility amplitudes and closure phases. The visibilities are nothing else than the Fourier Transform of the brightness distribution of the object taken at the spatial frequency observed $\mathbf{b}/\lambda$, where $\mathbf{b}$ is the interferometer baseline projected onto the sky at the object position. The VLTI is able to measure the visibility amplitudes, the differential phases between two wavelengths and for AMBER the closure phases. Closure phase is the addition of the phases measured simultaneously on three baselines when observing with three telescopes. It is free of atmospheric errors and is sensitive to objects which are not centrosymmetric. This is especially interesting when observing disks with a large vertical structure (see poster by Lachaume et al. for more information in these proceedings).

Figure 2 presents the results obtained on FU Orionis [4,8]. We observed this object both in H and K bands. The object is clearly resolved and is best interpreted by the presence of an accretion disk [4]. The color information can help us to constrain the radial distribution of the effective temperature $T(r)$.

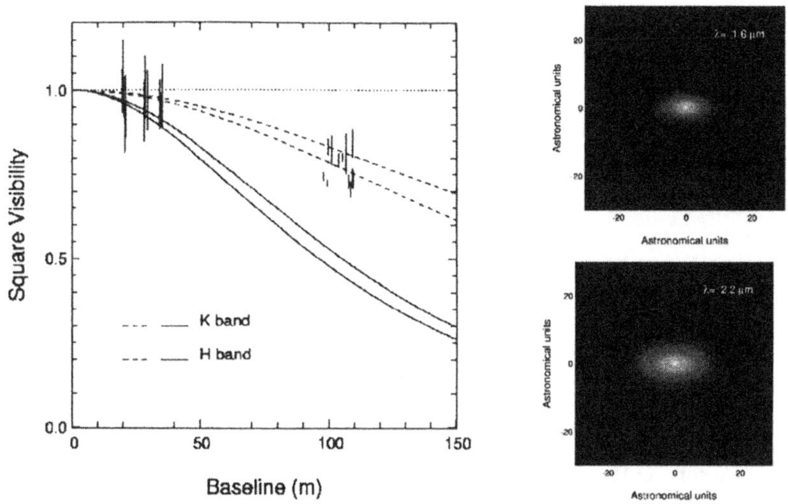

Fig. 2. Left panel: Observations of FU Orionis carried out on PTI and IOTA in H (blue) and K (red) bands. Right panels: Simulated disk brightness distribution in H (top) and K (bottom). Blue and red lines are the visibility curves derived from the simulations displayed in right panels. Dashed and solid lines corresponds to 2 perpendicular position angle of the disk

The disk emits light which is the sum of rings emitting like black bodies of temperature $T(r)$. The temperature radial distribution is modeled by a power law: $T(r) \propto r^{-q}$. Since the visibilities measured come mainly from the peak emission from the black-bodies, then the radius of a Gaussian brightness profile would be proportional to λ^{-q}. From the measurements both in H and K, we can estimate the power law coefficient q:

$$q = \frac{\log(\frac{V_K^2}{V_H^2})}{\log(\frac{\lambda_K}{\lambda_H})}$$

From our observations we derive $q \sim 0.4$ when the model fitting of spectral energy distributions leads to $q \sim 0.75$. We interpret this mismatch by the fact that the spectral energy distribution is sensitive to the disk mainly at $\lambda \geq 10$ μm, i.e. at radius larger than 1 AU when our interferometric observations probes the inner 1 AU of the disk. If it is the case then, this would mean that the energy dissipation due to the accretion in the inner part of disk is less than in the outer part and could come from the conversion of some energy in the launching of a strong bipolar jet as proposed by [9].

In conclusion, we constrain the physical conditions in disks with different techniques, but infrared interferometry is the only current technique which can constrain the inner 10 AU. The first detailed results on FU Ori are consistent with the model of an accretion disk but environments are usually more complex. We need to improve the data set: more spectral coverage and improved (u, v)

coverage. The VLTI will undoubtedly boost these studies because of its higher sensitivity, better (u, v) coverage, access to closure phases and maybe in the future imaging? It is also essential to prepare models in order to be ready to interpret the observations.

References

1. L. Hartmann, S.J. Kenyon: ARA&A **34**, 207 (1996)
2. F. Malbet, C. Bertout: A&AS **113**, 369 (1995)
3. G.H. Herbig, K.R. Bell: Lick Obs. Bull. **1111** (1988)
4. F. Malbet, J.-P. Berger, M.M. Colavita, C. Koresko, C. Beichman, et al.: ApJ **507**, L149 (1988)
5. R. Millan-Gabet, P. Schloerb, W. Traub, F. Malbet, J.-P. Berger, et al.: ApJ **513**, L131 (1999)
6. R.L. Akeson, D.R. Ciardi, G.T. van Belle, M.J. Creech-Eakman and E.A. Lada: ApJ **543**, 313 (2000)
7. R. Millan-Gabet, P. Schloerb, W. Traub: ApJ **546**, 358 (2001)
8. J.-P. Berger, F. Malbet F., M.M. Colavita, D. Ségransan, R. Millan-Gabet., W. Traub: 'New insights in the nature of the circumstellar environment of FU Ori'. In *Interferometry in Optical Astronomy, SPIE'2000, Munich, Germany* ed. by P. Léna, A. Quirrenbach A., vol. 4006, pp 597
9. J. Ferreira, G. Pelletier: A&A **295**, 807 (1995)

Interferometric Tunnel

Eike Guenther and Beate Stelzer

Preparing for the VLTI: A Search for Pre-Main Sequence Spectroscopic Binaries

Eike Guenther[1], Viki Joergens[2], Guillermo Torres[3], Ralph Neuhäuser[2], Matilde Fernández[4], and Reinhard Mundt[5]

[1] Thueringer Landessternwarte Tautenburg, 07778 Tautenburg, Germany
[2] Max-Planck-Institut für extraterrestrische Physik, 85741 Garching, Germany
[3] Harvard-Smithsonian Center for Astrophysics, Cambridge, MA 02138, USA
[4] Instituto de Astrofísica de Andalucia, Apartado 3004 18080 Granada, Spain
[5] Max-Planck-Institut für Astronomie, Königstuhl 17, 69117 Heidelberg, Germany

Abstract. The most fundamental parameter of a star is its mass. For low mass pre-main sequence (pms) stars this parameter is practically always derived by comparing the location of the star in the Hertzsprung-Russell diagram with theoretically calculated evolutionary tracks. In order to really test the tracks, determination of the masses of a number of pms-stars are necessary. Up to now, masses could only be derived for a few individual cases. In the near future the VLTI will allow us to determine the masses of many pms-stars with high accuracy. In order to prepare for the VLTI-observation, we have begun a survey for long period pms-spectroscopic binaries that should be resolvable with the VLTI. We find that 8% of the stars surveyed show significant radial-velocity variations, and are thus most likely spectroscopic binaries. In addition to these, 6% of the targets are double-lined spectroscopic binaries. The second aim of the survey is to find also all eclipsing binaries amongst these stars, and we have in fact found a highly promising candidate. An eclipsing binary announced previously by other authors turned out to be a spotted star.

1 Introduction

The most fundamental parameter of a star is its mass, which determines almost everything about its birth, life, and death. For low mass pms-stars this parameter is practically always derived by comparing the location of the star in the Hertzsprung-Russell diagram with theoretically calculated evolutionary tracks. The problem, however is that there are considerable differences for tracks published by different authors. The differences are especially notable for stars of lower mass. It is thus necessary to test the evolutionary tracks by determining the true masses of number of pre-main sequence (pms) stars. Direct determinations of the mass are possible in several ways:

- Determine the inclination from the light-curve in an eclipsing binary system and then measure $m\,sin^3\,i$ spectroscopically if it is an double-line spectroscopic binary (SB2) system.
- Determine the total mass of the system from the relative astrometric orbit and then derive the mass-ratio from spectroscopic observations.

- Determine the masses of the components of a visual binary system from the absolute astrometrical orbit.
- For classical T Tauri stars (cTTS) the inclination and rotation velocity of circumstellar disk can be determined from mm-interferometry. If it is then assumed that the mass of the disk is negligible, and if the distance of the object is known, the mass of the central star can be estimated.

The problem in determining the masses is that up to now only very few eclipsing binary system is known where the pms-stars have masses of about one solar mass (Covino et al. 2000, Joergens et al. 2001). Objects with masses much higher than this are not very suitable for testing the evolutionary tracks. The method for determining the masses of cTTS by measuring the rotational velocity of matter in the circumstellar disc is suitable for many objects but is not very precise. Measuring the masses of visual binaries has the problem that the distance to even relatively nearby star-forming regions is about 150 pc. Thus the periods of visual pms-binaries are decades, even if speckle interferometry or AO-systems are used. Thus, only the average mass of the whole sample can be estimated statistically, given the relatively small time-basis that is available up to now (Woitas et al 2001). With lunar occultations it is in principle possible to resolve systems with separations of 5 mas, bringing system with periods less than half a year into view. However, the problem with this method is that only a few nearby star-forming regions are within the path of the moon. Encouraging is that a number of young stars have recently been found that have distances that are much smaller than 150 pc. The TW Hydra association for example, has a distance of only 55 pc. Thus, Neuhäuser et al. (2001) were able to resolved a spectroscopic binary (TWA-5 A) in this association using Gemini North. The period of this system is presumably of the order of 10 years or so. Hipparcos has also proven to be an extremely powerful tool for determining stellar masses (Martin et al. 1998). If the astrometric motion of the photo-centre is combined with the spectroscopic orbits, it is possible to determine the masses down to the brown dwarf regime (Zucker & Mazeh 2000). However, for pms-stars the sensitivity, and accuracy of the future missions DIVA, FAME, or GAIA are needed.

2 Searching for More Eclipsing Binaries

Thanks to the success of the ROSAT mission, a large number of new pms-stars have been found. How many eclipsing binaries can be expected among these stars? Typically 4% of the pms-stars are spectroscopic binaries with periods of less than 10 days. Given the probability for an eclipse for a pms-binary of such a period is about 20%, we estimate that about 1% of the newly discovered pms-stars are eclipsing. However, since binaries are generally brighter in X-rays than single stars, the binary frequency for ROSAT sources is generally higher than for optically selected stars. As pointed out before, Covino et al. (2000) already found an eclipsing double-lined spectroscopic (SB2) binary amongst the young stars discovered by ROSAT in Orion. Wichmann et al. (1998) published indications for another eclipsing binary in Lupus. However, through spectroscopic and photometric observations we could demonstrate that this object is not an eclipsing binary but a spotted star (Joergens et al. 2001). Another very promising candidate was recently discovered by us (Fig. 1). This object clearly is a pms-star. The also shows large radial velocity (RV) variations, and the light curve resembles that of an eclipsing system.

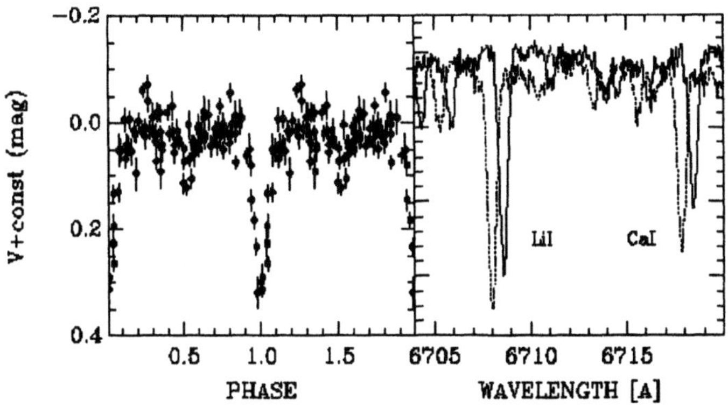

Fig. 1. Newly discovered eclipsing pms-star. The left panel shows the phased light-curve. The primary and secondary eclipse is seen. The right panel shows the part of the spectrum taken at different epochs (barycentric wavelength). The RV of the star differs by 26 km/s in the two spectra

3 Preparing for the VLTI: A Spectroscopic Survey

While photometric and spectroscopic observation possibly yield to a few more low-mass eclipsing pms-binaries, it is desirable to have accurate masses of a much larger sample of stars. This should be possible with the VLTI, as the VLTI will allow to resolve systems with a separation of 3 mas at 2.3 μm which brings pms-binaries with periods down to 100 days in nearby star-forming regions into view. The VLTI offers two possibilities for the mass-determination. Using AMBER the relative distance of the two components can be measured, which allows the determination of the total mass of the system. Given the typical $v \sin i$ values of pms-stars, binaries with periods of a few hundred days could also be resolved as SB2s. This would allow us to derive the mass ratio, and hence the masses of the stars individually. The second possibility is to determine the absolute astrometric orbit of stars by using PRIMA, which would then give the masses of the two stars right away. Since a fishing expedition for suitable binaries is too time-consuming on the VLTI, it is necessary to find the targets beforehand. For the AMBER

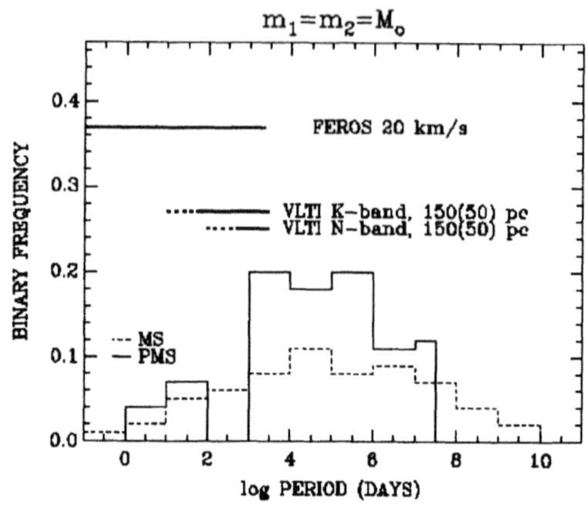

Fig. 2. Distribution of known pms-binaries in Taurus Auriga, and of old stars in the solar neighbourhood. The bars indicate, which of theses binaries could be resolved with the VLTI, and spectroscopically with FEROS and the Echelle spectrograph in Tautenburg

observations, we need SB2s with periods of more than 100 days in star-forming regions that are distances of not more than 150 pc. For two stars of one solar mass in a circular orbit with a period of two years, the semi-amplitude of the radial velocity variations would be 15 km/s for an inclination of 90 degrees. Given the typical *v sin* values of young stars of 10 to 20 km/s, the spectral lines of the two components could be disentangled. However, long period binaries always have elliptical orbits. For a typical eccentricity of 0.3, and a typical inclination of 45 degrees, the semi-amplitude would be about 11 km/s. Thus, binaries with periods up to years could at least at periastron be resolved spectroscopically. Fig. 2 shows the distribution of pms-binaries in Taurus Auriga and of old stars in the solar neighbourhood. From this diagram, we estimate that 10 to 20% of all stars are binaries of suitable periods. Binaries with periods between 100 days and a few years would not only be suitable for observations with AMBER but also for the observations with PRIMA. Since the brightness difference for objects of different mass is far less in the K-band than in the optical, most of the SB1-systems will be resolved into the two components when observed with the VLTI. The difference between objects suitable for AMBER and PRIMA is that the objects for the observations with AMBER have to be resolved in to SB2s. For PRIMA it is just necessary to know that this object is a binary with a suitable period.

How to find suitable targets for the project? Since long-period spectroscopic binaries have eccentric orbits, and since the periastron-crossing is only of very short duration, it is highly unlikely to actually see both components in the spectrum, when one spectrum is taken at a random moment in time. The only suitable way to find theses objects is to look for RV-variations. We thus have initiated a survey for pms-SBs located in nearby star-forming regions (Chamaeleon, Tucanae, Lupus, Scorpius-Centaurus, ρ Oph, TW Hydrae, Taurus-Auriga, Corona Australis) using the high resolution Echelle spectrographs on the ESO 1.5m-telescope (FEROS), and on the 2-m-Alfred-Jensch telescope in Tautenburg. Initially, the survey contained 250 pms-stars but 30% of the stars had to be disregarded, because the v sin i was to high. An other problem is that the intrinsic RV-variations of young stars due to stellar activity are typically 2 km/s. In order to be on the safe side, we thus consider only objects that show RV-variation of more than 10 kms^{-1} as SB candidates.

4 Results from the Spectroscopic Survey

In order to detect SB-candidates on the basis of the RV-variations, we have carried out 4 campaigns over a two year period using FEROS. Unfortunately, the fourth campaign was severely effected by bad weather. A grand total of 200 hours of observing time with the Echelle spectrograph on the 2-m-Alfred-Jensch telescope in Tautenburg was also invested in order to study sources that where accessible from there. With FEROS, we have detected 9 SB2, and 11 objects that show RV-variations larger than 10 km/s. Additionally, there are 22 objects which show RV-variations between 3 and 10 km/s. We expect that most of theses stars will also be binaries but it might be possible that some of them are just extremely active stars. The Tautenburg data-set additionally contains 5 new SB2s, and 7 SB1s. Additionally, there are two objects where only two spectra where taken, and where the RV measured is completely inconsistent with membership in that star-forming region. Given the time lag of the observation, and the fact that almost all objects were observed three times, we estimate that we found more than 90% of the binaries which have roughly equal masses, and have periods in the interesting regime.

Even before the start of the VLTI observations, interesting conclusions can be drawn from the spectroscopic data. Due to tidal interaction, short period binaries have circular orbits, whereas long-period binaries have elliptical ones. Theories of tidal interaction can be constrained if the position of the boundary between circular and elliptical orbits is measured for objects of different age. According to Zahn & Bouchet (1989), most of the orbital circularisation of low-mass stars occurs during the pms-phase of evolution. In contrast to this, Mathieu (1992) found that the transition period between circular and eccentric orbits is between 4 and 5 days for pms-stars, and thus much less than the 11 days of old stars. However, we recently found an a pms-binary with a circular orbit with a 7.5 day period, thus confirming that orbital circularisation happens quite early in the evolution of a binary system. The spectra can also be used in order to test whether the mass-ratio derived for SB2-systems are consistent with the evolutionary tracks. We have done this for RXJ1603.8-3938. For this object we derive a mass-ratio of 0.9266 ± 0.0063 from the orbit and 0.85 from the evolutionary tracks of D'Antona and Mazzitelli (1994) (Guenther et al. 2001).

5 What can be Expected from the VLTI Observations?

Before embarking on such a project, one should first have a look at the feasibility of the VLTI observation. Since almost all objects are in the brightness-regime between 6th and 10th magnitude in K, they could easily be observed with AMBER using only the auxiliary telescopes. If an off-axis reference star is available, it will even be possible to take spectra with a resolution of up to 10000 of most of the targets. Since the brightness difference of stars of different mass is less in the infrared than in the optical, most of the SB1s will be visual binaries when observed with the VLTI in the IR. We will thus certainly be able to get IR photometry of each of the stars individually, and possibly even spectra. The spectra would allow us to determine the veiling in the case of cTTS so that we will be able to place also these stars into the HR-diagram.

After a point source, a binary is the simplest object and it is comparatively easy to derive the separation and the position angle from interferometry. For observations with AMBER+spectroscopy we estimate that the errors of the determination of the masses will typically be between 2 and 5%. If PRIMA really reaches an accuracy of 10μas, then the masses can be determined with an accuracy of better than 1.0%. Thus, we hope that the outcome of this project will eventually be the accurate masses of 30 to 80 pms-stars. We could then perform strict tests of stellar evolution tracks, and even approach the question, whether or not the tracks are the same for all star-forming regions.

References

1. Covino, E., Catalano, S., Frasca, A., Marilli, E., Fernández, M., Alcalá, J. M., Melo, C., Paladino, R., Sterzik, M. F., Stelzer, B., 2000, A&A 361, L49
2. D'Antona, F., Mazzitelli, I., 1994 ApJS 90, 467
3. Guenther, E. W., Torres, G., Batalha, N., Joergens, V., Neuhäuser, R., Vijapurkar, J., Mundt, R., 2001, A&A 366, 965
4. Joergens, V., Guenther, E.W., Neuhäuser, R., Fernández, M., Vijapurkar, J., 2001, A&A in press
5. Martin, C.; Mignard, F.; Hartkopf, W. I.; McAlister, H. A., 1998, A&AS 133, 149
6. Mathieu, R.D., 1992, in Binaries as Tracers of Stellar Formation, Duquennoy, A., Mayor, M, (eds.), Cambridge University Press, Cambridge, p.155
7. Neuhäuser R., Potter D., Brandner W., 2001, 'Observing the planet formation time-scale by ground-based direct imaging of planetary companions to young nearby stars: Gemini/*Hōkūpa'a* image of TWA-5'. In: von Hippel T., Manset N., Simpson C. (eds.) Astrophysical ages and time-scales (held in Hilo, Hawai'i, Feb 2001), ASP Conf. Series, in press
8. Neuhäuser, R., Wolk, S. J., Torres, G., Preibisch, Th., Stout-Batalha, N. M., Hatzes, A. P., Frink, S., Wichmann, R., Covino, E., Alcala, J. M., Brandner, W., Walter, F. M., Sterzik, M. F., Köhler, R., 1998, A&A 334, 873
9. Wichmann, R.; Bouvier, J.; Allain, S.; Krautter, J., A&A 330, 521
10. Woitas, J.; Köhler, R.; Leinert, Ch., 2001, A&A 369, 249
11. Zahn, J.-P., Bouchet, L., 1989, A&A 223, 112
12. Zucker, S., Mazeh, T., 2000, ApJ 531, 67

Delay Line Carriage

Alex de Koter, Carsten Dominik, and Jeroen Bouwman

From Protoplanetary to Debris Disks

Carsten Dominik[1] and Cornelius P. Dullemond[2]

[1] Sterrenkundig Instituut "Anton Pannekoek", Kruislaan 403, NL-1098 SJ Amsterdam
[2] Max-Planck-Institut für Astrophysik, Postfach 1523, D-85740 Garching

Abstract. We discuss the physical difference between protoplanetary and debris (Vega-like) disks and consider the course of evolution which leads from the first to the latter. We discuss a simple evolutionary model for debris disks and predict 20μm fluxes for future observations with VISIR.

1 Protoplanetary and Debris Disks Differ in Kind

Most stars are born with circumstellar disks, and these disks decay over time. While the massive disks around protostars with strong 2μm excess and strong submillimetre fluxes [1] only last for about 10 Myrs, tenuous cold disks have been found around much older stars. These disks are called *debris disks* and were found around main sequence stars up to 400Myr old, in a few cases even much older [2]. When one studies the decay of disks, the transition from protoplanetary to debris disks is often assumed to be a smooth one, and the measured masses of both types of disk are plotted in a single diagram and fitted with a single power law [3]. This, however, is deceptive since the two types of disks are of different physical nature, and a single parameter will not be sufficient to describe the evolution of the disk. The following table lists the most important differences.

PMS disk	**Debris disk**
warm material (2μm excess)	cold material (60μm excess)
$0.01 - 0.1 M_\odot$ (dust+gas)	1 $M_{\rm moon}$ (dust)
gas rich	gas poor
decoupling size 1 m	decoupling size $< 1\mu$m
dust dynamics: gas drag	dust dynamics: collisions & radiation
growth environment	destructive environment
disappears by growth	disappears by depletion

PMS disks are dominated by gas. Most of the mass is gas, and interaction with gas also dominates the dynamical behavior of solid bodies up to a size of typically 1 m. Such drag controlled bodies move on circular orbits in or above the disk plane. Collisions between such bodies are controlled by Brownian motion, or by differential drift relative to the gas[4,5]. These velocities are so small (cm/s to m/s), that colliding particles will stick. Collisions will therefore lead to growth. The reason why eventually these disks become invisible has most likely to do

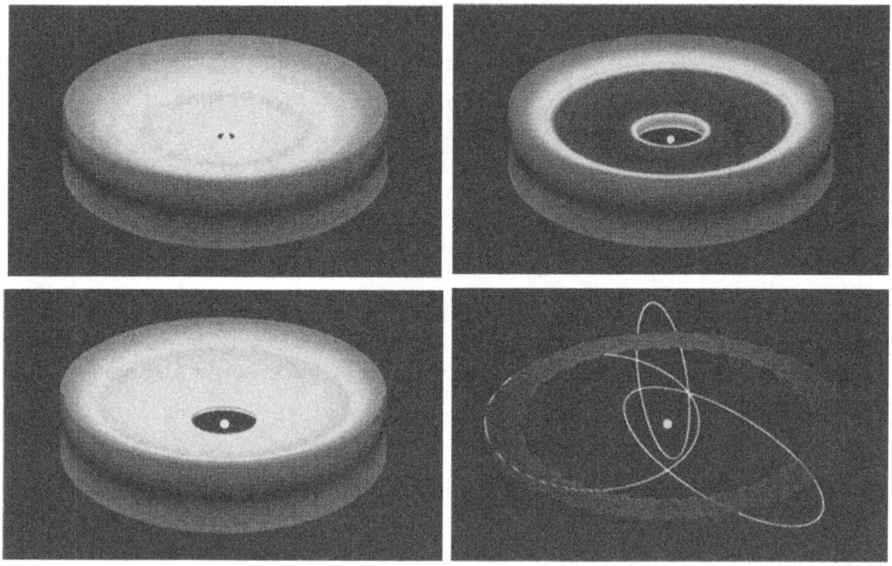

Fig. 1. Evolution of a disk from a flared irradiated disk via with gap opening, puff-up of the inner wall with shadowing to the final state of a debris disk.
Figure available in colour on the CD-ROM

with the fact that the dust growth process decreases the opacity of the solid material and therefore the ability of the disk to re-process stellar radiation.

Debris disks, on the other hand, are gas poor. Even though the gas-to-dust ratio may still be 100 as in the case of β Pic [6], the total amount of gas present in the disk is too small to be dynamically important. Dust particles and larger bodies move on modified (by radiation pressure) Keplerian orbits, and collision velocities can reach Keplerian velocities as well. Since such collisions will be destructive, the dust in such systems is not growing anymore. The reason why these disks become invisible eventually is the removal of the material and its sources (colliding planetesimals) from these systems.

The difference between protoplanetary and debris disks is therefore significant. The transition takes place when the gas in the disk is thinned out sufficiently to allow the dust particles to decouple from the gas.

2 When Accretion Stops

While a star is still actively accreting, the mass content is determined by the infall of fresh material onto the outer disk and by the viscous processes in the disk itself. When the infall of fresh material stops, material transport through the disk will still continue for some time and reduce the amount of gas in the disk. Starting with the outer regions of the disk, the energy budget of the disk will be dominated by irradiation from the star instead of the viscous dissipation. In this stage the disk will look as shown in the upper left panel of Fig. 1 [7]. In the innermost part of the disk, dust cannot survive, and eventually, as the gas mass decreases, this part will become transparent to the radiation emitted by the star. The star will then directly illuminate the dust sublimation boundary of the disk (lower left panel in Fig. 1). This inner boundary will therefore puff up and cast a shadow on part of the disk as shown in the upper right panel of Fig. 1. As we have shown [8–10], the inner boundary provides a very good match for the 3μm bump observed in the spectrum of most Herbig Ae stars. This model also makes strong predictions for the visibilities of interferometric observations with MIDI and AMBER.

Very little is known on how in detail the further development of the disk proceeds, and what observational consequences this will have. The 2μm and the submillimetre excess of these disks disappears on similar timescales [11] of about 7 Myrs. This simultaneous disappearance of both the short and long wavelength radiation is surprising. As we said above, the reason for the disappearance of PMS disk is likely the growth of particles which decreases the opacity of the disk material and therefore its ability to reprocess stellar radiation. Dust growth is very slow as long as the particles are small and suspended in the gas [4]. An important parameter for the dust growth is the dust settling time, i.e. the time needed by a dust grain to reach the midplane of the disk. In a laminar nebula it is given by [12]

$$t_{\mathrm{set}} = \frac{\pi}{2} \frac{\Sigma(r)}{\rho_{\mathrm{d}}} \frac{1}{\Omega_{\mathrm{k}(r)}} \tag{1}$$

where $\Sigma(r)$ is the surface density of the disk at distance r, ρ_{d} is the specific density of the grain material, a is the grain radius and $\Omega_{\mathrm{k}}(r)$ is the Kepler frequency. If the surface density varies as $r^{-1.5}$ as derived from the mass distribution in our solar system, then the settling time becomes independent of distance. For typical surface densities at 1 AU in a T Tauri disk of 10^3 cm/s, the settling time for a 0.1μm grain is 2.5×10^7 years. However, if the disk starts to thin out and gas is removed, the settling time drops quickly. At 10 g/cm^2, the typical time is 2.5×10^5 yr, and the removal of the small grains and growth of large bodies can proceed quickly.

When the gas is largely removed from the disk, the orbits of the formed planetesimals in the disk can become increasingly non-circular, due to gravitational stirring of the largest bodies formed in the system [13]. Small grains will be removed from the system by radiation pressure [14], and the dust particles still observed must be replenished from a population of colliding larger bodies.

3 Evolution of Debris Disks: A Simple Model

We assume that the dust produced in the system is the tail of a collisional equilibrium size distribution which is ultimately due to the destruction of a group of large bodies which we will call *comets* for convenience. Let N_p be the initial number of such bodies with collisional cross section σ_p and mass M_p. Let these bodies occupy a volume V where they move around and collide with

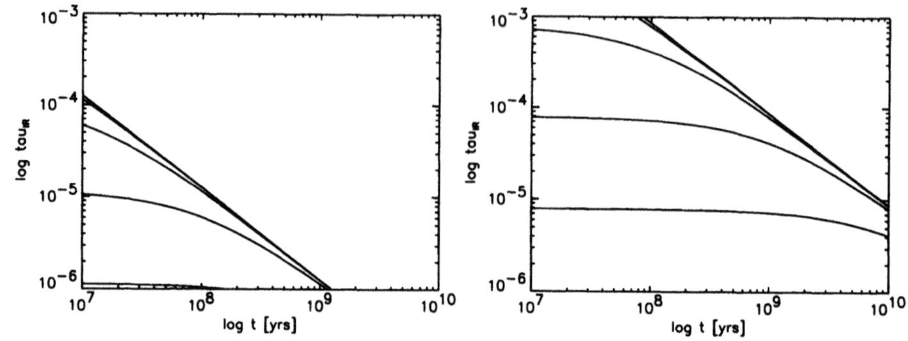

Fig. 2. τ_{IR} as a function of time. Left panel: a system with $\nu = 1$, i.e. highly eccentric orbits. Right panel: a system with $\nu = 0.1$, i.e. low-eccentricity orbits. Different curves are for different initial masses of the comet cloud. From bottom to top 0.1,1,10,100,1000 $M_\oplus$

relative velocities $v_{\rm coll} = \nu v_K$ with v_K being the Kepler velocity in the regions where the comets exist. ν will be zero for circular orbits in a single plane and approach one for highly excited orbits. If all collisions are destructive, the time derivative of the number of comets in the system is given by

$$\dot{N}_p = -\frac{N_p^2}{t_s} \tag{2}$$

where t_s is the sweeping time, the time the by the cross section of a comet moving at velocity $v_{\rm coll}$ to sweep the entire volume V. The solution of eq. (2) is obviously

$$N_p(t) = \frac{N_0}{1 + \frac{N_0}{t_s} t} \tag{3}$$

where N_0 is the initial number of comets. For times $t \gg \frac{t_s}{N_0}$, the solution becomes independent or N_0.

$$N_p(t) \rightarrow \frac{t_s}{t} \quad \text{for} \quad t \gg \frac{t_s}{N_0} \quad . \tag{4}$$

We may now proceed to compute the amount of dust grains produced by the collisional cascade of the comets. It has been show that collisional equilibrium produces a size distribution $f(m) \propto m^{-11/6}$ [15]. At the small end of the size distribution, particles are removed. The main removal processes are radiation pressure for grains which are small enough to receive an acceleration close to or higher than the gravitational acceleration of the star. Somewhat bigger grains can still be removed from the disk by Poynting Robertson drag. Which process dominates is dependent upon the dust density in the disk. If the density is high, collisional destruction dominates all the way down to small particles which are removed by radiation pressure. If the disk contains less mass, particles of

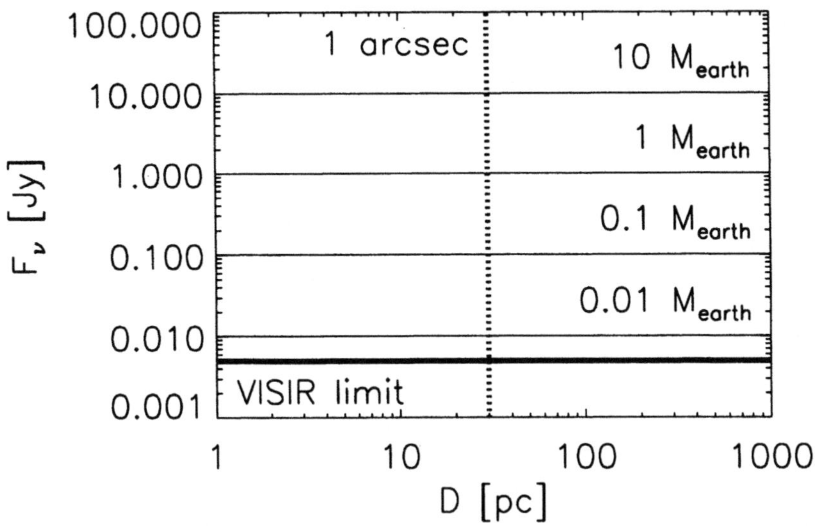

Fig. 3. Predicted fluxes levels of spatially resolved Vega-like disks with different dust masses. The 1 arcsec line marks the 30 pc distance where a typical disk (50 AU) can just be resolved at 20μm. The fluxes are computed per 0.5″ beam

a certain size have Poynting Robertson timescales which are smaller than the collisional time and are being pulled out of the cometary region towards the star. It turns out that in a system with as little mass as the solar Kuiper Belt (estimates indicate about 0.1 $M_\oplus$), Poynting Robertson drag dominates for 1μm sized grains. In the denser Vega-like systems collisions always dominate.

One can show that in a collisionally dominated disk the number of small grains is proportional to N_p while in a PR-drag dominated disk, the number of grains is proportional to N_p^2. The number of grains in the disk can then be calculated analytically and the infrared/bolometric luminosity τ_{IR} be evaluated. For a detailed discussion, see [16].

4 Predictions for VLT/VISIR

Can we observe the changes in the structure of a disk and the later slow decay of debris disks with the VLT? We have already mentioned above, that the interferometers AMBER and MIDI are very well suited to probe the structural changes in the inner disk. Historically, the first detection of debris disks was in mid-IR wavelength with the IRAS satellite [17]. Since the contrast between the stellar photosphere and the disk emission is largest at long wavelength, IR and sub-mm are the best ways to study the total emission from these disks. A few disks have also been imaged in the near-IR, in particular β Pic [18]. However this still requires excellent coronographs to remove the bright star from the image. The large spacial resolution offered by 8m-class telescopes like the VLT together with 10μm and 20μm capabilities opens another possibility to study these disks by direct imaging, unhindered by the stellar light. At 20μm, a typical Vega-like disk and the central star will be equally bright. If the system is nearby and the disk can be resolved, direct images can be obtained. The region directly around the star can be studied with only the resolution and not the size of the coronograph as the limit.

What will be possible with VISIR at 20μm in this respect is shown in Fig. 3. We have plotted the predicted flux per pixel from a dust disk between 30 and 50 AU from the star. The VISIR limit at these wavelength will be 7 mJy in one hour of integration [19]. With this detection limit, all disks with dust masses of 1% of an Earth mass can be directly detected out to a distance of 30 pc.

Acknowledgments. CD acknowledges the financial support from NWO Pionier grant 6000-78-333. CPD and CD acknowledge support from the European Commission under TMR grant ERBFMRX-CT98-0195 ('Accretion onto black holes, compact objects and protostars'). We thank Vincent Icke for help with Figure 1.

References

1. C. Lada. this volume.
2. H. J. Habing, C. Dominik, M. Jourdain De Muizon, M. F. Kessler, R. J. Laureijs, K. Leech, L. Metcalfe, A. Salama, R. Siebenmorgen, and N. Trams. Nature, 401:456–458, 1999.
3. W. S. Holland, J. S. Greaves, B. Zuckerman, R. A. Webb, C. MCCarthy, I. M. Coulson, D. M. Walther, W. R. F. Dent, W. K. Gear, and I. Robson. Nature, 392:788–790, 1998.
4. S. Kempf, S. Pfalzner, and T. K. Henning. *Icarus*, 141:388–398, October 1999.
5. S. J. Weidenschilling. MNRAS, 180:57–70, July 1977.
6. W. F. Thi, G. A. Blake, E. F. van Dishoeck, G. J. van Zadelhoff, J. M. M. Horn, E. E. Becklin, V. Mannings, A. I. Sargent, M. E. van den Ancker, and A. Natta. Nature, 409:60–63, January 2001.
7. E. I. Chiang and P. Goldreich. ApJ, 490:368+, November 1997.
8. C.P. Dullemond, C. Dominik, and A. Natta. A&A, 2001. submitted.
9. C.P. Dullemond, C. Dominik, and A. Natta el al. Passive irradiated circumstellar disks with an inner hole: Predictions for VLTI. poster paper, this volume.
10. A. Natta, T. Prusti, R. Neri, D. Wooden, and V. P. Grinin. A&A, 371:186–197, May 2001.
11. C. L. Clarke. this volume.
12. K. Miyake and Y. Nakagawa. ApJ, 441:361–384, March 1995.
13. S. J. Kenyon and B. C. Bromley. AJ, 121:538–551, January 2001.
14. P. Artymowicz. ApJ, 355:L79–82, 1988.
15. J. W. Dohnanyi. *J. Geophys. Res.*, pages 2531–2554, 1969.
16. C. Dominik. A&A, 2001. in preparation.
17. H. H. Aumann, F. C. Gillett, C. A. Beichmann, T. deJong, J. Houck, F. J. Low, G. Neugebauer, R. Walker, and P. R. Wesselius. ApJ, 278:23–27, 1984.
18. B. A. Smith and R. J. Terrile. *Science*, 226:1421–1424, 1984.
19. R. Siebenmorgen. priv. communication.

Turbulent Radial Mixing in the Solar Nebula as the Source of Crystalline Silicates in Comets

Dominique Bockelée-Morvan[1], Daniel Gautier[1], Franck Hersant[1], Jean-Marc Huré[1,2], and François Robert[3]

[1] Observatoire de Paris, F-92195, Meudon, France
[2] Université de Paris 7, 2 place Jussieu, F-75251, Paris Cedex 05, France
[3] Muséum d'Histoire Naturelle, 61 rue Buffon, F-75005, France

Abstract. We present a time-dependent model which investigates the thermal annealing of amorphous silicates in the hot regions of the solar nebula and their diffusive transport to the comet formation zone by turbulence. The model uses pseudo-time-dependent temperature and surface density profiles of the solar nebula generated with a 2-D accretion disk model based on the α-prescription for turbulent viscosity. We show that some of the solar nebulae which explain the D/H ratios measured in meteorites, long-period comets, Uranus and Neptune, namely the warmest, provide a relative mass fraction of crystalline silicates in the Jupiter-Saturn region in agreement with that measured in comet C/1995 O1 (Hale-Bopp).

1 Introduction

The origin of cometary silicates is a matter of debate. Infrared spectra show that Mg-rich olivines and pyroxenes are present in cometary grains in both amorphous and crystalline form [6][7][12][13][24]. Since the formation of crystalline silicates requires high temperatures, in contrast to the amorphous variety, the puzzling question is how comets could have incorporated both high- and low temperature materials, including ices.

Crystalline silicates are found in interplanetary dust particles (IDP) and coexist with amorphous silicates in the IDPs which are thought to originate from comets [2]. Silicates in chondritic meteorites, formed in the inner hot nebula, are entirely crystalline. Crystalline silicates have been detected in disks around pre-main-sequence Herbig AE/Be stars and young main sequence stars (e.g. [17], [16]). On the other hand, there is no spectral evidence for their presence in the diffuse ISM or molecular clouds, nor in young stellar objects [11]. Likely, the crystalline silicates present in cometary grains were produced during or after the collapse of the presolar cloud. Crystalline silicates can form at high temperature by direct condensation or thermal annealing of amorphous silicates. This latter process requires temperatures of at least 800 K.

Chick and Cassen [4] studied the thermal processing of silicates during the collapse of the presolar cloud, combining results of envelope radiative heating with heating in the accretion shock at the surface of the disk. From this work, it can be deduced that, during this stage, thermal annealing of amorphous silicates was inefficient at distances larger than 2–3 AU from the proto-Sun. The thermal regime experienced in the mid-plane of the accretion disk, namely the solar nebula, was more severe. However, the temperature was too cold in the outer nebula, where comets formed, for the interstellar silicate grains to be thermally processed prior to their incorporation into comets. Therefore, either crystallization was initiated by a low temperature process, as proposed for disks around evolved stars [18], or there was a significant radial mixing between the warm and cold parts of the solar nebula.

We investigate here the thermal annealing of amorphous silicates in the hot regions of the solar nebula and their diffusive transport out to the comet formation zone, by using a turbulent evolutionary model of the nebula. In fact, a large-scale radial mixing in the nebula has been shown to be a requisite in order to fit measurements of the D/H ratio in water in LL3 meteorites and Oort cloud comets [8][14][20]. We take advantage here of the recent 2-D evolutionary model of [14] based on the accretion disk model of [15], which accounts for the vertical structure of the disk. The model generates time-dependent temperature, density and pressures profiles throughout the nebula. Hersant *et al.* [14] have studied the radial distribution and time evolution of the deuterium enrichment in water in the solar nebula due to isotopic exchange with H_2 (the main reservoir of deuterium) and diffusive turbulent transport. The interpretation of the D/H value in LL3 meteorites and Oort cloud comets allow [14] to constrain the ranges of input model parameters describing the nebula which fit the deuterium data. The goal of our study is to investigate whether these nebulae are also able to explain the amount of crystalline silicates present in cometary dust.

2 Model

The model treats both the amorphous-to-crystalline phase transition of small silicate particles and their diffusive transport throughout the nebula by turbulence as a function of time. Only two dust entities are considered, regardless their mineralogy: amorphous and crystalline silicates. Following [19], it can be shown that, in cylindrical coordinates, the relative mass concentrations of amorphous and crystalline silicatic grains with respect to H_2 (C_a and C_c, respectively) obey the following coupled transport equations in the disk midplane:

$$\frac{\partial C_a}{\partial t} + v_r \frac{\partial C_a}{\partial r} = \frac{1}{\Sigma r}\frac{\partial}{\partial r}\left(r\Sigma\kappa\frac{\partial C_a}{\partial r}\right) - \frac{C_a}{t_{hop}} \tag{1}$$

$$\frac{\partial C_c}{\partial t} + v_r \frac{\partial C_c}{\partial r} = \frac{1}{\Sigma r}\frac{\partial}{\partial r}\left(r\Sigma\kappa\frac{\partial C_c}{\partial r}\right) + \frac{C_a}{t_{hop}} \tag{2}$$

The second term in the left part of Eqs 1 and 2 is the advection term describing inward mass transport towards the central object at the velocity v_r. We assume that the dust particles are well mixed with the gas, so that their radial velocity is the same as that of gas. In the case of a thin Keplerian disk, v_r is equal to $-3\nu_t/2r$, where ν_t is the turbulent viscosity and r is the radial distance. The effect of turbulence is to smooth out any concentration gradient. This diffusion is represented by the first term in the right part of Eqs 1 and 2: the turbulent diffusity κ is set to the ratio of turbulent viscosity ν_t to the Prandtl number P_r, taken equal to 0.7 [14]. Σ is the local surface density of the disk.

The characteristic time scale for the amorphous-crystalline transition t_{hop} is expressed as $t_{hop} = \nu^{-1} e^{E_a/kT}$. E_a has been measured for various silicatic materials, assuming ν equal to 2–2.5 × 10^{13} s^{-1}. For Mg-rich silicates, E_a/k is found between 39100 and 47500 K [3][10]. We present calculations made with $E_a/k = 39100$ K, as determined for Mg_2SiO_4 (olivine) smoke particles [10].

Sublimation is taken into account in the inner boundary condition by setting $C_a = C_c = 0$ at the distance r where the temperature of the nebula is high enough to allow efficient sublimation. We used the data on the stability of Mg-rich silicates against vaporization published by [9]. We do not consider the re-formation of silicates which subsequently occurs as the hot gas is cooling down.

Solving Eqs 1 and 2 requires an evolutionary solar nebula model giving surface density and temperature radial profiles as a function of time. We used that described in length by [14], in which the temporal evolution of the solar nebula is modelled by a sequence of stationary solutions provided by the 2-D accretion disk model of [15]. This model uses the well known α prescription [21], which consists in writing the turbulent viscosity ν_t, which governs the transport of angular momentum, as $\nu_t = \alpha C_s^2/\Omega$, where C_s is the sound velocity at the midplane, Ω is the Keplerian angular velocity, and α is a non-dimensional parameter whose value depends on the nature of the turbulence and is taken as constant with r and time. The model of [15] provides stationary solutions for a given accretion rate $\dot{M}$, which governs the infall of matter from the disk onto the proto-Sun. The temporal evolution is obtained taking into account that, with time, the accretion rate diminishes and the radius of the nebula (R_D) spreads out viscously according to:

$$\dot{M}(t) = \dot{M}_0(1 + t/t_0)^{-1.5}, \tag{3}$$

$$t_0 = \frac{R_{D0}^2}{3\nu_t(R_{D0})}, \tag{4}$$

Table 1. Solar nebula models

Type	α	$\dot{M}_0$ [$M_\odot yr^{-1}$]	R_{D0} [AU]
cold	2×10^{-2}	4.0×10^{-6}	27
nominal	9×10^{-3}	5.0×10^{-6}	17
warm	8×10^{-3}	9.8×10^{-6}	12.8

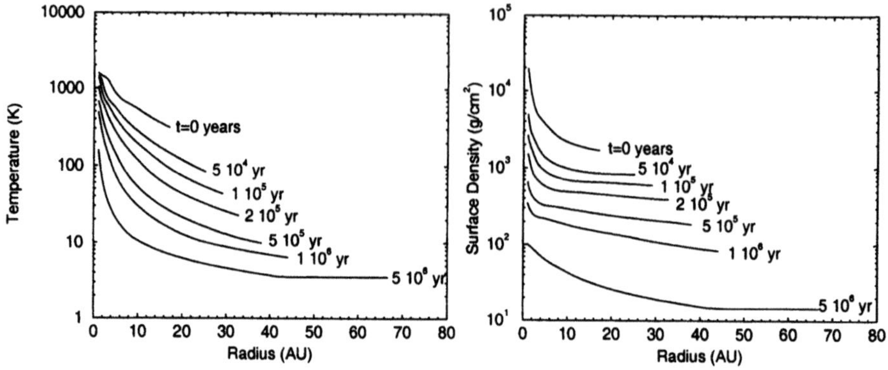

Fig. 1. Temperature and surface density profiles for the nominal solar nebula

$$\frac{dR_D}{dt} = \frac{3}{2}\frac{\nu_t}{R_D}. \tag{5}$$

where $\dot{M}_0$ is the initial accretion rate at $t = 0$, R_{D0} is the initial radius of the nebula and $\nu_t(R_{D0})$ the turbulent viscosity at R_{D0}. The model begins when the formation of the Sun is almost complete. The model input parameters are R_{D0}, $\dot{M}_0$, and α. From the analysis of D/H enrichments in comets and LL3 meteorites, [14] conclude that $\dot{M}_0$ was in between 2.2×10^{-6} and 10^{-5} $M_\odot \mathrm{yr}^{-1}$, R_{D0} between 12.8 and 39 AU, and α between 0.006 and 0.02. Among all nebulae fitting observational constraints, they selected a nominal nebula, one of the coldest and one of the hottest, named cold nebula and warm nebula, respectively, characterized by the parameters given in Table 1. Radial profiles of Σ and T are shown on Fig. 1 for the nominal nebula.

3 Results

We assumed that, at $t = 0$ yr, silicates are in amorphous form throughout the nebula and infall of matter onto the disk has almost ceased. The temporal and radial evolution of the relative mass fraction of crystalline silicates $f_c = C_c/(C_a + C_c)$ is shown in Fig. 2, for the nominal nebula. Profiles computed with the cold and warm nebulae show the same behaviour. At $t \approx 0$ yr, silicate grains are vaporized or in crystalline form within some distance ($\sim$ 5–7 AU) which depends on the initial thermal structure of the nebula. As time goes on, amorphous silicates, present in the outer regions, are transported towards the inner regions by advection and turbulent diffusion, and then thermally annealed. This inward flux of amorphous material inside the region where crystallisation takes place balances the loss of crystalline silicates by advective transport inside the evaporation zone. As seen on Fig. 2, crystalline silicates are progressively

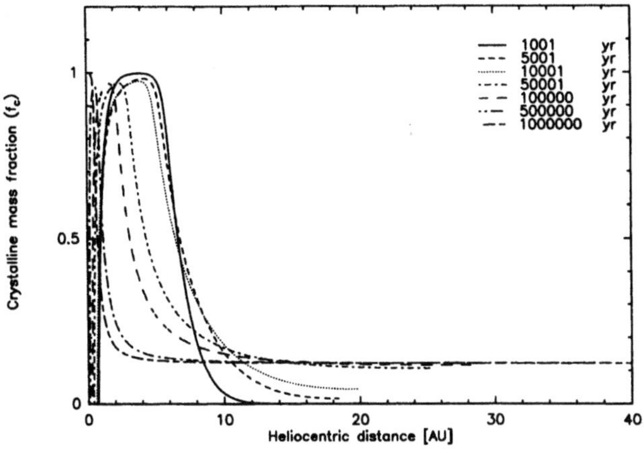

Fig. 2. Crystalline mass fraction f_c for the nominal solar nebula

mixed, by turbulent diffusion, to the amorphous grains present in the outer regions, so that f_c continuously increases in the outer regions to finally reach a plateau. In turn, the crystalline front moves towards the Sun. The value of the plateau is $f_c = 0.12$ for the nominal nebula, and $f_c = 0.02$ and $= 0.6$ for the cold and warm nebulae, respectively. The time scale for radial mixing is typically a few 10^4 years. At times larger than $\sim$ 5–10 $\times 10^4$ yr, the mass fraction of crystalline silicates in the outer nebula ($r > 10$ AU) no more evolves.

The relative mass fraction of crystalline silicates in comet Hale-Bopp is estimated between 30–50 % and 90 % [6][5][24]. The high value of 90 % determined by [24] is somewhat debated [13]. Dynamical considerations suggest that Oort cloud comets, such as Hale-Bopp, formed mainly in the Uranus-Neptune zone. Our model shows that microscopic grains made of crystalline silicates present in the inner regions of the Solar Nebula are efficiently transported outwards to the comet formation zone by turbulence. The model succeeds in explaining a crystalline mass fraction of 30–50 % in Hale-Bopp, provided the warm nebula is used. A still warmer nebula is required for interpreting the high concentration of 90 % inferred by [24].

4 Discussion and Perspectives

The present analysis shows that it is possible to explain consistently, with the same model, the crystalline/amorphous silicate mass ratio and the D/H in water measured in comet Hale-Bopp. This requires warm nebulae with initial mass accretion rates $\sim 10^{-5}\ M_{\odot}\mathrm{yr}^{-1}$, corresponding approximately to Class 0 sources. In this model, transport of crystalline silicatic grains to the comet formation zone is made by turbulent mixing. This mechanism is efficient as long as the grains are well coupled to the gas, that is for sizes less than ~ 1 m [23] or even much smaller [22]. Therefore, a critical point for this scenario is that time scales for grain growth should not be too short with respect to diffusion time scales. Dust coagulation in protoplanetary disks is a complex topic which is not yet well understood [1]. Estimated time scales for grain growth are typically 1000 orbital periods, which compare well, within a factor of a few, to our diffusion time scales given by R^2/ν_t ($\sim 5 \times 10^3$ yr to 5×10^4 yr from 1 to 10 AU).

So far, the crystalline-to-amorphous mass ratio has only been measured in a very limited number of comets coming from the Oort cloud. Measurements in a larger sample of comets, including short-period comets originating from the Kuiper Belt, could provide additional tests to the proposed interpretation. This will be possible thanks to the mid-infrared observational capabilities of the VLT.

References

1. S.V.W. Beckwith, T. Henning, Y. Nakagawa: In: *Protostars and Planets IV*, ed by V. Manning, A.P. Boss, S.S. Russel (University of Arizona Press 2000) pp. 533–558
2. J.P. Bradley, L.P. Keller, T. Snow, G.J. Flynn, J. Gezo, D.E. Brownlee, M.S. Hanner, J. Bowie: Science **285**, 1716–1718 (1999)
3. J.R. Brucato, L. Colangeli, V. Mennela, P. Palumbo, E. Bussoletti: Astron. Astrophys. **348**, 1012–1019 (1999)
4. K.M. Chick, P. Cassen: Astrophys. J. **477**, 398–409 (1997)
5. L. Colangeli, J.R. Brucato, L. Ferrini, V. Mennela, E. Bussoletti, P. Palumbo, A. Rotundi: Adv. Space Res. **23**, 1243–1252 (1999)
6. J. Crovisier, T.Y. Brooke, K. Leech, D. Bockelée-Morvan, E. Lellouch, M.S. Hanner, B. Altieri, H.U. Keller, T. Lim, T. Encrenaz, A. Salama, M. Griffin, T. de Graauw, E. van Dishoeck, R.F. Knacke: In: *Thermal emission spectroscopy and analysis of dust disks, and regoliths* ed. by M.L. Sitko, A.L. Sprague, D.K. Lynch, Astron. Soc. Pac. Conf. Ser. **196**, 109–117 (2000)
7. J. Crovisier, K. Leech, D. Bockelée-Morvan, T.Y. Brooke, M.S. Hanner, B. Altieri, H.U. Keller, E. Lellouch: Science **275**, 1904–1907 (1997)
8. A. Drouart, B. Dubrulle, D. Gautier, F. Robert: Icarus **140**, 129–155 (1999)
9. W.J. Duschl, H.-P. Gail, W.M. Tscharnuter: Astron. Astrophys. **312**, 624-642 (1996)
10. D. Fabian, C. Jäger, Th. Henning, J. Dorschner, H. Mutschkle: Astron. Astrophys. **364**, 282–292 (2000)
11. M.S. Hanner, T.Y. Brooke, A.T. Tokunaga: Astrophys. J. **502**, 871–882 (1998)
12. M.S. Hanner, D.K. Lynch, R.W. Russel: Astrophys. J. **425**, 274–285 (1994)
13. T.L. Hayward, M.S. Hanner, Z. Sekanina: Astrophys. J. **538**, 428–455 (2000)
14. F. Hersant, D. Gautier, J.-M. Huré: Astrophys. J., in press (2001)
15. J.-M. Huré: Astron. Astrophys. **358**, 378–394 (2000)
16. R.F. Knacke, S.B. Fajardo-Acosta, C.M. Tedesco, J.A. Hackwell, D.K. Lynch, R.W. Russel: Astrophys. J. **418**, 440-450 (1993)
17. K. Malfait, C. Waelkens, L.B.F.M. Waters, B. Vandenbussche, E. Huygen, M.S. de Graauw: Astron. Astrophys. **332**, L25–L28 (1998)
18. F.J. Molster, I. Yamamura, L.B.F.M. Waters, A.G.G.M. Tielens, Th. de Graauw, T. de Jong, A. de Koter, K. Malfait, M.E. van den Ancker, H. van Winckel, R.H.M. Voors, C. Waelkens 1999: Nature **401**, 563–565 (1999)

19. G.E. Morfill, H.J. Völk: Astrophys. J. **287**, 371–395 (1984)
20. O. Mousis, D. Gautier, D. Bockelée-Morvan, F. Robert, B. Dubrulle: Icarus **148**, 513–525 (2000)
21. N.L. Shakura, R.A. Sunyaev: Astron. Astrophys. **24**, 337–355 (1973)
22. T.F. Stepinski, P. Valageas: Astron. Astrophys. **309**, 301–312 (1996)
23. S.J. Weidenschilling: Icarus **127**, 290-306 (1997)
24. D.H. Wooden, D.E. Harker, C.E. Woodward, H. Butner, C. Koike, F.C. Witteborn, C.W. McMurtry: Astrophys. J. **517**, 1034–1058 (1999)

Patrick Roche, João Alves, and Thierry Montmerle

X-Rays from Star-Forming Regions in the VLT Era

Thierry Montmerle[1] and Nicolas Grosso[2]

[1] Service d'Astrophysique, CEA Saclay, 91191 Gif-sur-Yvette, France
[2] Max-Planck Institut für Extraterrestrische Physik, D-85741 Garching, Germany

Abstract. The association between star-forming regions and X-ray emission was discovered over 30 years ago. We now know that essentially all young stellar objects, T Tauri stars and protostars, are X-ray emitters, although the case of the youngest, Class 0 protostars, is less clear. The paper highlights X-ray emission and absorption mechanisms, and summarizes X-ray observations of young stellar objects. The impact of the VLT on the characterization of the new X-ray sources is also briefly discussed.

1 From Past to Present

The first evidence that X-rays could reach the Earth from outer space, and more specifically from the Sun, was obtained in 1949 by means of a V2 rocket launched from White Sands, New Mexico, by a team of physicists from the US Naval Research Laboratory; the flight lasted a mere 336 s (Friedman, Lichtman, & Byram 1951). The real astronomical start took place with the launch of the *Uhuru* satellite, in 1970. This satellite, operating for three years in scanning mode with collimators, obtained the first X-ray all-sky survey, with the discovery of 339 sources, with error boxes up to several $10'$ on a side. One of them included M42, the Orion nebula (Giacconi et al. 1972): ***the knowledge that star-forming regions are emitting X-rays is nearly 30 years old!*** The discovery of the X-ray emission from M42 was confirmed by the ANS satellite (den Boggende 1978). The second major leap was the launch of the second "High-Energy Astronomical Observatory", *Einstein*, in 1978, which carried X-ray focussing mirrors able to obtain the first pointed images of astronomical X-ray sources, with a field-of-view of $\sim 1^\circ \times 1^\circ$. Orion and other star-forming regions were successfully detected, with a combination of point sources identified with OB stars and/or T Tauri stars (Ku & Chanan 1979). Launched in 1990, ROSAT undertook an all-sky survey, which resulted in a catalog of several 10^5 sources, including ~50,000 stars in all evolutionary stages, from PMS to evolved. Launched in 1993, the Japanese satellite ASCA was the first imaging satellite using CCD detectors and reaching the hard X-ray range. To this date, four X-ray satellites are in orbit: BeppoSAX, RossiXTE, and the two "workhorses" of the present decade, the US *Chandra* and the European *XMM-Newton*, both launched in 1999. These two major observatories currently have significant programs dedicated to the study of young clusters and star-forming regions. Tables 1 and 2 give the current list of accepted targets, updated from Sciortino (2001).

Table 1. SFRs and young stellar clusters in the *Chandra* AO-1 Guaranteed Time Observation (GTO) and Guest Observation (GO) AO-1, AO-2 [left] and AO-3 [right] programs (June 2001)

Target Name	Exposure Time	Target Name	Exposure Time
NGC 1333	50 ks	IRAM04191	20 ks
IC 348	50 ks	L1527	20 ks
Orion Trapezium	50 ks + 69 ks	α Tau	20 ks
Orion region	several obs	ONC Flanking Field S	50 ks
R CrA Cloud Core	20 ks	ONC Flanking Field N	50 ks
ρ Oph	mosaic of 6 $\times$ 5 ks	σ Ori	100 ks
ρ Oph	2 $\times$ 100 ks	HH 24-26	70 ks
NGC2516	20 ks + 20 ks	Maddalena's Cloud	30 ks
NGC2516	20 ks + 50 ks	NGC2264	50 ks + 100 ks
M16	80 ks	NGC2362	47 ks + 48 ks
NGC 6530	61 ks	IRAS16293	30 ks
W3B	40 ks	Trifid Nebula	60 ks
HH 1	20 ks	HH 80/81	40 ks

Table 2. SFRs and young stellar clusters in the *XMM-Newton/EPIC* GTO and AO-1 GO program (June 2000).

Target Name	Exposure Time	Target Name	Exposure Time
NGC 2024	45 ks + 25 ks	L1448-C	30 ks
NGC 2023	25 ks	NGC 1333	50 ks
IC 2602	45 ks	Orion P1795	20 ks
IC 2391	45 ks	OMC2/3	100 ks
alpha Per	50 ks	L1641-N	50 ks
Upper Sco-Cen	2 $\times$ 50 ks	NGC 2264	2 $\times$ 35 ks
R CrA	20 ks	Chamaeleon	28 ks
ρ Oph	25 ks + 50 ks	Serpens	50 ks
IC 348	40 ks	NGC 2362	50 ks
L1551	50 ks	NGC 2547	50 ks
Blanco 1	50 ks	Pleiades	2 $\times$ 50 ks

The latest review of pre-*Chandra/XMM-Newton* results on Young Stellar Objects (YSOs) is by Feigelson & Montmerle 1999 (FM); the review by Glassgold, Feigelson & Montmerle (2001; hereafter GFM) puts more emphasis on the various physical processes at work in YSOs and in their vicinity. The reader is referred to these reviews for details. In the present short contribution, we only give some key background information, summarize and update the main X-ray results, and emphasize the role of the VLT in the problem of the characterization of the new X-ray sources.

2 X-Ray Emission and Absorption Mechanisms

• *YSO X-ray emission* comes from the thermal bremsstrahlung emission of an optically thin, "coronal" plasma at temperatures $T_X \sim 10^6$–10^8 K and densities $n_e \sim 10^{10}$–$10^{12}\,\mathrm{cm}^{-3}$ (FM). It can be shown that the emitting plasma must be confined in magnetic loops, like on the Sun, but much larger. Stellar X-ray spectra also indicate the presence of lines of heavy elements (see examples in, e.g., Audard et al. 2001), so that abundances and metallicities can be determined from χ^2-type fits. This is also the case for the X-ray brightest young stars.

• *X-ray absorption* is an important parameter in all studies of young stars (GFM). It is due to the photoelectric effect, in which inner-shell (mostly K and L) electrons of an atom are ejected by the incoming photon. However, these photoelectrons basically carry away the energy of the original X-rays (in the ~ keV range), minus the shell energy, so are themselves initially highly energetic compared to the energy of the outer shells of the ambient atoms. A "shower" results, in which many secondary electrons are ejected from ambient atoms as the primary electrons collide with them and loose their initial energy. Other secondary electrons may be produced as "Auger electrons", i.e., from internal rearrangement of the energy levels of the original atoms hit by the X-ray photons.

Fig. 1, adapted from Ryter (1996), gives in graphic form the total extinction cross-section all the way from the far-IR to the hard X-ray range, assuming a normal gas-to-dust ratio and cosmic abundances. First, it can be noted that *the cross-section has comparable values in the IR range* (where it is dominated by the dust) *and in the X-ray range* (where it is dominated by the gas): in other words, the "penetrating power" of X-rays is the same as in the IR, a key point allowing to characterize embedded sources such as young T Tauri stars (hereafter TTS) and protostars (see the detailed discussion by Casanova et al. 1995). Second, the total photoelectric cross-section, which for each element has the form $\sigma \propto E_X^{-2.5}$ beyond the Lyman limit (H), displays discontinuities. With increasing X-ray energy, each "jump" corresponds to the inner-shell ionization of a new atomic species present in the ISM: He, C, N, O below 1 keV, up to Fe and Ni above ~ 7 keV. Although these cross-section jumps generally are small, they accumulate to the point that above 0.5 keV the absorption comes predominantly from heavy atoms, and is > 10 times larger than from H+He beyond ~ 7 keV (for details, see Wilms, Allen, and McCray 2000).

For $E_X = 1\,\mathrm{keV}$ and solar abundances, one finds $\tau(1\,\mathrm{keV}) = N_H/4.41 \times 10^{21}\,\mathrm{cm}^{-2}$. If the usual conversion between N_H and A_V for diffuse interstellar clouds is used (e.g., Ryter 1996), then $\tau(1\,\mathrm{keV}) = 1$ occurs for $A_V \simeq 2$. It is also a useful rule-of-thumb that the penetrability of X-rays varies like $\sim E_X^{-2.5}$: the absorption decreases by a factor ~ 3000 between 0.4 keV and 10 keV. Note that the contribution of dust grains (into which certain heavy atomic species are condensed) to the X-ray opacity is small, and depends on grain models only at the lowest energies ($E_X \lesssim 0.3\,\mathrm{keV}$) (see discussion in Wilms, Allen, and McCray 2000).

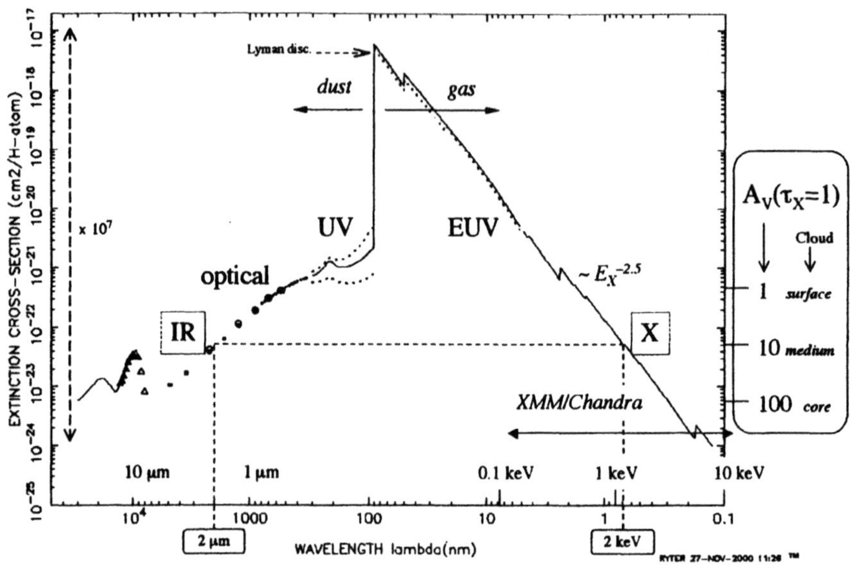

Fig. 1. Total extinction cross-section, from the IR range ($1\,\mu m \equiv 1000\,nm$), to the X-ray range ($1\,keV \equiv 0.81\,nm$, using $E = h \times \lambda^{-1}$, with h the Planck's constant) (see Ryter 1996). The sharp peak in the middle is the Lyman discontinuity (almost four orders of magnitude!)

3 X-Ray Observations of Low-Mass Star-Forming Regions

To date, all major star-forming regions within $\sim 500\,pc$ of the Sun have been observed in X-rays, plus a couple of more distant ones ($d \geq 1\,kpc$) (see FM). Perhaps the most spectacular observation to date is that of the Orion Trapezium region and its surrounding M42 nebula (Garmire et al. 2000), in which a single $17' \times 17'$ *Chandra* field has revealed nearly 1,000 sources. As shown on Fig. 2, almost every X-ray source has a stellar counterpart (position accuracy $\sim 0.1''$), all the way from the central massive stars down to lower-mass stars. A full analysis is in progress (Feigelson et al., in preparation).

Here we restrict our discussion to the numerous X-ray detected low-mass, solar-like stars. The Taurus-Auriga, R Coronae Australis, Chamaeleon, and ρ Ophiuchi regions have been particularly well studied for many years. For example, ROSAT observations of the central region of the ρ Ophiuchi dark cloud have revealed several dozen X-ray sources associated with young stellar objects (Grosso et al. 2000). This region has now been also observed in X-rays by *Chandra* (Imanishi et al. 2001), and by *XMM-Newton* (Grosso et al., in preparation: see Fig. 3). ROSAT being sensitive in the soft X-ray band only ($< 2.4\,keV$), the sources are mainly identified with objects having low to inter-

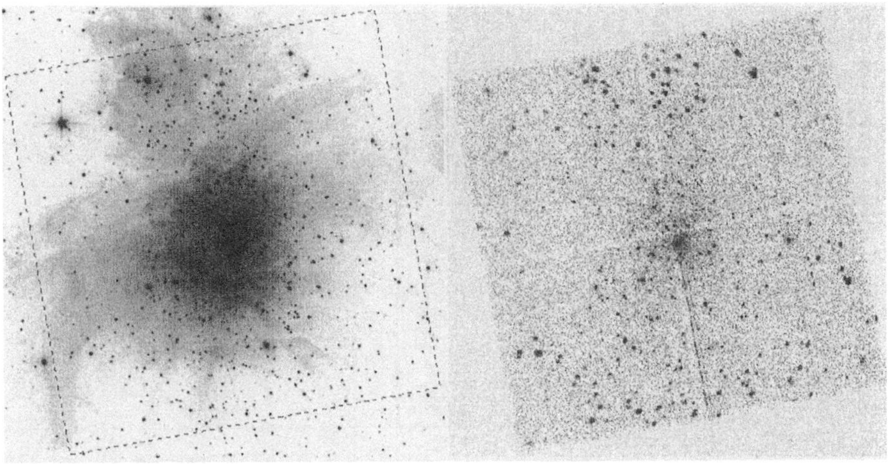

Fig. 2. *Left:* The Orion Nebula Cluster seen at 2 μm, from the 2MASS survey. The Trapezium O stars exciting the M42 nebula are located at the center of the image. *Right:* The same region seen by *Chandra/ACIS.* Almost 1000 X-ray sources are detected (Garmire et al. 2000); there is nearly a one-to-one subarcsecond correspondence between the near-IR and X-ray sources in both images. Field of view of *Chandra/ACIS*: 17′ × 17′

mediate extinction (see Fig. 1), i.e., TTS. Nevertheless thanks to X-ray flares, which combine an increase of X-ray luminosity with an increase of plasma temperature hence a better visibility, ROSAT caught X-rays from two protostars (see Grosso et al. 1997 and Grosso 2001). In contrast ASCA was sensitive up to 10 keV, hence was able to penetrate deep into dense material, revealing more easily X-rays from protostars. The past generation of X-ray satellites detected protostars only during their high activity level. Today, the new generation of X-ray satellites, *Chandra* and *XMM-Newton*, with an increased sensitivity up to 10 keV, gives access to the quiet coronae of protostars.

Let us now summarize and update briefly some key X-ray properties of TTS and protostars.

- *There is a strong correlation between the X-ray and bolometric luminosities of TTS,* which differs somewhat from region to region: $L_X/L_{bol} \sim 10^{-4}$, as compared with $\sim 10^{-6}$ for the active Sun. Typical TTS X-ray luminosities are $L_X \sim 10^{30}$–10^{31} erg s^{-1} (e.g., Grosso et al. 2000). The presence of disks does not seem to influence the TTS X-ray properties, but as TTS lose their disks while contracting towards the main sequence, they become more X-ray luminous, presumably because of the increased rotation velocity and resulting enhanced dynamo (Stelzer & Neuhäuser 2001). X-ray surveys of star-forming regions are thus very efficient to make a reliable census of the TTS population. Nevertheless IR follow-ups are needed to confirm the nature of the X-ray selected objects, and their young age.
- *TTS X-ray variability is generalized.* Peak X-ray luminosities may be as high as $L_{X,peak} \sim 10^{32}$ erg s^{-1} or more. The light curves are strongly suggestive

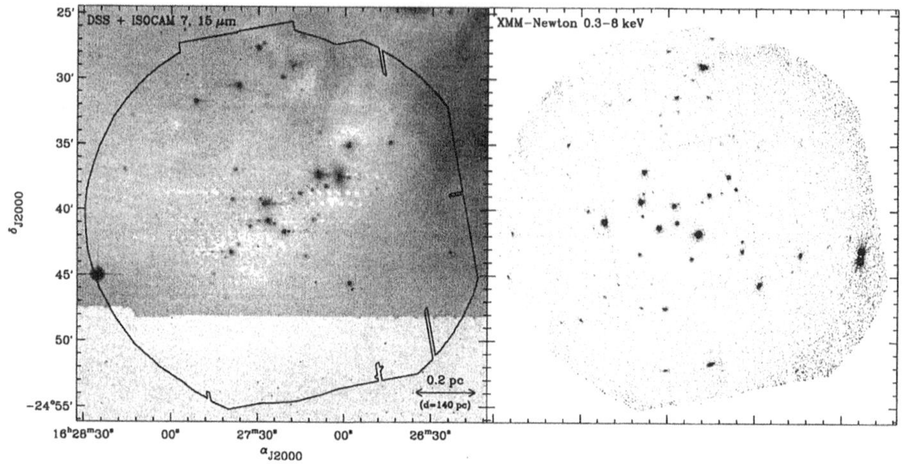

Fig. 3. *Left:* Field centered on the core F region of the ρ Ophiuchi dark cloud. Background: DSS optical image. Foreground: ISOCAM 7+15 μm image (Abergel et al. 1996). Quasi-circular contour: *XMM-Newton/EPIC* field-of-view. Most of the bright ISOCAM sources are protostars, the other sources are T Tauri stars (see Bontemps et al. 2001). *Right: XMM-Newton* image of the same region. With a few exceptions, almost all the IR sources, including protostars, are detected (Grosso et al., in preparation)

of flares. In the best cases, flare spectroscopy can be time-resolved, showing a clear decline in temperature (from T_X as high as $\sim 10^8$ K down to a few 10^7 K : see the example of V773 Tau, Tsuboi et al. 1998).

• *X-ray detections have reached the brown dwarf limit.* A few *bona fide* brown dwarfs have been detected in X-rays, already by ROSAT (Neuhäuser et al. 1999), and recently by *Chandra* (Rutledge et al. 2000; Imanishi et al. 2001) and *XMM-Newton* (Grosso et al., in preparation). Their L_X/L_{bol} ratio is found to be $\gtrsim 10^{-4}$, i.e., very similar to that of TTS. It is reasonable to think that a number of new brown dwarf candidates will soon be *X-ray detected* among *Chandra* and *XMM-Newton* sources in star-forming regions. An IR spectroscopy confirmation will be necessary, and will require large telescopes.

• *Deeply embedded sources like protostars are more difficult to detect in X-rays.* The number of protostars detected remained small (about a dozen) in the ROSAT/ASCA era, for lack of access to the hard X-ray range (ROSAT), or limited sensitivity (ASCA) (FM). The new *Chandra* and *XMM-Newton* results on the ρ Ophiuchi dark cloud (see above), are very promising, with a ~ 70 % detection rate, including at their non-flare, quiescent level.

• *Almost all of the* $\sim$ 15 bona fide *protostars detected so far in X-rays are evolved ("Class I") protostars.* Although magnetism also plays a central role in their X-ray emission, significant differences exist with that of TTS : most detected Class I protostars have X-ray luminosities and temperatures somewhat higher than those of most TTS. Variability is as ubiquitous as for TTS, with many examples of flares. In a few cases $L_{X,peak}$ reaches very high values of several $10^{32}\,\mathrm{erg\,s^{-1}}$ or more, which may be due to star-disk magnetic interactions (Tsuboi et al. 2000, Montmerle et al. 2000). In contrast, the X-ray detection of young ("Class 0") protostars is open to discussion. Tsuboi et al. (2001) recently reported the *Chandra* discovery of two weak X-ray sources in the direction of candidate Class 0 protostars in the Orion OMC 2/3 clouds, but their extinction, measured from the X-ray spectrum itself, is nearly one order of magnitude smaller than the typical extinction towards the central object of a Class 0 source across its own envelope. They may therefore be somehow associated with the Class 0 protostars, but not be strictly identified with them.

4 The Role of the VLT

Thanks to their much improved sensitivity and resulting improved performance in spectral and spatial resolution, *Chandra* and *XMM-Newton* allow detailed studies of previously known, X-ray bright YSOs like TTS and protostars. But a new frontier appears: *low-luminosity objects.* With a few exceptions, the $L_X/L_{bol} \sim 10^{-4}$ correlation approximately holds empirically for all YSOs, so that in general weak X-ray sources (now detected at the level of $L_X \sim 10^{28}\,\mathrm{erg\,s^{-1}}$, i.e., only 10 times the active Sun, at $d \sim 150\,\mathrm{pc}$) are expected to be intrinsically low-luminosity sources, i.e., of low mass and/or low temperature. Very Low-Mass objects and brown dwarfs are typical examples.

However, contrary to the situation holding with the past generation of X-ray satellites, *low-luminosity sources are not necessarily YSOs!* The reason is that the sensitivity of *Chandra* and *XMM-Newton* is such that weak, hard X-ray sources like AGNs start to be visible through molecular clouds. From the log$N$-log$S$ curve of the Hubble Deep Field seen by *Chandra*, the contamination is estimated to be ≈ 20 background extragalactic sources in the $17' \times 17'$ field of view (Garmire et al. 2001). Of course, such sources must be among the weakest and most absorbed of the sample, mimicking very low-mass stars deeply embedded or at the back of the cloud. Note that the same situation holds in the IR, where background stars and galaxies routinely contaminate the young star sample, except in the densest regions.

It is therefore clear that one needs *both* X-ray and IR observations to disentangle the source sample, and large ground-based telescopes are mandatory in almost all cases. More generally, we have also seen (§2) the deep connection existing between the X-ray and IR ranges. Therefore, the VLT, with its large choice of IR cameras, is certainly the facility of choice for follow-ups of the sensitive X-ray observations in progress.

A foretaste of the discoveries to come is offered by the preliminary results of an NTT/SOFI campaign to search for counterparts of *XMM-Newton* new X-ray sources without 2MASS counterpart in the ρ Ophiuchi dark cloud (see Fig. 3; Grosso et al., in preparation). In the course of this study, two spectacular new embedded Herbig-Haro objects were found, one probably related to the peculiar X-ray emitting protostar YLW15, the other to the weaker X-ray emitting protostar, IRS54 (Grosso et al. 2001). The campaign has also yielded very low-luminosity IR counterparts for faint *XMM-Newton* sources, the nature of which can be elucidated only with follow-up IR spectroscopy such as provided by ISAAC in the near-IR or VISIR in the mid-IR. Could these be "free-floating planets" so well appreciated by Mark McCaughrean? (After all, Jupiter is also an X-ray source, although of auroral origin...)

N. G. acknowledges financial support from the European Union (HPMF-CT-1999-00228).

References

1. Abergel, A., et al. 1996, A&A, 315, L329
2. Audard, M., et al. 2001, A&A, 365, L329
3. Bontemps, S. et al. 2001, A&A, 372, 173
4. Casanova, S., Montmerle, T., Feigelson, E.D., & André, P. 1995, ApJ, 439, 752
5. den Boggende, A.J.F., Mewe, R., Gronenschild, E.H.B.M., Heise, J. & Grindlay, J. E. 1978, A&A, 62, 1
6. Feigelson, E.D., & Montmerle, T. 1999, Ann. Rev. Astr. Ap., 37, 363 [FM]
7. Friedman, H., Lichtman, S.W., & Byram, E.T. 1951, Phys. Rev., 83, 1025
8. Garmire, G., Feigelson, E. D., Broos, P., Hillenbrand, L. A., Pravdo, S. H., Townsley, L., & Tsuboi, Y. 2000, AJ, 120, 1426
9. Garmire, G., et al. 2001, ApJ, submitted
10. Giacconi, R., Murray, S., Gursky, H., Kellogg, E., Schreier, E. & Tananbaum, H. 1972, ApJ, 178, 281
11. Glassgold, A.E., Feigelson, E.D., & Montmerle, T. 2000, in *Protostars & Planets IV*, Eds. V. Mannings, A. Boss, & S. Russell (Tucson: U. of Arizona Press), p. 429 [GFM]
12. Grosso, N. 2001, A&A, 370, L22
13. Grosso, N., Alves, J., Neuhäuser, R. & Montmerle, T. 2001, A&A, submitted
14. Grosso, N., Montmerle, T., Feigelson, E.D., et al. 1997, Nature, 387, 56
15. Grosso, N., Montmerle, T., Bontemps, S., André, P. & Feigelson, E.D. 2000, A&A, 359, 113
16. Imanishi, K., Koyama, K. & Tsuboi, Y. 2001, ApJ, 557, 747
17. Ku, W.H. & Chanan, G.A. 1979, ApJ, 234, L59
18. Montmerle, T., Grosso, N., Tsuboi, Y. & Koyama, K. 2000, ApJ, 532, 1097
19. Neuhäuser, R., Briceño, C., Comerón, F., Hearty, T., Martín, E.L., Schmitt, J.H.M.M., Stelzer, B., Supper, R., Voges, W. & Zinnecker, H. 1999, A&A, 343, 883
20. Rutledge, R. E., Basri, G., Martín, E.L., & Bildsten, L. 2000, ApJ, 538, L141
21. Ryter, C.E. 1996, Astr. Sp. Sci., 236, 285
22. Sciortino, S. 2001, in *From Darkness to Light*, Eds. T. Montmerle & P. André, ASP Conf. Ser. 243, in press
23. Stelzer, B. & Neuhäuser, R.N. 2001, A&A, in press
24. Tsuboi, Y., Koyama, K., Murakami, H., Hayashi, M., Skinner, S., & Ueno, S. 1998, ApJ, 503, 894
25. Tsuboi, Y., Imanishi, K., Koyama, K., Grosso, N., & Montmerle, T. 2000, ApJ, 532, 1089
26. Tsuboi, Y., Koyama, K., Hamaguchi, K., Tatematsu K., Sekimoto, Y., Bally, J. & Reipurth, B. 2001, ApJ, 554, 734
27. Wilms, J., Allen, A., & McCray, R. 2000, ApJ, 542, 914

RCW 38
(Antu/UT1 + ISAAC)

Michael Meyer

The Formation and Evolution of Planetary Systems: SIRTF Legacy Science in the VLT Era

M.R. Meyer[1], D. Backman[2], S.V.W. Beckwith[3], T.Y. Brooke[4], J.M. Carpenter[5], M. Cohen[6], U. Gorti[7], T. Henning[8], L.A. Hillenbrand[5], D. Hines[1], D. Hollenbach[7], J. Lunine[9], R. Malhotra[9], E. Mamajek[1], P. Morris[10], J. Najita[11], D.L. Padgett[10], D. Soderblom[3], J. Stauffer[10], S.E. Strom[11], D. Watson[12], S. Weidenschilling[13], and E. Young[1]

[1] Steward Observatory, The University of Arizona, Tucson, AZ 85721–0065, U.S.A.
[2] Franklin & Marshall College, P.O. Box 3003, Lancaster, PA 17604–3003, U.S.A.
[3] STScI, 3700 San Martin Drive, Baltimore, MD 21218, U.S.A.
[4] JPL, MS 169–237, 4800 Oak Grove Dr., Pasadena, CA 91109, U.S.A.
[5] Caltech, Astronomy Program, MS 105–24, Pasadena, CA 91125, U.S.A.
[6] Radio Astronomy Laboratory, UC–Berkeley, Berkeley, CA 94720–3411, U.S.A.
[7] NASA–Ames, MS 245–3, Mountain View, CA 94035, U.S.A.
[8] Astrophysikalisches Institut, Schillergäßchen 2–3, Jena, D–07745, Germany
[9] LPL, The University of Arizona, Tucson, AZ 85721, U.S.A.
[10] SIRTF Science Center, Caltech, MS 314–6, Pasadena, CA 91125, U.S.A.
[11] NOAO, P.O. Box 26732, Tucson, AZ 85726, U.S.A.
[12] University of Rochester, Rochester, NY 14627–0171, U.S.A.
[13] Planetary Science Institute, 620 N. Sixth Avenue, Tucson, AZ 85705–8331, U.S.A.

Abstract. We will utilize the sensitivity of SIRTF through the Legacy Science Program to carry out spectrophotometric observations of solar-type stars aimed at (1) defining the timescales over which terrestrial and gas giant planets are built, from measurements diagnostic of dust/gas masses and radial distributions; and (2) establishing the diversity of planetary architectures and the frequency of planetesimal collisions as a function of time through observations of circumstellar debris disks. Together, these observations will provide an astronomical context for understanding whether our solar system – and its habitable planet – is a common or a rare circumstance.

Achieving our science goals requires measuring precise spectral energy distributions for a statistically robust sample capable of revealing evolutionary trends and the diversity of system outcomes. Our targets have been selected from two carefully assembled databases of solar-like stars: (1) a sample located within 50 pc of the Sun spanning an age range from 100-3000 Myr for which a rich set of ancillary measurements (e.g. metallicity, stellar activity, kinematics) are available; and (2) a selection located between 15 and 180 pc and spanning ages from 3 to 100 Myr. For stars at these distances SIRTF is capable of detecting stellar photospheres with SNR >30 at $\lambda \leq 24\mu$m for our entire sample, as well as achieving SNR >5 at the photospheric limit for over 50% of our sample at $\lambda = 70\mu$m. Thus we will provide a *complete* census of stars with excess emission down to the level produced by the dust in our present-day solar system.

SIRTF observations obtained as part of this program will provide a rich Legacy for follow–up observations utilizing a variety of facilities including the VLT. More information concerning our program can be found at **http://gould.as.arizona.edu/feps**.

1 Introduction

The **S**pace **I**nfra**R**ed **T**elescope **F**acility (SIRTF) is a key element of NASA's *Origins* program [20] . The 85 cm cryogenic space telescope will be launched into an earth–trailing orbit in 2002. There are three science instruments on–board: IRAC, IRS, and MIPS which will provide imaging and spectroscopy from 3.6–160 μm for an estimated mission lifetime of $\sim$ 5 yrs. The *SIRTF Legacy Science Program* was established to provide access to large coherent datasets as rapidly as possible in support of general observer (GO) proposals. In addition to the program described here there are complementary programs to survey nearby star–forming (Evans et al.), the inner galactic plane (Churchwell et al.), star–formation in nearby galaxies (Kennicutt et al.), a wide–field extragalactic survey (Lonsdale et al.), and a deep pointed survey (Dickinson et al.). For more information concerning the SIRTF Legacy Science Program please visit `http://sirtf.caltech.edu`.

Our modern understanding of the ubiquity of dust disks associated with young stars began with the revelations provided by SIRTF's ancestor IRAS [15]. Later, ISO produced a more complete census of optically–thick disks within 200 pc and revealed the rich dust mineralogy and gas content of these disks (see [13] for review). Understanding the evolution of young circumstellar dust and gas disks as they transition through the planet–building phase requires the $\times 100$ enhancement in sensitivity and increased photometric accuracy offered by SIRTF at far-infrared wavelengths. Concerning dust surrounding main sequence stars, IRAS discovered the prototypical debris disks [1] and ISO made additional limited-sample surveys [6]. Neither IRAS nor ISO were sensitive enough to detect dust in solar systems older than a few hundred Myr for any but the nearest tens of stars. SIRTF will detect orders of magnitude smaller dust masses, down to below the mass in small grains inferred for our own present-day Kuiper Belt (6×10^{22} g) surrounding a solar–type star at 30 pc!

We will probe circumstellar dust properties around a representative sample of primordial disks (dominated by ISM grains in the process of agglomerating into planetesimals) and debris disks (dominated by collisionally generated dust) over the full range of dust disk optical–depths diagnostic of the major phases of planet system formation and evolution. Our Legacy program is designed to complement those of Guaranteed Time Observers (GTOs) such that a direct link between disks commonly found surrounding pre–main sequence stars < 3.0 Myr old and our 4.56 Gyr old solar system can be made. Together, these data will help guide studies of the formation and evolution of planetary systems undertaken with facilities such as the VLT.

2 Science Strategy

We take advantage of the efficacy of infrared observations in revealing evidence for planetary systems embedded in dust distributions. In three coordinated modules we will: 1) conduct a survey of post–accretion circumstellar dust disks in order to understand evolution of disk properties (mass and radial structure) and dust properties (size and composition) during the main phase of planet–building and early solar system evolution for 150 F–G–K stars aged 3–100 Myr; 2) conduct a sensitive search for warm molecular hydrogen in a sub–sample of 50 targets from our dust disk survey, to constrain directly the time available for embryonic planets to accrete gas envelopes; and 3) trace for 150 F–G–K stars aged 100 Myr to 3 Gyr the evolution of dust disks generated through collisions of planetesimals and thereby infer the locations and masses of giant planets through their action on the remnant disk.

Understanding gas–dust dynamics is crucial to our ability to derive timescales important in planet formation and evolution. Modules (1) and (2) have an important synergy in furthering this understanding because dust dynamics are controlled by gas drag rather than radiation pressure when the gas-to-dust mass ratio is >10, while it is the presence of dust that mediates gas heating (and therefore detectability). Module (3) investigates epochs of terrestrial and ice-giant (Uranus- and Neptune-like) planet formation and the subsequent dynamical evolution of planetary systems. Our combined program will help place the formation and evolution of our own solar system in context, by providing the first estimates of the diversity of planetary architectures over the full range of radii relevant for planet formation. Our large sample will enable us to measure the *mean properties* of evolving dust disks, discover the *dispersion* in evolutionary timescales, and provide a database against which future studies can measure how various evolutionary paths might be *related to stellar properties.*

2.1 Formation of Planetary Embryos

Our experiment begins as the disks are making the transition from optically thick to thin, the point at which all of the disk's mass first becomes detectable through direct observation [17]. The goals are to: 1) constrain the initial structure and composition of $\tau < 1$ post-accretion disks; 2) measure changes in the dust particle size distribution due to coagulation of interstellar grains and shattering associated with high-speed planetesimal collisions; 3) characterize the timescales over which primordial disks dissipate and debris disks arise; and 4) infer the presence of newly formed planets at orbital radii of 0.3-30 AU.

Photometric observations from 3.6-160μm probe temperatures (radii) encompassing the entire system of planets in our solar system [2]. Detailed spectrophotometry from 5.3–40 μm will permit a search for gaps in disks caused by the dynamical interaction of young gas giant planets and the particulate disk from 0.2–10 AU [3]. Mid–infrared spectroscopic observations are sensitive to dust properties including size distribution and composition which in turn probe the physical conditions in the disk [7]. We will determine, for example, the relative importance of broad features attributed to amorphous silicates (ubiquitous in the ISM) compared to numerous narrow features throughout the 5.3–40 μm region due to crystalline dust (observed only in circumstellar environments [12]). In this way, we can diagnose radial mixing in the disk because the temperature required to anneal grains (> 1500K) is substantial higher than the inferred temperature of the emitting material (~ 300K). Further, the shape and strength of spectroscopic features can provide constraints on the fractional contribution of each grain population to the total opacity; a necessary ingredient to estimate dust mass surface densities.

2.2 Growth of Gas Giants

We will undertake the most comprehensive survey to date of H_2 gas in post–accretion disk systems in order to characterize its dissipation and to place limits on the time available for giant planet formation. We plan to survey 50 stars selected from our dust disk survey sample at high spectral resolution (R=600) from 10–37 μm, including both the S(0) 28 μm and S(1) 17 μm H_2 lines. We focus on the post-accretion epochs from 3–100 Myr to examine whether gas disks do indeed persist after disk accretion onto the star has ceased [19] and planetesimal agglomeration has provided "nucleation sites" for gas giant planet formation [8]. We estimate that if the gas and dust temperatures are within 40K and both are optically thin, the dust emissivity in the continuum can be suppressed enough to enable gas detection while maintaining the gas temperature via collisions. We expect to be sensitive to >2×10^{-4} $M_{\odot}[H_2]$ at 70–200 K.

2.3 Mature Solar System Evolution

We complement our investigation of the initial decay of both dust and gas signatures in first generation "primordial" disks with a comprehensive study of second generation "debris disks". The presence of *any* observable circumstellar dust around stars older than the maximum lifetime of a primordial dust disk (the sum of the to–be–determined gas dissipation timescale and the characteristic Poynting–Robertson drag timescale) provides compelling evidence not only for large reservoirs of planetesimals colliding to produce the dust, but also for the existence of massive planetary bodies that dynamically perturb planetesimal orbits inducing frequent collisions [10].

We will undertake the first comprehensive survey of solar-type stars with ages 100 Myr to 3 Gyr sensitive to dust disks comparable to that characteristic of our own solar system throughout its evolution from 100–300 Myr (the last phase of terrestrial and ice giant planet–building in our solar system) through 0.3–1 Gyr (bracketing the "late heavy bombardment" impact peak in our own solar system) and finally to 1.0–3.0 Gyr (examining the diversity of evolutionary paths from early activity to mature planetary system). Spectroscopic observations from 5.3–40 μm enable diagnosis of gaps caused by giant planets [9] and estimates of dust size and composition which translate directly into constraints on the mass opacity coefficients for the dust [14] as well as Poynting-Robertson drag timescales [1].

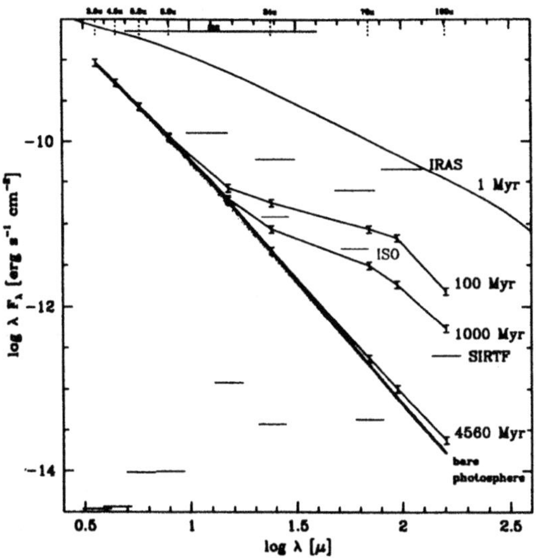

Fig. 1. Model SED for a hypothetical solar system based on a G2V stellar photosphere at 30 pc, asteroid belt zodiacal dust, and Kuiper belt dust for ages 4560, 1000, and 100 Myr along with an optically thick disk SED characteristic of <1-2 Myr old stars. Also shown are the IRAS, ISO, and SIRTF sensitivity limits. IRAS detected optically thick disks out to 160 pc. At 30 pc, ISO would have detected this young solar system at ages of a few hundred Myr while SIRTF will detect it at ages as old as the Sun

3 Observing Strategy

3.1 Sample Characteristics

To derive statistically meaningful results on the dust properties, we will observe ~50 stars in each of 6 logarithmically spaced age bins from 3 Myr (connecting our Legacy program to that of Evans et al.) to 3 Gyr (beyond which there is strong emphasis by GTO's). Our targets span a narrow mass range (0.8-1.2 $M_\odot$) and are proximate enough to enable a complete census for circumstellar dust comparable to our model solar system as a function of age ([11]; [16]). We will measure the stellar photosphere at SNR>30 for $\lambda \leq 8\mu$m at ages < 100 Myr, and SNR>20 for $\lambda \leq 70\mu$m at ages > 100 Myr (subject to calibration uncertainties). To identify gaps in the dust distribution created by the presence of giant planets from 0.2–10 AU, we require relative spectrophotometry with SNR>30 from 5.3–40 μm.

For the gas evolution module, we have chosen 10 stars for first-look observations with the high resolution mode of the IRS. This sample spans a range of spectral type (F3–K5), age (3–100 Myr), L_x/L_{bol} ratios (10^{-3} – 10^{-5}). Because the line-to-continuum ratio starts to limit our detectable H_2 line flux at R=600 when the continuum at 20μm is >100 mJy these sources are chosen to have optically-thin mid-infrared excess emission on the basis of IRAS and ISO observations. These observations will enable us to explore the limits implied by null results and guide our choice of follow–up observations for 40 more stars drawn from our dust disk survey. Our goal is a quantitative limit on the lifetime of gas–rich disks capable of forming giant planets.

Table 1. Proposed SIRTF Observations

Instr.	**# stars**	**Total Time inc. overhead**	**SNR on Photosphere**	**Objectives**
IRAC	300	40 hrs [419s/star]	>30 all bands	3.6, 4.5, 5.8 and 8.0 μm photometry. **Measure photosphere** and hot dust excess.
MIPS	300	135 hrs [1100-1900s/star]	>30 at 24μm >5 at 70μm	24, 70, and 160 μm phot. **Complete census** for dust at 24 μm and most of sample at 70 μm.
IRS (Lo)	300	125 hrs [740-2020s/star]	>30 at 5.3-14.2μm >10 at 14.2-40μm	R $\sim$ 60-120 spectra. **Detailed SED** and spectral analysis.
IRS (High)	50	50 hrs [720-5180s/star]	>3 to detect 2×10^{-4} $M_\odot$ of H_2	R $\sim$ 600 spectra 10-37μm. **Measure H_2 gas mass** & resolve dust features.
SUM		**350 hrs**		

3.2 SIRTF Data

The SIRTF data, in conjunction with the ancillary observations described below, will be used to: (1) establish the contribution of the stellar photosphere to the observed spectral energy distribution; (2) measure any excess infrared emission from estimates of the opacity of the dust as a function of wavelength; and (3) determine the amount, distribution, and composition of the circumstellar material through mid–infrared spectroscopy. Our goal is to collect data capable of realizing the fundamental limits imposed by instrument stability and systematic calibration uncertainties. Integration times are chosen according to each star's distance, age and spectral type to reach uniform SNR at the photospheric limits. Table 1 summarizes the observations, their most basic objectives, and the total amount of observing time requested including all overheads.

3.3 Ancillary Data

While SIRTF observations alone are an extremely valuable dataset, the scientific impact of our program will be enhanced when these data are combined with those at shorter and longer wavelengths. The ancillary data that will be assembled and provided to the community for each star in our sample is summarized in Table 3. We will search for dust located at large radii and too cold to radiate strongly in the MIPS 160 μm band by obtaining sub–mm continuum observations for every source in our sample. These observations will provide us with measurements of (or constraints on, in the case of nondetections) the spectral slope ($F_\nu \sim \nu^{-\alpha}$) which can be used to constrain particle size distributions. We also plan a limited campaign of 5 μm, 10 μm and 20 μm imaging using the MMT, Keck, Magellan, and the VLT for the brightest objects in our sample. Intermediate–band photometry in carefully chosen atmospheric windows (e.g. 5.3 and 11.7 μm) will provide a useful check on the SIRTF calibration when tied to the same photometric standards of Cohen et al. [5].

Table 2. Ancillary Data

Instrument	# stars	Observing Time	Objective
Tycho/Hipparcos	300	Public	Proper Motions
Tycho/Hipparcos	300	Public	B_T, V_T, and H_p photometry
2MASS	300	Public	J, H, K_S photometry
Optical Spectrographs	300	In hand	Spectral classification
Mid-Infrared Imagers	30	4 runs	5-25μm imaging photometry
SMT/CSO/SEST	300	17 weeks	sub–mm photometry

4 The Legacy

The combined SIRTF + Ancillary Data catalog along with specific tools for reduction, calibration, and interpretation of data will create a rich Legacy in science and in services provided to the community.

First, we will construct 3.6-160μm + sub–mm spectral energy distributions for a representative sample of ~300 F-G-K stars aged 3 Myr - 3 Gyr within 15-160 pc of the Sun, providing a complete set of ancillary data characterizing the properties of the stars.

Second, we will combine these data with model calculations to discern the timescales for gas giant and terrestrial planet formation in circumstellar disks, and the evolution of these systems on Myr to Gyr timescales.

Third, we will provide a database of targets for follow–up with SIRTF and other platforms. Our results will support GO programs with similar scientific aims and help in selecting samples (by age, mass, metallicity, etc.) for additional work. Follow-up with MIPS in SED mode and with IRS at high spectral resolution are areas in which GO's can directly exploit the Legacy database. Our SIRTF+Ancillary results will also have a strong impact on future space–based infrared surveys, as well as programs enabled by new ground-based facilities such as those soon available on the VLT. Follow–up observations with VISIR are capable of directly resolving the thermal disk emission from 10–20 μm [18]. Further, high velocity resolution spectroscopic observations could resolve the line profiles of warm circumstellar gas indicating its radius of origin in a Keplerian disk [4]. Additional studies utilizing ISAAC, CONICA, and VLTI with MIDI will yield further insights.

Fourth, due to the nature of our program (large sample of uniform observations) we hope to assist the SIRTF Science Center and instrument teams in improving the photometric accuracy of SIRTF from the initial projections of 20% to target values of 5 % for the benefit of the entire astronomical community.

Fifth, a second fundamental data product of this Legacy project derives from the serendipitous discoveries made as part of our primary survey. Our program involves observing ~300 fields with $38'' \times 38''$ field of view to the limiting sensitivity of SIRTF at 3.6, 4.5, 5.8, & 8 μm and these same ~300 field centers with $5' \times 5'$ field of view at 24, 70 & 160 μm, over galactic latitudes ranging from $b = 20°$ to $b = 90°$.

References

1. Backman, D.E., & Paresce, F.: In: *PP III*, ed. by E. Levy, J. Lunine (University of Arizona Press, Tucson 1993) pp. 1253
2. Beckwith, S.V.W.: In *Origins of Stars and Planetary Systems*, ed. by C. Lada and N. Kylafis (Kluwer Academic Press, Dordrecht 1999) pp. 579
3. Bryden, G., Rozyczka, M., Lin, D.N.C., & Bodenheimer, P.: ApJ **540** 1091 (2000)
4. Carr, J.S., Mathieu, R.D., & Najita, J.: ApJ **551** 454 (2001)
5. Cohen, M. et al.: AJ **117** 1864 (1999)
6. Habing, H.J., et al.: A&A **365** 169 (2001)
7. Hanner, M.S., Lynch, D.K., & Russell, R.W.: ApJ **425** 274 (1994)
8. Hollenbach, D.J., Yorke, H.W, & Johnstone, D.: In *PP IV* ed. by V. Mannings, A.P. Boss, and S.S. Russell (University of Arizona Press, Tucson 2000) pp. 533
9. Liou, J.C. & Zook, H.A.: AJ **118** 580 (1999)
10. Malhotra, R., Duncan, M.J. & Levison, H.F.: In *PP IV* ed. by V. Mannings, A.P. Boss, and S.S. Russell (University of Arizona Press, Tucson, 2000) pp. 1231
11. Mamajek, E.E. & Feigelson, E.D.: In *Young Stars Near Earth: Progress and Prospects* ed. by R. Jayawardhana and T.P. Greene (A.S.P. Conference Series, San Francisco, 2001) in press
12. Meeus, G. et al.: A&A **365** 476 (2001)
13. Meyer, M.R. & Beckwith, S.V.W.: In *ISO Surveys of a Dusty Universe* ed. by D. Lemke, M. Stickel, & K. Wilke (Springer–Verlag, Heidelberg, 2000) pp. 347
14. Miyake, K., Nakagawa, Y.: Icarus **106** 20 (1993)
15. Rucinski, S.M.: AJ **90** 2321 (1985)
16. Soderblom, D.R., King, J.R., & Henry, T.J.: AJ **116** 396 (1998)
17. Spangler, C. et al.: ApJ **555** 932 (2001)
18. Telesco, C. et al.: ApJ **530** 329 (2000)
19. Thi, W. F., et al. : Nature **409** 60 (2001)
20. Werner, M., et al.: In *The Extragalactic Infrared Background and its Cosmological Implications* IAU Symp. **204** 439 (2001)

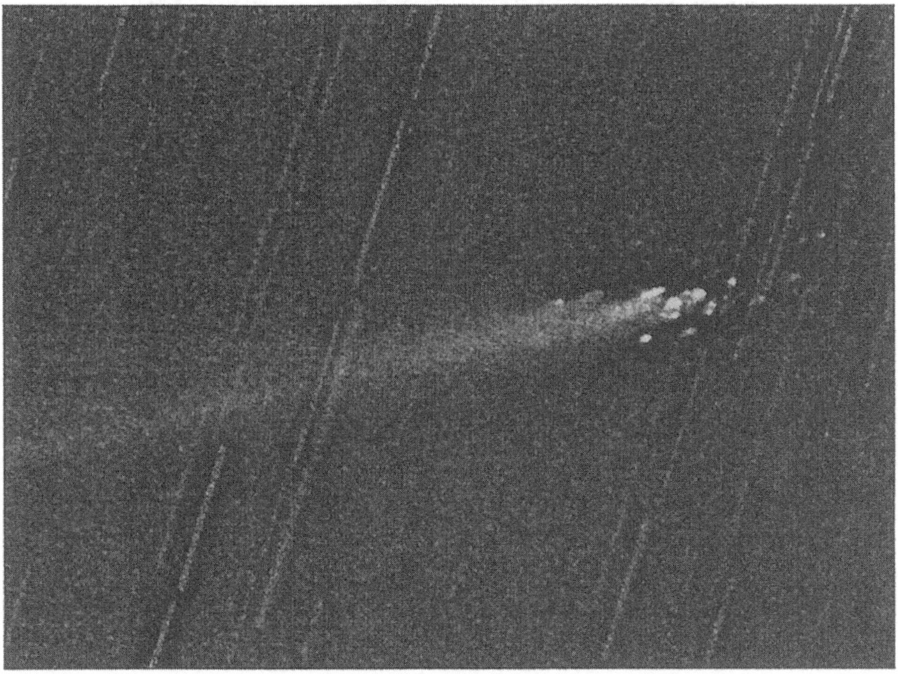

Comet LINEAR (C/1999 S4). Break-up of nucleus
(Antu/UT1 + FORS1)

Philippe André

The Herschel Space Observatory and the Earliest Stages of Star Formation

Philippe André

CEA Saclay, Service d'Astrophysique,
F-91191 Gif-sur-Yvette, France

Abstract. Despite recent progress, both the earliest stages of protostellar collapse and the origin of stellar masses remain poorly understood. Since prestellar condensations and young protostars have $T_{bol} < 30$ K and emit the bulk of their luminosity between $\sim 80\ \mu$m and $\sim 350\ \mu$m, a large submillimeter space telescope such as *FIRST/Herschel* will undoubtedly yield significant breakthroughs in this area. In particular, *Herschel* will provide a unique probe of the energy budget and temperature structure of pre-/proto-stellar condensations. With an angular resolution at 90–300 μm comparable to, or better than, the largest ground-based millimeter radiotelescopes, *Herschel* will make possible complete surveys for such condensations in all the nearby ($d < 1$ kpc) cloud complexes of the Galaxy down to the proto-brown dwarf regime. These surveys should greatly help develop a satisfactory theory for the origin of the IMF.

1 Introduction: The Herschel Space Observatory

Herschel, formerly known as *FIRST* (for "Far InfraRed and Submillimeter Telescope"), is the fourth cornerstone mission in the ESA science programme (see [27]). It was renamed *Herschel Space Observatory* (*HSO*) at the end of 2000 to celebrate William Herschel's discovery of infrared light in 1800. To be launched in 2007, *Herschel* will be the first large space facility to completely cover the entire far-IR and submillimeter range from $\sim 60\ \mu$m to $\sim 700\ \mu$m.

With a 3.5-m diffraction-limited telescope, *Herschel* will provide good ($\sim 7''$–$17''$) angular resolution for the first time in the far-IR (at $\lambda \sim$ 90–250 μm). It will be equipped with two imaging instruments: a Photodetector Array Camera and Spectrometer (PACS – e.g. [28]), providing simultaneous broad-band photometry (with full beam sampling) at 110 μm (or 75 μm) and 170 μm over an instantaneous field of view of $\sim 1.75' \times 3.5'$; and a Spectral and Photometric Imaging REceiver (SPIRE – e.g. [16]), able to perform simultaneous broad-band photometry at 250 μm, 350 μm, and 500 μm over $4' \times 8'$. In mapping mode, the expected 5σ sensitivity in 1 hr of integration time is ~ 3 mJy/beam for PACS and ~ 7 mJy/beam for SPIRE (over their respective fields of view). Note that this is a factor ~ 10 more sensitive (i.e., a factor ~ 100 faster) than SOFIA (cf. [7]). The third instrument is a single-pixel heterodyne spectrometer (HIFI – e.g. [12]) for high-resolution spectroscopy ($\lambda/\Delta\lambda \lesssim 10^7$ or $\Delta v \sim 0.03$–$300\,\mathrm{km\,s^{-1}}$) with complete wavelength coverage from $\sim 240\ \mu$m and $\sim 625\ \mu$m and between $\sim 157\ \mu$m and $\sim 213\ \mu$m.

Once operational *Herschel* will offer a minimum of 3 years of routine observations, with roughly 2/3 of the available observing time being open to the general astronomical community. More details about the mission may be obtained from the *Herschel* web page (`http://astro.estec.esa.nl/FIRST/`) maintained by Göran Pilbratt, the ESA Project Scientist.

The two main science objectives of *Herschel* are 1) the study of primordial galaxies in the early Universe, and 2) the study of star formation in the interstellar medium (ISM) of our Galaxy as well as nearby galaxies. The first topic is itself closely related to star formation since a key goal here is to characterize the evolution of the star-formation rate in primordial galaxies as a function of redshift. Both primordial galaxies and young protostars have spectral energy distributions (SEDs) that peak around 200 μm (e.g. Fig. 1), which is precisely the prime wavelength range of *Herschel*.

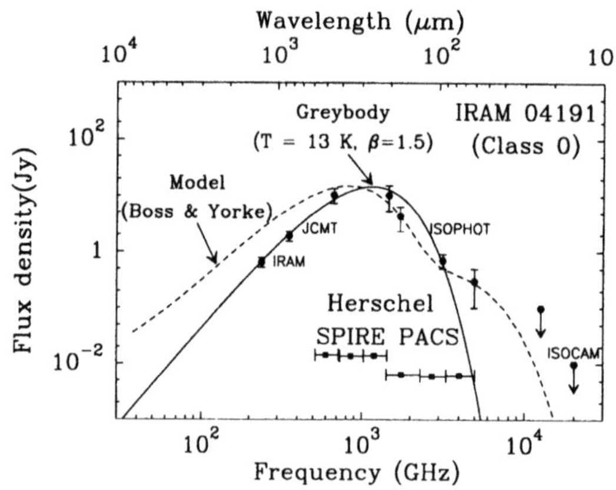

Fig. 1. Spectral energy distribution of the Class 0 protostar IRAM 04191 (cf. [3]). This object is at $d = 140$ pc and has $L_{bol} \approx 0.15\,L_{\odot}$, $T_{bol} \approx 18$ K, $M_{env} \approx 0.5\,M_{\odot}$ (in a $1'$-diameter aperture). The six photometric bands of SPIRE and PACS on *Herschel* are shown, along with the corresponding (10σ, 1hr) sensitivities (in mapping mode). The solid curve is a greybody dust spectrum which fits the SED longward of 90 μm; the dashed curve shows the model SED of a single "first protostellar core" seen along its rotational axis ([10]). *Herschel* is ideally suited for detecting and characterizing all such cold protostars to $M_{proto} \sim 0.03$–$0.1\,M_{\odot}$ and $d \sim 1$ kpc in the Galaxy

2 Prestellar Cores and Young Protostars

Herschel will be particularly well suited for studying the earliest stages of star formation and cloud collapse. These stages are of crucial interest as they can differentiate between collapse models and, to some extent at least, must govern the origin of stellar masses. So far, observational progress in this area has been seriously hindered by two main factors: the associated timescales are short ($\lesssim 10^4$-10^5 yr) and the corresponding SEDs peak around $\lambda \sim 100$–300 μm (e.g.

Fig. 1), i.e., in a wavelength range which has been inaccessible with good resolution and sensitivity up to now. While *IRAS*, *ISO*, and ground-based infrared studies have provided a fairly complete census of evolved protostars and pre-main sequence objects (i.e., near-IR sources of Classes I to III) in nearby clouds (e.g. [35], [29], [9]) no such census exists yet for younger (Class 0) accreting protostars and cold prestellar cores/condensations. Only about thirty Class 0 protostars are known to date ([4]), which all are relatively massive ($M \gtrsim$ 0.5-1 $M_\odot$) and were discovered either serendipitously (e.g. [11]) or through their powerful outflows (see [5]). With present ground-based (sub)-millimeter telescopes, systematic surveys for prestellar condensations and cold protostars are possible only down to $\sim 0.1\,M_\odot$ in nearby ($d \sim$ 150 pc) clouds such as the ρ Ophiuchi cloud (cf. [23], [18]). Even in the Taurus cloud complex where stars are known to form in relative isolation, the angular resolution of *ISO* around $\sim$ 100–200 μm was barely sufficient to probe the emission from individual cores and protostars (cf. [33]). Furthermore, the recent millimeter discoveries of the Class 0 object IRAM 04191+1522 ([3] - see Fig. 1) and of the cold $H^{13}CO^+$ condensation MC 27 ([26]) clearly show that the current census of protostars in Taurus is incomplete and that there may exist a significant, as yet unknown, population of cold protostars with $L_{bol} \lesssim 0.1\,L_\odot$ in this cloud. This example emphasizes the need for high-resolution, extensive surveys of molecular clouds in the submillimeter band.

3 Wide-Field Surveys of Nearby Molecular Clouds

Equipped with the PACS and SPIRE bolometer arrays (e.g. [28], [16]), *Herschel* will have the ability to carry out deep, wide-field imaging surveys of nearby molecular clouds at 75–170 μm and 250–500 μm, respectively. These surveys should tremendously improve our knowledge of the first phases of protostellar collapse, on both individual (§ 3.1) and global (§ 3.2) scales in the Galaxy.

3.1 Structure and Energy Balance of Individual Cloud Cores

Unbiased submillimeter continuum surveys with SPIRE and PACS will detect large, *complete samples of young protostars and prestellar condensations*, down to much smaller masses ($M \lesssim$ 0.03 $M_\odot$) than is possible from the ground (see § 3.2 below). The much better angular resolution of *Herschel* ($\lesssim$ 7″ at 90 μm with PACS) compared to *IRAS*, *ISO*, or *SIRTF* in the far-IR will be sufficient to separate the main individual members of nearby ($d \lesssim$ 800 pc) embedded clusters (which all have stellar surface densities < 2000 stars pc^{-2}, except the Trapezium). *Herschel* will provide, for the first time, *accurate bolometric luminosities* (down to low values $\lesssim 0.01\,L_\odot$) for many cold protostellar sources, thanks to a good sampling of the SEDs with six photometric bands between $\sim 75\,\mu$m and $\sim 500\,\mu$m (see, e.g., Fig. 1). This will make it possible to fully exploit the potential of the M_{env}–L_{bol} and L_{bol}–T_{bol} diagrams (e.g. [2], [30], [25]) as practical evolutionary diagrams for embedded protostars.

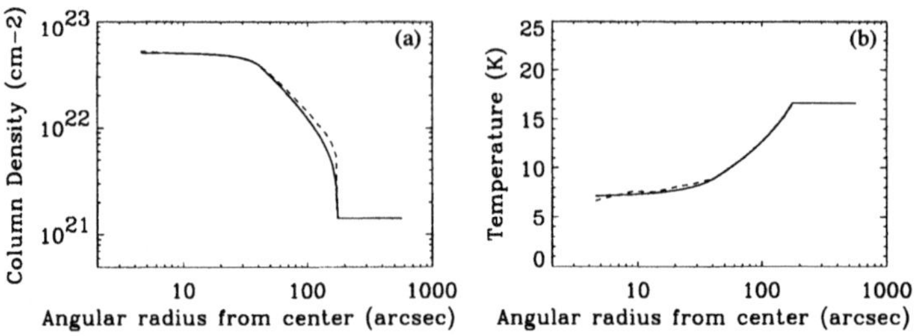

Fig. 2. Simulated *Herschel* observations of the radial structure of prestellar cores. The solid curves show the column density profile (a) and the temperature distribution (b) in a model cloud core at $d = 160$ pc, consistent with recent ISOCAM absorption results ([6]) and radiative transfer calculations ([14], [38]). The dashed curves (almost indistinguishable from the solid curves) show the reconstructed distributions that would be derived from *Herschel* maps at 6 bands between 75 μm and 500 μm (plus ground-based data at 850 μm and 1.3 mm), assuming the 3-D core geometry is known.

Second, using combined SPIRE and PACS images to construct 75–500 μm SED maps for at least the nearest (spatially resolved) sources, it will be possible to derive the *temperature distribution* within both prestellar condensations/cores and protostellar envelopes. Recent modeling of the thermal energy balance suggests that starless (externally-heated) cloud cores are not isothermal but significantly colder in their central regions (with T as low as $\sim$ 5–7 K) than in their outer parts ([22]; [14]; [38]). *Herschel* will directly measure the magnitude of this effect (see Fig. 2).
Using complementary ground-based dust continuum observations at longer submillimeter wavelengths, the *column density structure* and the *dust properties* (e.g. the opacity index β) of the same sources will also be measurable with unprecedented accuracy. Comparison between the structure of prestellar condensations and that of the envelopes surrounding the youngest protostars will give insight into the *initial conditions of individual protostellar collapse.* These initial conditions hold the key to understanding early protostellar evolution and, in particular, control the history of the mass accretion rate at the Class 0 and Class I stages (e.g. [15]; [17]).
Promising results have been obtained in this area using 800–1300 μm emission maps (e.g. [34]), as well as mid-IR or near-IR absorption maps (e.g. [34], [6], [1]). However, because both the temperature and the density distribution affect the dynamics of cloud collapse, the only way to reach definitive conclusions is to constrain both distributions simultaneously through multi-band imaging around the peak of the emission spectrum.

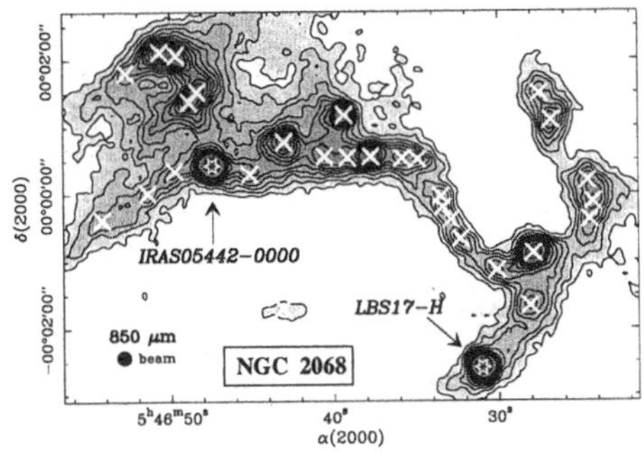

Fig. 3. Dust continuum map of the NGC 2068 protocluster at 850 μm extracted from a SCUBA mosaic of NGC 2068/2071 ([24]). The mean rms noise level is $\sim$ 22 mJy/13″-beam. A total of 30 compact starless condensations (cf. crosses), with masses between $\sim 0.4\,M_\odot$ and $\sim 4.5\,M_\odot$, are detected in this field. *Herschel* will provide images of comparable angular resolution around 200 μm, with a much better mass sensitivity (by $\sim$ 1 order of magnitude) over much wider fields (square degrees)

3.2 Origin of the Stellar Initial Mass Function

On a more global level, wide-field submillimeter imaging of both active and quiescent regions with *Herschel* will allow us to better understand the origin of stellar masses and the *nature of the fragmentation process in molecular clouds*, for which we still have no satisfactory theory (e.g. [13]). Sensitive submillimeter dust emission maps have the remarkable property that they can probe cloud structure, prestellar condensations, collapsing/accreting protostars, and post-collapse circumstellar envelopes/disks, *simultaneously*. Thus, they make it possible to investigate the genetic link between cloud cores and protostars, hence the origin of the stellar initial mass function (IMF).

This point is illustrated by the results of recent ground-based dust continuum surveys around 1 mm. In particular, a 1.3 mm imaging survey of the central $\sim$ 480 arcmin^2 portion of the ρ Oph cloud ($d \approx 150$ pc) with the MPIfR bolometer array on the IRAM 30 m telescope ([23]) led to the identification of 58 starless 'condensations' (undetected by ISOCAM in the mid-IR – cf. [9]) with characteristic scales of $\sim$ 2500–5000 AU. Follow-up molecular-line observations ([8]) showed that most of these starless condensations were gravitationally bound and thus likely prestellar in nature. The mass distribution of these 58 prestellar condensations, complete down to $\sim 0.1\,M_\odot$, was found to be remarkable in that it mimicked the shape of the stellar IMF ([23], see Fig. 4a). It is also very similar in shape to the YSO mass function recently determined down to $\sim 0.06\,M_\odot$ for the Class II sources of the ρ Oph cluster from ISOCAM 7 μm and 15 μm observations ([9], see Fig. 4a). Such a resemblance to the IMF suggests that *the starless condensations detected in the (sub)millimeter dust continuum on the same spatial scales as protostellar envelopes are the direct progenitors of*

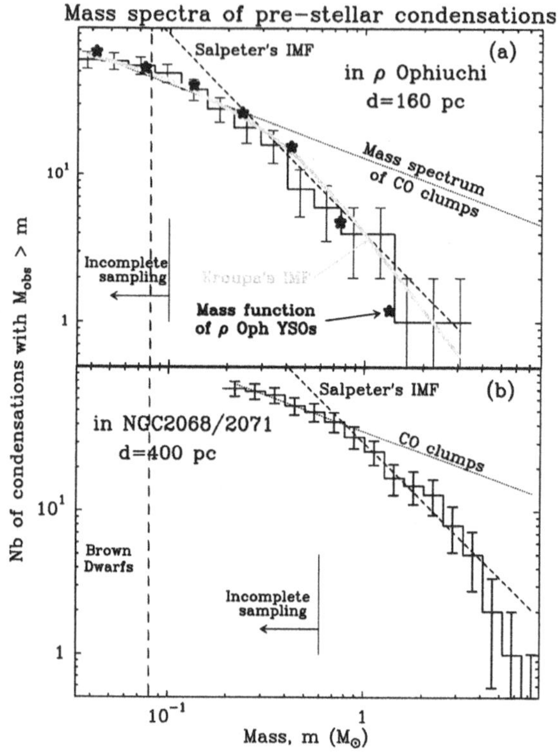

Fig. 4. Cumulative mass distributions of the prestellar condensations of the ρ Oph (**a**) and NGC 2068/2071 (**b**) protoclusters ([23], [24]). The dotted and dashed lines show power-laws of the form $N(>m) \propto m^{-0.5}$ (mass spectra of CO clumps, see [36]) and $N(>m) \propto m^{-1.35}$ (Salpeter's IMF), respectively. The solid curve in (**a**) shows the shape of the field star IMF ([21]), and the star markers represent the mass function of ρ Oph YSOs derived from an extensive mid-IR survey with ISOCAM ([9])

individual stars or systems. In agreement with this view, some of the ρ Oph condensations exhibit spectroscopic evidence of collapse ([8]).
By contrast, recall that the typical clump mass spectra found by large-scale molecular line studies (cf. [36]) are much shallower than the stellar IMF above $\sim 0.5\,M_\odot$ (see Fig. 4). The difference presumably arises because, up to now, line studies have been primarily sensitive to transient, unbound structures (cf. [20]) which are not immediately related to star formation.

Other studies have found prestellar mass spectra consistent with the IMF. An OVRO interferometer mosaic of the inner ~ 30 arcmin^2 region of the Serpens cloud core at 3 mm revealed 26 starless condensations above $\sim 0.5\,M_\odot$ with a mass spectrum close to the Salpeter IMF ([31]). An 850 μm SCUBA survey of the ρ Oph main cloud with JCMT essentially confirmed the above-mentioned IRAM results ([18]). SCUBA was also used to image a ~ 575 arcmin^2 field at

850 μm and 450 μm around the NGC 2068/2071 protoclusters in Orion B ([24]). The images (e.g. Fig. 3) reveal a total of $\sim$ 70 compact starless condensations whose mass spectrum is again reminiscent of the IMF between $\sim 0.6\,M_\odot$ and $\sim 5\,M_\odot$ (see Fig. 4b).

These recent findings on the mass spectrum of protocluster condensations are very encouraging because they support scenarios according to which *the low-mass end of the IMF is at least partly determined by turbulent fragmentation at the prestellar stage of star formation* (see, e.g., [13] and [19]). It is nevertheless clear that present ground-based dust continuum studies are limited by small-number statistics due to insufficient sensitivity and mapping speed. Submillimeter surveys with *Herschel* will be $\sim$ 2 orders of magnitude faster and can probe $\sim$ 1 order of magnitude deeper into the mass distributions of prestellar condensations and young protostars. For instance, one would need only $\sim$ 15 days to survey 100 deg^2 with SPIRE/*Herschel* down to the estimated cirrus confusion limit ($\sigma_{250\mu} \sim$ 10 mJy/17″-beam) in a region like the Taurus complex. Such a sensitivity would be sufficient to detect (and spatially resolve) proto-brown dwarfs of temperature $T_{proto} = 10K$ and mass $M_{proto} \sim 0.03\ M_\odot$ at the 5σ level in the nearest clouds ($d = 150$ pc). Furthermore, the mass uncertainties will be much reduced since coordinated SPIRE and PACS observations between $\sim 80\,\mu$m and $\sim 500\,\mu$m will strongly constrain the temperature and emissivity of the dust, as well as the nature of the objects (see § 3.1 above).

4 Conclusions

Large-scale, multi-band continuum surveys of molecular clouds at $\lambda \sim$ 75–500 μm with the SPIRE and PACS imaging instruments on *Herschel* will yield a complete census of prestellar condensations and young accreting protostars in the nearby ($d < 1$ kpc) ISM. This will set direct constraints on the lifetimes of the earliest phases of star formation and will provide the corresponding luminosity/mass functions. Follow-up, high-resolution spectroscopy with both the HIFI heterodyne instrument and ALMA will give quantitative constraints on the dynamical and chemical states of the most interesting objects.

Other major star formation projects envisaged with *Herschel* include: (i) an unbiased mapping of the richest part ($\sim 5° \times 360°$) of the Galactic plane at 250–500 μm with SPIRE to probe the distribution of massive star-forming regions as a function of galacto-centric radius; and (ii) complete spectral line surveys between $\sim 100\,\mu$m and $\sim 600\,\mu$m with HIFI and PACS for a broad sample of objects to determine the cooling rates of the gas, as well as the abundances of the principal atoms and molecules in a variety of ISM conditions and evolutionary states (cf. [32]).

References

1. Alves, J.F., Lada, C.J., & Lada, E.A. 2001, Nature, 409, 159
2. André, P., Montmerle, T. 1994, ApJ 420, 837

3. André, P., Motte, F., Bacmann, A. 1999, ApJL, 513, L57
4. André, P., Ward-Thompson, D., Barsony, M. 2000, in Protostars and Planets IV, eds. V. Mannings, A.P. Boss, & S.S. Russell (Univ. of Arizona Press, Tucson), p. 59
5. Bachiller, R. 1996, ARA&A, 34, 111
6. Bacmann, A., André, P., Puget, J.-L., Abergel, A., Bontemps, S., & Ward-Thompson, D. 2000, A&A, 361, 555
7. Becklin, E.E. 1997, in The Far InfraRed and Submillimetre Universe, ESA SP-401, p. 201
8. Belloche, A., André, P., Motte, F. 2001, in From Darkness to Light, Eds. T. Montmerle & P. André, ASP Conf. Ser., 243, p. 313
9. Bontemps, S., André, P., Kaas, A.A., Nordh, L., Olofsson, G. et al. 2001, A&A, 372, 173
10. Boss, A.P., & Yorke, H.W. 1995, ApJ 439, L55
11. Chini, R., Reipurth, B., Sievers, A., Ward-Thompson, D., et al. 1997, A&A, 325, 542
12. de Graauw, Th., & Helmich, F.P. 2001, in The Promise of FIRST, ESA SP-460, in press
13. Elmegreen, B.G. 2001, in From Darkness to Light, Eds. T. Montmerle & Ph. André, ASP Conf. Ser., 243, p. 255
14. Evans, N. J., Rawlings, J. M. C., Shirley, Y. L., & Mundy, L. G. 2001, ApJ, in press
15. Foster, P.N., Chevalier, R.A. 1993, ApJ 416, 303
16. Griffin, M.J., Swinyard, B.M., & Vigroux, L. 2001, in The Promise of FIRST, ESA SP-460, in press
17. Henriksen, R.N., André, P., & Bontemps, S. 1997, A&A, 323, 549
18. Johnstone, D., Wilson, C. D., Moriarty-Schieven, G., et al. 2000, ApJ, 545, 327
19. Klessen, R. S., & Burkert, A. 2001, ApJ, 549, 386
20. Kramer, C., Stutzki, J., Rohrig, R., Corneliussen, U. 1998, A&A, 329, 249
21. Kroupa, P., Tout, C. A., Gilmore, G. 1993, MNRAS, 262, 545
22. Masunaga, H., Inutsuka, S. 2000, ApJ, 531, 350
23. Motte, F., André, P., Neri, R. 1998, A&A, 336, 150
24. Motte, F., André, P., Ward-Thompson, D., & Bontemps, S. 2001, A&A, 372, L41
25. Myers, P.C., Adams, F.C., Chen, H., & Schaff, E. 1998, ApJ, 492, 703
26. Onishi, T., Mizuno, A., Fukui, Y. 1999, PASJ, 51, 257
27. Pilbratt, G.L. 2000, Proc. SPIE, Vol. 4013, 142
28. Poglitsch, A., Waelkens, C., & Geis, N. 2001, in The Promise of FIRST, ESA SP-460, in press
29. Prusti, T. 1999, in The Universe as seen by ISO, eds. P. Cox & M.F. Kessler, ESA SP-427, p. 453
30. Saraceno P., André P., Ceccarelli C., Griffin M., Molinari S. 1996, A&A, 309, 827
31. Testi, L., Sargent, A.I 1998, ApJL, 508, L91
32. van Dishoeck, E.F. 1997, in the Far Infrared and Submillimetre Universe, ESA SP-401, p. 81
33. Ward-Thompson, D., André, P., & Kirk, J.M. 2001, MNRAS, in press
34. Ward-Thompson, D., Motte, F., André, P. 1999, MNRAS, 305, 143
35. Wilking B.A., Lada C.J., Young E.T. 1989, ApJ 340, 823
36. Williams, J. P., Blitz, L., McKee, C. F. 2000, in Protostars & Planets IV, ed. V. Mannings, A. Boss, & S. Russell (Tucson: Univ. Arizona Press), p. 97
37. Wootten, A. 2001, Science with the Atacama Large Millimeter Array, ASP Conf. Ser., Vol. 235
38. Zucconi, A., Walmsley, C.M., & Galli, D. 2001, A&A, in press

Kueyen structure

Mark McCaughrean, Tom Ray, and Takuya Fujiyoshi

A Look Forward to Star and Planet Formation with the NGST

Mark J. McCaughrean

Astrophysikalisches Institut Potsdam,
An der Sternwarte 16, 11482 Potsdam, Germany

Abstract. The Next Generation Space Telescope is being built by NASA, ESA, and the CSA, to be launched to the Sun-Earth L2 point in late 2008. The present NGST design has a deployable 6.5 metre diameter primary mirror, diffraction-limited at 2 μm and passively cooled to 50K. Free of the bright background emission, poor transmission, and atmospheric seeing suffered by groundbased telescopes, the NGST will achieve extraordinary sensitivity for imaging and low/medium-resolution spectroscopy across the near- and mid-infrared, with excellent spatial resolution across wide fields-of-view. Here, I outline the present state of development of the NGST, its instrument complement, and a few key areas where it should be making unique contributions to star and planet formation studies in the decade 2010–2019.

1 The NGST Basic Architecture

The *Next Generation Space Telescope* (NGST) is under study by NASA, the European Space Agency (ESA), and the Canadian Space Agency (CSA), as a successor to the extraordinarily successful *Hubble Space Telescope* (HST), with a presently scheduled launch date of late 2008, and a goal lifetime of 10 years. The basic NGST architecture includes a deployable $\sim$6.5-m diameter primary mirror (cf. the monolithic 2.4-m diameter HST primary), and will be optimised for observations in the near-infrared and mid-infrared ($\sim$ 1–30 μm), with diffraction-limited performance at 2 μm (Figure 1).

To be optimised for infrared wavelengths, the NGST must be cold, in order to reduce its thermal self-emission background to levels below that of the zodiacal light. The large aperture of the NGST precludes the use of cryogens, as in ISO and SIRTF for example, and thus passive cooling techniques will be employed. This requires that the NGST be located far from the Earth to avoid being heated by it, and the mission scenario places the NGST behind a large sunshield at the Earth-Sun L2 Lagrange point, some 1.5 million kilometres outside the Earth's orbit. The primary mirror should end up at 50 K, and thus the observatory will be zodiacal-light limited out to 12 μm at least.

These specifications ensure that in the decade 2010–2019, when the first ELTs (e.g., CELT, GSMT, OWL) may be (coming) online with MCAO, the NGST will be uniquely sensitive for wide-field, high-resolution imaging surveys and low- to medium-resolution multiobject spectroscopy at near-infrared (1–2.5 μm) wavelengths, and greatly more so at thermal wavelengths beyond 2.5 μm, where the NGST thermal background is $\sim 10^4$–10^6 less than that seen on the ground.

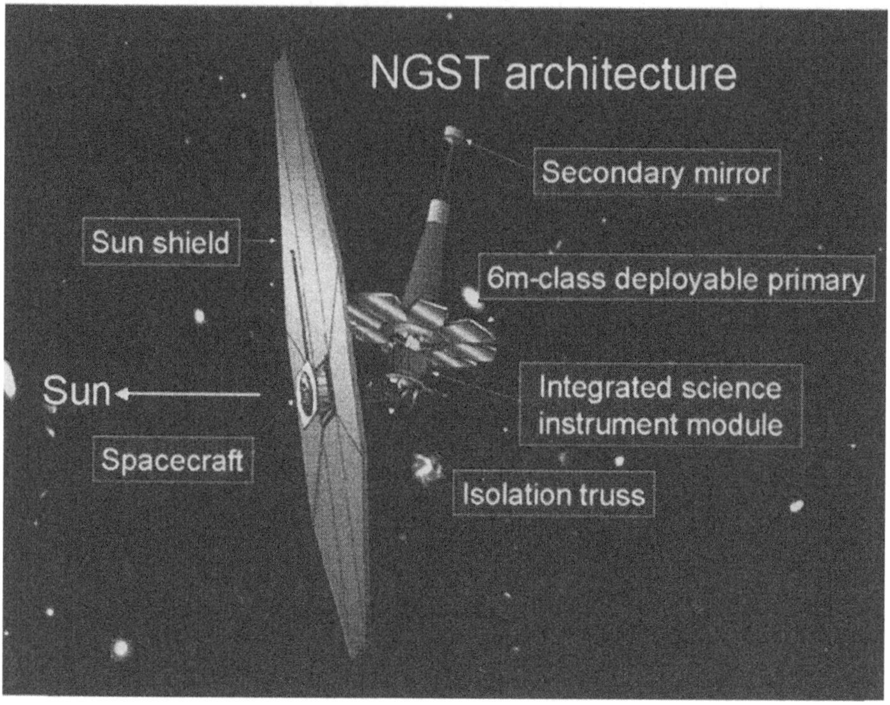

Fig. 1. A schematic representation of the NGST and its key components, using the early baseline design from NASA Goddard Space Flight Center.

The initial scientific justification for making the NGST an infrared telescope was that sources in the early Universe, including the first galaxies, will have much of their radiation redshifted from optical to infrared wavelengths (Dressler et al. 1997). However, there are obviously significant other science interests at the same wavelengths, and the NGST is now under development as a general purpose observatory at the core of NASA's Origins programme (`http://origins.jpl.nasa.gov`). The NGST Ad Hoc Science Working Group (ASWG) drew up a so-called *Design Reference Mission* (DRM) that describes a broad sweep of key projects defining the science goals of the mission, and details of these projects can be found at `http://www.ngst.stsci.edu/drm`. The five basic themes of the DRM are: cosmology and the structure of the universe; the origin and evolution of galaxies; the history of the Milky Way and its neighbours; the birth and formation of stars; the origins and evolution of planetary systems.

Of greatest interest to the present audience, of course, are the latter two themes. Stars and planets are both cool and heavily obscured at optical wavelengths by dust extinction when young. NGST will cover the shorter, infrared wavelengths, while ALMA will provide directly complementary coverage in terms of imaging resolution and sensitivity at the longer, millimetre wavelengths.

2 Instrumentation Package

The NGST observatory includes a package of three instruments, the basic parameters of which were defined by the ASWG, working to the science goals outlined in the DRM, as follows.

2.1 Near-Infrared Camera (NIRCam)

The near-infrared camera will be a PI-led instrument developed in the US, with additional Canadian participation, with the following basic characteristics:

- Wavelength coverage: 0.6–5 μm (InSb or HgCdTe detectors)
- Field-of-view: 4×4 arcmin
- Image scale: 0.04 arcsec/pixel (fully-sampled diffraction-limit of 6.5-m telescope at 2 μm)
- Detector size: 3×3 mosaic of 2048×2048 pixel arrays
- Provision for single object / long-slit R=100 spectroscopy
- Provision for simple coronographic imaging

Predicted Performance The nominal performance of the NIRCam can be encapsulated in a single sensitivity figure, namely that in broad-band imaging (R=3) at any wavelength from 1–5 μm, point sources as faint as ~2 nanoJanskys (nJy) should be detectable at S/N=10 in 10^5 seconds. This corresponds to an AB magnitude of $30\overset{\mathrm{m}}{.}7$, which in turn corresponds to normal Vega magnitudes of $J=29\overset{\mathrm{m}}{.}8$, $H=29\overset{\mathrm{m}}{.}3$, $K=28\overset{\mathrm{m}}{.}8$, $L=27\overset{\mathrm{m}}{.}9$, and $M=27\overset{\mathrm{m}}{.}2$. These numbers can be scaled accordingly for other integration times under the assumption that the NIRCam will be background-limited at all wavelengths.

2.2 Near-Infrared Multiobject Spectrometer (NIRSpec)

The near-infrared spectrometer will be developed by ESA and European industrial contractors, with NASA delivering the detectors and MEMS devices, according to the following characteristics:

- Wavelength coverage: 1–5 μm (InSb or HgCdTe detectors)
- Field-of-view: 3×3 arcmin
- Image scale: 0.1 arcsec/pixel
- Spectral resolution: R=100 and 1000
- Number of simultaneous objects: >100 (up to 1000 possible with MEMS)
- Minimum slit width: 0.1 arcsec

Predicted Performance The nominal performance for NIRSpec can again be characterised by a single sensitivity figure, namely that in 10^5 seconds, an R=1000 spectrum covering the 1–5 μm region at S/N=20 should be achievable for a point source with a flux of 420 nJy or $K=22\overset{\mathrm{m}}{.}9$. At this resolution, NGST is read-noise limited and thus its sensitivity is dependent on detector parameters. To illustrate how important good detectors are to the NGST, it should be noted that this sensitivity figure assumes the NGST *requirement* detectors with a read-noise of 7 e^- RMS, a dark current of 0.05 e^-/sec/pixel, and a 2 μm QE of 0.6. For NGST detector *goal* performance of 2.1 e^- RMS read-noise, 0.0045 e^-/sec/pixel dark current, and QE of 0.7, the sensitivity limit improves to 129 nJy or $K=24\overset{m}{.}2$.

The Challenge of MEMS The baseline technology for slit selection over a wide field-of-view is MEMS, or microelectromechanical systems, which uses microchip techniques to create small integrated mechanical systems. In the case of the NIRSpec, the plan is to use an array of microshutters or micromirrors (similar to Texas Instruments' Digital Micromirror Device, used in computer projectors; `http://www.dlp.com`) to redirect light from a randomly reconfigurable series of slits into the entrance aperture of the spectrograph.

This is a challenging technology to implement at cryogenic temperatures, and there are several key performance issues involving contrast and fringing that may yet prove problematic. Thus fallback options are also being studied by ESA in case the MEMS-based approach fails to mature appropriately, including the use of mechanical slits (with a much reduced number of slits per exposure) and an integral field based approach.

2.3 Mid-Infrared Camera/Spectrometer

The mid-infrared instrument (MIRI) will be developed jointly by NASA (led by JPL) and a consortium of European member states (including the UK, Germany, France, the Netherlands, and Italy), with the following characteristics:

- Wavelength coverage: 5–28.3 μm (Si:As detectors)
- Camera field-of-view: 1.5×1.5 arcmin
- Camera image scale: 0.1 arcsec/pixel (fully-sampled diffraction-limited of 6.5-m telescope at 5 μm)
- Detector size: 1024×1024 pixels
- Low-res spectroscopy: R=100 at 7.5 μm in camera, 5 arcsec long slit
- Integral field spectrometer field-of-view: 2×2 arcsec (goal 3×3 arcsec)
- Integral field spectrometer image scale: 0.2 arcsec/pixel
- High-res spectroscopy: R=1000–3000 in integral field spectrometer
- Provision for coronograph (simple or phase mask)

Predicted Performance The MIRI covers a wide range of parameter space, from 5–28.3 μm in wavelength and 5–3000 in spectral resolution. In addition, the zodiacal light and thermal background from the telescope become much more significant, and rise sharply to longer wavelengths. As a result, it is difficult to give simple reference sensitivities for the MIRI. It is nevertheless instructive to note that for broad-band (R=3) imaging, 10^5 seconds will yield S/N=10 detections of point sources at 100 nJy and 2 μJy at 10 μm and 20 μm, respectively.

The equivalent point source detection limits predicted for VISIR on the VLT are roughly 150 μJy and 2 mJy, i.e., factors of 1500 and 1000 worse at 10 μm and 20 μm, respectively (`http://www.eso.org/instruments/visir`).

The important point here is that the MIRI will be enormously more sensitive than ground-based mid-infrared instrumentation due to the huge reduction in thermal background. Furthermore, even cold space-borne mid-infrared missions such as ISO, SIRTF, and ASTRO-F all have small apertures, between 60–85 cm diameter, and since the flux limit of a background-limited, diffraction-limited telescope is inversely proportional to D^2, NGST goes ~ 100 deeper than them in a given time, or is $\sim 10{,}000$ faster at reaching a given sensitivity limit. This advantage comes coupled with a spatial resolution improvement of 10 in addition. Thus it is at thermal-infrared wavelengths that the so-called 'discovery space' of the NGST will be largest, with the potential for completely new and unique experiments.

3 The NGST and the Origins of Stars and Planets

The NGST will contribute to the study of the origin and evolution of stars, circumstellar disks, and planetary systems in various ways:

- Very earliest stages of star formation: evolution of protostars and the determination of stellar mass
- Initial mass function at substellar masses: physics of fragmentation and the effects of environment
- Evolution of protoplanetary and debris disks: timescales for converting gas and dust into planets
- Detection and characterisation of extrasolar planetary systems
- Physical and chemical conditions in the ISM: feedback, processing of proto-organic materials

Detailed science programmes covering each of these areas can be found in the DRM, and further ideas may also be found elsewhere (e.g., ESA NGST Science Study Team 2000; Robberto & Beckwith 2000). In the limited space available, I will discuss just two topics, and the reader is encouraged to think of additional projects that will be enabled by the instrumental capabilities detailed above.

3.1 The IMF at Substellar Masses

The NGST will be able to probe the substellar mass IMF in young clusters across the galaxy, from the hydrogen-burning limit at $0.072\,M_\odot/72\,M_{\rm Jup}$, to the deuterium-burning limit at $13\,M_{\rm Jup}$, down to planetary-mass objects at 3–$5\,M_{\rm Jup}$. The form of the IMF needs to be examined over this substellar regime in a variety of clusters, to test for the effects of local environment (ionising flux and winds from high-mass stars, cluster density) and of location (galactocentric radius, metallicity). One important goal is to search for empirical evidence in the IMF for a minimum mass to formation by fragmentation, predicted by theory to lie at $\sim$1–10 $M_{\rm Jup}$ (cf. Low & Lynden-Bell 1976; Boss 2001). Dynamical evolution of clusters can be probed, searching for evidence of mass segregation at birth, evaporation of low-mass systems from the cluster, and the destruction of binaries. Finally, these same low-mass sources can be examined for evidence of the evolution of associated circumstellar disks and potential planetary systems.

In order to do so, imaging surveys of young clusters from 1–10 μm are required to identify sources down to 1 $M_{\rm Jup}$ and define their spectral energy distributions, from which extinction, accretion signatures, and excess flux from the terrestrial planet forming region of a disk can be deduced. Repeat imaging observations can be used to measure proper-motions and thus ascertain membership and dynamical properties. Medium-resolution (R=1000) spectroscopy will then be carried out to provide effective temperatures, in order to create an HR diagram from which the IMF and other properties can be determined.

The imaging section of this programme is relatively straightforward for the NGST. For example, a 1 Myr old, 1 $M_{\rm Jup}$ source at 0.5 kpc (e.g., Orion), and embedded in A_V=10$^{\rm m}$ will have a K-band flux of $24\overset{\rm m}{.}4$ (Burrows et al. 2001; Saumon, personal communication). Such an object will detectable to 3.5 kpc in 10^5 seconds in the K-band, while an equivalent 1 Myr, 10 $M_{\rm Jup}$ object will be visible in 10^5 seconds out to at least the LMC.

Spectroscopy is more challenging, however. Our fiducial 1 Myr, 1 $M_{\rm Jup}$ object will be measurable at 0.5 kpc, with some slight spectral binning to (say) R=500 to yield S/N=20 in 10^5 seconds with the NIRSpec, assuming NGST goal detector parameters. Objects of 10 $M_{\rm Jup}$ are much easier, at K=$18\overset{\rm m}{.}2$ at 0.5 kpc under the same assumptions, and thus readily amenable to spectroscopy there. Even at the Galactic Centre at 8 kpc, such a source would have K=$24\overset{\rm m}{.}2$ and thus still be reachable by NGST spectroscopy: many young embedded clusters lie at intermediate distances.

Thus NGST requirements for this programme include wide-field near-/mid-infrared imaging and wide-field multiobject spectroscopy to cover hundreds of sources in clusters typically a few arcmin in size at $\lesssim$5 kpc, and high spatial resolution to detect faint sources against the nebular background present in these regions. Diffraction-limited imaging at K (0.08 arcsec) will resolve binaries with 40 AU separation at 0.5 kpc, near the peak of the field star binary distribution.

3.2 From Disks to Planetary Systems

Circumstellar disks are the raw material of planetary systems. It is important to study their evolution as a function of environment, since interactions between star-disk systems, heating and evaporation of disks by nearby massive stars, and tidal truncation may all help determine the frequency and characteristics with which planetary systems form. NGST will permit a detailed study of circumstellar disks spanning a wide range of ages (1–300 Myr) in several key ways.

HST and ground-based AO studies have shown that the outer structure of disks can be traced via scattering and absorption due to cold dust (McCaughrean, Stapelfeldt, & Close 2000). NGST will extend this imaging work to greater sensitivity and higher spatial resolution, to further probe the structure and dust composition of outer disks in a variety of star-forming environments. It may be possible to detect gaps in the disks where planets have already formed, and the temporal evolution of dust grains can be traced as they settle and agglomerate, converting sub-micron ISM grains into millimetre-sized particles, the seeds of planetesimal formation (cf. Beckwith et al. 2000; Throop et al. 2001).

Direct emission from warm (100–2000 K) dust in the inner few AU of disks can be observed at 1–28 μm. Its frequency and form as a function of stellar mass, age, and environment can be used to infer the typical timescales for dust agglomeration in the inner disk. Images of disks from 5–28 μm and spatially-resolved (integral field) mid-infrared spectroscopy in dust solid state features will probe their temperature and density distributions, search for evidence of inner holes due to ongoing planet formation (cf. Jayawardhana et al. 1998; Koerner et al. 1998), and study the time-dependent growth and processing of materials such as amorphic and crystalline silicates.

However, dust is a minor constituent of young disks: 99% of the mass is in gas, which is essential for the formation of Jovian-type planets. NGST will enable deep searches for the 28.2 μm S(0) and 17 μm S(1) lines of H_2, the dominant molecule in disks, to constrain the disk temperature structure, gas mass, and gas/dust ratio (see, e.g., Thi et al. 2001). Unlike CO, H_2 does not freeze-out onto grains, and thus NGST measurements will complement those of the gas in the outer disk traced in CO with ALMA. NGST will be sensitive to gas with masses of 10^{-3} to 10^{-7} $M_\odot$ at T_{gas}=50–200 K, and such data will provide constraints on the timescale for gas dissipation and important tests of theories of gas-rich giant planet formation.

Finally, mid-IR (5–28 μm) spectroscopy of other warm molecular gas species (CO, H_2O, CH_4, C_2H_2) in disks will be used as a probe of organic chemistry, to kinematically identify cleared gaps, and to measure temperature structure.

4 Project Status

Progress on the NGST has been enormous over the past few years, with the project moving from a concept to one of the prime space missions for the coming decade. The US National Academy of Sciences decadal review selected the NGST as its top priority project for the coming decade (McKee & Taylor 2000), and in October 2000, ESA committed itself to participating in the project at the level of an F-class mission ($\sim$ 176 MEuro) in return for guaranteed access to 15% of the NGST observing time, as is presently the case with the HST.

At the time of writing (October 2001), the NGST project is about to enter a major new phase, moving from wide-ranging scientific and technical studies, to the hard work of actually developing, building, and flying the observatory. Proposals are due soon from the two potential prime contractors, namely Lockheed-Martin and TRW/Ball, and after selection, we will know the likely mission architecture. NASA has also issued an AO soliciting proposals for a US team (with Canadian participation) to build the NIRCam, members of the MIRI science team, and members of the Flight Science Working Group, including interdisciplinary scientists. On the ESA side, a series of technology studies looking at various ways of implementing the NIRSpec are nearing completion, and two competitive Phase A studies will soon commence. A Flight Science Team for the NIRSpec will replace the present ESA NGST Science Study Team in late 2002. Finally, funding is being sought from several ESA member states, in order to cement the European contribution to the MIRI.

All in all, things are moving quickly, and by the beginning of the next decade, we can certainly hope to see the NGST in orbit, providing us with extraordinary new capabilities for star and planet formation research.

Acknowledgements The NGST project involves a very large international community of scientists and engineers, and in the limited space available, I have only been able to give a brief overview of the project and its goals. Any errors in this paper are purely my responsibility, and when in doubt, the truth can always been found from the official NGST website at `http://ngst.gsfc.nasa.gov`. I would like to wish all my colleagues involved in the project the best of luck bringing the NGST to fruition.

References

1. S. V. W. Beckwith, T. Henning, Y. Nakagawa: In *Protostars & Planets IV*, ed. by V. Mannings et al. (Univ. Arizona Press, Tucson, 2000), pp533–558
2. A. P. Boss: Astrophys. J. (Letts) **551**, L167 (2001)
3. A. Burrows, W. B. Hubbard, J. I. Lunine, J. Liebert: Rev. Mod. Phys. in press (astro-ph/0103383) (2001)
4. A. Dressler, editor: *Exploration and the Search for Origins: A Vision for Ultraviolet-optical-infrared Space Astronomy, Report of the HST & Beyond Committee* (AURA 1996) `http://ngst.gsfc.nasa.gov/project/bin/HST_Beyond.PDF`
5. ESA NGST Science Study Team: 'European participation in the Next Generation Space Telescope' (ESA-SCI(2000)9, 2000) `http://astro.estec.esa.nl/SA-general/Projects/NGST/study_rep.pdf`
6. R. Jayawardhana, S. Hartmann, L. Hartmann, C. Telesco, R. Pina, G. Fazio: Astrophys. J. (Letts) **503**, L79 (1998)
7. D. Koerner, M. E. Ressler, M. W. Werner, D. E. Backman: Astrophys. J. (Letts) **503**, L83 (1998)
8. C. Low, D. Lynden-Bell: Mon. not. R. astr. Soc. **176**, 367 (1976)
9. M. J. McCaughrean, K. R. Stapelfeldt, L. M. Close: In *Protostars & Planets IV*, ed. by V. Mannings et al. (Univ. Arizona Press, Tucson, 2000), pp485–507
10. C. F McKee, J. H. Taylor, editors: *Astronomy and Astrophysics in the New Millennium, Report of the Astronomy and Astrophysics Survey Committee* (Washington, National Academy Press, 2000) `http://books.nap.edu/html/astronomy_and_astrophysics`
11. M. Robberto, S. V. W. Beckwith: 'Understanding star formation with the NGST'. In: *Mid- and far-infrared astronomy and future space missions*, ed. by T. Matsumoto, H. Shibai (Sagamihara, Japan, ISAS, 2000), pp11–18
12. W. F. Thi, G. A. Blake, E. F. van Dishoeck et al.: Nature **409**, 60 (2001)
13. H. B. Throop, J. Bally, L. Esposito, M. J. McCaughrean: Science **292**, 1686 (2001)

Aerial view of Paranal

Virginia Trimble

Origin of Stars and Planets: The View from 2021

Virginia Trimble

Dept. of Physics, Univ. of California, Irvine, CA 92697, USA
and
Dept. of Astronomy, Univ. of Maryland, College Park, MD 20742, USA

Abstract. What (or who) made the Earth, and how? What made the Sun, and are there others? Thus mankind has wondered for eons. Not surprisingly, the answers have evolved a good deal over the decades, centuries, and millenia. So, perhaps more importantly, have the meanings of the questions. Here is an attempt to explore, "as far as thought can reach", both back into the past and forward into the future, the territory defined by such questions, illuminated by light (and infrared) from the present workshop. If you decide to come along on the quest, it is probably worth carrying in your baggage the related, if strange-sounding, question, is star formation important?

1 Introduction

Thirty-some years ago, Jesse L. (for Leonard) Greenstein, who will appear again in a few pages, instructed the women graduate students in astronomy at Caltech not to use their middle names or initials, on the grounds that the space would soon be needed for our "maiden" names or initials, when we married and adopted our husbands' surnames. (His wife was born a Kittay.) It will not surprise you to hear, if you know any of us, that Judith G. (for Gamora) Cohen, Donna E. (for Etta) Weistrop, and I all immediately began using our middle names and initials on any and all possible occasions. A decade or two later, it dawned on us that there were not really enough astronomers in the world to make this necessary, and (as you can check by looking at our recent papers), we mostly stopped using said middle initials, as incidentally did Allan Sandage, who has not inserted R. (for Rex) in many years.

Nevertheless, it remains true that I was born Virginia L. (for Louise) Trimble, and so share the initials of the VLT, whose present and future accomplishments we gathered to celebrate. Whether this is other than the purest of coincidences[1], only The Shadow (no middle initial, but J. for Jansen would not be inappropriate) knows. It was clear from the chair's introduction of this presentation that the correspondence had at least been noted by the SOC.

An alternative explanation for my being asked to give this talk can be found on a postcard [4], which says, "You can get a great reputation for wisdom just by telling people what they already know". This is precisely the function of the concluding speaker or reviewer of a conference, and one which I have been exercising for at least 25 years [18] or perhaps 50, if you count appearances as narrator in the annual Christmas pageant at Toluca Lake Grammar School.

[1] The cube of 26 is, after all, a large number. On the other hand, at least three astronomical entities share the acronym AIC [20].

2 The Beginnings

"Vaya'as Elohim et shney hame'orot hag'dolim ... And God made two great lights, the greater light to rule the day and the lesser light to rule the night. V'et hakokhavim. And also the stars. Vayehi erev, vayehi boker, yom r'vi-i. And there was evening, and there was morning. A fourth day." This, from the author of the first chapters of Genesis, is the earliest published description of star formation.

There are three things to be noted about this scenario. First, all the stars will be the same age. Second star formation is not really a separate, important phenomenon, but is more or less incidental to the creation of the universe and the formation of the solar system. We will encounter both ideas again. Third, stars did not appear until the fourth day, and so a four day conference is barely long enough to solve the associated problems.

There is a contrasting tradition, at least as old, in which star formation is an on-going process, most familiar perhaps in the form of Greek myths, in which various heroes and demi-gods are placed among the stars as a reward for meritorious behavior. The ancient Egyptians also held this view, with the souls of Pharaohs being entitled to ascend to the heavens and join the stars, and shafts out of their tombs were sometimes provided for the purpose of facilitating their journey to the Indestructable, or circumpolar, constellations [2]. Frequently-reproduced texts describe the events, e.g. *Sinuhe*, "The King of Upper and Lower Egypt (Sehetepibre) flew up to heaven.", and, a sort of converse, in the *Story of the Shipwrecked Sailor*, "A star fell, and they (75 serpents, children and brothers and a little daughter) went up in flames through it."

The Middle Egyptian word for star, conventional pronunciation sebah, (like writers of Hebrew and Arabic they generally left you to guess the vowels), consists of five glyphs, the individual ones for s, b, and aleph, then a five-pointed star, with one point straight up and looking very much like a modern child's drawing, and last the determinative meaning sun, which is a circle with a dot in it, like the medieval astrological symbol and our subscript in solar mass, $M_\odot$, etc. The short form is just the star, sometimes with the dotted circle at the center, but if the idea "stars are suns" ever occurred to any of them, I have not seen it. Other words with the same pronunciation but different meanings often include the star. Thus it developed into a triliteral, implying sound. Such words have an additional determinative; a simplified house for sebah = door, a man carrying a stick in threatening fashion for sebah = teach, and a sort of arc and compass for sebah = surveying instrument.

We can, therefore, identify two questions with very long pedigrees. First, is star formation an on-going or one shot process? This was not finally settled until about 1950. And, is star formation important? If you think the answer to this is a hearty, unambiguous "yes", take a look at how many pages it occupies in the reference and review volumes of Cox [7], Lang [13], and Bahcall and Ostriker [3] and in issues of *Annual Reviews* since the paper of Shu et al. [17] appeared.

There is a comparably old question connected with the formation of planets (explored from Greek times down to the present by Brush [5] and references therein), phrased most succinctly as monadic vs. dyadic. That is, did our (and by extension other) solar systems arise from a single entity like the Swedenborg/Kant/Laplace rotating nebula or out of some interaction between two or more previously existing entities, as in the Buffon/Chamberlin-Moulton/Jeans-Jeffreys-Lyttleton tidal encounter scenarios?

In case you might have forgotten, some form of tidal encounter picture was the main-stream, respectable point of view from the 1890s until the 1940s. How did they manage to get it so wrong, to the point where Russell et al. [14] (in the textbook that introduced English speakers to astronomy for two generations) put "The Nebular Hypothesis" in small print and "The Hypothesis of Dynamic Encounter" in large type? They were misled by the concentration of angular momentum in planetary orbits around a slowly-rotating sun, and concluded that "the angular momentum of the planets must have been put into the system from outside". In other words, there was important physics – magnetic fields and the solar wind – missing from their world view.

If half of all three-sigma results turn out eventually to be wrong (which is roughly so), it is very frequently just because there is physics missing from the discussion, and John Bahcall has used this as part of his private definition of what is really meant by a 3σ result – if it is wrong, we have left out something important. The next section looks at the sense in which this might still be a problem in understanding planet and star (vs. galaxy and large scale structure) formation.

Incidentally, the monadic/dyadic issue is still with us, in the form of the question of the extent to which cloud contraction and star formation are triggered. White described observational evidence for triggering in S185 and the Eagle. Triggering by a nearby supernova was assumed in the talk by Adams on the sun's parent star cluster. But poster presenter Zavagno mentioned that she had been warned off giving too much emphasis to triggered star formation in her discussion of Sharpless 128, on the grounds that it was not currently fashionable.

3 Missing Physics

There seem to be four issues here, three of which affect calculations of star and planet formation one way or another. First is physics that you already know is likely to be important, but do not or cannot yet include in your code for lack of space or time. Magnetic fields in N-body simulations (of nearly anything) are an obvious example. Bate's talk addressed the problem of dynamic range: it takes many particles to watch one protostar and its disk form, limiting substructure in his calculations to things bigger than about 1 AU and clouds to at most 50 $M_\odot$. Thus he forms no 100 $M_\odot$ stars. Heating and cooling of gas by dissipation and by the turn-on of the first stars are often also not treated adequately in simulations of formation of galaxies and large scale structure, where the dynamic range problem is also severe, with smaller galaxies sometimes represented by only about one particle. Among the ways of tackling missing physics in this sense are (a) deliberately make your code worse in resolution or some other parameter and see how much difference it makes, (b) draw on observations for your initial conditions (talks by Alves, Kramer, Lombardi, Walmsley), and (c) wait for computers to become faster.

Second is essential physics that is currently missing from human knowledge, not just from your calculation. I believe star formation does not suffer from this. In contrast, models of large scale structure and streaming must begin with assumptions about the kinds and amounts of dark matter, dark energy, topological singularities, and so forth.

Third is the curse of the variable or adjustable parameter, sometimes including the initial conditions for a simulation. Gamow is supposed to have said that, with five parameters, you could fit an elephant. This is roughly true if you think of a Fourier decomposition of the outline. Burkert, speaking on formation of binary stars, remarked that, historically, what you got out was what you put in, and this was to a certain extent still the case.

All scenarios for galaxy formation have this problem in spades – there are infall and outflow of primordial or enriched gas, star formation as a function of gas density, composition, and whatever else matters, the initial mass function and its dependence on gas density, composition, etc., and, at some level (especially if you care about nucleosynthesis by Type Ia supernovae) binary star populations to be considered. Notice that, in principle, solutions to some of the problems in understanding star formation discussed at the workshop would be, in turn, improved input for galactic calculations.

Fourth is dimensionality. Klessen noted that star formation logically proceeds from 3-d (turbulence) to 2-d (disks) to 1-d (main sequence stars). In practice, the physicist's first approximation is the spherical cow. The next step is, of course, the bipolar cow (Fig. 1), and Shu et al. [17] noted as one of the principal advances of the decade preceding their review the recognition of the ubiquity of bipolar outflows in young stellar objects. The talks by Bourke, Bacciotti, Garcia,

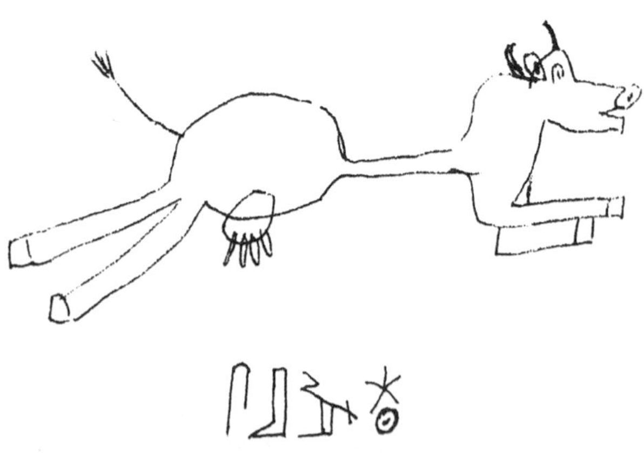

Fig. 1. The bipolar cow (F.C. Adams) jumps over the hieroglyphs for star (in a modern Egyptological hand, V. Trimble)

and Romaniello and many of the posters made clear that such outflows are even more in evidence (and less purely bipolar) when your angular resolution and wavelength coverage improve. The thrust of Li's talk was that it is possible to model what we see, assuming suitable amounts of rotation and magnetic fields, but that this is not quite the same as understanding what causes the outflows, or, in my opinion, how they know how much rotation and field need to be removed, and when, to leave the range of ZAMS stars we see (Wring, Hogerheijde).

Disks are the other extreme of two-dimensionality. The talks of Brandner, Lada, Launhardt, and Wilner showed that disks are like belly buttons, that is, everybody has one when young, but they are not all the same shape.

Finally, it came as a surprise to hear from Clarke that disks may be more difficult to get rid of on the right time scale than they were to form in the first place. Her discussion left many of us uncertain about whether this was a case of "all of the above" (we will meet more shortly) or a case of physics missing from the calculation, in the sense of the dominant process not yet having been identified. Some of this physics may be preserved in our own Kuiper belt (talk by Jewitt). Or perhaps there is an octupole cow lurking out there somewhere.

4 Some More Recent History

William Herschel, beginning in 1789 and perhaps guided by Kant's and later Laplace's nebular hypothesis, arranged the various sorts of nebulae he had observed and drawn into what he regarded as an evolutionary sequence, beginning with the diffuse and/or planetary nebulae, going on to bright/dense stellar nebulae, then to blurry or hazy stars and clusters, and finally to isolated stars and clusters. Well, he had some of the details wrong, but he was sure that star formation was an on-going process that observers could reasonably study, and so essentially right.

Another William, Huggins, who in 1865 was the first to see emission lines from some of Herschel's nebulae (and thus prove them to be truly diffuse, not just unresolved star clusters) began to gum up the works when he declared that Orion and other bright nebulae could not be Herschel's "nebulous fluid" from which stars could be built, because he could see only a hydrogen, a nitrogen, and one other line, while uniformity of nature required the presence of all the elements found in the sun, earth, and stars, dominated by calcium, sodium, iron, and so forth. That the strongest lines did not always come from the most abundant elements was not clarified until the 1925 work of Cecilia Payne on spectra of K giants.

Then, strangely it now seems to us, came an epoch in which on-going star formation ceased to be regarded as a part of the real universe or as a topic on which serious astronomers should work. It came in the wake of Hubble's 1929 linear relation between galaxy distances and velocities, implying a cosmic age close to 2 billion years, about the same as that of the solar system, and about the same as the time a typical star might plausibly live on "subatomic energy". And, from many things having similar measured ages to the dictum that everything was the same age seems to have been a short step. Thus one finds many of the great and good of the astronomical community in the 1920s and 30s and even later making snide remarks about the whole subject of star formation. (Most of the references are in Trimble [19], where also is to be found a discussion of the discoveries of general interstellar extinction, H I, and all that led the community back to something like Herschel's view.) Here are some of my favorites:

"In former times, when theories of cosmic evolution were taken somewhat more seriously than they are at present ... " (Baker 1932)

"A good many arguments weigh against the hypothesis of recent creation" (of S Dor, Jeans 1929)

The alternative of on-going star formation is "not very attractive", and bright stars live on some much more generous source of energy than synthesis of heavy elements. (Atkinson 1936)

"Most stars were created at the beginning of the universe some 2 or 3 thousand million years ago." (Goldberg and Aller 1943)

" ... origin of stars and stellar systems at the epoch of catastrophe" (Bok 1936)

"A time of turbulence in which various things happened which are very difficult to account for otherwise – such as the origin of the solar system and of double stars" (Russell, Dugan and Stewart 1938 edition)

Hoyle and Lyttleton were led specifically to consider accretion on to a previously existing solar type star as being responsible for the more massive ones. This is foreshadowed in Eddington's 1926 *Internal Constitution of the Stars*. Hoyle and Lyttleton (1939 etc) [11] are still cited not because anyone still supposes that this is where OB stars come from, but because their formulae are useful for estimating accretion rates in interacting binary systems and such.

A last published quote from Jesse Greenstein in 1951 (in the Hynek astrophysics compendium) says, "Like the accretion hypothesis, the condensation hypothesis now seems suitable only for creation of a few exceptional objects rather than all the stars." And he leaves the problem unsolved.

In a preliminary effort to find out more about this epoch, I wrote to Fred Hoyle and to Lyman Spitzer, asking for their "takes" on the period. Hoyle wrote (9 Jun 1997) saying, "You ask why in 1939 we did not think of ongoing star formation. Because at the time the astronomical world did not believe in an interstellar medium containing hydrogen." And he went on to mention that, soon after, he and his colleagues had other things (like radar and proximity fuzes to worry about.

Spitzer responded (November 26, 1996), "My keen interest in the problem (star formation) dates from 1939 ... supergiant stars must be younger than any likely age for the Galaxy ... dusty clouds ... were the stellar birthplace. Later that year, I submitted a paper to which these theoretical suggestions provided an introduction ... Several astronomers objected to this introduction as too speculative, and much to my regret the paper was published without (it)." To my subsequent letter asking who the objectors had been, he responded that it was not appropriate to say because one of them was still living. I thought it likely that he meant Greenstein, but had not attempted to verify this. At any rate, the original, "too speculative", version of Spitzer's 1939 paper now appears in his collected works (edited by him and J.P. Ostriker), though it is now too late to ask any further questions of Spitzer. A phone call to Greenstein this afternoon (13 May 2001) yielded the information that it wasn't him and he hasn't a clue who it might have been.

5 The Poster Session

More than 60 presentations, with authors ranging from Apai to Ziegler, covered almost as many topics, including some just a little outside the main thrust of the conference, occasionally indeed way out. My favorites included:

- The youngest protostar? Well possibly not, but Belloche et al. indicated that IRAM 04191 is no older than the neolithic revolution.
- The most distant dust that reaches us? Krivova suggested that certain radar-detected meteoroids represent dust from the disk of Beta Pictoris, expelled by a Jupiter-like protoplanet.
- The longest correlation length? Ibrahimov displayed light curves of FU Orionis and V1057 Cygni and showed that they share several symmetries (one only obvious when you turn the light curve upside down).
- The most deviant IMF? Stolte presented evidence that the young galactic center cluster, the Arches, is particularly rich in high mass stars (or deficient in low mass ones).

6 Current Conditions: Observational

I will take the (by no means original) point of view that the status of a field like origins of stars and planets can be described by the questions that the community is trying to answer. Many of these, as Mark McCaughrean made clear in his multi-faceted (and multi-media) introduction, have been alive and well for 20 years and more, in the case of stars. Names in parentheses are those of speakers who presented some part of an answer or a better-posed version of the question.

6.1 Planets

Some of my "questions about planets" would be regarded by others as part of the definition of a planet. The definition assumed here is things less than about 10 times the mass of Jupiter, currently in orbit around stars.

Statistical questions (Mayor, Neuhäuser, Hatzes): How many are there as a function of planet mass, composition (gas, ice, rock, metal) and orbit parameters (a, e, and i to the equator of the parent star; angles involving the node are not a property of orbits but only of our point of view)? How are the several properties of planets correlated with each other (where the solar system example includes correlation of composition with semi-major axis)? What are the minor constituents of systems like, including comets, asteroids, and moons (Schulz)?

Host star questions: What are their masses, compositions, locations in the Milky Way, rotation periods, magnetic fields, kinematics, and ages, and how are these correlated with the sorts of planets they have? Do the averages and distributions of host properties differ from those of stars in general? It was once advertized that OBA stars still had their birth quotas of angular momentum, while most GKM stars had formed planets. I think no one would still argue this.

System questions: Are multiple planets the norm or the exception? How typical is our mix of Jovian and terrestrial? Are all planet systems coeval with their stars? (No – one pulsar has some, but it remains a unique example, despite accurate timing data for many pulsars.) Are all planet systems roughly co-planar? You don't expect them to be, said a theorist during discussion, even if they formed that way, because of subsequent interactions (Monin).

Evolution: can we see disks giving way to planets? How? And what do they look like? (Brandner, Launhardt, Looney, Wilner, Wolf).

Individual planet questions: Is chemical differentiation common among Jupiter-mass planets outside the solar system? Do other terrestrial planets have non-equilibrium atmospheres? This last is, of course, perilously close to "Is there extraterrestrial life?".

Of these, the issues of initial co-planarity and chemical differentiation are part of some experts' definitions of planets, as is the question of whether there are free-range ones in Orion or elsewhere (Barrado y Navascués, Roche).

Data reduction: How do you turn measured fluxes, line profiles (etc.) into masses, temperatures, compositions, and all the rest? This is important both for planets and for low mass stars (Baraffe, Guenther, Testi).

6.2 Stars

Statistical questions: What is the IMF? What is it correlated with (gas composition, turbulence, density, cloud mass, presence/absence of spiral arms or other observable triggers ...) and how? (Bontemps, Eislöffel, Zinnecker).

Binary questions: How many of them are there? (Bouvier, Joergens). What is the initial distribution of semi-major axis, eccentricity, and mass ratio? (Sterzik). How are these correlated among themselves and with gas composition, density, degree of ordering of magnetic fields, etc.?

Evolution questions: Which of these features can already be discerned among pre-main-sequence stars and young stellar objects, or even cloud cores, and how do statistics, binary populations, and planetary companions change as stars age? (Chini, Greene, Jayawardhana, Kaper, Monin, Nürnberger, Petr-Gotzens, Prusti).

Collective questions: How efficient is star formation in a given gas mass? Not very, noted Shu et al. [17] as the second major surprise in the decade leading up to their work. How does that efficiency depend on composition, presence or absence of triggering, and all the rest?

Cosmic questions: What was the total star formation rate as a function of time or redshift and environment? What are the entities in which most of it occurred at each epoch? How have the statistical, binary, and collective properties changed with time? Here we need to know not only about star formation per se, but about the assemblage of gas clouds dense enough to participate in the process. Blitz made it clear that we have no theoretical understanding of GMC formation, while Alves and Lombardi pointed out that observations can at least tell us what conditions the assembly process needs to aim at to make clouds like real ones.

7 Current Conditions: Theoretical

It is easy to make insulting remarks about theorists (e.g. the back end of the bipolar cow), so in the interests of fairness, I quote two theorems and their corollaries.

A. Redman's theorem: A competent theorist can explain any set of observations using any theory. Longair's corollary: In some cases he need not even be competent.

B. Redman's anti-theorem: A competent observer, given a large enough telescope, can find an object that matches any theoretical prediction and another which falsifies it. Corollary: In these days of virtual observatories, she may not even need a telescope.

7.1 Planet Questions

If we accept that planets form in circum (proto)stellar disks, then what are the dominant processes, gravitational instability, coagulation of dust to planetesimals and beyond, something else? (Wuchterl)

What patterns of post-formation dynamical evolution are possible? Which are common? Migration, it seems, including in our own system (Jewitt). Expulsion? Mutual tilting of orbits?

If free-range planets exist, were they formed that way and how, or liberated? (Reipurth)

7.2 Star Questions

What (all) supports molecular clouds against premature dynamical collapse? All perhaps of turbulence, magnetic fields, rotation, and thermal pressure?

Is contraction/collapse always, sometimes, or never triggered? By what? At what stage do dense cores form, and is the dominant process fragmentation or something else?

What makes the IMF? Turbulence is a relatively new player in this game.

What makes binaries and is responsible for the observed distribution of separations (from two stellar radii to a tenth of a parsec: at least we know what is responsible for the end points, just as we know what sets the upper and lower limits to masses of single stars)? What determines the initial distributions of eccentricities from 0 to 1 and of mass ratios from 1 to 0, and how do they evolve? How are the initial values of these correlated?

How do clouds, on all scales from entire giant molecular assemblages down to individual Bok globules and disks, get rid of excess angular momentum and magnetic fields, and how do they know when to stop, since even very young stars do not rotate at break-up or have fields of Megagauss? Is collimated, bipolar, magnetic outflow the only game in town? What about ambipolar diffusion?

7.3 All of the Above?

Several of the items in the two previous sections and others mentioned during the workshop were phrased as A vs. B (sometimes vs. C). That is, which description of the observations is most accurate? Which physical process is responsible or dominant? Such dichotomization and the defense of an alternative by its propounders are a customary and often useful part of the scientific landscape. But I believe we have heard about several cases where the correct answer has to be "both please" (Winnie the Pooh), "all of the above" (multiple choice on the GRE), "some of them are and some of them aren't" (Franklin Pangborn), or "that depends".

Clarke tentatively put disk destruction into this class. I suspect that whether you see rotation or turbulence in molecular clouds depends a good deal on the length scale you examine and whether you reduce your data to the rest frame of a chap living inside the cloud near a dense core. Another example must surely be the IMF. Is it determined by fragmentation, mergers, when accretion stops, the efficiency of outflows? Surely they all matter.

Fritz Zwicky was a great believer in "all of the above", which he called the morphological method. He had little to say about star formation. The strongest contrast I can think of is with the work of Viktor A. Ambartsumian, who had very definite views on star formation which could not be combined with any of the ideas discussed at the meeting. He was indeed among the first to revive the idea of on-going star formation after its period of disrepute. But he decided very early (see Ambartsumian [1] for a retrospective) that stars were born by expansion out of very dense, "prestellar" matter, so that planetary nebulae were an early stage, not a late one, and expanding OB associations a natural product. The idea colored work in Armenia and, to a certain extent, throughout the USSR for decades, and without knowing about it one is puzzled to find Shklovskii in non-technical books working very hard to demonstrate that planetaries are old. Chapter 6 of Shklovskii and Sagan [16] should also be read keeping that history in mind, and it is appropriate to quote from them, "It was once believed that all the stars in the heavens were formed at about the same time, several billions of years ago. But there are now a number of lines of evidence that stars are being formed continuously, by condensation of the interstellar gas and dust."

On-going vs. one-shot star formation and expansion from pre-stellar matter vs. condensation from gas and dust are cases where "both, please" is not a possible resolution. We think, however, we now know the right answer in both those cases!

7.4 Potentially Useful Analogies

It is much easier to tell a story after it is over than while you are in the middle of it. In this spirit, some events of early 20th century astronomy may be enlightening. The astrophysicist of 1900 would surely have said that the most important unsolved problem was the source of stellar energy. Well he would have said it in German, but that is another story. The answer, nuclear reactions, led directly to considerations of nucleosynthesis in stars, from Atkinson and Houtermans (1929) to B^2FH [6]. The nuclear reactions, their energetics and products, and how they interface with stellar structure and evolution can be regarded as largely solved problems.

In 1920, a (again perhaps the) most divisive observational issue was superficially about distance scales, but really about the existence of external galaxies. Curtis [8] said yes, and Shapley [15] said no, in a 1920 debate that is still remembered by cognoscenti. There was a definite observational answer, coming from Hubble [12] in the form of the discovery of Cepheid variables in NGC 6822 and M31. The answer then led directly to Hubble's use of Cepheids as an extragalactic distance indicator, calibrating all the rest, and so to the discovery of his linear distance-velocity relation, that is, the expansion of the universe. Notice that the customary inclusion of a K term in studies of stellar motions means that, if the galaxy had been expanding, it would have been discovered in time to save Einstein from his cosmological constant! Oh, Hubble's answer was "yes", in case you weren't sure.

It seems plausible that there is a corresponding key question for planet and star formation whose answer will lead us in some new, interesting direction. Unfortunately, I do not know what it is (or I would be working on it instead of writing this manuscript). But it is probably not, "Are there other solar systems?" or "How do stars form?".

8 The View from 2021

"I find it difficult to make predictions, especially about the future", so supposedly said Enrico Fermi. In fact, only the future is safe. Predictions about the past, variously called hindcasts, retroposals, and postdictions (the words came from Thomas Gold, an expert), are all too easily falsified.

Easiest to summarize is the computational front, where astronomers will continue to benefit from smaller hardware and bigger software driven primarily by commercial and military goals (though occasionally our algorithms, especially from radio interferometry, are borrowed back the other way into telecommunications, medicine, and so forth). This means, of course, more and faster, including many more ways to link, ask, and communicate. Perhaps there will be ubiquitous PixelPrintPads (pronounced "3-Ps") against which you touch your thumb to log into anywhere, from anywhere, and can carry on by talking, scribbling, or even feeding in punch cards. Reading of brain waves is probably a couple more revolutions ahead.

It is still, however, a good bet that you and your students will continue to strain computing capacity with magnetic fields, gas flows, dust, multi-wavelength, and dynamic range problems. Twenty years is, after all, only about 13 Moore's law doubling times, or 10^4, and the star formation code that now stretches from a 50 $M_\odot$ cloud down to 1 AU details won't yet quite reach, simultaneously, a 10^5 $M_\odot$ cloud and planet sizes. Cleverness will, therefore, still be required and rewarded. And, of course, you will still spend more time debugging then running, because you are a scientist, not an accountant.

On the observational front, someone will probably be writing just about the last proposal for observing time on NGST (McCaughrean), if it is launched a little late (what else is there) into a 10–15 year mission. SOFIA should still be available, but SIRTF (Meyer) long gone, unless it has not yet been launched. Other things relevant to origins of planets and stars that you might reasonably find in space are SIM and GAIA (American and European astrometric missions, somewhat oriented to planet searches); conceivably TPF (Terrestrial Planet Finder, the next generation) funded if not flown; Astro-F (which, like all Japanese missions will get a proper name only when it is safely in orbit to do its infrared photometry and spectroscopy); and HerschelFIRST (André). SAFIR, the Single Aperture Far InfraRed Observatory, or whatever it morphs into, might be under construction. Potentially available to look for X-rays as a signature of young stars (Montmerle), and surely renamed, will be Spectrum-X, MAXIM, Const-X, and perhaps others.

On the ground, there will have been some sacrifices, probably the Lick 120″ and plausibly some of the four-meter class instruments at less than ideal sites, perhaps the Greenbank Telescope (and, of course, NSF will still be looking for a partner to operate the 12-meter Kitt Peak millimeter telescope). But the total area of glass covered by aluminum or silver or – surely this is an area of materials science ready for exploitation with a substance that reflects all wavelengths penetrating to ground and not in need of frequent recoating? – will still be larger, and larger per astronomer, than at any time in the past. Already with us, or nearly so, and good for at least 20 more years are the VLT and its outriggers (Lagage, Malbet, Montmerle, Moorwood, Richichi, Siebenmorgen), Kecks and their sideKecks (Boden), Gemini, Subaru, the Large Binocular Telescope, GranTeCan, Magellan, HET, SALT, and the MMT (for single-mirror telescope).

At longer wavelengths, the VLA should long since have become the Expanded VLA (followed presumably by the Grossly Expanded VLA), and ALMA (Henning, van Dishoeck) be a major, multi-user facility, well-tuned to many aspects of star formation near and far. If the Square Kilometer Array does not yet have a filler aperture adding up to a square kilometer, it has the advantage of being accomplishable in steps (unlike most space missions).

What about the next step from 10-meter apertures up to 30, 50, or 100? A Giant Segmented Mirror Telescope is the second highest priority in the US

National Research Council decade report for the 2000's. Competing concepts in the US include MAXAT (Gemini consortium), ELT (Univ. of Texas), and CELT (Univ. of California and Caltech). There is the Swedish XLT and the largest of all, OWL (Gilmozzi). Will they all happen? Not by 2021, and, because the main showstopper for both CELT and OWL (Gilmozzi; G.A. Chanan, private communication) is money, perhaps not at all except as the "AOL-OWL", that is, with major private sponsorship. I believe the two syllables of that compound are pronounced identically, but I am not sure.

All of these, from 6-meters on up, are being planned with some form of active and adaptive optics (which is like setting your windmill to blow the breeze back the way it came, before it gets to you, but which also enormously improves sensitivity by reducing noise within an image as well as angular resolution) and some 4-meters are being retrofitted. Many also form or will form parts of interferometers, whose analysis is like being led blindfolded to the top of a mountain and then being required to describe not only the entire mountain but its Fourier Transform. This is even more true in optical than in radio astronomy (the description came from the later Peter Scheuer), because there are not, so far (and may never be) optical superheterodynes that let you keep dividing your supply of photons in half and still have the same number you started with, limiting optical interferometers to a dozen or so baselines. NGST is, in contrast, like throwing a butterfly up at 20 or 30 g and then expecting it to spread its wings and fly away.

Theory is even harder to project into the future, because the retooling time (and also the half life) can be very short. It seems unlikely that anyone will be doubting that the dominant physics in star and planet formation is gravity and electromagnetism, but this does not seem particularly helpful.

It may be a bit more helpful to remember that there have been in the past several cycles, in which angular momentum (à la Laplace) gave way to gravitational processes (à la Chamberlin/Moulton and Jeans), which gave way to dynamically important magnetism (à la Mestel and Woltjer and Parker), and back to angular momentum again (Larson, Shu) and gravitational instability (Boss, for planets) and magnetic fields (Mouschovias) again, and so forth, as the bandwagon process of the year. This will surely happen again.

At each return, naturally, the considerations become more sophisticated (or anyhow harder for a novice to understand). Thus last year a couple of my Maryland colleagues published a paper tabulating no fewer than 11 kinds of instabilities in accretion disks, most of which have not yet been applied to the problems of planet and binary formation. The process of the year for star formation in 2001 is probably turbulence, and it could come and go several times, though I would bet about 50-50 that the process of the year in 2021 will be one not currently in the inventory, because it hasn't yet been thought of at all or hasn't been recognized as relevant.

Another theoretical territory comprises chemical bonds, stickiness, fluffiness, and other "materials" properties of matter (Bockelée-Morvan, Dominik, Wright) that must surely be important in planet formation, and here is one jurisdiction in which Space Station experiments may actually be a useful guide to theory.

Finally, any science progresses by a continuous cycling around data, the models they inspire, the predictions of those models, collection of confirming or conflicting data, improvement of the models, deducing the consequences of the improved models, and so forth. You can enter the cycle at any point (notice that numerical simulations often belong to the "deducing the consequences of ideas" stage). Some phases of the scientific process have the stark elegance of chess. Others – and studies of star and planet formation appear to be at this stage – are more like mud wrestling.

9 Changing the Course of History

Just as only the future is safe when making predictions, so in the case of influencing the course of history, only the future is possible. What can we do (beyond the obvious of producing the best science possible with the resources already in hand) to ensure the future health of work on origins of stars and planets, astronomy as a whole, or indeed scientific research in general? Diamond [9] has synthesized an enormous sweep of human history and prehistory, identifying a handful of factors that have led to widely different rates of economic progress in the range of human tribes, clans, nations, and empires on the several inhabited continents. Four of these are directly applicable to science, to wit (1) surplus food or productivity, (2) population size, (3) ease of diffusion of ideas and artifacts, and (4) concentration of power.

Surplus productivity may seem too obvious even to mention. Societies of hunter-gatherers build very few large telescopes; we and our graduate students are fed on grain and grapes that we do not grow; and the number of members of the International Astronomical Union in a country is a fairly good proxy for gross national product. If you take the number of IAU members in a country and divide first by total population and then by per capita annual income in dollars, you get an index on which India, Japan, and Portugal rank nearly equal at 1.3–1.5 $\times 10^{-10}$ and the two Chinas are nearly even at $8 \pm 0.4 \times 10^{-11}$. Indeed 44 of the 62 adhering countries fall between 0.8 and 8 $\times 10^{-10}$ on this scale. The outliers on the high side are former portions of the Soviet Union and its Eastern European neighbors (as are many of the "high normals") who got into the habit of living scientifically above their incomes during the cold war (plus Vatican City at $R = 2.9 \times 10^{-8}$!).

The outliers on the low side are (a) large and still climbing out of poverty, with science not yet very high on the list of priorities (like Malaysia and Indonesia) and (b) not necessarily poor so recently, but with little recent tradition of scientific research of any kind (Algeria, Iran). And the 120 or so UN member nations that do not adhere to the IAU are, on average, both considerably poorer and considerably smaller than the ones that do. A better normalization might include both per capita income above some threshold and population over some threshold (the quota for Vanuatu and St. Kitts and Nevis is rather less that one astronomer per country, and considerably less than one per island), which leads us toward the second item.

I am not quite sure what you are supposed to do with these numbers except to recognize that if you vote simultaneously for slower total economic growth (for whatever reason, however meritorious), lower taxes, and more scientific research, you are asking your fellow country persons to do the improbable, if not the impossible.

Second, population size, is already a more complex variable and leads to an "action item". If all human beings fall in a single gaussian curve of ingenuity (which may well be true) and there were no benefits or penalties from interactions, then the creation of new scientific ideas should be linearly proportional to numbers working in a field, independent of the size of the interacting community. This is the equivalent of star formation rate being linearly proportional to gas density and is almost certainly wrong for human creativity. Mozart and van Rijn (though not Vermeer) flourished among unusually large numbers of other not-quite-so-good composers and painters.

What is the real relationship? (a) A threshold, above which productivity scales as some power of population size? This is the philosophy behind the concept of "critical mass" that university departments are forever trying to achieve for their new research initiatives. It seems to mean either "two more than we have now" or "two more than you have in your group". (b) Next simplest is a power law, $P(n) \propto n^x$ and x larger than unity, like a Schmidt law for star formation. (c) Or there could be several thresholds for different levels of innovation, each followed by some proportionality. These are the optimistic ideas. Alternatively, however, the truth may be more like (d) a broken power law, with the benefits of interaction decreasing as the population becomes too large for each member to know all the others. Or (e) indeed at some point perhaps there is a saturation, where, like rats in a too-crowded cage, we spend most of our time biting each other (i.e. writing and reviewing proposals) and very little on eating and reproducing, let alone thinking how to improve the cage, or, like the servants in a very large household, we spend almost all our time looking after each other (writing letters of recommendation, sitting on committees) and almost none serving the mistress, astronomy.

Many American astronomers feel that we are already perilously close to condition (e) and so, I suspect, do other national and transnational communities. We all know colleagues who have had 30 or 60 or even 100 PhD students in their careers (60+ for Dennis Sciama; 100 for the mathematical physicist Elliott Montroll, who also had 10 biological children) and it may be time for our admiration and envy of such colleagues to be tinged with caution.

The numbers in Table 1 are provided for you to try testing your own feelings or hypotheses, if you feel so inclined. The total American astronomical community (based on AAS membership with minor corrections) is about three times the number shown as IAU members. This may not be typical. The smallest populations represented at the workshop seem to have been Hungary, Ireland, Portugal, and Uzbekistan.

Table 1. Astronomy as a measure of GNP

Country	Members	R	Country	Members	R
Algeria	3	2.5×10^{-11}	Japan	448	1.4×10^{-10}
Argentina	90	2.5×10^{-10}	Korea RP	51	8.1×10^{-11}
Armenia	31	3.3×10^{-9}	Latvia	8	7.9×10^{-10}
Australia	191	4.8×10^{-10}	Lithuania	12	7.9×10^{-10}
*Austria	31	1.8×10^{-10}	Macedonia	0	—
**Belgium	88	3.7×10^{-10}	Malaysia	7	3.0×10^{-11}
Bolivia	0	—	Mexico	83	1.0×10^{-10}
Brazil	109	1.0×10^{-10}	**Netherlands	167	8.7×10^{-10}
Bulgaria	50	1.5×10^{-9}	New Zealand	26	4.1×10^{-10}
Canada	199	3.0×10^{-10}	Norway	22	1.8×10^{-10}
Chile	46	2.2×10^{-10}	Peru	1	8.9×10^{-12}
China, PR	368	8.6×10^{-11}	*Poland	117	4.2×10^{-10}
China, ROC	23	7.4×10^{-11}	**Portugal	17	1.3×10^{-10}
Croatia	13	6.2×10^{-10}	Rumania	37	3.1×10^{-10}
*Czech Rep.	71	6.4×10^{-10}	Russia	344	5.0×10^{-10}
**Denmark	52	4.2×10^{-10}	Saudi Arabia	11	5.3×10^{-11}
Egypt	39	1.3×10^{-10}	*Slovak Rep.	27	5.8×10^{-10}
*Estonia	22	2.4×10^{-9}	South Africa	46	1.7×10^{-10}
*Finland	37	3.6×10^{-10}	*Spain	204	3.6×10^{-10}
**France	609	4.6×10^{-10}	**Sweden	95	5.4×10^{-10}
Georgia	19	2.4×10^{-10}	**Switzerland	70	4.0×10^{-10}
**Germany	488	2.9×10^{-10}	Tajikistan	8	6.9×10^{-9}
*Greece	89	6.4×10^{-10}	Turkey	53	1.4×10^{-10}
*Hungary	41	5.4×10^{-10}	UK	535	4.3×10^{-10}
Iceland	4	7.8×10^{-10}	Ukraine	119	9.5×10^{-10}
India	227	1.4×10^{-10}	Uruguay	6	2.0×10^{-10}
Indonesia	13	1.3×10^{-11}	USA	2235	2.7×10^{-10}
Iran	15	3.9×10^{-11}	Uzbekistan	8	1.3×10^{-10}
Ireland	33	4.9×10^{-10}	Vatican City	5	2.9×10^{-8}
Israel	45	9.6×10^{-10}	Venezuela	11	5.8×10^{-11}
**Italy	409	3.3×10^{-10}			

R = Number of members/(population × per capita income)

Membership data from IAU records just prior to 2000 General Assembly

Population and economic data from The New York Times Almanac (1999)

* Participants in A&A
2617 IAU members

** Also ESO members
1998 IAU members

Third of the conditions considered by Diamond is the ease of diffusion of various things. His examples include (a) cultivars, easy in longitude, much harder in latitude because the climate changes, and also hard across barriers of deserts, mountain ranges, and dense forests, hence some of the advantages Eurasia has had over the pre-Columbian Americas and pre-Vascan Africa, and (b) technologies, easier across wide continents than isthmuses, still harder across oceans, and often requiring a minimum population for maintenance. You will immediately think of some scientific/astronomical analogies, like the sudden implementation of adaptive optics when the barriers of military classification fell (even though the idea came originally out of the astronomical community and wheels were slowly being reinvented within it). Another case is the largely independent development of ideas about both high energy radiation mechanisms and cosmology (dark matter, origin of large scale structure, limits on neutrino masses) east and west of the Oder-Neisse line.

The action item under this third heading is, clearly, to tell everybody anything that you think might be of use to them. This unfortunately contradicts the action item from heading two, which is not to drown your colleagues in your preprints, emails, and all the rest.

Fourth and last is the issue of concentration of power. Even more than items 2 and 3 this is a two-edged sword. Only a community with a great deal of centralized decision making can declare "we will build and deliver an atomic weapon" or "we will put a man on the moon in this decade" and make it happen. Although the primary drivers were quite different, science including astronomy has benefitted at some level from these, even study of formation of stars and planets, since at least we now know that the oldest moon rocks are a bit younger than the meteorites but a bit older than the oldest earth rocks, and quite a lot about some kinds of shock propagation. But converse examples are also easy to accumulate. Ming China began a series of open ocean voyages in the early 15th century, to which a new government faction brought a screeching halt in 1433. Contract the case of Columbus who (in case you had forgotten) had already been turned down by three funding agencies in three different statelets before he approached Ferdinand and Isabella. The Japanese decision to stop gun production in the late 1600s was also made to stick because of a centralized government isolated from competition.

It is not in any sense that the West has been immune from such errors. The words Luddite and saboteur are our very own. Other examples [10] are edicts forbidding the use of Arabic numerals by Florentine bankers (1299), the use of machines by tailors in Cologne (1397) and ribbon looms in Danzig (1579). But, because there were and are many European states, treading many different paths, competition soon forced adoption of the more efficient technology. Just knowing that there is another, better way of doing things can be a powerful incentive to change. Contrast the delayed (but not much delayed) electrification of London street lights around 1888 (the Gas company objected) with the persistence of the QWERTY keyboard on which we write. Yes, there are almost certainly better arrangements, but have you ever seen one? Whole books have been written about the subject.

What astronomical lessons can we learn here? Chances are, you will agree with some but not all of the following:
1. Britain perhaps performed a public service by staying out of ESO as long as possible and carrying on with independent development of 4-meter class telescopes and focal plane instrumentation, though not necessarily a service to her own astronomical community.
2. Very few things of the CELT/ELT/OWL classes will be built, so they will probably not be the best possible of their type. Electric cars would work a lot better if there had been as many manufacturers and generations of them as of internal combustion vehicles.
3. An historic strength of the US scientific community has been the multiplicity of funding sources and decision processes, including several large states (especially California and Texas) and several government agencies (NASA, NSF, DOE, and DOD and its ancestors the Army, Navy, and Air Force Offices of Scientific Research). Thus the possibility of restricting large projects to NASA should be Viewed With Alarm.
4. The number of different journals in which a fairly respectable astronomer can publish has fallen enormously over the years, with the loss of something like 100 independent observatory publications that existed near the turn of the previous century, the decline of national academy publications (at least for astronomy; remember Hubble's law first appeared in the Proceedings of the US National Academy of Sciences), and, more recently, the mergers that led to A&A, and the demise of QJRAS, Irish Astronomical Journal, Comments on Astrophysics, and others. While this has presumably yielded economies of scale, it has also concentrated the power of saying yeah or nay into fewer editorial hands and encouraged the remaining ones to adopt more nearly uniform standards.

The action item that comes out of this fourth, concentration of power, consideration could be phrased as "celebrate diversity" (a singularly politically correct sentiment) or "don't always vote for mergers".

Acknowledgements

It is the traditional prerogative of the last speaker at a conference to express the collective thanks of the participants to those who have made it possible. First comes Catherine Cesarsky for her gracious hospitality. She may or may not be the best director ESO has ever had, but she is definitely the best looking (consult the corridor outside the refreshment room if you doubt this). The assemblage of the scientific program was in the capable hands of João Alves of ESO and Mark McCaughrean of Potsdam. The considerable work involved in the local arrangements for food, lodging, transport, and so forth was largely carried out by Christina Stoffer and Martino Romaniello of ESO. I am personally grateful to all of them, to Pamela Bristow for manuscript assistance above and beyond the call of duty, and also (for various factoids) to Peter Shaver, David Barrado y Navascués, Rosina Iping, and Dick Ballard. Conference participants included native speakers of more than 20 languages (not to mention broken English, the international language of modern science, according to Zeldovich) who kindly provided for my final overhead suitable expressions of farewell and hopes for speedy next meetings in them all. Unfortunately not even Springer has that complete a set of type fonts, and so only "Auf Wiedersehen, até à próxima, and Ankh Wdjah Seneb, the Ancient Egyptian, May you Live, Prosper, and be Healthy".

References

1. Ambartsumian, V.A. 1960. QJRAS 1, 152
2. Badawy, A.M. 1954. History of Egyptian Architecture (Gizah), p. 153
3. Bahcall, J.N. & Ostriker, J.P. (eds.) 1997. Unsolved Problems in Astrophysics, Princeton Univ. press
4. Brilliant, A. 1985. Pot Shots No. 3253, Brilliant Enterprises, Santa Barbara
5. Brush, S.G. 1996. A History of Modern Planetary Physics, Cambridge Univ. Press
6. Burbidge, E.M., Burbidge, G.R., Fowler, W.A. & Hoyle, F. 1957. RMP 29, 547
7. Cox, A.N. (ed.) 2000. Allen's Astrophysical Quantities, Fourth Edition, Springer
8. Curtis, H.D. 1921. Bull. NRC 2, 194
9. Diamond, J. 1998. Guns, Germs, and Steel, W.W. Norton
10. Economist, The 1999. Issue of 31 December
11. Hoyle, F. & Lyttleton, R.A. 1939. Proc. Cam. Phil. Soc. 35, 405, 595, 608
12. Hubble, E.P. 1924. NY Times, 23 November, p. 6
13. Lang, K.R. 1980. Astrophysical Formulae, 2nd Ed., Springer
14. Russell, H.N., Dugan, R.S., & Stewart J.Q. 1926. Astronomy (Boston: Ginn & Co.)
15. Shapley, H. 1921. Bull. NRC 2, 171
16. Shklovskii, I.S., Sagan, C. 1966. Intelligent Life in the Universe, Holden-Day
17. Shu, F.H., Adams, F.C., & Lizano, S. 1987. ARA&A 25, 23
18. Trimble, V. 1975. RMP 47, 877
19. Trimble, V. 1997. In S.S. Holt & L.G. Mundy (eds.) Star Formation Near and Far, AIP Conf. Ser. 393, p. 15
20. Trimble, V. & Aschwanden, M. 2001. PASP, 113 (Sept. issue)

Author Index

GPSR Compliance

The European Union's (EU) General Product Safety Regulation (GPSR) is a set of rules that requires consumer products to be safe and our obligations to ensure this.

If you have any concerns about our products, you can contact us on ProductSafety@springernature.com

In case Publisher is established outside the EU, the EU authorized representative is:

Springer Nature Customer Service Center GmbH
Europaplatz 3
69115 Heidelberg, Germany

Batch number: 10704711

Printed by Printforce, the Netherlands